Sustainable Engineering Forum 2017

Core Programming Area at the 2017 AIChE Annual Meeting

Minneapolis, Minnesota, USA
29 October – 3 November 2017

ISBN: 978-1-5108-5809-1

Printed from e-media with permission by:

Curran Associates, Inc.
57 Morehouse Lane
Red Hook, NY 12571

Printed by Curran Associates, Inc. (2018)

AIChE
120 Wall Street, FL 23
New York, NY 10005-4020

Phone: (800) 242-4363
Fax: (203) 775-5177

www.aiche.org

Additional copies of this publication are available from:

Curran Associates, Inc.
57 Morehouse Lane
Red Hook, NY 12571 USA
Phone: 845-758-0400
Fax: 845-758-2633
Email: curran@proceedings.com
Web: www.proceedings.com

TABLE OF CONTENTS

Some abstracts in this book refer to images that were not made available by the publisher

499972 Bioprocess Model and Economic Analysis of Microalgae Production in Flat-Panel Photobioreactors Taking into Account Geospatial Factors

Abstract:

Microalgae are a promising feedstock for production of fuels and value-added chemicals. In order to assess the economic, energetic and environmental performance of algae based system in an industrial scale biorefinery, it is important to model algae productivity accurately. In this work, a first principle based growth model of microalgae is developed for flat panel photobioreactors taking into account the transport and biological processes. The primary factors that affect microalgae growth, such as solar irradiance, temperature, specific characteristics of species, uptake rates of nutrients and carbon dioxide, are incorporated in the bioreactor process model. Year-round algae productivity potential is computed for different locations to elucidate the implications of climate variations for the locations. Based on the yearly microalgal productivity for different regions, economic analysis is conducted to quantify the microalgae production cost for the different locations. As there are significant uncertainties associated with the different parameters of the growth model and economic analyses, detailed sensitivity analyses are performed. Results show that microalgae production cost is heavily influenced by climatic factors like solar irradiance and air temperature as well as by different equipment and process design factors. Results from the dynamic model can be used to optimize and scaling up of algae systems for efficient production of fuels and chemicals for variety of algae species as well as geospatial locations.

497873 Continuous Photobioreactor Cultivation of *Nannochloropsis oculata* to Isolate Cosmetical Grade Phospholipids

Abstract:

Microalgae are a promising source of natural ingredients as their cell mass contains polysaccharides, proteins, lipids, pigments, and vitamins. Phospholipids, in particular, are of interest to the cosmetics industry for various formulations, such as emulsifiers, liposome formers, solubilizers, wetting agents, and bioactive compounds. They are currently extracted from food sources, but this practice raises sustainability questions because of the competition with food. Microalgae can serve as a source of phospholipids on a more sustainable basis. To advance the status of microalgae bioproducts for cosmetics applications, the marine microalga *Nannochloropsis oculata* was cultivated in a vertical flat panel photobioreactor (VFPPBR) in batch and continuous mode. Batch cultivation was employed to optimize the growth parameters of *N. oculata*, calculate its maximum specific growth rate (μ_{max}), and determine its phospholipid productivity. The measured μ_{max} of 0.015 hr^{-1} was then used to select the dilution rate for designing and operating a VFPPBR-based continuous cultivation that is deemed more scalable and productive from a commercialization standpoint. Cell harvesting was followed by organic solvent extraction to solubilize and recover the algal phospholipids. The experimental findings in continuous cultivation of *N. oculata*, the robustness of the strain, and its polar lipid productivity, bioreactor designs, performance of the strain in batch and continuous reactors, and green pathways for downstream processing to isolate, concentrate and fractionate algal phospholipids were analyzed and discussed.

498781 Effects of pH on Cell Growth, Lipid Production and CO_2 Demand of Microalgae *Chlorella sorokiniana*

Abstract:

Microalgae have emerged as one of the most promising alternative energy feedstocks. Some advantages include the simple cellular structure, short production cycle, high lipid content, and fast growth. However, high production costs due to high CO_2 usage and low lipid productivity have been one of the major challenges impeding the commercial production of algal biofuels, Here, cell growth and lipid contents of *Chlorella sorokiniana* DOE 1412 were tested at different pH in flask cultivation. Culture pH was manipulated by CO_2 addition. The optimal pH for DOE1412 is 5.8-6.0 when only taking into account the cell growth and lipid production. A flat panel airlift photobioreactor (PBR) was used for scale-up cultivation at five different pH levels (6.5, 7.0, 7.5, 8.0 and 8.5). Data of pH value and the CO_2 addition was collected by a data logger. Biomass productivity increased with decreasing pH. By taking into account not only the cell growth and lipid production but also CO_2 demand, the lowest value of g CO_2/g dry biomass was achieved at pH 8. Also, the fatty acid profiles and biodiesel properties in different culture pH were investigated. Cetane number (CN) of biodiesel in the pH 6.5, 7 and 7.5 treatments satisfied the US standard ASTM D6751; among them, the pH 6.5 treatment met the Europe standard EN 14214. Finally, protein content in algae biomass increased with increasing pH, while C/N ratio in cells decreased.

484990 Design of Marine Macroalgae Photobioreactor Integrated into Building with Natural Sun Illumination

Abstract:

Seagriculture, which can provide offshore grown macroalgae biomass for food, platform chemicals and biofuels production will play a significant role in bioeconomy and biorefineries. Nevertheless, seagriculture development has been hindered by the lack of laboratory photobioreactors that enable fundamental macroalgae research. Here we report on the indoor macroalgae photobioreactor (MPBR) system integrated into the LEED certified building that enables maintenance and cultivation of marine macroalgae at controlled environment under natural illumination. Depending on specific reactor location, illumination varied between reactors in 238-348 µmole photons m^{-2} s^{-1} . The total direct energy consumption of the system was 8.15-11.45kWh during the 24 hours of operationThe total operation volume of the system was 6,400 L of seawater (salinity 3.9%, pH 8.2). Initial water stock was provided from the Tel Aviv Reading marine area and was partially (1/4-1/3) exchanged every 3 months. The compensation for evaporation was done by adding deionized water to the system to maintain the salinity at 3.9%.

The MPBR operation was been demonstrated for 6 months on continuous maintenance and cultivation of three green macroalgae species *Cladophora sp.*, *Ulva compressa* and *Ulva rigida* isolated form the Eastern Mediterranean coast. The growth rates, protein, rhamnose, xylose, arabinose, glucose, galactose and glucuronic acid content of the cultivated species were quantified. *Cladophora sp.*, isolated from Rosh-ha-Nikra sampling site was cultivated for 122 days. The growth rates changed from -3.66 $\%day^{-1}$ to 9.78 $\%day^{-1}$, with an average 3.47 $\%day^{-1}$. The maximum growth was for 12 continuous days. *Cladophora sp.* isolated from Mikhmoret sampling site was cultivated for 122 days. The growth rates changed from 9.32$\%day^{-1}$ to 95.5 $\%day^{-1}$, average 36.63 $\%day^{-1}$. The maximum growth was for 12 continuous days. *Cladophora sp.* isolated from Reading sampling site was cultivated for 97 days. The growth rates were stable at 95.47 $\%day^{-1}$ during the entire period of 38 cultivation days.

Carbohydrate profiling was done for *Cladophora sp.* harvested at Mikhmoret site and cultivated at the system at the growing rate of 63.9$\%^{-1}$ during harvesting. Rhamnose content was 3.89 µg mg_{DW}^{-1}, arabinose 30.46 µg mg_{DW}^{-1}, galactose 26.59 µg mg_{DW}^{-1}, glucose 61.72 µg mg_{DW}^{-1}, xylose 17.79 µg mg_{DW}^{-1} and glucuronic acid 0.98 µg mg_{DW}^{-1}. The total carbohydrate yield was 14.10%. The protein content in the biomass was 11%.

U. compressa isolated from Rosh-ha-Nikra sampling site was cultivated for 187 days. The growth rates changed from -0.06% day^{-1} to 3.5 $\%day^{-1}$, average 1.01 $\%day^{-1}$. The maximum growth was for 9 continuous days. *U. compressa* isolated from Mikhmoret sampling site was cultivated for 172 days. The growth rates changed from 1.46 $\%day^{-1}$ to 5.51 $\%day^{-1}$, average 3.65 $\%day^{-1}$.The maximum growth was for 27 continuous days. *U.compressa* isolated from Reading sampling site was cultivated for 143 days. The growth rates were mostly negative, with the biomass loss at -82.34 $\%day^{-1}$, growth rate of only 0.83 $\%day^{-1}$ was observed 40 continuous days.

Carbohydrate and protein profiling was done for *U. compressa* harvested at Rosh-ha-Nikra site and cultivated at the system at the growing rate of 3.5% day^{-1} at the time of a harvest. Rhamnose content was 48.58 µg mg_{DW}^{-1}, arabinose 0.29 µg mg_{DW}^{-1}, galactose 9.03 µg mg_{DW}^{-1}, glucose 75.13 µg mg_{DW}^{-1}, xylose 23.83 µg mg_{DW}^{-1} and glucuronic acid 18.13 µg mg_{DW}^{-1}. The total carbohydrate yield was 17.50%. The total protein content of this biomass was 25.5%.

U. rigida isolated from Rosh-ha-Nikra sampling site was cultivated for 89 days. The growth rates changed from -3.13 $\%day^{-1}$ to 19.51 $\%day^{-1}$, average 4.61 $\%day^{-1}$. The maximum growth was for 16 continuous days. *U.* rigida isolated from Haifa sampling site was cultivated for 155 days. The growth rates changed from -0.09 $\%day^{-1}$ to 9.4 $\%day^{-1}$, average 2.01 $\%day^{-1}$. The maximum growth was for 64 continuous days.

Carbohydrate and protein profiling was done for *U. rigida* harvested at Haifa site and cultivated at the system at the growing rate of 1.4%/day at the time of a harvest. Rhamnose content was 37.29 µg mg_{DW}^{-1}, arabinose 0.46 µg mg_{DW}^{-1}, galactose 3.60 µg mg_{DW}^{-1}, glucose 54.84 µg mg_{DW}^{-1}, xylose 9.11µg mg_{DW}^{-1} and glucuronic acid 16.62 µg mg_{DW}^{-1}. The total carbohydrate yield was 12.20%. The protein content from the biomass harvested was 29%.

This work provides detailed insights on the design of practical indoors photobioreactors for isolation, selection, cultivation maintenance and analysis of macroalgae species required for biorefinery feedstock development.

488865 An Integrated Approach for Bioenergy Production from Microalgae Using Solar Energy, CO_2 and Wastewater

Abstract:

New biomass sources for alternative fuels has grown to be a subject of increasing importance as the world struggles to resolve the economic and strategic impacts of limited fossil fuel resources on environment and global climate. Microalgae are among the most promising non-food-crop-based biomass feedstocks. Microalgae are photosynthetic microorganisms with simple growing requirements (light, CO_2 and nutrients) that can produce lipids, proteins and carbohydrates in large amounts over short periods of time. These products can be processed into both bioenergy and valuable co-products. On the other hand, carbon dioxide (CO_2) is one of the most important contributors for the increase of the greenhouse effect. CO_2 concentrations are increasing in the last decades mainly due to the increase of emissions due to human activities. A promising technology is the biological capture of CO_2 using microalgae. These microorganisms can fix CO_2 using light with efficiency ten times greater than terrestrial plants.

To be economically competitive with carbon capture/sequestration methodologies, an intensive research is needed to integrate all microalgal culture benefits: flue gas and wastewater treatment and biomass production. In this research, an integrated process is introduced to produce bioenergy from microalgal biomass combined with simultaneous CO_2 capture/sequestration and bioremediation of wastewater nutrients. Also, this research investigates different processes and operations to optimize microalgal biomass production using open ponds or photobioreactors. Finally, this research focuses on the potential of producing biogas from microalgal biomass through anaerobic digestion and further upgrading of biogas into higher alkenes through catalytic processes

The main objective of this research is to investigate the growth kinetics of the appropriate microalgae species at different feed composition of CO_2 with air to optimize the growth rate of microalgae at maximum CO_2 consumption and in various wastewater media to reduce wastewater nutrients concentrations (COD, ammonia, nitrate and phosphate) to the potable clean water level. Also, this research aims to study the growth kinetics of the microalgae at other important parameters such as the effect of concentrated solar energy, pH, temperature, wastewater dilution and algae circulation rate. The anaerobic digestion of the harvested algae is investigated at different operating conditions such as liquid/solid ratio, temperature and pH. Methane concentration in the outlet stream will be optimized as function of these operating conditions.

There is a number of challenging issues in the realization of the proposed technology. One of the major challenges is to find appropriate microalgae species tolerant to high concentrations of CO_2. The flue gas of a power plant contains up to 500 times higher CO_2 content than that in the atmosphere. Therefore, different microalgae species will be studied under similar composition of CO_2 in the flue gas from a fossil fuel-based power generating station to find the proper type of microalgae adapting to higher CO_2 concentrations. Another major challenge is to find the proper technology for microalgal biomass production. There are two main technologies; open ponds and photobioreactors. Even that the open ponds are relatively cheap and easy to clean and maintain, they have poor biomass productivity, mixing, light and CO_2 utilization and they require large area of land. On the other hand, photobioreactors have different designs such as tubular, flat plate and column. Usually, the photobioreactors produce high microalgal biomass. Therefore, this research will explore different designs of open ponds and photobioreactors to optimize the growth of desired microalgae species. Finally, integrated processes propose different challenges that need to be faced. One of the powerful tools to meet these challenges is modelling and simulation software. For example, an essential part of any integrated process is heat and energy integration. Aspen simulation packages proved their capabilities to design and simulate such integration. In this research, modelling and simulation packages will be used to optimise the overall proposed integrated process along with analysing and modelling collected data from the experimental work.

It is expected that the innovation and implementation of the integrated approach for microalgal bioenergy production CO_2 will offer a solution of providing sustainable energy demand and addressing the global commitment to reduce greenhouse gases emissions by CO_2 capture/sequestration.

491438 Bio/Catalytic Synthesis of Methacrylic Acid

Abstract:

2-methylpropenoic acid, also known as methacrylic acid, is an organic acid with one carboxyl group as the main functional group. This organic acid is a significant monomer in the polymer industry, due to its application as a precursor in production of methyl methacrylate (MMA) and poly methyl methacrylate (PMMA). Polymers synthesized from methacrylic acid and methyl methacrylate are chemically resistant and transparent such that they can be used as glass substitutes and in optical fibers. Industrially, methacrylic acid is produced using the hazardous acetone cyanohydrin process involving large amounts of sulfuric acid.

In this work, methacrylic acid was synthesized from the bio-based substrates citric acid, itaconic acid, and 2-hydroxyisobutyric acid. Hydrotalcite, a solid base catalyst, was employed to form methacrylic acid through decarboxylation of itaconic acid and citric acid. The effect of varying temperatures, catalyst mass, residence times and substrate concentrations on methacrylic acid concentration and yield was determined. Comparing yields of methacrylic acid in each series of reactions resulted in finding an optimum substrate to catalyst mass ratio where methacrylic acid yield reaches to its highest value (9.6 g-substrate/g-catalyst and 22.7% yield for citric acid, 6.4 g/g and 22.9% yield for itaconic acid). Catalyst reusability experiments resulted higher methacrylic acid yields. Methacrylic acid was also formed from 2-hydroxyisobutyric acid in a single-step dehydration reaction. Among these three substrates, the highest yield of methacrylic acid (71.48 %) was achieved at 275 °C and a 1 minute residence time using 2-hydroxyisobutyric acid and subcritical water.

495780 Green Processes to Use Extract from Citrus Peel Waste for Novel Applications (Direct Polystyrene Recycling to Natural Solvent to Source of Carbon)

Abstract:

Recycling and reuse through green engineering is always an open challenge for sustainable environment. Here, we present some innovative ways to recycle and reuse two different kinds of waste: plastic (polystyrene) waste and agriculture waste (citrus peel). Worldwide, Citrus fruits production is more than 31 million tons annually out of which nearly 50% is waste in terms of citrus peel. This amounts to a large quantity of citrus peel waste as a by-product in fruit juice industry. Orange contribute nearly half of this citrus peel waste. Although orange peel waste is further processed to yield some chemicals such as limonene and pectin which are used a fragrance, cosmetics, cleansing agents and sometimes as solvents, these processes (distillation, microwave heating) are highly energy consuming and uneconomical and therefore have limited industrial acceptability. We have developed cost effective, low-energy novel and green approaches to directly use the orange peel extract for a variety of applications such as in print transfer and natural solvent for biopolymers. Print transfer technique is used to prepare beautiful patterns or artistic designs on cloth, fabric, paper, walls etc. in order to enhance the appearance of respective objects creatively. However, conventionally used solutions for print transfer such as alcohols, xylene, acetone, acrylic mediums or Citra-solv are known to be toxic and harmful, if concentrated or overused. Therefore, the use of orange peel extract as the transfer solution is not only advantageous because it is natural, non-toxic, and inexpensive. In this process, orange peel extract facilitates in deteriorating the adhesive forces between the ink and the paper and thus transfers the ink on to the other substrates. Interestingly, this technique of print transfer is not limited to any particular choice of substrate.

Further, we centrifuged orange peel extract to yield three distinct layers. Bottom layer was then used as a solvent to prepare the solution for biopolymers such as gelatin. This gelatin solution made by bottom layer of orange peel extract as a solvent was then electrospun to prepare gelatin nanofibers followed by its successful demonstration in drug delivery. Again, the use of natural solvent in these biopolymer formulations minimizes any toxic effect associated with conventionally used organic solvents especially for healthcare applications. The middle part of the orange peel extract was comprised of small sized cellulosic flakes which can be pyrolyzed into nano-structured hard carbon and thus may potentially be used as an electrode material for rechargeable batteries.

The top layer of the orange peel extract was used for direct recycling of Polystyrene waste. Polystyrene is one of the most widely used thermoplastic polymer. It is used in various forms like solid in applications such as electronics, construction, house and medical ware, disposal food services etc. whereas, for protective packaging, in electrical, pharmaceutical and retail industries etc., it is used in form of foam, because of its light weight, shock resistance, cushioning properties, and flexibility in design possibilities. Obviously due to huge applications, polystyrene is one of the major solid waste. It is major component of plastic debris in ocean, where it becomes hazardous to marine life. Also, Styrofoam (another form of polystyrene) blows in wind and floats on water, making it difficult to collect and control the waste. Polystyrene packaging products are usually discarded in dumps, landfills or simple litter after their

useful application. As the waste plastic material has become a serious problem, recycling is getting attention to save environment and resource recovery. The three main alternatives for PS recycling are, mechanical, chemical and thermal recycling. Mechanical recycling involves relatively simple technologies for converting scrap polystyrene into new product, by compressing and melting. However it is quite labour- or energy-intensive, depending on whether the process is manual or automated. Chemical recycling can have high capital cost such as in catalytic degradation of polystyrene whereas, toxic emission in thermal recycling refrains its use. All recycling methods involves energy input as the preliminary step either to crush PS objects into granules or thermally degrade them. Therefore, it is not surprising that recycling of polystyrene waste gets the worst rating. Recycled plastic is always a lower grade than the "virgin" material, so some applications, such as that for food packaging, cannot use recycled plastic. According to a 2004 study by the California Integrated Waste Management Board, of the 377,580 tons of polystyrene produced in the state, only 0.8% is recycled. Of that, only 0.2% (310 tons) of polystyrene food service packaging is recycled.

As stated above, we present a novel and innovative way to recycle objects made up of polystyrene directly into sub-micron, aligned fibers using citrus fruitâ€™s peel extract (top layer). Recycled PS fibers were then structurally characterized and were found to be hydrophobic as well as lyophilic. Based on this property, these fibers were then tested as a sorbent material for oil in terms of absorption capacity and retention capacity. Furthermore, we also figured out that same fabrics can be re-used at least for five time with half of their original absorption capacity and therefore these fabrics are also useful not only for oil spills cleaning but also for oil recovery. We also measured buoyancy properties of as-fabricated recycled Ps fabrics in static as well as in turbulent conditions as it is important for their practical applications in oil spills remediation in large water bodies. In summary, the novel processes developed here in this work are perfect example of green engineering to make use of waste material into useful products.

498508 Development and Applications of Profitable Pollution Prevention Technologies

Abstract:

To achieve industrial sustainability, environmental, social, and economic factors need to be integrated to reduce resource consumption, while increasing health, equity, and quality of life for all stakeholders. Profitable pollution prevention (P3) technologies seek the simultaneous realization of waste reduction and production improvement. It can be realized through fundamental understanding of the process operation and waste generation mechanism. This paper will present the concept of P3 and an array of profitable pollution prevention technologies developed, including clean combustion technologies, recycle and reuse strategies for the petrochemical and electroplating technologies. Significant economic gain and environmental benefit were achieved.

500513 Kinetics, Yield and Rate Limiting Processes in the Biosynthesis of Colloidal Silver Nanoparticles By a Fresh Water Microalga

Abstract:

In the current study, a fresh water microalga, *Chlamydomonas reinhardtii* has been used to synthesize colloidal AgNPs through bio-reduction, and a mathematical model is developed to study the reaction kinetics and yield. The properties of silver nanoparticles (AgNPs) make them excellent candidates for fabricating biosensors, therapy and diagnostic tools, medical imaging agents and biomaterials. The prevailing synthesis methodologies consume imprudent amounts of energy and use harmful chemicals such as sodium borohydride hence creating a need to look for sustainable technologies. The present set up uses live culture of *C. reinhardtii* therefore reducing the energy, cost and doing away with toxic chemicals. The biosynthesis takes place inside the thylakoids that have abundance of reducing equivalents, and through reduction catalyzed by the cellâ€™s exopolymeric substances (EPS). The production of AgNPs was confirmed by the characteristic SPR band in the range of 415- 430 nm using UV-vis spectrophotometry. Crystallinity of nanoparticles was determined using X-ray diffraction (XRD) and the mean particle size was calculated from Scherrer equation. Micrographs from scanning and transmission electron microscopy (SEM and TEM) were analyzed to study the shape and the size of AgNPs. Ag^{+} to AgNP conversion kinetics and yields were determined by measuring the composition of the solid and liquid phases by inductively coupled plasma atomic emission spectroscopy (ICP-AES). The AgNPs were stable over time as determined through negligible change in the intensity of the SPR band. The results of this study will be used evaluate the prospects of a scalable, fully automated, and permanent photobioreactor for the continuous production of different biogenic inorganic nanoparticles.

499120 Microwave-Assisted Synthesis of Silver Nanoparticles Using Glucose and Starch

Abstract:

In this study, we have developed a rapid, easy and novel method for the microwave assisted, â€œgreenâ€ synthesis of silver nanoparticles (AgNPs) using glucose and starch as the reducing and capping agents, respectively. The properties of AgNPs make them excellent candidates for antibacterial and antitumoral agents, biosensors, supercapacitors and theranostic vehicles amongst others. Different shapes and sizes of the nanoparticles impart them these multifarious applications. Many traditional synthesis routes use toxic chemicals and imprudent amounts of energy and therefore pose a potential hazard to health and the environment. Microwave synthesis of silver nanoparticles is a faster and more energy efficient route. In addition, microwave assisted synthesis can be a robust and efficient platform to synthesize various shape-controlled inorganic nanoparticles. A simple synthesis method for AgNP formation using glucose and starch as the reducing and capping agents, respectively, has not been well understood. AgNP production was optimized as a function of reaction time, temperature and molar concentrations of the reactants. The size, shape and crystallinity of the silver nanoparticles were characterized by UVâ€“vis spectroscopy, scanning electron microscopy (SEM), transmission electron microscopy (TEM), X-ray diffraction (XRD) and inductively coupled plasma atomic emission spectroscopy (ICP-AES). We also carried out thermal gravimetric analyses (TGA) to assess the thermal stability of the nanoparticles and to verify the formation of AgNPs. Furthermore, the AgNP yield, energy balance and reaction mechanism, which has not been well elucidated in the literature, was studied. There was a change of less than 2% in the absorbance of the SPR band of the as prepared AgNPs after 2 weeks, thereby validating that the particles were uniform and stable over time. In the future, we will study the antimicrobial and electronic properties of the AgNPs by looking into its cytotoxic effects and through cyclic voltammetry (CV), respectively.

Keywords: Silver nanoparticles, microwave synthesis, green chemistry

501820 Enhancing Energy Efficiency in Saccharide-HMF Conversion with Core/Shell Structured Microwave Responsive Catalysts

Abstract:

Core/shell structured microwave responsive catalysts with carbon nanotube core and TiO_2/SO_4^{2-} shell were synthesized in this work for catalytic conversion of saccharide to 5-Hydroxymethylfurfural (HMF). The microstructure and surface property of these hybrid composites were carefully characterized. With such a structure of inner microwave absorber and outer catalyst shell, the heat generated at the core area by microwave radiation can be translated to the shell catalyst directly. Such localized heating allows maximum heat utilization in the reaction and uplifts the energy efficiency of the catalytic reaction by up to six times compared to pure TiO_2. The effects of TiO_2 shell thickness, annealing temperature, catalyst concentration and microwave power on the saccharide conversion and HMF yield were systematically investigated. These catalysts have been demonstrated effective in the hydrolysis conversion of various saccharides such as fructose, glucose and sucrose.

486988 Impacts of CO_2 Supply Systems for Algal-Based Biorefineries on Biofuel Life Cycle Assessments

Abstract:

Biofuels have great potential as low-carbon alternatives to fossil-based transportation fuels. This talk will focus on utilization of CO_2 in an advanced biofuel system and the integration of that system with CO_2 generation sources. The biofuel system is the Algenol Direct to Ethanol® technology, which provides an efficient, cyanobacteria-based system for producing ethanol and bio-crude. The Algenol process consumes CO_2 in a process that yields a biofuel with much lower greenhouse gas emissions than gasoline.[1-3] In this talk, different options for CO_2 capture and utilization (CCU) are considered and their impact on the overall system operation evaluated in terms of carbon footprint reduction. The boundary of the life cycle analysis includes the CO_2 source. Overall 13 scenarios are described including an artificial reference case with CO_2 arriving at the boundary limits of a 2000 acre, model biorefinery with zero carbon footprint burden and zero CO_2 cost at a 10% IRR. Most scenarios involve combined heat and power (CHP) systems with no power supplied from the grid. Both natural gas and cellulosic fuel sources are considered for the CHP system. Five of the scenarios are "stand-alone" in that they do not require co-location with an anthropogenic CO_2 source. This is achieved by generating enough CO_2 on site to supply the needs of the biorefinery and selling excess power back to the grid. Four of the five stand-alone systems have carbon footprints and economics that are fully competitive with the commonly considered coal flue gas case. Overall carbon footprint reductions range from 23% to 116%. Equivalent CO_2 costs (based on comparisons to the reference case) range from \$35/-\$96/tonne with coal flue gas being \$45/tonne.

[1]D. Luo et al, "Lifecycle Energy and Greenhouse Gas Emissions from Ethanol Produced by Algae", *Environmental Science and Technology* **44**, 8670 (2010).

[2]Notice of EPA pathway approval: http://epa.gov/otaq/fuels/renewablefuels/new-pathways/approved-pathways.htm

[3]R. P. Lively et al, "Anthropogenic CO_2 as a feedstock for the production of algal-based biofuels", *Biofuels, Bioproducts, and Biorefining* **9**, 72-81 (2015)

498698 Carbon Fixation By Rubisco-Nanostructure Complex to Produce 3-Phosphoglyceric Acid: A Life Cycle Assessment

Abstract:

As an effort to eliminate greenhouse gases (GHGs), biochemists, microbiologists, and engineers are working toward systems that mimic the plant cellular environment's ability to convert GHGs into higher value products [1]. In the Calvin cycle of photosynthesis, ribulose-1,5-bisphosphate carboxylase oxygenase (RubisCO), which is an enzyme present in chloroplasts, catalyzes the conversion of ribulose 1,5-bisphosphate (RuBP) to 3-phosphoglyceric acid (3-PGA) while incorporating atmospheric CO_2 into the organic molecule [1,2]. It has been said that RubisCO is the most abundant protein in the world and accounts for most of the biological CO_2 fixed on earth [3]. Satagopan et al. have developed multienzyme-nanostructure complexes that catalyze cascade reactions in the Calvin cycle to enhance the catalytic properties of enzymes [1]. The complexes are made of three enzymes that include Rubisco and a nanostructure to support the enzymes.

In this work, a carbon footprint and other environmental impacts of 3-PGA produced through carbon fixation by the multienzyme-nanostructure complexes are evaluated using the Life Cycle Assessment (LCA) approach [4]. Two types of nanostructures for the complex are investigated as follows: Camptothecin (CPT)-dipeptide nanotube [1,5] and fluorenylmethyloxycarbonyl (Fmoc) tetrapeptide nanofiber. To examine the effectiveness of carbon fixation by those emerging technologies, another route that doesn't fix CO_2 to prepare 3-PGA is investigated [6] and the carbon footprint of each route is compared. Also, the potential use of 3-PGA and its impacts are examined to account for a cradle-to-grave life cycle boundary. From the results of life cycle impact assessment, processes that show the highest contribution to the footprint are identified.

Many challenges exist in applying the LCA to emerging technologies due to the nascent nature of these technologies [7]. If any life cycle inventory (LCI) data are not available from existing database [8,9], the data are estimated from laboratory experiments or obtained from literatures and patents. To account for the uncertainty of those data, a sensitivity analysis is performed. Three cases are considered based on the amount of fugitive emissions and the recyclability of unreacted reactants and solvents. The reusability of the complexes is also taken into consideration through the sensitivity analysis because they function as catalysts for the carbon fixation. Also, most of energy flows are excluded from the inventory analysis because most of biological processes take place around room temperature. If product yields are unknown, they are determined by a stoichiometry calculation.

CPT is extracted from the stem of a specific tree species, *Camptotheca acuminate* [10]. Since its extraction yield from the tree and the subsequent product yield of the CPT-dipeptide nanotube are low, the total carbon footprint is affected significantly depending on whether biogenic CO_2 emissions are included in the analysis or not. In this work, therefore, the effects of the biogenic CO_2 emissions on the footprint are examined and discussed.

Even though this work accounts for some data uncertainty via sensitivity analysis, other sources of uncertainty are still left for a more robust analysis. Some data, such as labor, equipment, and other unavailable data, are excluded from the analysis boundary due to the lack of information. Also, inventories from different LCI database [8,9] often base on different location. Moreover, since many unavailable data from the LCI database are obtained from literatures and patents, there may be errors in the life cycle network as well as the inventory data. Those potential uncertainty and errors are mostly attributed to the nature of emerging technologies. If a more complete analysis that accounts for those issues is developed, it would present significant benefits for evaluating the impacts of emerging technologies.

References

[1] Sriram Satagopan, Yuan Sun, Jon R. Parquette, and F. Robert Tabita. Self-assembled nanostructures for capturing atmospheric CO_2. Unpublished manuscript.

[2] Spencer M. Whitney, Robert L. Houtz, and Hernan Alonso. Advancing our understanding and capacity to engineer nature's CO_2-sequestering enzyme, Rubisco. Plant Physiology. 155(1): 27-35, 2011.

[3] John A. Raven. Rubisco: still the most abundant protein of Earth? New Phytologist. 198(1): 1-3, 2013.

[4] ISO. Environmental management—Life cycle assessment—Principles and framework (ISO 14040:2006). ISO. Geneva, Switzerland, 2016.

[5] Se Hye Kim et al. The self-assembly of anticancer camptothecin–dipeptide nanotubes: A minimalistic and high drug loading approach to increased efficacy. Chemistry - A European Journal. 21(1): 101-105, 2015.

[6] Ines Mandl and Carl Neuberg. Preparation of D(-)-3-phosphoglyceric acid. Methods in Enzymology. 3:208-214, 1957.

[7] Marcelle C. McManus et al. Challenge clusters facing LCA in environmental decision-making–what we can learn from biofuels. The International Journal of Life Cycle Assessment. 20(10):1399-1414, 2015.

[8] U.S. Life Cycle Inventory Database. National Renewable Energy Laboratory, 2012.

[9] Rolf Frischknecht et al. The ecoinvent database: Overview and methodological framework. International Journal of Life Cycle Assessment, 10(1):3–9, 2005.

[10] J. George Buta and Michael J. Novak, Isolation of camptothecin and 10-methoxycamptothecin from camptotheca acuminate by gel permeation chromatography. Industrial and Engineering Chemistry Product Research and Development. 17(2): 160-161, 1978.

496672 Life Cycle Assessment of Biofuel and Bioenergy Production from Brown Algae through Biochemical Pathways

Abstract:

Brown algae *Laminaria Japonica* is the most cultivated type of seaweed with a large potential for sustainable biofuel and bioenergy production. This study aims to evaluate the environmental impacts of seaweed cultivation and utilization for biofuel production through biochemical conversion routes including the sugar platform (SP), methane platform (MP), and volatile fatty acids platform (VFAP). In the SP, bioethanol is produced through saccharification of carbohydrates and fermentation of produced sugars. In the MP and VFAP, methane and volatile fatty acids (VFA) are produced through anaerobic digestion of seaweed. In the MP, methane is utilized as fuel source in a boiler/turbogenerator for heat and power production. In VFAP, VFAs are recovered through extraction/distillation and then hydrogenated to mixed alcohols. All three processes are simulated in Aspen Plus v.8.8 to obtain heat and material balances. Simulation results along with data from literature and Ecoinvent V3 databases for seaweed cultivation, chemicals, nutrients, and waste treatment operation are used to obtain life cycle inventory. Then, life cycle models are developed in SimaPro V8.2 software for each process. The environmental impact categories considered include global warming potential, acidification, eutrophication, fresh water consumption, and non-renewable energy use. Results are used to identify bottlenecks, limitations, and main contributors to the environmental profile of each route and assess the potentials to reduce the emissions and enhance the sustainability of the production process.

497439 Integrated Design and Analysis of Chemical Production from Biomass Feedstocks

Abstract:

Integrated Design and Analysis of Chemical Production from Biomass Feedstocks

Abhay Athaley[1], Praneeth Annam[1], Basudeb Saha[2], Marianthi Ierapetritou[1]

[1] Department of Chemical and Biochemical Engineering, Rutgers - The State University of New Jersey

[2]. Catalysis Center for Energy Innovation & Department of Chemical & Biomolecular Engineering, University of Delaware

This work principally concentrates on the integrated design, techno-economic and the life-cycle analysis of the production of chemicals from the lignocellulose biomass. The priority for the development of variate fine chemicals from biomass feedstock stepped-up over the changing time. The idea of bio-refinery has been proposed that uses different conversion technologies to produce multiple products. The success of bio-refinery is highly dependent on its mature technology and its applicability to produce not only high volume low cost fuels but also to produce low volume high cost chemicals.[1]

Bio-based products acceptance in the market depends on the competitiveness of economics and sustainability when compared to oil-based chemicals and products. In our previous work, [2]we investigated a novel hydrolysis process[3] for the productions of sugars and the subsequent production of para-Xylene and compared it with two other hydrolysis process (Concentrated acid hydrolysis and Dilute acid hydrolysis) using techno-economic and life cycle assessment. In our present work, we integrate the novel hydrolysis process to production of other high value chemicals such as surfactants,[4] butadiene, 2-methyl styrene etc. and high value jet fuels and combine it with the production process of para-Xylene and phthalic anhydride.[5] The aforementioned chemicals are lucrative and also it furthers our goal of designing an integrated bio-refinery.

This work uses techno-economic analysis and life cycle assessment to design and evaluate the production alternatives. The detailed process flowsheet is developed and simulated using Aspen Plus. The steps for the production of different chemicals and fuels are integrated with the novel hydrolysis process into a single flowsheet. Next, heat integration and economic analysis is carried out to calculate the minimum selling price of each of the product. Furthermore, sensitivity analysis is performed to analyze the bottlenecks of entire process and to compare the cost of the chemicals with fluctuating cost of raw materials and capacity of the plant. It is also seen that the overall cost of raw materials increases considerably for the production of surfactants as compared to other chemicals such as butadiene or phthalic anhydride. To check the competitiveness of the above process in terms of sustainability Life Cycle Analysis is carried out on SimaPro®. The results with respect to carbon production and water consumption are shown and compared to the current oil based processes.

References:

1. Lin, Z., V. Nikolakis, and M. Ierapetritou, *Alternative Approaches for p-Xylene Production from Starch: Techno-Economic Analysis.* Industrial & Engineering Chemistry Research, 2014. **53**(26): p. 10688-10699.

2. Athaley A., S.B., Ierapetritou M.,, *TECHNO-ECONOMIC ANALYSIS OF DIFFERENT TYPES OF HYDROLYSIS PROCESS FOR THE PRODUCTION OF P-XYLENE.* 2017.

3. Sadula, S., Athaley, A., Zheng, W., Ierapetritou, M., Saha, B., *Process Intensification for Cellulosic Biorefineries.* ChemSusChem, 2017. **10**.

4. Park D., J.K., Koehle M., Krumm C., Ren L., Damen J., Shete M., Lee H., Zuo X., Lee B., Fan W., Vlachos., Lobo R., Tsapatsis M., Dauenhauer P.,, *Tunable Oleo-Furan Surfactants by Acylation of Renewable Furans.* ACS Catalysis, 2016. **2**: p. 820-824.

5. Lin, Z., Ierapetritou, M., Nikolakis, V.,, *Phthalic anhydride production from hemicellulose solutions: Technoeconomic analysis and life cycle assessment.* AIChE Journal, 2015. **61**: p. 3708-3718.

499590 Life-Cycle Analysis (LCA) of Bio-Derived Terephtalic Acid (TPA) and Bio-Derived-Hexamethylenediamine (HMDA)

Abstract:

Increasing efforts to reduce environmental impacts while boosting U.S chemical production have been an important driver in the growing bioeconomy. These efforts not only stimulate economic development but also promote competitiveness in the bioeconomy. Under the analysis led by the Clean Energy Manufacturing Analysis Center (CEMAC), screening of the impact of initial 170 chemicals derived from biomass were conducted, resulting in three relevant bioproducts: 3- hydroxypropionic acid (HP), terephtalic acid (TPA), and hexamethylenediamine (HMDA). These bio-derived chemicals can be used in the production of superabsorbent polymers, coatings, adhesives, plastics, fibers, paints, etc. In this work, we use the Greenhouse gases, Regulated Emissions and Energy use in Transportation (GREET®) model to conduct a life-cycle analysis (LCA) on the last two bioproducts: TPA and HMDA. We evaluate the key drivers of the GHG emissions and water consumption. We present thermochemical and biochemical pathways as alternative production pathways for these chemicals. These biomass-derived chemicals can displace fossil-derived products potentially and reduce GHG emissions, water and fossil fuel consumption as compared to their fossil-derived counterparts.

495175 Sustainability Analysis of Condensed-Phase n-Butanol Production from Ethanol

Abstract:

Bioethanol is one of the most valuable replacements for or additives to fossil fuels. However, ethanol is not without its limitations that include low energy density, corrosivity, and contamination issues when mixed with water. In contrast, n-butanol has a notable advantage in these aspects, as it is more miscible in gasoline, has higher energy density, and has an air to fuel ratio comparable to that of gasoline. As a result, there has been growing interest in the production of biobutanol and higher alcohols as biofuels as well as for use in the chemical industry.

The condensation of ethanol to butanol and higher alcohols is known as Guerbet chemistry [1]. The reactions are well known and have been substantially investigated. The path to Guerbet chemistry commercialization, however, has been hindered by substandard yields, catalyst stability concerns, and high capital and energy costs.

Our research group has studied the condensed-phase Guerbet reaction of ethanol to higher alcohols [2], [3]. Recently, we have optimized a condensed phase continuous flow reactor using multi-metallic catalysts on γ-Al_2O_3 support modified with La_2O_3. We have achieved higher alcohols selectivity of 80% at ethanol conversions of 40%. These encouraging results have prompted us to explore the economic feasibility and environmental impact of industrial-scale higher alcohol production using Guerbet chemistry.

An economic and environmental assessment of a first-generation facility producing 25 million gallons of n-butanol per year has been performed for a range of butanol and higher alcohol selectivities. Simulation of the catalytic reactor and the recovery and purification processes have been conducted using AspenPlus. Economic parameters such as capital costs, operating costs, annual profit, and return on investment have been determined for several scenarios. The results obtained show n-butanol required selling prices are reasonably competitive for conversion and selectivity combinations achieved in laboratory reactions. Interestingly, analysis of the separation train indicate significant economic benefit by carrying out distillations at elevated pressures. Environmental impacts such as climate change, resource depletion, water use, and eutrophication are also assessed.

References

[1] D. Gabriëls, W. Y. Hernández, B. Sels, P. Van Der Voort, and A. Verberckmoes, "Review of catalytic systems and thermodynamics for the Guerbet condensation reaction and challenges for biomass valorization," *Catal. Sci. Technol.*, vol. 5, pp. 3876–3902, 2015.

[2] T. L. Jordison, L. Peereboom, and D. J. Miller, "Impact of Water on Condensed Phase Ethanol Guerbet Reactions," *Ind. Eng. Chem. Res.*, vol. 55, no. 23, pp. 6579–6585, 2016.

[3] T. L. Jordison, C. T. Lira, and D. J. Miller, "Condensed-Phase Ethanol Conversion to Higher Alcohols," *Ind. Eng. Chem. Res.*, vol. 54, no. 44, pp. 10991–11000, 2015.

494548 Power Capacity Expansion Planning Considering Endogenous Technology Cost Learning

Abstract:

Integrated modelling can guide structural and operational decision-making in energy systems. Long-term models can assess pathways complaint with future demands and environmental targets and optimal investment timing for power generation, energy storage, and transmission capacity. Short-term operational models can elucidate the role and value of specific technologies within the power system.

We develop a mixed-integer linear formulation of a hybrid power generation capacity expansion and unit commitment model with a high level of technical detail and granular time discretisation [1]. The model considers up to 2000 units of 16 different power generation and storage technologies, including international interconnectors for electricity import and export, and grid-level energy storage. The national-scale single-node formulation includes the consideration of endogenous learning for technology capital cost. In line with previous approaches, we determine a piece-wise linear formulation of the exponential learning curve model [2, 3]. A reformulation of the non-convex term of specific unit cost installed capacity is implemented in form of the cumulative investment cost as a function of cumulative installed capacity.

The model is applied to the power system of the United Kingdom for the years 2015 to 2050, subject to electricity demand and ancillary service requirements, CO_2 emission targets, technology specific performance parameters, and learning rates [4]. Wefind that the consideration of cost learning for an individual technology moves the optimal investment timing to earlier planning years and increases the economic amount of capacity deployment of the respective technology. Other unsupported technologies, however, might experience deployment delays and become less competitive. These systemic implications on the profitability of technologies is again observed in scenarios where all technology types experience cost learning. We compare a case of local (national) and global technology learning curves, andfind more pronounced effect in the case of global technology diffusion assumptions.

Specifically in the case of offshore wind capacity, wefind that the omission of cost reduction effects leads to an underestimation of the optimal economic deployment level in 2050 by 32 % to 50 %. Similarly, coal-fired power generation with Carbon Capture and Storage capacity which is crowded out by less costly low-carbon technologies under static cost assumptions, becomes an attractive investment when cost reduction is taken into account. System-wide effects prove that early investments are able to drive technology learning and reduce total system cost by mid-century. In addition, wefind that the model results are highly sensitive to the build rate parameter which limits the maximum annual capacity additions. A high build rate can reduce cumulative total system cost by a multiple of the cost learning effects.

[1] C. F. Heuberger, I. Staffell, N. Shah, and N. Mac Dowell. Levelised Value of Electricity - A Systemic Approach to Technology Valuation. In 26th European Symposium on Computer Aided Process Engineering, volume 38, pages 721-726, 2016.

[2] S. Messner. Endogenized technological learning in an energy systems model. Journal of Evolutionary Economics, 7(3):291-313, 1997.

[3] A. Gruebler and S. Messner. Technological change and the timing of mitigation measures. Energy Economics, 20(5-6):495-512, 1998.

[4] E. S. Rubin, I. M. L. Azevedo, P. Jaramillo, and S. Yeh. A review of learning rates for electricity supply technologies. Energy Policy, 86:198-218, 2015.

496532 Data-Driven Battery Sizing and Market Participation

Abstract:

High-profile utility-scale battery storage systems have been deployed during the past few years to improve the grid reliability, ease renewable integration, curtail emissions, and reduce overall systems cost [1]. The financial soundness of these investments has yet to be proven because energy storage economics heavily depends on electricity market dynamics, energy policies, and long-term performance degradation.

We propose a multiscale optimization framework to co-optimize battery system designs and electricity market participation [2]. This includes mathematical models for California market rules, battery physics, and operational constraints. The models involve multi-year investment horizons with 5-minutes timesteps (to capture fast fluctuations of real-time prices), giving rise to linear programs with up to tens of million variables and constraints. We use a high-performance computing framework to analyze over 1 trillion historical price data points from the California ISO market from year 2015 and address key questions that drive battery storage economics:

1. Which market products (energy, ancillary services) and market timescales (day-ahead, real-time) offer the greatest revenue potential?
2. How much does battery degradation hinder market revenue? What is the potential impact for new battery technology breakthroughs?
3. What are the optimal sizing, battery replacement strategies, and locations for storage systems? What is the expected payback period for recent investments?

Our analysis shows that most market revenues come from the fastest timescales (real-time energy markets, regulation ancillary services), indicating a strong need for fast flexibility in the California grid. Surprisingly, we find that market participation strategies are an order of magnitude more important that degradation effects. Moreover, our analysis reveals that small energy to power ratios are optimal, suggesting that current battery systems are often oversized. This is because revenue collected from market participation is often limited by the power rating of batteries and not by the available storage capacity. We also find strong spatial variability of the revenues in the California grid: the 1% most lucrative nodes (locations) provide revenues that are 10 times larger than the mean. This indicates that battery placement is key. Finally, we demonstrate the complex relationship between battery placement, sizing, degradation, and market participation by analyzing the Tesla Powerpack battery system and recent installation near Chino, CA [1,3].

References:

[1] Cardwell and Krauss (2017). A Big Test for Big Batteries. *New York Times*. https://www.nytimes.com/2017/01/14/business/energy-environment/california-big-batteries-as-power-plants.html

[2] Dowling, Kumar, Zavala (2016). A Multi-Scale Optimization Framework for Electricity Market Participation. *Applied Energy* 190, pg. 147-164.

[3] https://www.tesla.com/powerpack

490433 Techno-Economic and Environmental Optimization of Wind Farm with Energy Storage Systems

Abstract:

The growing concern of climate change induced by energy consumption form fossil fuels, has led to a worldwide focus on the use of renewable energy sources such as solar and wind to fulfil the growing energy demand. Wind energy is one of the most promising renewable energy sources with an estimated total of 2-6 TW globally accessible, which can fully satisfy the current electricity consumption of the whole world which is estimated to be around 2.3 TW [1]. However, the discrete generation and inherent uncertainty in wind power cannot be overlooked. The integration of an Energy Storage System (ESS) is being considered as a promising solution to overcome the problem of intermittency.

Facing challenges in designing and operating wind farms with energy storage systems, we propose a systematic methodology to minimize the relevant costs. A receding horizon optimization, as a supervisory layer of economic model predictive control (EMPC), provides a macro-scale scheduling of the whole wind farm and energy storage components, which regulates the trade-off between power output and life-cycle performance. Furthermore, individual pitch controllers of wind turbine are designed to mitigate effects from vibrations and fatigue loads on its main flexible components such as the blades, tower, and drive shaft. A management system for ESS is also integrated for flexible adaptation of energy storage technologies.

In this paper we detail the holistic approach from the wind farm and ESS modeling and optimization, to wind turbine aerodynamic and structural models with appropriate control strategies. A case study is discussed thoroughly combining an offshore wind farm (OWF) and multiple innovative low-cost battery technologies. OWF is an emerging technology to utilize wind energy on a large-scale, but faces more complicated weather conditions than onshore systems leading to higher operational and maintenance costs. The rich offshore wind resources are located far away from the end load and require coupling to high capacity transmission lines. Additional ESS can mitigate transmission congestion, perform frequency control, and defer or avoid transmission and distribution upgrades. The wind farm with ESS needs to be deployed as an integrated system taking its operational, economic and environmental impacts into account. It is demonstrated how the proposed methodology is applied to an OFM constituted by 5-MW and 1.5 MW wind turbines in a real environment, with a properly sized ESS, in various scenarios of feed-in-tariff or other incentive schemes. The results show significant benefits in the overall economic, environmental and social performance for not only the wind farm owners, but also the overall energy supply structure. The simulation results using the Fatigue, Aerodynamics, Structures, and Turbulence (FAST) wind turbine simulation tool developed by NREL [2] prove the adopted control strategy is able to further reduce the unbalanced loads acting on the rotor compared to the traditional PI control method with reduced computational efforts, making it suitable for real time implementations. The environmental issues in wind farm and ESS development are also discussed.

[1] International Energy Agency (IEA). Key World Energy Statistics. 2016.

[2] Jonkman, J., Butterfield, S., Musial, W. and Scott, G., 2009. Definition of a 5-MW reference wind turbine for offshore system development. *National Renewable Energy Laboratory, Golden, CO, Technical Report No. NREL/TP-500-38060.*

495110 Hybrid Renewable Energy System Evaluation for Industry Applications

Abstract:

Chilean economy is known to be closely tied to ore processing capabilities. Ore-processing is known for involving energyâ€“intensive processes, such as comminution, concentrating, and cathodic processes. Given these energy needs, and the fact that electrical energy supply in the north of Chile is based on fossil fuels, it is important to consider the use of renewable energies as environment-friendly and economic alternatives. The aim of this work is to evaluate, by an optimized techno-economic analysis, the use of photovoltaic and a novel electrolytical wind-based offshore technology to supply at least 10% of the current and the predicted electrical energy requirements of the mining industry in the Antofagasta region. A combination of optimization problems was performed to obtain the optimal number of solar and electrolytical wind-based technology units. Total energy generation from a novel technology unit is 67,616 MWh/year. Considering a 65% efficiency of the combined cycle fed with hydrogen and oxygen, 28,133 MWh/year of electrical energy would be obtained. For the cases studied the cost of energy from the combined system was estimated to be between 0.255 US$/KWh and 0.273 US$/KWh, slightly higher than the average energy regional cost. According to the analysis, the renewable energy system could be a sustainable alternative to supply economic green energy to the mining industry in Chile.

490335 Sustainable Planning of Urban Energy-Water-Food Nexus through Decision Making Tools

Abstract:

The United Nationâ€™s Sustainable Development Goals (UN SDGs) condense the major challenges facing human society into 17 categories, many of which are interlinked and related to the energy-water-food (EWF) and climate change nexus. Integrated system analysis is more and more important for sustainable development by addressing challenges in natural resources, economic sectors, and human behaviors in a systematic manner. However, the main infrastructure and service of energy, water and food sectors are delivered in an isolated manner, causing investment and operational strategies limited to a single system and neglecting potential synergies across sectors.

To address the urgent challenges society is facing in urbanization and its associated energy demand and supply, we developed a holistic methodology and platform to support the resilient and sustainable planning at city-region level for multiple sectors, combining agent-based modelling to simulate and forecast resource demands on spatial and temporal scales, with resource network optimization incorporating capital expenditures, operational costs, and environmental impacts to find optimal allocations of resources and technologies given these demands [1]. In addition to a comprehensive study of water and sanitation sector study in a developing country context [2], we will also demonstrate how this framework can be applied to urban energy systems and the energy-water-food nexus.

The results suggest environmentally friendly and cost effective plans for sustainable city development by 2030. Both energy demand side simulation and supply side optimization are used to test innovative energy deployment policies especially to address the provision of clean energy for every citizen. A set of scenarios consisting of technological evolvement, effects of climate change and related policies are created to illustrate the application of the model especially in energy sector development. The cost of greenhouse gas emissions and the opportunity cost of land use based on the feasible food production are evaluated and optimized in various scenarios. Some typical scenarios are discussed in details including climate change impacts and policy interventions, such as how the capacity factor reduction for large-scale hydropower, feed-in electricity tariffs or agricultural intensification with yield increasing change the optimal power generation mix in future. Specific to the studied region (e.g., Ghana), it is found that the reduced power generation by large-scale hydropower is compensated for by the introduction of additional natural gas, solar thermal and photovoltaic capacity. Agricultural intensification is observed to promote the use of both solar thermal and solar photovoltaic power with a preference for solar photovoltaic power. Not only specific case and scenario analysis, the major contributions of this work also include the creation of a generalized algorithm within a uniform platform that can effectively handle and dispatch different types of technology in a variety of essential sectors.

[1] Xiaonan Wang, Koen H. van Dam, Charalampos Triantafyllidis, Rembrandt H.E.M. Koppelaar, and Nilay Shah. "Improving energy and water management in sustainable urban development by a systematic decision-making platform" (under review).

[2] Charalampos Triantafyllidis, Rembrandt H.E.M. Koppelaar, Xiaonan Wang, Koen H. van Dam, and Nilay Shah. "Resilience.io: An integrated optimisation platform for sustainable resource and infrastructure planning" (under review).

504143 Theory-guided Data Science: A New Paradigm for Scientific Discovery in the Era of Big Data

Abstract:

The potential of data science methods, that have found tremendous success in the commercial arena, is increasingly being recognized for advancing scientific discovery, especially in applications that are critically linked to sustainability. However, the "black-box" application of data science methods in oblivion to the domain context and knowledge has met with limited success in scientific domains, where complex physical phenomena are insufficiently represented using scarcely available data samples. This talk will introduce a novel paradigm for scientific discovery, termed as theory-guided data science, that uses the unique capability of data science methods to automatically learn patterns and models from big data, but without ignoring the treasure of accumulated scientific knowledge. An overarching goal of theory-guided data science is to learn data science models that not only make accurate predictions on the available labeled data, but are also interpretable and consistent with scientific theories and principles, thus translating to scientific advancements. This talk will describe several strategies for integrating scientific knowledge in conventional data science frameworks, using illustrative examples of applications from a diverse range of scientific domains such as material science, hydrology, turbulence modeling, bio-medical engineering, and neuroscience. The talk will conclude with a detailed case study on mapping the dynamics of freshwater bodies at a global scale using data from Earth observing satellites.

504245 Deepmetabolism: A Deep Learning System to Predict Phenotype from Genome Sequencing

Abstract:

High-throughput sequencing technology has brought life science into a "big data" era with an unrivaled explosion in the amount of genomic and transcriptomic data. The falling cost (<$1,000 per human genome) and increasing speed (<1 day per human genome) of high-throughput sequencing lead to the snowballing data at petabyte level. However, it is still difficult to transfigure such "Big Data" to valuable biological insights such as cell growth rate and metabolic pathway activities. The gap between genome sequencing and cell phenotypes is one of the biggest challenges that need to be solved to achieve "Data-to-Insight". In recent five years, the rapid development of artificial intelligence, especially deep learning, provides a novel option to overcome this challenge. Deep learning is found to be extremely effective in learning and modeling complex systems based on the graphic processing unit computation. In this study, we developed DeepMetabolism, a deep learning system that predicts cell phenotypes from genome sequencing data such as transcriptomics data. DeepMetabolism uses biological knowledge to design a neural network model and integrates unsupervised learning with supervised learning for predicting multiple phenotypes. In a prototypic application on *E. coli*, DeepMetabolism is able to predict phenotypes with high accuracy (PCC>0.92), high speed (<30 min for >100 GB data using a single GPU), and high robustness (tolerate up to 75% noise). Such outstanding features can be used to facilitate the discovery of natural products via rapid analysis based on DeepMetabolism. We envision DeepMetabolism to bridge the gap between genotype and phenotype and serve as a springboard for applications in synthetic biology and precision medicine.

501863 Comparison of Ozone Analyses between Camx and Denfis for Selected Monitoring Sites in an Ozone Nonattainment Area

Abstract:

Ground-level ozone is a secondary air pollutant produced by photochemical processes of nitrogen oxides (NOx) and volatile organic compounds (VOC). The Houston-Galveston-Brazoria (HGB) is identified by the USEPA as an ozone nonattainment area, and it is important to identify and quantify essential factors affecting ozone formation to improve the State Implementation Plans (SIP) for effective ozone reduction. In this study, site-specific ozone analyses were performed using a CAMx (Comprehensive Air Quality Model with Extensions) model and a DENFIS (Dynamic Evolving Neural-Fuzzy Inference System) algorithm at several selected ozone monitoring stations in the HGB area. The CAMx is a multi-scale, three dimensional photochemical grid model which simulates dynamic ozone concentrations under specific meteorological and emission inventory conditions. The DENFIS algorithm can generate data clusters along with the corresponding Takagi-Sugeno fuzzy rule set to predict dynamic ozone concentrations. Raw datasets for the fuzzy system were downloaded from the Texas Air Monitoring Information System on the TCEQ website. They were processed through a series of filters to ensure the quality of data before modeling. Each fuzzy system takes one year of hourly measurements as inputs for ozone modeling, which includes 11 meteorological variables and 5 types of air pollutants. The DENFIS algorithm uses the ECM (Evolving Clustering Method) for data clustering. Cluster center generation is distance-based, which has two modes: offline constrained optimization mode, and online one-pass mode. In the offline mode, pre-processed data were sampled every 17 hours for model testing, and the rest were used for model training; whereas in an online mode, new fuzzy rules can be dynamically inserted to update the model in real-time. The adaptive training is controlled by: (a) threshold value that determines number of cluster centers; (b) maximum number of iterations; (c) step size for least square optimization; and (d) width of the triangular membership function. Defuzzification of the system is a standard first order Takagi-Sugeno type.

The objectives of the study were to compare the effectiveness of the two models for ozone prediction and investigate the corresponding advantages and disadvantages of the two models for sensitivity analysis for the studied sites in the HGB area based on the 2012 Ozone Episode recently made available by TCEQ (Texas Commissions on Environmental Quality). Additional fuzzy inference analyses based on the 2013 through 2016 ozone data were also performed to establish the year-by-year trends of essential factors affecting ozone concentrations at selected monitoring sites in the HGB area.

511618 Importance of interdisciplinary frameworks and data in the Sustainable Healthy Cities SRN

Abstract:

This presentation will speak about how data is integrated with interdisciplinary projects addressing environmental sustainability, health and well-being drawing from work of the NSF-funded Sustainability Research Network: Integrated Urban Infrastructure Solutions for Environmentally Sustainable, Healthy, and Livable Cities. Integration of data with interdisciplinary sustainability projects will be illustrated through two example projects addressing the food-energy-water nexus. Examples will demonstrate the need for frameworks to organize data, the quantity and diversity of data needed for sustainability and systems analysis, partnerships for data collection, and application to policy.

The Waste Plastic Challenge

- In 2010, 270 million metric tons (MMT) of plastic produced
  - 99.5 MMT discarded by coastal populations
  - 31.9 MMT mismanaged on land
  - 4.8-12.7 MMT entered the oceans
- 2050 Projections:
  - 33 Billion MT produced
  - 3,900 MMT mismanaged
  - 6,00-1,600 MMT will enter oceans
  - 122 fold increase in 40 years!

UK UNIVERSITY OF KENTUCKY College of Engineering

2

Need for Decentralized Solutions

- Lack of infrastructure and capital resources to manage municipal solid waste (MSW)
- Openly dumped on streets, drainage channels, near slums, or on undeveloped plots
- Leads to leaching, soil impregnation and contamination of underground waters
- Incineration converts waste plastic into CO_2, NOx, SOx, dusts, dioxins and similar toxins
- Effective waste abatement strategies must be decentralized
  - Targeting where generated rather than focusing on centralized processing
  - Empowers individuals to adapt solutions
  - One potential solution: waste plastic to fuel

UK KENTUCKY

3

Metric Indicators

Categories
Economic Factors
Global Economic
Gross Domestic Product (GDP) 2016 (Billion USD)
GDP Per Capita 2016 (USD)
Social Factors
Global Social
Population 2016
Population Below Poverty Line 2016 (%)
Population Density (capita/sq km)
Environmental Factors
Global Environmental
Estimated MSW Generation 2016 (tonnes/day)
Environmental Stress (tonnes MSW/sq km)
Estimated Plastic Waste in MSW 2016 (tonnes/day)
Estimated Unsound Plastic Waste Disposal (tonnes/day)

- Goal: Critical locations for implementing decentralized plastic waste solutions
- Analyzed:
  - 200 countries
  - 9 indicators – representing 3 pillars of sustainability
  - A holistic approach to understand contributions to waste generation
  - Metric base year: 2016

UK KENTUCKY

4

Calculational Methodology

- World Bank reported 2005 and 2025 projections for *Estimated MSW Generation (tonnes/day)*
  - Data linearly interpolated to obtain 2016 values
  - Similar case for *Estimated Plastic Waste in MSW (tonnes/day)*
- Land Area
  - Used to obtain *Population Density (capita/ km^2)*
  - *Environmental Stress (tonnes MSW/km^2)*
- *Estimated Unsound Plastic Waste Disposal (tonnes/day)*
  - Based on MSW unsound disposal from Waste Atlas

5

Accounting for Missing Data

Income Level
Low Income (LI)
Low Middle Income (LMI)
Upper Middle Income (UMI)
High Income Countries (HIC)

Region
East Asia Pacific (EAP)
Europe & Central Asia (ECA)
Latin America & the Caribbean (LCR)
Middle East & North Africa (MENA)
South Asia (SAR)
Sub-Saharan Africa (AFR)
Organization for Economic Co-operation & Development (OECD)

- Data classifications: Income Level and Region
- Average simultaneously grouped classifications to replace missing data
- Groups with insufficient information replaced with income level averages only

6

Methodology

- Determining country performance:
  - Assign weighting factors to individual indicators
  - Lowest performing nation ranked 1
- Global weighting factors – Balance 3 Pillars of Sust.
- All weights sum to 100%
- Indicators equally balanced for base case

$$\sum_{i=1}^{n} G_{EC}\left[(I_{EC1} * W_{EC1}) + (I_{EC2} * W_{EC2})\right] +$$

$$G_{SC}\left[(I_{SC1} * W_{SC1}) + (I_{SC2} * W_{SC2}) + (I_{SC3} * W_{SC3})\right] +$$

$$G_{EV}\left[(I_{EV1} * W_{EV1}) + (I_{EV2} * W_{EV2}) + (I_{EV3} * W_{EV3}) + (I_{EV4} * W_{EV4})\right]$$

Where:
I = Indicator
G = Global weighting factor
EC = Economic; SC = Social; EV = Environmental
W = Single weighting factor corresponding to an indicator

UK KENTUCKY

7

Results

Categories	Weighting
Economic Factors	
Global Economic	**33.33%**
Gross Domestic Product (GDP) 2016 (Billion USD)	50.00%
GDP Per Capita 2016 (USD)	50.00%
Social Factors	
Global Social	**33.33%**
Population 2016	33.33%
Population Below Poverty Line 2016 (%)	33.33%
Population Density (capita/sq km)	33.33%
Environmental Factors	
Global Environmental	**33.33%**
Estimated MSW Generation 2016 (tonnes/day)	25.00%
Environmental Stress (tonnes MSW/sq km)	25.00%
Estimated Plastic Waste in MSW 2016 (tonnes/day)	25.00%
Estimated Unsound Plastic Waste Disposal (tonnes/day)	25.00%

Top 10 Results

Country	Value
Bangladesh	44.6
Burundi	46.0
Haiti	47.7
Pakistan	49.3
Malawi	49.4
Nigeria	51.0
Rwanda	51.2
Syria	53.4
Vietnam	54.8
India	57.3

Countries with lowest total sum considered to be most feasible locations for implementing decentralized solutions to plastic waste accumulation

UK KENTUCKY

8

Effects of Weighting Factors

Categories	Weighting
Economic Factors	
Global Economic	**50.00%**
Gross Domestic Product (GDP) 2016 (Billion USD)	50.00%
GDP Per Capita 2016 (USD)	50.00%
Social Factors	
Global Social	**25.00%**
Population 2016	33.33%
Population Below Poverty Line 2016 (%)	33.33%
Population Density (capita/sq km)	33.33%
Environmental Factors	
Global Environmental	**25.00%**
Estimated MSW Generation 2016 (tonnes/day)	25.00%
Environmental Stress (tonnes MSW/sq km)	25.00%
Estimated Plastic Waste in MSW 2016 (tonnes/day)	25.00%
Estimated Unsound Plastic Waste Disposal (tonnes/day)	25.00%

Top 10 Results

Country	Value
Burundi	39.2
Malawi	43.6
Haiti	44.4
Rwanda	47.2
Comoros	51.7
Togo	51.8
Syria	54.5
Bangladesh	56.1
Gambia, The	58.2
Congo, Democratic Repulbic of the	59.2

UK UNIVERSITY OF KENTUCKY

9

Effects of Weighting Factors

Categories	Weighting
Economic Factors	
Global Economic	**33.33%**
Gross Domestic Product (GDP) 2016 (Billion USD)	50.00%
GDP Per Capita 2016 (USD)	50.00%
Social Factors	
Global Social	**33.33%**
Population 2016	33.33%
Population Below Poverty Line 2016 (%)	33.33%
Population Density (capita/sq km)	33.33%
Environmental Factors	
Global Environmental	**33.33%**
Estimated MSW Generation 2016 (tonnes/day)	15.00%
Environmental Stress (tonnes MSW/sq km)	15.00%
Estimated Plastic Waste in MSW 2016 (tonnes/day)	15.00%
Estimated Unsound Plastic Waste Disposal (tonnes/day)	55.00%

Top 10 Results

Country	Value
Bangladesh	43.4
Burundi	43.9
Haiti	46.0
Malawi	46.0
Pakistan	47.3
Nigeria	47.8
Rwanda	49.2
Syria	50.3
Vietnam	53.4
Congo, Democratic Repulbic of the	54.1

UK UNIVERSITY OF KENTUCKY

10

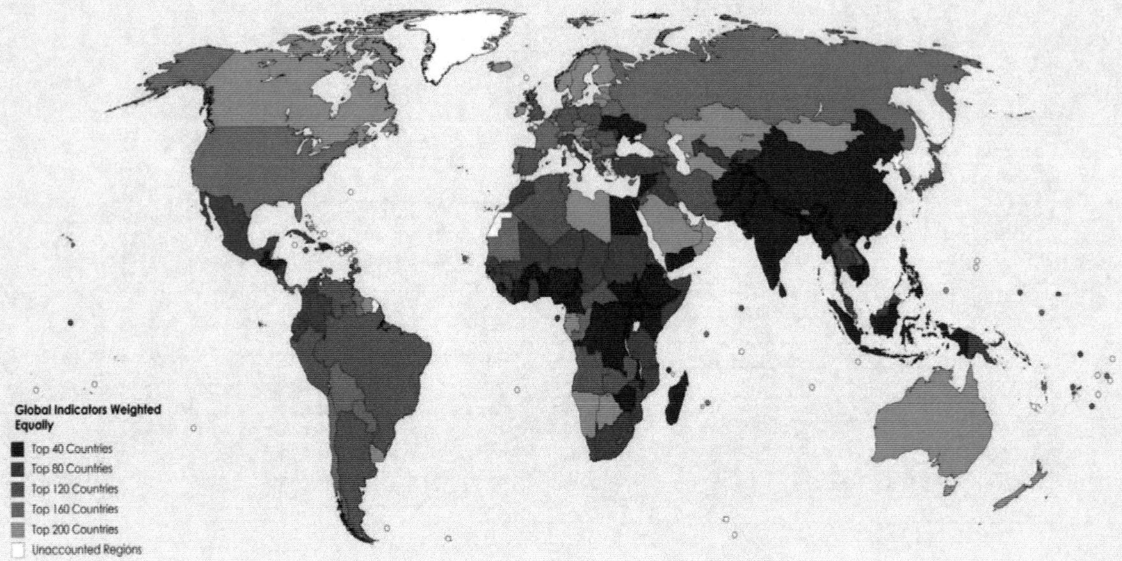

Heat Map Overview

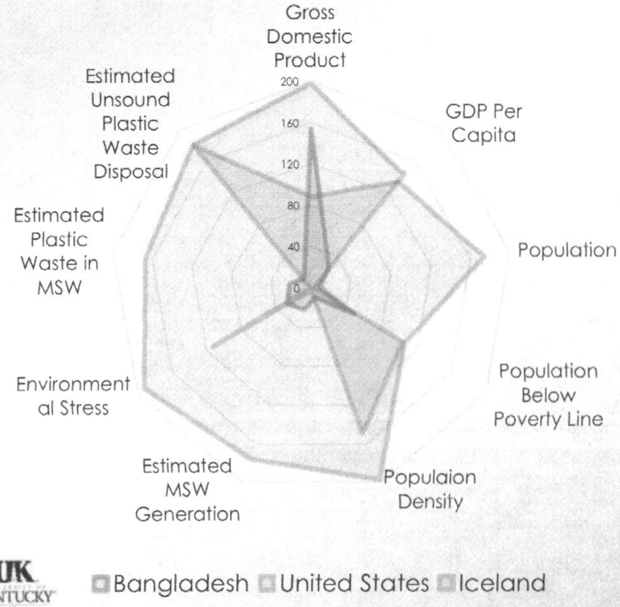

Results Spotlight

- Bangladesh:
  - Population – 156 million
  - 31.5% of population living below poverty line
  - 60-87% reliable access to drinking water
- Iceland:
  - Population – 334,000
  - 2.7% unemployment rate
  - 98% access to drinking water
- United States:
  - Highest GDP – $18.57 trillion
  - 99.2% access to drinking water
  - 2nd in MSW generation with 12% plastic

12

Kampala, Uganda Case Study

- Kiteezi Landfill Waste Accumulation Analysis:
  - Serves the city's 5 divisions (210 sq km radius)
  - 1300-1500 metric tons of MSW per day
  - Accounts for half of MSW generated, rest unsoundly disposed
  - 500 waste-pickers employed to manage recyclable materials
    - Pick 40 kg of plastic at $5.54 USD/day
  - Soft plastics, grocery bags not recycled – 3.8% of total MSW
  - 49,400 kg of plastic landfilled per day
  - 1,235 additional jobs created

13

Conclusion & Future Work

- Waste Plastic Data Analysis is a tool for researcher and policy makers to:
  - Collect in-depth understanding of a region's waste management outlook
  - Determine feasible locations to implement decentralized plastic waste solutions
  - Reduce accumulation on land and eventual migration to oceans
- Future Work:
  - Development of a metric to assess sustainability of appropriate technology based solutions used to target waste accumulation

14

References

- Patni, N., Shah, P., Agarwal, S., & Singhal, P. Alternate Strategies for Conversion of Waste Plastic to Fuels. *ISRN Renew. Energy.* **2013**, 902053 (2013)
- Rochman, C. *et al.* Policy: Classify plastic waste as hazardous. *Nature* **494**, 169-171 (2013)
- Wilcox, C., Sebille, E. V., & Hardesty, B. D. Threat of plastic pollution to seabirds is global, pervasive, and increasing. *Proc. Natl. Acad. Sci. U.S.A.* **112**, 11899-11904 (2015)
- Li, W. C., Tse, H. F., & Fok, L. Plastic waste in the marine environment: A review of sources, occrence and effects. *Sci. Total Environ.* **566-567**, 333-349 (2016)
- Jambeck, J. R. *et al.* Plastic waste inputs from land into the ocean. *Science* **347**, 6223, 768-771 (2015)
- Jayasiri, H. B., Purushothaman, C. S., & Vennila A. Quantitative analysis of plastic debris on recreational beaches in Mumbai, India. *Marine Poll. Bull.* **77** (1-2), 107-112 (2013)
- Barnes, D. F. K., Galgani, F., Thompson, R. C., & Barlaz, M. Accumulation and fragmentation of plastic debris in global environments. *Phil. Trans. R. Soc. B* **364**, 1985-1998 (2009)
- Joshi, C. & Seay, J. An Appropriate Technology Based Solution to Convert Waste Plastic into Fuel Oil in Underdeveloped Regions. *J. Sust. Dev.* **9** (4), (2016)
- DeNeve, D., Joshi, C., Samdani, A., Higgins, J., & Seay, J. Optimization of an Appropriate Technology Based Process for Converting Waste Plastic in to Liquid Fuel via Thermal Decomposition. *J. Sust. Dev.* **10** (2), (2017)
- A-Z list of country and other area pages. https://www.state.gov/misc/list/ (2017)
- The world factbook, field listing: GDP (official exchange rate). https://www.cia.gov/library/publications/the-world-factbook/fields/2195.html (2017)
- The world factbook, field listing: GDP – per capita (PPP). https://www.cia.gov/library/publications/the-world-factbook/fields/2004.html (2017)
- The world factbook, field listing: Population below poverty line. https://www.cia.gov/library/publications/the-world-factbook/fields/2046.html (2017)
- Hoornweg. D, & Bhada-Tata, P. What a waste: A global review of solid waste management. The World Bank, Washington, DC. https://openknowledge.worldbank.org/handle/10986/17388 (2012)
- Waste Atlas, field listing: Unsound disposal. http://www.atlas.d-waste.com/ (2017)
- How are the income group thresholds determined? The World Bank Group. https://datahelpdesk.worldbank.org/knowledgebase/articles/378833-how-are-the-income-group-thresholds-determined (2017)
- mapchart.net. https://mapchart.net/detworld.html (2017)

UK KENTUCKY

15

WENQIN LI

1329 Mayfield, Unit 202
Ames, IA, 50014

515-708-8112
wenqin@iastate.edu

EDUCATION

- **Ph.D. Iowa State University** 2013-2018
  Major: Mechanical Engineering Co-Major: Bio-renewable Resource and Technology
  Minor: Chemical Engineering
- **Bachelor China Agricultural University** 2009-2013
  Major: Vehicle Engineering

RESEARCH

- **Process modeling of multiple chemical processes**

- Built detailed chemical process models for chemical conversion of biomass using Aspen Plus/Hysys and ChemCAD, processes include fast pyrolysis, fractionation, hydrotreating, distillation, H_2 reforming, liquefaction, fermentation, anaerobic digestion etc.;
- Calculated Mass and energy balance to evaluate the energy efficiencies of the processes;
- Did heat integration and pinch analysis to improve the heat usage efficiency;
- Used Aspen Plus Simulation Workbook to run fast pyrolysis process of over 400 types of different feedstocks to evaluate how biomass properties affect the biofuel and biochar yield.

- **Techno-economic analysis to evaluate economic performance of different new technologies**

- Evaluated the capital and operating costs for novel chemical processes using Aspen Plus Economic Analyzer (APEA), scaling factors, installation factors etc.;
- Wrote Mathematica programming codes to apply discounted cash flow calculation to determine the break-even selling price of final products (biofuel or chemicals), to further evaluate the market competitiveness of the products;
- Evaluated the impacts extent of different parameters using sensitivity/uncertainty analysis;
- Calculated minimum fuel selling prices for fast pyrolysis of over 400 types of biomass to assess how biomass properties affect the platform economic performance.

- **Life cycle analysis to evaluate environmental performance of different new technologies**

- Designed customized biofuel production pathway in GREET and SimaPro, drew the system boundary, chose the functional unit and different allocation methods, and created life cycle inventory;
- Wrote Mathematica codes to calculate GHG emissions for fast pyrolysis of over 400 different types of biomass to evaluate how biomass properties affect the environmental performance;
- Compared GHG emissions of stand-alone fast pyrolysis plant vs. plant integrated with refineries, as well, compared emissions when producing different byproducts.

SKILLS

- **Programming:** C, MATLAB, Mathematica, Fortran, Visio Basic
- **Applications:** Aspen Plus/Hysys, ChemCAD, GREET, SimaPro, Biograce, GHGenius, AutoCAD, MS Office, Microsoft Access

WORKING EXPERIENCE

- **Renewable Energy Group, Internship in EHS** Summer 2017

- Developed a database tool using Microsoft Access to automatically calculate carbon intensity (CI) for biodiesel produced from different refineries and feedstocks, broadened the impacts of the CI calculation tool by educating plant engineers of how to use it;
- Involved in designing and submitting novel product pathways to California Air Resources Board (CARB) and EPA;
- Investigated CARB and EPA's future programs to strategically prepare REG for the policy adjustments and maximize the RIN and LCFS credits.

TEACHING EXPERIENCE

- **ME 270: Introduction to Mechanical Engineering Design** Fall 2013

- Guided students through the assignments of each class;
- Answered students in-class questions to help them complete the design to improve people's lives in African developing countries;
- Organized the final design expo for over 150 students.

AWARDS

- **Third place prize in poster competition** April 2017

- Poster competition was held by the Center for Crops Utilization Research (CCUR) and BioCentury Research Farm (BCRF)

PUBLICATIONS

- **Wenqin Li**, Dang, Q., Smith, R., Brown, R. C. & Wright, M. M. Techno-economic analysis of the stabilization of bio-oil fractions for insertion into petroleum refineries. *ACS Sustain. Chem. Eng.* (2017). **(Special Issue Paper)**

- **Wenqin Li**, Qi Dang, Robert C. Brown, David Laird, M. M. W. The impacts of biomass properties on the pyrolysis yields, economic and environmental performance of the pyrolysis-bioenergy-biochar platform to carbon negative energy. *Bioresour. Technol.* (2017). (**Front Cover Paper**)

492233 Process Synthesis and Economical Evaluation of Torrefied Wood Pellet Production Processes

Abstract:

Torrefied wood pellet is a solid biofuel derived from forestry, agriculture, or industrial woody residues. The torrefied wood pellets, with properties close to coal [1], can be potentially used to replace coal for power generation. The design and optimization of the wood pellet production process can help save energy consumption, reduce CO_2 emissions, and increase economic benefits.

A torrefied wood pellet production process involves several unit operations including drying, grinding, torrefaction, pelletizing and cooling, with major heat consumption in drying and torrefaction, and power mainly consumed for grinding and pelletization. During torrefaction combustible organic volatiles are also generated from biomass decomposition, which can be used to provide heat for the drying and torrefaction to make the overall process auto-thermal, thus improve process energy and material efficiency [2]. The heating value of organic volatiles and their mass fraction can be adjusted by controlling torrefaction reactor conditions, such as reactor temperature and mean solid residence time. There are several potential configurations of the above biomass torrefaction and densification processes based on desired pellet product quality and energy efficiency, such as 1) drying " torrefaction " grinding " pelletization; 2) drying " grinding " torrefaction " pelletization, 3) drying " grinding " pelletization " torrefaction. Previous studies of the torrefied wood pellet production process have focused on either single energy balances or single reactor models. Our study combined both considering detail reactor equipment design with particle mechanics, thermal dynamics, and kinetics etc., as well as different heat integration strategies.

In this study, we conducted a systematic investigation of the above three configurations for the production of torrefied pellets using biomass feedstocks of different physical properties. Different biomass feedstocks of different properties require the use of different equipment for drying and torrefaction. For wood chips, rotary dryers and torrefiers are considered with combined direct and indirect heat transfer from combustion flue gases. For sawdust powders, fluidized bed dryer and torrefier with or without build-in heat exchangers are applied. The equipment types and heat transfer mode are selected based on the constraints of operating conditions e.g. inlet gas temperature, gas velocity and availability of gas blowers for high temperature flue gases etc., as well as the overall flowsheet heat integration strategies. Sensitivity analysis was also performed to search for the optimum operating conditions. After the flowsheet synthesis, we carried out economical evaluation to identify the most promising configurations for making desired products. Different strategies were proposed and discussed for different purpose e.g. high energy qualified product, low energy demand, or low production costs. All the synthesized processes were modeled and simulated using Aspen Plus under steady state.The developed flowsheet simulation packages could be applied to assist the development and scale-up of commercial biomass torrefaction and densification process.

References

[1] WH Chen, J Peng, XT Bi., 2015. A state-of-the-art review of biomass torrefaction, densification, and applications. Renewable and Sustainable Energy Reviews, 44, 847-866.

[2] Prins, M.J., 2005. Thermodynamic analysis of biomass gasification and torrefaction. Ph.D. Thesis. Eindhoven University of Technology.

492924 Implementation of Six Sigma in a Methane Generation Process

Abstract:

Distillery process waste water contains organic and inorganic compounds with high COD strength, which causes negative impact on environment. Instead of discharging this waste water to sewer system, it can be pre-treated and used for energy generation before discharging it to the environment. The pre-treatment that possibly used is the anaerobic digestion process, several groups of bacteria will feed on this waste water to generate methane gas. An expanded granular sludge bed reactor has been utilized as the main reactor for this process. The biogas produced will be mainly consists of methane and carbon dioxide. In this paper we studied, how to improve the biogas production and improving methane composition in the generated gas using Six Sigma methodology. DMAIC (Define Measure Analyze Improve Control) method has been implemented in this biogas production process so, process failure could be identified, improve the gauges used in measurement and finally, enhance the yield and composition of methane.

Six Sigma DMAIC methodology behaves as a roadmap to understand various unexplored areas in this process that could help in organizing and updating the standard operating procedure for the experiments. The improvement was observed after introducing six sigma tools and concepts in our experiment, the yield was increased from 11 gallons to 28 gallons in 60 hours which is an impressive 154.5% increase in yield and increase in the percentage composition of methane in the yield from 50% to 76%.

495569 Performance of Fed-Batch Relative to Batch Processes in Unsteady State Fermentation and *in situ* Gas Stripping Simulations Under Different Conditions

Abstract:

The production of biofuels and chemicals from lignocellulosic biomass on the fermentation platform is considered a sustainable energy alternative to fossil fuels mainly because of the mitigation of environmental impacts and energy independence. Butanol and ethanol have properties that make them excellent liquid transportation fuel alternatives. Fermentation is frequently operated as a batch or semi-batch process. The batch fermentation using microorganisms is characterized by low reactor productivities, yield and total solvent produced because of product toxicity and substrate inhibition to the microorganisms. The integrated fed-batch fermentation and *in situ* gas stripping circumvents substrate inhibition by using a substrate feeding strategy that utilizes a lower initial substrate concentration at the beginning of fermentation. As fermentation proceeds and the substrate is used up, fresh substrate is added at concentrations that sustains microbial growth and product formation, and below substrate concentrations inhibitory to the microorganisms. Furthermore, product inhibition is alleviated by removing solvents as they are produced to prevent the build-up of solvents to inhibitory concentrations.

In this study, the batch reactor, RBatch block, in Aspen Plus is linked to a Fortran user kinetics subroutine to simulate an unsteady state fed-batch fermentation process. A fed-batch and *in situ* gas stripping process will be simulated by continuously feeding nitrogen gas to an RBatch block. With this simulation approach, the advantages of the integrated fed-batch and *in situ* gas stripping to improve the productivity, yield and total solvent production by reducing substrate and product inhibitions relative to a batch fermentation will be highlighted under different conditions. An important feature of the procedure developed in this study will be to show that the trends predicted from the model are consistent with general trends and microbial growth kinetics observed in fermentation systems that use microorganisms in the face of high substrate and product inhibitions. This study provides a platform that can be used to as guide to the fermentation experimentalist and assess the accuracy of literature data. Additionally, a universally accepted process simulator, Aspen Plus, can be used in the simulation of fermentation processes linked with separation unit operations to provide strategies for the design and optimization of bioprocesses.

496926 Technoeconomic Assessment of Biofuel and Biochemical Production Using New Pretreatment Processes for Lignocellulosic Biomass and Lignin Valorization

Abstract:

Recently developed extractive ammonia (EA) and copper–bipyridine catalyzed alkaline hydrogen peroxide (AHP) pretreatments are capable of solubilizing a large portion of lignin in lignocellulosic biomass. Liquefied lignin can be valorized through electrochemical processes to produce value-added biochemicals such as cyclohexanol which is a Nylon 6,6 precursor. In addition, lignin removal improves saccharification by increasing the carbohydrate accessibility to the enzymes. This study evaluates the potential of two pretreatment processes for bioethanol and biochemical production from lignocellulosic biomass, namely: 1) extractive ammonia (EA) pretreatment and 2) copper–bipyridine catalyzed alkaline hydrogen peroxide (AHP) pretreatment. Both processes are simulated in Aspen Plus v8.8 to obtain heat and material balances required for energy and economic assessments. Technoeconomic models are developed to calculate the minimum ethanol selling price considering a 10% internal rate of return and 30 years of plant life. Results showed that high ammonia-to-biomass ratio in EA makes the ammonia recovery highly energy intensive. Similarly, high hydrogen peroxide and bipyridine demand in AHP makes it a costly process. Therefore, reducing the amount of pretreatment chemicals is required to improve the economic viability of both processes. Finally, sensitivity analyses on different process parameters were performed to identify the pitfalls and potential of each process for further development and economic profitability.

498923 Technoeconomic and Strategic Insights on Cellulosic Biofuel Production with Ionic Liquid Pretreatment

Abstract:

Biomass pretreatment is a key step in the biochemical conversion of cellulosic biomass to biofuels. Some ionic liquids (ILs) are very efficient solvents for biomass pretreatment (1-4). Given the possibility of numerous anions and cations, there is a significant opportunity to *design* ILs and *tune*their properties to improve sugar yields. Due to the highly-integrated nature of the cellulosic biorefineries, however, the pretreatment process needs to be developed not only to improve sugar yields but also to minimize any negative impacts on the other parts of the biorefinery (including enzymatic hydrolysis, microbial fermentation wastewater treatment, and on-site cogeneration facility to produce steam and electricity). Design of such processes requires understanding of the underlying material & energy balances, mass & heat transfer effects, thermodynamics, and kinetics. To this end, detailed model based technoeconomic analysis (TEA) offers valuable guidance. Our recent work (1-4) highlighted the importance of biorefinery processes with reduced water and energy footprint, and high IL recovery to ensure economic viability. With further advancements, some of the recently developed processes could be cost-competitive, with a minimum selling price of <$4/gal of advanced biofuel. This presentation will discuss TEA driven insights and challenges that must be addressed to realize an economical biomass conversion process at industrial scale .

References:

1. Konda et al. (2014). Understanding cost drivers and economic potential of two variants of ionic liquid pretreatment for cellulosic biofuel production. Biotechnology for Biofuels 7: 86.
2. Konda et al. (2015). Towards a cost-competitive ionic liquid based pretreatment: Insights from technoeconomic analysis. AIChE Annual Meeting, November 2015, Salt Lake City, UT.
3. Sun et al. (2016) CO_2 enabled process integration for the production of cellulosic ethanol using bionic liquids." Energy & Environmental Science 9.9: 2822-2834.
4. Xu et al. (2016) Transforming biomass conversion with ionic liquids: process intensification and the development of a high-gravity, one-pot process for the production of cellulosic ethanol." Energy & Environmental Science 9.3: 1042-1049.

500238 Process Simulation of Integrated Biomass Torrefaction and Pelleting Plant

Abstract:

Process simulation is a useful technique to perform a detailed mass and energy balance of a system using different operating conditions. design parameters and equipment specifications. Simulation is beneficial to scale-up the process and estimation of the investment costs. The process of torrefied pellet production including different unit operations such as combustion, drying, torrefaction, grinding, and pelletizing was simulated with Aspen Plus software platform. In this study, a process simulation of converting wood chips into torrefied pelleting plant was developed. Drying section was simulated through a solid conventional dryer applying the experimental data for drying kinetics. In the torrefaction section, a two-step kinetic model consists of intermediate products was employed and the solid product, as well as volatile products, were produced at various torrefaction temperature. A volatile stream was considered as a combination of ten major components with fixed composition deriving from the torrefaction experiments of forest biomass. At last, a pinch analysis was implemented, and several scenarios were proposed based on different capacities. The size reduction section and pelletizing were also modeled based on experimental particle size distribution and industrial energy consumption data. This simulation model might be applied to a wide range of feedstock types through woody biomass to energy crops, and the effects of various process parameters on the production cost and environmental impacts can be evaluated.

Keywords: process simulation, torrefaction, drying, pelletizing, Aspen Plus

497748 Producing Biocrude from Renewable Feedstocks through Hydrothermal Liquefaction

Abstract:

To maintain sustainability with respect to people, prosperity, and the planet, we must produce liquid biofuels from renewable and sustainable resources at a price that is competitive to petroleum derived fuels. Based on this motivation and market need, we have successfully developed a straightforward but highly efficient process to produce biodiesel from an oleaginous yeast strain, *Cryptococcus curvatus* grown on hydrolysate of sweet sorghum bagasse. The simple pretreatment of using 0.5% sulfuric acid at 121 °C for 1 h released 92.2% of theoretically available sugars in sorghum bagasse. The yeast strain thrived on the hydrolysates and accumulated lipids to an extent of 48% of dry cell weight. Through a one-step *in-situ* transesterification reaction using microwave energy, 92.0% of lipids in wet cells of *C. curvatus* were converted to crude biodiesel under an optimal condition.

The bagasse-to-biodiesel pathway is certainly ready to be scaled up. However, using this pathway, yeast cell residues (YCR) after transesterification and the pretreated and washed bagasse (PWB) following pretreatment are generated. To utilize these two materials, we seek to empoly hydrothermal liquefaction (HTL) to convert both to biocrude considering two reasons: (1) compared with other thermochemical processes, HTL is conducted at a relatively low temperature of 250-350°C; and (2) liquefaction can handle materials with any levels of water content. Thus, for the two target materials with moisture content of around 80%, HTL is an excellent choice. In this talk, we will present our recent results on HTL of these two biomaterials with different catalysts. Yields of biocrude and biochar from each material will be presented. In addition, the properties of the biocrude samples will be reported. The characterization of the resulting aqueous phase samples will be detailed, too. Our results have shown that biofuels can be developed successfully from renewable and domestic feedstocks through a combination of biochemical and thermochemical processes.

490011 Cell Lysis, Lipid Recovery, and Lipid Hydroprocessing from Oleaginous Yeast

Abstract:

Single-cell oils, or lipids derived from oleaginous microbes, have seen recent interest as a potential renewable source of transportation fuels. However, one of the main technical challenges involved in single-cell-oil-derived products is the need to lyse the cells and extract the lipids from the resulting complex matrix. We have screened a series of cell lysis approaches using the oleaginous yeast *Lipomyces starkeyi*, and have found that a dilute acid pretreatment compares favorably to high-pressure homogenization, alkaline pretreatment, bead beating, enzymatic hydrolysis or thermolytic approaches. In particular, acid pretreatment allows recovery of lipids in yields greater than 90% after hexane extraction. We have also shown that the acid pretreatment is effective on strains of *Cryptococcus curvatus* and *Rhodosporidium toruloides*, is readily scalable, and the extracted lipids can be hydroprocessed to a renewable diesel blendstock.

501451 Wood Pellet Boiler Heating Systems Evaluation and Optimization

Abstract:

The use of wood pellet boilers for residential space heating has significantly increased over the past decade. Wood pellets are a biomass-based renewable energy made by pelletizing debarked wood fiber. Compared to log wood and wood chips, wood pellets have higher energy density, relatively uniform fuel quality, easier to automate their use, and therefore, are being more widely used.

A typical wood pellet heating system has three components: the boiler (energy generation unit), the thermal energy storage (TES) tank (energy storage unit), and the building (energy consumption unit). The three components form into two loops: the boiler to TES tank loop and the TES tank to building loop.

Three modern wood pellet boilers were installed and monitored in this research. Two 25 kW boilers (PB and WPB) were installed in the end of 2014 and a 50 kW boiler (LGB) was installed in March of 2016. PB is used solely for radiant floor heating and WPB is used for traditional baseboard/cast iron radiators space heating as well as providing domestic hot water (DHW) supply. LGB boiler is used for concrete slab heating. The three boiler installations provide different methods of heating and different levels of building heat demand.

This research first evaluated the performance of residential scale wood pellet boilers (25 â€“ 50 kW) in terms of boiler operation characteristics, thermal efficiency, boiler emissions, TES tank stratification and discharge efficiency, etc. Comparisons were also made among different boilers with suggestions for improvements were made.

A process dynamic simulation using VMGSim was built up based on the field monitoring data. Good agreement between the simulation and field data was found. The model was then used to size the TES tank with respect to different boiler capacities and heating demands. The results will be used to compose a guidance document for wood pellet boiler heating systems.

UNIVERSITY OF BATH

An MILP model for integrated carbon-free heat networks

André Prates Pereira (*afpp20@bath.ac.uk*)
PhD student
Department of Chemical Engineering
University of Bath
United Kingdom

Supervisor: Dr Sheila Samsatli (*s.m.c.samsatli@bath.ac.uk*)

2017 AIChe Annual Meeting
Minneapolis Convention Center, 101C
October 30, 2017

UNIVERSITY OF BATH

A world-class University in a World Heritage City

50 UNIVERSITY OF BATH Established 1966

TOP 5 UK UNIVERSITY The Guardian 2018

TEF Gold

Department of Chemical Engineering

- **Top 3** for course satisfaction *(Guardian University Guide 2018)*
- **Top 5** in the UK for Chemical Engineering overall *(2018 Complete University Guide and Guardian University Guide 2018)*
- **Top 5** for teaching satisfaction *(Guardian University Guide 2018)*
- **89%** of our research is graded as either world-leading or internationally excellent *(Research Excellence Framework 2014)*

Outline

1. Introduction
2. Problem statement
3. MILP model
4. Great Britain case studies
   - Wind speed and heat demands
   - Site suitability for wind turbines (GIS analysis)
   - Results
5. Conclusion and future work

Introduction

- UK committed to reducing carbon emissions by 80% of 1990 levels until 2050.
- Decarbonising UK heat sector by replacing the use of fossil fuels with renewable energy.
- Challenges: intermittency of renewable energy, expensive energy storage and lack of infrastructure for hydrogen.
- H_2 as an energy carrier can present advantages:
  - Low transmission losses;
  - Large-scale and long-term storage, tackling the intermittency problem of renewable energy.
- Aim: To optimise strategic design and tactical operation of wind-electricity-hydrogen network to satisfy GB's heat demands

Problem Statement

UNIVERSITY OF BATH

Given:

- Suitable sites for onshore and offshore wind turbines
- Spatio-temporal onshore and offshore wind speed
- Spatio-temporal domestic heat demands
- Characteristics of each technology/infrastructure

Determine:

- The number, location and capacity of conversion and storage technologies
- Transmission infrastructure network
- Which is the best combination of technologies to use
- When and how to transport the resources/products
- When to store the resources/products

Subject to:

- The availability of onshore and offshore area to install wind turbines
- Technological constraints (e.g. production rates,
- Demand satisfaction

Objective function:

- Maximise profit

Temporal representation

UNIVERSITY OF BATH

- **Planning horizon: 1 year**
- **4 different seasons in a year**
- **13 identical weeks in a season**
- **5 identical weekdays and 2 identical weekend days in a week**
- **24 different hours in a day**

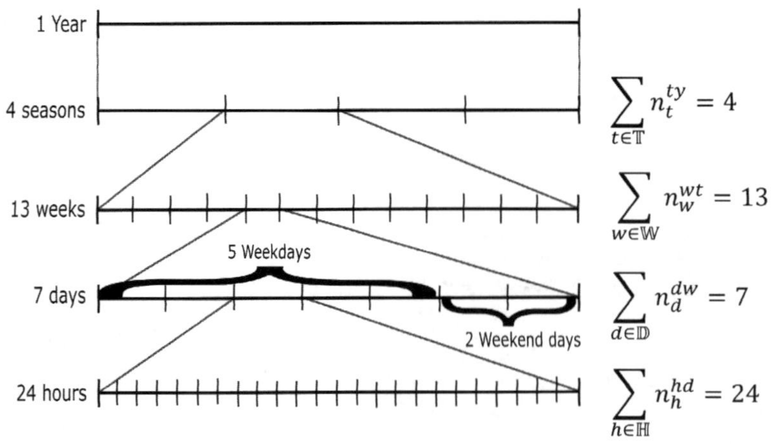

Spatial Representation

UNIVERSITY OF BATH

- **Great Britain is divided into 16 zones**
  (based on National Grid study zones
- **Each zone:**
  - **Characterised by:**
    - **available suitable sites for onshore and offshore wind turbines**
    - **variable wind speeds**
    - **variable heat demands**
  - **May host**
    - **Wind turbines**
    - **Conversion technologies**
    - **Storage technologies**
    - **Electricity transmission lines or hydrogen pipelines**

Integrated energy pathways

UNIVERSITY OF BATH

- Represented using Resource-Technology Networks (RTN)
- Able to model the most general situation
  - convert any resource to any other, including recycles (circular chains)
  - store and transport resources at any stage in the chain
- Consider a wide range of different feedstocks and technologies
  - different operating modes to generate different energy vectors
    - heat, electricity, transport fuel etc.

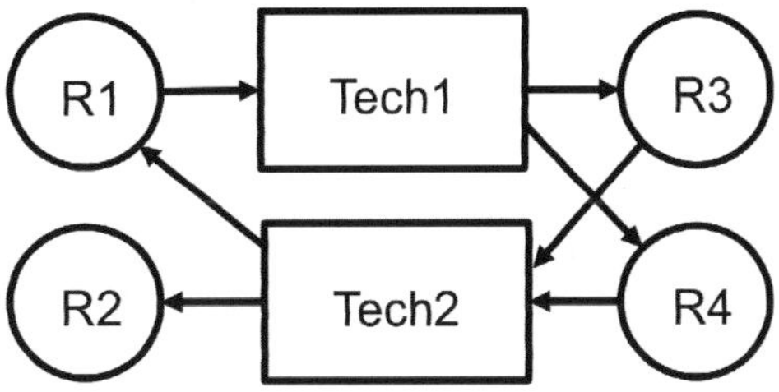

S. Samsatli, N.J. Samsatli (2015). Computers & Chemical Engineering 80, 155-176, 0098-1354.

Resource-Technology Network (RTN)

UNIVERSITY OF BATH

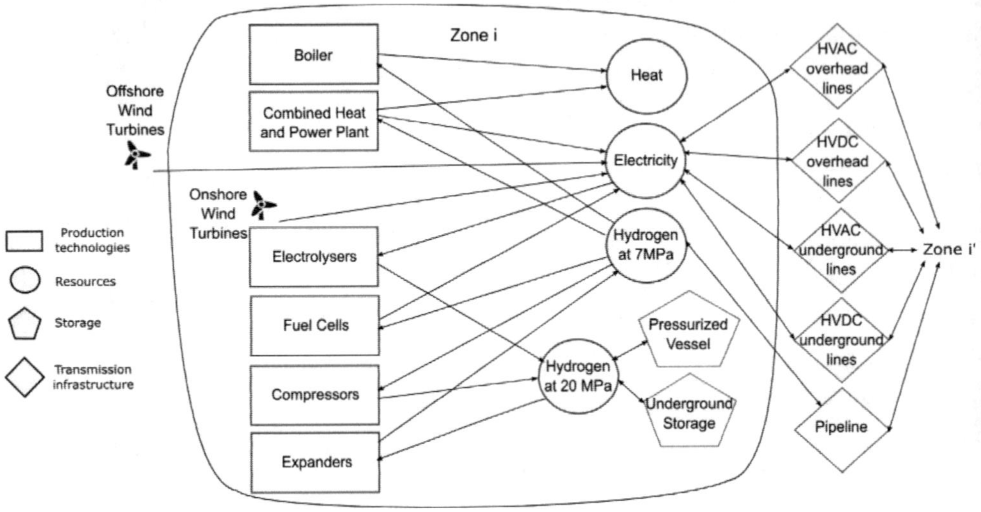

A. Prates Pereira and S. Samsatli. Optimal design and operation of heat network utilising hydrogen as an energy carrier. *Computer Aided Chemical Engineering*, 2017. 27th European Symposium on Computer Aided Process Engineering

Modelling Storage

UNIVERSITY OF BATH

Set of **storage tasks** to store resource r_1.

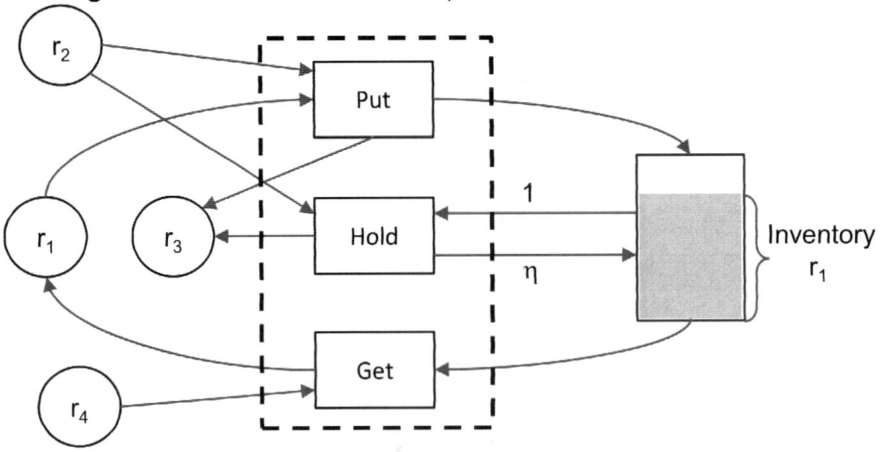

The "put" task transfers r_1 from the zone to the store, requiring some r_2 and producing some wastes r_3 (e.g. CO_2). The "hold" task maintains r_1 in storage which could be at less than 100% efficiency, the losses being converted to r_3 ; this task may also require some r_2. Finally, the "get" task retrieves r_1 from storage and delivers it to the zone, requiring some r_4.

S. Samsatli, N.J. Samsatli (2015). Computers & Chemical Engineering 80, 155-176, 0098-1354.

Modelling Transport

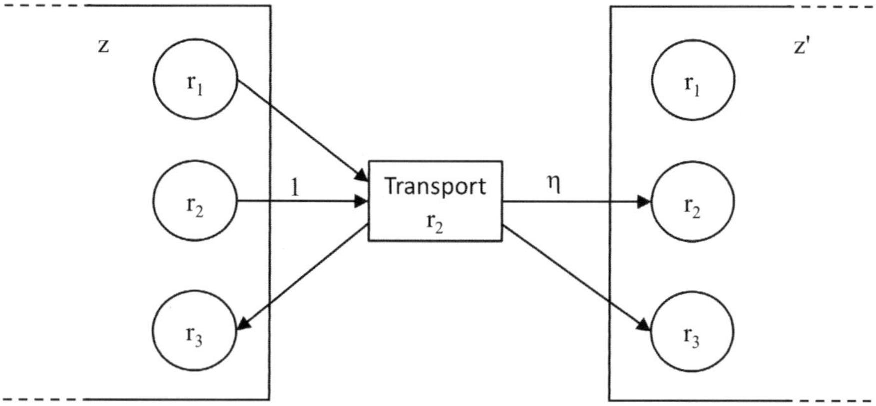

Resource r_2 is transported from zone z to zone z', which requires r_1 from zone z and results in waste r_3 being generated in both zones

S. Samsatli, N.J. Samsatli (2015). Computers & Chemical Engineering 80, 155-176, 0098-1354.

Resource Balance

$$U_{rzhdty} + M_{rzhdty} + P_{rzhdty} + S_{rzhdty} + Q_{rzhdty} \geq D_{rzhdty} + X_{rzhdty}$$

$$\forall r \in \mathbb{R},\ z \in \mathbb{Z},\ h \in \mathbb{H},\ d \in \mathbb{D},\ t \in \mathbb{T},\ y \in \mathbb{Y}$$

S. Samsatli and N.J. Samsatli. A multi-objective MILP model for the design and operation of future integrated multi-vector energy networks capturing detailed spatio-temporal dependencies. Applied Energy (In Press).

Onshore and offshore wind turbines constraints

Resource utilisation: $U_{rzhdty} \leq u^{\max}_{rzhdty}$ $\forall r \in \mathbb{R}, z \in \mathbb{Z}, h \in \mathbb{H}, d \in \mathbb{D}, t \in \mathbb{T}, y \in \mathbb{Y}$

Where: $u^{max}_{Elec,zhdty} = u^{IWT,max}_{Elec,zhdty} + u^{OWT,max}_{Elec,zhdty}$

	Onshore wind turbines	**Offshore wind turbines**
Wind power potential	$u^{IWT,max}_{Elec,zhdty} = 0.5 \times 10^{-6} \eta^{IWT} \rho^{air} \pi \left[N^{IWT}_{zy}(R^{IWT})^2 + N^{exist,IWT}_{zy}\left(R^{exist,IWT}\right)^2\right]\left(v^{i}_{zhdty}\right)^3$	$u^{OWT,max}_{Elec,zhdty} = 0.5 \times 10^{-6} \eta^{OWT} \rho^{air} \pi \left[N^{OWT}_{zy}(R^{OWT})^2 + N^{exist,OWT}_{zy}\left(R^{exist,OWT}\right)^2\right]\left(v^{o}_{zhdty}\right)^3$
Utilised existing WT	$N^{exist,IWT}_{zy} \leq N^{exist,IWT,total}_{zy}$	$N^{exist,OWT}_{zy} \leq N^{exist,OWT,tot}_{zy}$
Land footprint contraints	$\pi(5R^{IWT})^2 N^{IWT}_{zy} \leq A^{IWT,max}_{zy}$	$\pi(5R^{OWT})^2 N^{OWT}_{zy} \leq A^{OWT,max}_{zy}$
Wind turbines balance	$N^{IWT}_{zy} = N^{IWT}_{z,y-1} + NI^{IWT}_{zy} - NR^{IWT}_{zy}$	$N^{OWT}_{zy} = N^{OWT}_{z,y-1} + NI^{OWT}_{zy} - NR^{OWT}_{zy}$
Wind turbines retirement	$NR^{IWT}_{zy} = \sum_{yy}\left(Rf^{IWT}_{yy,y} \times NI^{IWT}_{z,yy}\right)$	$NR^{OWT}_{zy} = \sum_{yy}\left(Rf^{OWT}_{yy,y} \times NI^{OWT}_{z,yy}\right)$

Conversion technologies constraints

Net production of resource r : $P_{rzhdty} = \sum_{p} \mathscr{P}_{pzhdty} \alpha_{rpy}$

Min/max rate of operation $N^{\mathrm{P}}_{pzy} p^{\min}_{p} \leq \mathscr{P}_{pzhdty} \leq N^{\mathrm{P}}_{pzy} p^{\max}_{p}$

Technology balance: $N^{P}_{pzy} = N^{P}_{pz,y-1} + NI^{P}_{pzy} - NR^{P}_{pzy}$

Retirements: $NR^{\mathrm{P}}_{pzy} = \sum_{y'} RF^{P}_{py'y} NI^{P}_{pzy'}$

Max build rate: $\sum_{z} NI^{\mathrm{P}}_{pzy} \leq BR_{py}$

S. Samsatli and N.J. Samsatli. A multi-objective MILP model for the design and operation of future integrated multi-vector energy networks capturing detailed spatio-temporal dependencies. Applied Energy (In Press).

Transmission infrastructure constraints

UNIVERSITY OF BATH

Net transport of resource r:

$$Q_{rzhdty} = \sum_{z'|\nu_{z'z}=1} \sum_{l \in \mathbb{L}} \left[\left(\bar{\tau}_{lr,\text{dst},y} + \hat{\tau}_{lr,\text{dst},y} d_{z'z} \right) \mathscr{Q}_{lz'zhdty} \right] + \sum_{z'|\nu_{zz'}=1} \sum_{l \in \mathbb{L}} \left[\left(\bar{\tau}_{lr,\text{src},y} + \hat{\tau}_{lr,\text{src},y} d_{zz'} \right) \mathscr{Q}_{lzz'hdty} \right]$$

Max rate of operation: $\mathscr{Q}_{lzz'hdty} \leq \sum_{b \in \mathbb{B}} q_l^{\max} N^{\text{B}}_{bzz'y} |_{LB_{lb}=1 \wedge \nu_{zz'}=1}$

Max infrastructure capacity: $\sum_{l \in \mathbb{L}} \mathscr{Q}_{lzz'hdty} LB_{lb} \leq b_b^{\max} N^{\text{B}}_{bzz'y}$

Bi-directional links:

$$N^{\text{B}}_{bzz'y} = N^{\text{B}}_{bz'zy} \qquad \forall b \in \mathbb{B} | \beta_b = 1,\ z \neq z' \in \mathbb{Z}$$

S. Samsatli and N.J. Samsatli. A multi-objective MILP model for the design and operation of future integrated multi-vector energy networks capturing detailed spatio-temporal dependencies. Applied Energy (In Press).

Storage technology constraints

UNIVERSITY OF BATH

Max charging and discharging rates:

$$\mathscr{S}^{\text{put}}_{szhdty} \leq N^{\text{S}}_{szy} s_s^{\text{put,max}} a_{sz}$$

$$\mathscr{S}^{\text{get}}_{szhdty} \leq N^{\text{S}}_{szy} s_s^{\text{get,max}} a_{sz}$$

Net inflow of resource r due to storage operations:

$$S_{rzhdty} = \sum_s \left(\mathscr{S}^{\text{put}}_{szhdty} \sigma^{\text{put}}_{sr,\text{src},y} + \mathscr{S}^{\text{hold}}_{szhdty} \sigma^{\text{hold}}_{sr,\text{dst},y} + \mathscr{S}^{\text{get}}_{szhdty} \sigma^{\text{get}}_{sr,\text{dst},y} \right)$$

Inventory levels:

$$I_{szhdty} = n_h^{\text{hd}} \sum_r \left(\mathscr{S}^{\text{put}}_{szhdty} \sigma^{\text{put}}_{sr,\text{dst},y} + \mathscr{S}^{\text{hold}}_{szhdty} \sigma^{\text{hold}}_{sr,\text{src},y} + \mathscr{S}^{\text{get}}_{szhdty} \sigma^{\text{get}}_{sr,\text{src},y} \right)$$

Changes in inventory:

Daily: $\delta^{\text{d}}_{szdty} = I_{sz,|\mathbb{H}|,dty} - I^{0,\text{sim}}_{szdty}$ Weekly: $\delta^{\text{t}}_{szty} = \sum_d \delta^{\text{d}}_{szdty} n_d^{\text{dw}}$

S. Samsatli and N.J. Samsatli. A multi-objective MILP model for the design and operation of future integrated multi-vector energy networks capturing detailed spatio-temporal dependencies. Applied Energy (In Press).

Storage technology constraints (cont.)

UNIVERSITY OF BATH

Link inventories between different time levels:

$$I^{0,\mathrm{act}}_{szdty} = I^{0,\mathrm{act}}_{sz,d-1,ty} + n^{\mathrm{dw}}_{d-1}\delta^{\mathrm{d}}_{sz,d-1,ty}$$

$$I^{0,\mathrm{act}}_{sz,1,ty} = I^{0,\mathrm{act}}_{sz,1,t-1,y} + n^{\mathrm{wt}}_{t-1}\delta^{\mathrm{t}}_{sz,t-1,y}$$

Shift inventory levels to match average:

$$I^{0,\mathrm{sim}}_{szdty} = I^{0,\mathrm{act}}_{szdty} + \left[\left(n^{\mathrm{dw}}_{d} - 1\right)\delta^{\mathrm{d}}_{szdty} + \left(n^{\mathrm{wt}}_{t} - 1\right)\delta^{\mathrm{t}}_{szty} + \left(n^{\mathrm{yy}}_{y} - 1\right)\delta^{\mathrm{y}}_{szy}\right]/2$$

$$\mathscr{S}^{\mathrm{hold}}_{sz,1,dty} = I^{0,\mathrm{sim}}_{szdty}/n^{\mathrm{hd}}_{1} \qquad \mathscr{S}^{\mathrm{hold}}_{szhdty} = I_{sz,h-1,dty}/n^{\mathrm{hd}}_{h}$$

Storage capacity constraints:

$$s^{\mathrm{hold,min}}_{s} N^{S}_{szy} a_{sz} \le I_{szhdty} \pm \left[\left(n^{\mathrm{dw}}_{d} - 1\right)\delta^{\mathrm{d}}_{szdty} \pm \left(n^{\mathrm{wt}}_{t} - 1\right)\delta^{\mathrm{t}}_{szty} \pm \left(n^{\mathrm{yy}}_{y} - 1\right)\delta^{\mathrm{y}}_{szy}\right]/2 \le s^{\mathrm{hold,max}}_{s} N^{S}_{szy} a_{sz}$$

S. Samsatli and N.J. Samsatli. A multi-objective MILP model for the design and operation of future integrated multi-vector energy networks capturing detailed spatio-temporal dependencies. Applied Energy (In Press).

Objective function

UNIVERSITY OF BATH

Minimise $Z = (\mathcal{I}^{iw} + \mathcal{I}^{ow} + \mathcal{I}^{p} + \mathcal{I}^{s} + \mathcal{I}^{q} + \mathcal{I}^{m} + \mathcal{I}^{x}) + (\mathcal{I}^{IW} + \mathcal{I}^{OW} + \mathcal{I}^{P} + \mathcal{I}^{S} + \mathcal{I}^{Q})/\gamma - V$

Impacts	Capital	O&M
OnshoreWind turbines	$\mathcal{I}^{IW}_{iz} = \varsigma \sum_{z} C^{IWT}_{iy} NI^{IWT}_{zy}$	$\mathcal{I}^{iw}_{iz} = \varsigma \sum_{z} F^{IWT}_{iy} N^{IWT}_{zy}$
OffshoreWind turbines	$\mathcal{I}^{OW}_{iz} = \varsigma \sum_{z} C^{OWT}_{iy} NI^{OWT}_{zy}$	$\mathcal{I}^{ow}_{iz} = \varsigma \sum_{z} F^{OWT}_{iy} N^{OWT}_{zy}$
Conversion techs	$\mathcal{I}^{P}_{iy} = \varsigma \sum_{pz} C^{P}_{piy} NI^{P}_{pzy}$	$\mathcal{I}^{p}_{iy} = \varsigma \sum_{pz} F^{P}_{piy} N^{P}_{pzy}$
Transport techs	$\mathcal{I}^{Q}_{iy} = 0.5\varsigma \sum_{pz} C^{B}_{biy} NI^{B}_{bzz'y} d_{zz'}$	$\mathcal{I}^{q}_{iy} = 0.5\varsigma \sum_{pz} F^{B}_{biy} N^{B}_{bzz'y} d_{zz'}$
Storage techs	$\mathcal{I}^{S}_{iy} = \varsigma \sum_{sz} C^{S}_{siy} NI^{S}_{szy}$	$\mathcal{I}^{s}_{iy} = \varsigma \sum_{sz} F^{S}_{siy} N^{S}_{szy}$
Export/import	$\mathcal{I}^{x}_{iy} = \varsigma \sum_{rzhdt} V^{X}_{r} X_{rzhdt} n^{hd}_{h} n^{dw}_{d} n^{wt}_{w}$	$\mathcal{I}^{m}_{iy} = \varsigma \sum_{rzhdt} V^{M}_{r} M_{rzhdt} n^{hd}_{h} n^{dw}_{d} n^{wt}_{w}$

Revenue $V = \varsigma \sum_{rzhdt} V^{Rev}_{riy} D_{rzhdt} n^{hd}_{h} n^{dw}_{d} n^{wt}_{w}$

Wind speed and heat demands

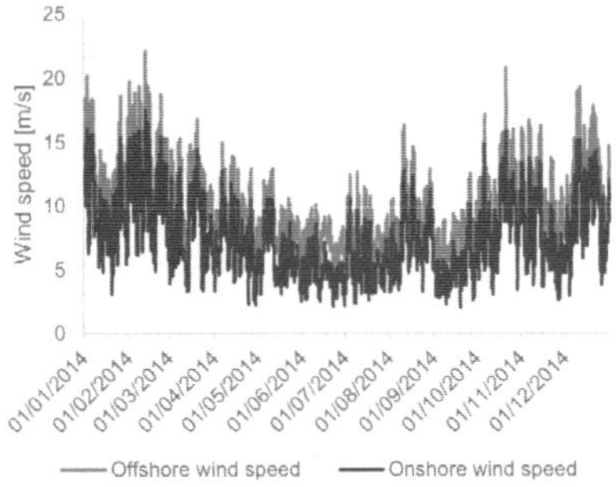

Offshore vs onshore wind speed variability at an hourly interval for an entire year

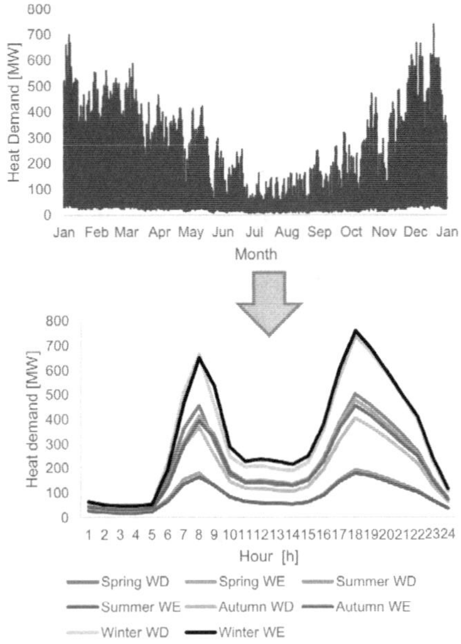

Hourly heat demands during the year

Available onshore area

Onshore available area limited by 10 criteria:

- Average wind speed of at least 5 m/s at above 45m above ground level
- Land slope of less than %
- A minimum distance of 500m from minor roads
- At least 200m but not more than 1500m from major road network
- Not a Site of Special Scientific Interest
- At least 500m from developed population áreas
- At least 200m away from rivers
- At least 250m from woodland
- At least 5 km from airports
- Not occupied by existing wind turbines

Suitable onshore area: 4,255 km^2

S. Samsatli, I. Staffell, and N.J. Samsatli (2016). International Journal of Hydrogen Energy, 41(1), 447 – 475, 0360-3199

Available offshore area: GIS analysis

UNIVERSITY OF BATH

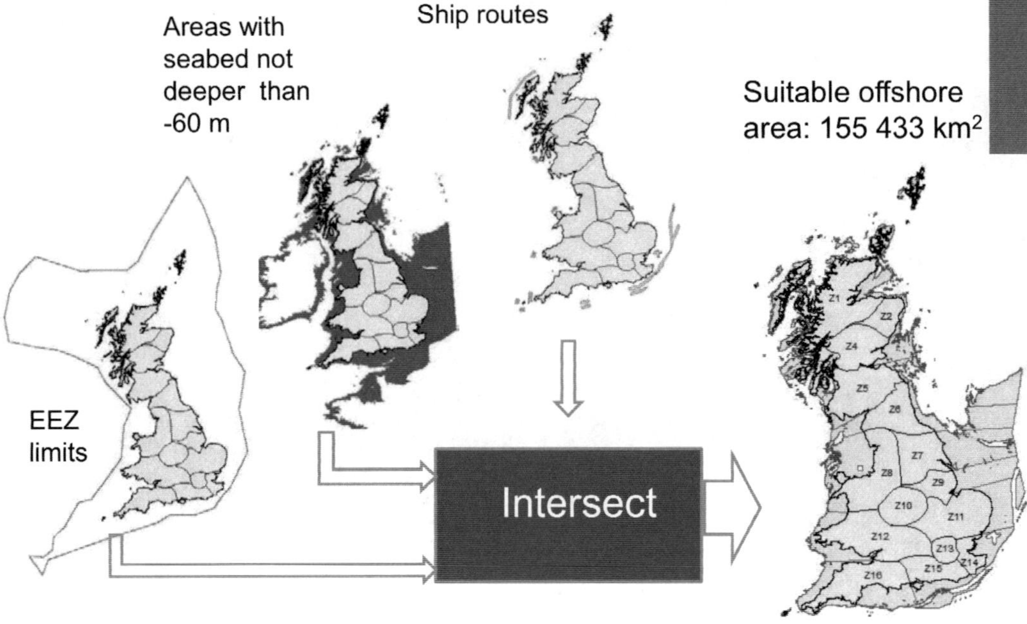

Results

UNIVERSITY OF BATH

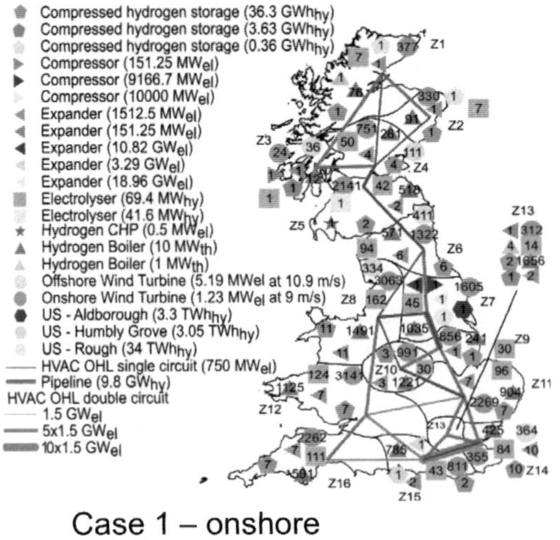

Case 1 – onshore and offshore

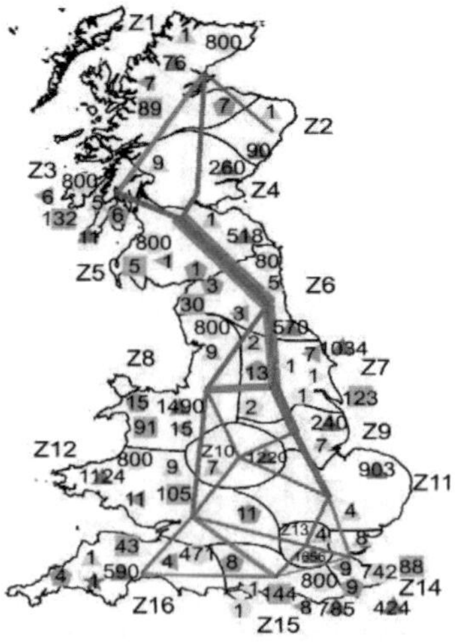

Case 2 – Offshore

Case 3 - Onshore

Infeasible –the available onshore area is not enough to build the number of wind turbines required to satisfy the heat demands

Results

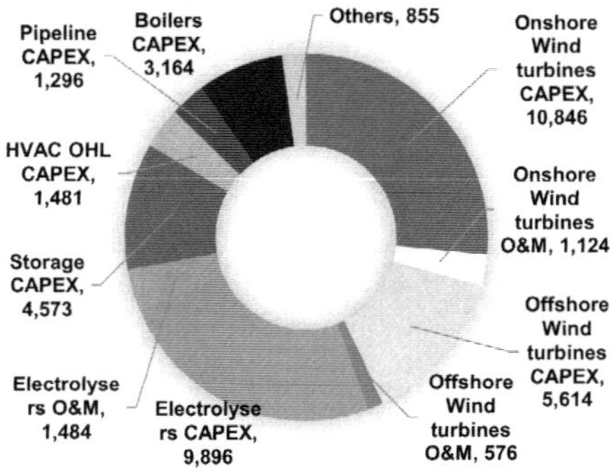

Case 1 – Breakdown of cost (values in £M). Total cost: £ 37,057 M

	Onshore WT	Offshore WT	Objective function (£M)
Case 1	20,622	1,244	37,057
Case2	-	5,003	44,924

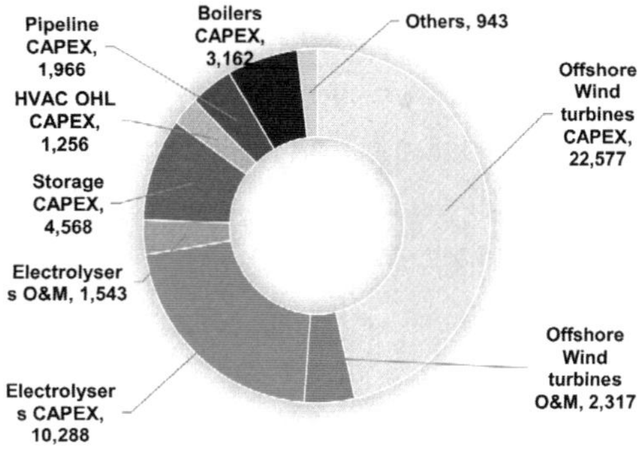

Case 2 – Breakdown of cost (values in £M). Total cost: £ 44,924 M

Conclusions

UNIVERSITY OF BATH

- GB domestic heat demands can be decarbonised using wind power
- Mixed onshore & offshore wind turbines most cost-effective strategy
- New investment in: hydrogen storage, compressors, expanders, electrolysers, hydrogen boilers, hydrogen CHP, onshore and offshore wind turbines, pipeline and high voltage alternating current lines
- Large hydrogen pipeline network required to transport energy from the production sites to storage facilities and the location of demands
- Not possible to satisfy GB heat demands using only onshore wind power
- Decarbonising the heat sector using hydrogen produced from wind power would prevent 47 Mt of CO_2 annual emissions and would cost £795 per avoided tonne of CO_2

Future work

- Extend the planning horizon to 2050.
- Incorporate more technologies and, consequently, more pathways to the model (SMR, CCS, electric heaters, district heating vs domestic heating).
- Include existing infrastructure (e.g. existing wind turbines).
- Satisfy new demands, such as household electricity and domestic transports.

Thank you for your attention.

Comments/questions/suggestions?

The authors would like to thank the University of Bath for the PhD studentship of André Prates-Pereira.

498955 Characterization of Used Cooking Oils and Their Supply Chain for the Exploitation As Raw Materials in Urban Biorefineries

Abstract:

In recent years, the chemical industry has focused its efforts on the sustainable generation of chemical products derived from renewable and sustainable resources (biobased chemicals) (1). This trend has pushed the oleochemical industry to exploit alternative raw materials that are not suitable for human consumption (second generation) such as used cooking oils (UCOs). Because these are waste materials, they do not compete in the market with edible oils and fats (reducing pressure on food prices and land use), are low cost, have a large volume of production, and can maintain a sustainable supply under a circular economy concept. In particular, UCOs are generated in large volumes, mainly in urban centers, making them suitable feedstock for urban biorefineries.

After impurities removal and refining, the UCOs become an oleochemical platform for the production of commodity biobased fuels (e.g. biodiesel (2,3)), but more importantly, valued added products. UCOs can be used as building block for the synthesis of valuable biobased products such as monomers, additives for polymers and resins, surfactants, etc (4,5). In particular, production of polyurethane monomers (polyols and non-isocyanate moieties) is of major interest in the near future(6). However, there is a major limitation when using UCOs as feedstock associated to the chemical heterogeneity. UCOs composition largely varies with the nature of original virgin oil, the conditions during the cooking process (processed food, fryer type, temperature, processing time, etc.), and the refining process (7,8).

In order to assess the feasibility of using UCOs as feedstock for urban biorefineries in the production of value-added chemicals, this work developed a method to characterize UCOs and their supply chain in a typical urban center (Bogota, Colombia). The studied supply chain corresponded to food establishments because they are larger UCOs generators compared with industrial users (e. g. snack producers). Initially the UCOs supply spots were classified according to the following categories: institutional food, fast food, casual food, and tablecloth restaurants (9,10). UCOs were obtained from the categories with the largest outlets across the city. The physicochemical characterization of representative samples collected around the city was done by evaluation of iodine value, acid value, moisture, density, saponification value, peroxide value and total polar compounds.

A pilot test was performed to evaluate the chemical heterogeneity of UCOs obtained from each category around the city. According to results, the density of samples presented low coefficient of variation (0,5%) as this property barely changes during oil processing. On the other hand, properties such as iodine value and saponification value presented a coefficient of variation around 14%, indicating that these properties do not change much among the different categories. Finally, properties such as moisture, acid value, peroxide value and total polar compounds presented high coefficient of variation (55%). This indicate that these properties largely depend on the type of food establishment, and that UCOs originated from each supplier might need a specific refining process. Properties characterized with high coefficient of variation were used to stablish the required processing conditions for UCOs refining (11).

Iodine and saponification values were of special interest because they are surrogate variables that determines the potential use of UCOs as oleochemical feedstock. Thus, the ranges of iodine values and saponification index of UCOS collected in the city of Bogota were 85-105 g I_2/100 g, and 130-201 mg KOH/g respectively. According to these results UCOs can be used as raw materials for epoxidized oils and subsequently for the ring-opening reaction to biobased polyols (12,13). As some suppliers produced highly saturated UCOs, it is necessary to avoid their blending with UCOs from different categories in the central collection depots. Instead, they can be separately processed and be used for the production biofuels, surfactants, or other additives for commodity plastics (such as PVC) (14,15).

References

1. Commission E. Bio-based products [Internet]. [citado 13 de marzo de 2017]. Disponible en: http://ec.europa.eu/growth/sectors/biotechnology/bio-based-products/index_en.htm

2. Mandolesi de Araújo CD, De Andrade CC, De Souza e Silva E, Dupas FA. Biodiesel production from used cooking oil: A review. Renew Sustain Energy Rev. 2013;27:445-52.

3. Talebian-Kiakalaieh A, Amin NAS, Mazaheri H. A review on novel processes of biodiesel production from waste cooking oil. Appl Energy [Internet]. 2013;104:683-710. Disponible en: http://dx.doi.org/10.1016/j.apenergy.2012.11.061

4. Cai C, Dai H, Chen R, Su C, Xu X, Zhang S, et al. Studies on the kinetics of in situ epoxidation of vegetable oils. Eur J Lipid Sci Technol. 2008;110:341-6.

5. Ríos LA, Cardeño F, Guzmán AF. EPOXIDACIÓN DE ACEITES VEGETALES PARA LA PRODUCCIÓN DE POLIURETANOS SIN ISOCIANATOS (NIPUS).

6. Alam M, Akram D, Sharmin E, Zafar F, Ahmad S. Vegetable oil based eco-friendly coating materials: A review article. Arabian Journal of Chemistry. 2014.

7. Sanli H, Canakci M, Alptekin E. Characterization of Waste Frying Oils Obtained from Different Facilities. World Renew Energy Congr. 2011;479-85.

8. Choe E, Min DB. Chemistry of deep-fat frying oils. J Food Sci. 2007;0(0).

9. Ortegón DG. El sector gastronómico creció 22% en el último año con 90.000 restaurantes. La republica [Internet]. 16 de junio de 2015; Disponible en: http://www.larepublica.co/el-sector-gastronómico-creció-22-en-el-último-año-con-90000-restaurantes_266206

10. Garzón RL. Las cadenas de restaurantes que más venden en el país. El tiempo. 10 de junio de 2015;

11. Serjouie A, Tan CP, Mirhosseini H, Man YBC. Effect of vegetable-based oil blends on physicochemical properties of oils during deep-fat frying. Vol. 5, American Journal of Food Technology. 2010. p. 310-23.

12. Gamage PK, O'Brien M, Karunanayake L. Epoxidation of some vegetable oils and their hydrolysed products with peroxyformic acid - Optimised to industrial scale. J Natl Sci Found Sri Lanka. 2009;37(4).

13. Hazmi ASA, Aung MM, Abdullah LC, Salleh MZ, Mahmood MH. Producing Jatropha oil-based polyol via epoxidation and ring opening. Ind Crops Prod. 2013;50:563-7.

14. Aisyah N, Nor B. Bio-Surfactant Syhnthesis From Waste Cooking Oil. 2015;(January).

15. Mozejko J, Ciesielski S. Saponified waste palm oil as an attractive renewable resource for mcl-polyhydroxyalkanoate synthesis. J Biosci Bioeng. 2013;116(4):485-92.

501441 Electrochemical Activity of Non-Noble Metal Alloy As Catalyst Towards Oxidation of Glycerol in Acidic Media: A Case for theÂ Conceptual Glycerol/Ferric Redox Flow Battery

Abstract:

Electrochemical activity of non-noble metal alloy as catalyst towards oxidation of glycerol in acidic media: A case for the Conceptual Glycerol/Ferric Redox Flow Battery

James Akrasi, Egwu Kalu and Yaw Yeboah

FAMU-FSU College of Engineering

Department of Chemical and Biomedical Engineering, Tallahassee, Florida. 32310

The electrochemical oxidation of glycerol has been studied and known to be a source of several chemicals of premium industrial value and has also been utilized in the concept of direct glycerol fuel cells to produce electrical energy additionally[1]. In the current study however, this concept is extended to cover both electrochemical oxidation and reduction of glycerol as a substitute for the hydrogen ion in redox flow battery with a Fe^{2+}/Fe^{3+} couple as cathode [2]. A predictive thermodynamic analysis has been developed and applied to the study and proof of this conceptual glycerol redox flow battery. Reactions toward electrochemical formation of several partial oxidation and reduction products of glycerol in acidic media under discharge and charging conditions of the proposed glycerol flow battery are presented. Free energy values of the reaction species as estimated by the Group Additivity Value (GAV) method were used to determine the feasibility of these proposed reaction towards cogeneration of electricity and commodity chemicals. The equilibrium voltages representing the maximum electrical work under these conditions are also computed. Experimental validation of this theoretical concept has also been investigated using electrochemical methods to study the electro-catalytic activity of electrolessly deposited NiCuSnP supported on carbon towards glycerol oxidation in acidic medium as anode in a proton exchange membrane-glycerol redox flow battery.

References

1. Li W., Xin L., Zhang Z., Wang Z.,. (2012) *Simultaneous Generation of Mesoxalic Acid and Electricity from Glycerol on a Gold Anode Catalyst in Anion-Exchange Membrane Fuel Cells* ChemCatChem Vol 4 1105-1114

2. Khalid Fatih K, Wilkinson D., Moraw F., Ilicic A and Girarda F *Advancements in the Direct Hydrogen Redox*

Fuel Cell Electrochemical and Solid-State Letters, 11 2 B11-B15 (2008)

490758 Fecralloy Catalysts Partially Oxidize Methane to Syngas Selectively

Abstract:

Micro Gas-to-Liquids technology (GtL) reduces flared natural gas and CO_2emissions while producing valuable diesel. Integrating a high pressure syngas step with Fischer-Tropsch (FT) in a single vessel reduces investment and operating costs to synthesize GtL liquids. Catalytic methane partial oxidation (CPOX) to produce syngas for FT is an economic opportunity for micro-refineries. The CPOX catalysts should combine thermal resistance to high temperature (~1000 °C) and mechanical strength to high pressure (~2 MPa). FeCralloy knitted fibres resist high temperature, is highly conductive, and can be molded into shapes.
It is promising support that we overlaid washcoat of Al_2O_3 via spraying pyrolysis and of MgO via solution combustion synthesis (SCS). It is also highly reactive, which partially oxidizes methane to syngas. We explored a larger range of FeCralloy catalysts by depositing Pt/Rh and/or CeO_2 over its surface. The selectivity to syngas of 1% Pt/Rh FeCralloy catalyst was better than other catalysts at 900 °C and from 0.1 MPa and 2 MPa. Here, we characterized the methane oxidation behavior versus temperature of 5 FeCralloy catalysts, in micro gauze reactor operated up to 950 °C at ambient pressure. Temperature higher than 550 °C activates the CPOX reaction. Oxygen reacted with methane to form H_2, CO, and CO_2 while in pure Argon methane decomposes to H_2 and coke. We developped a kinetic model to predict its behavior.

491918 Rapid Wood Fractionation ≤ 80°C for Sustainable and Economic Biorefinery

Abstract:

Here we report a rapid and low temperature (≤ 80°C) wood fractionation process for producing wood pulp, cellulose and lignin nanomaterials, chemicals, and sugar/biofuels. Approximately 90% of wood lignin can be dissolved at 80°C for 20 min. Equivalent delignification using alkaline pulping can be achieved only at 170°C for 2 h. The novel process fractionated wood into two fractions: (1) a primarily carbohydrate-rich water-insoluble solid fraction that can be used for sustainable production of fibers such as dissolving pulp, lignocellulosic nanomaterials, and/or sugars through subsequent enzymatic hydrolysis, and (2) a spent acid liquor stream containing mainly dissolved lignin that can be easily precipitated as lignin nanoparticles simply by diluting the spent acid liquor. The dissolved hemicellulose sugars can be used to produce furfural after lignin precipitation. The chemical used has a low water solubility, which facilitates efficient recovery simply using commercially proven crystallization technology by cooling concentrated spent acid solution to ambient temperatures to achieve environmental sustainability.

502673 Insights into Biomass Recalcitrance

Abstract:

The utilization of biomass for biofuels and bio-derived chemicals and materials based on the biological platform is contingent on efficient separation, deconstruction and conversion. Todayâ€™s cellulosic ethanol plants, especially those based on the cellulase deconstruction, have demonstrated that the conversion of plant polysaccharides to ethanol can be accomplished on an industrial scale but current process challenges are hindering wide-spread adaption. One of the key issues is the recalcitrance of biomass and the need to utilize aggressive pretreatment technologies. Recent advances in our understanding of the plant cell wall structure and our ability to reduce biomass recalcitrance by altering these constituents by controlling genetic features of the plant and innovations in pretreatment chemistries provides innovative pathways to facilitate the biorefining of plant polysaccharides to biofuels and yield a lignin rich process stream that can be valorized to chemicals and/or materials. This seminar review our contributions in this field and highlight future opportunities.

492489 Low Temperature and High Efficiency Biomass Fuel Cell and Bio-Hydrogen Production

Abstract:

Title: Low temperature and high efficiency biomass fuel cell and bio-hydrogen production

Name: Yulin Deng

School of Chemical & Biomolecular Engineering, Georgia Institute of Technology, Atlanta, GA 30302-0620, yulin.deng@rbi.gatech.edu

A novel fuel cell which can directly use native polymeric biomasses, such as starch, cellulose, lignin, and even switchgrass and wood powders will be discussed. This fuel cell combines some features of solar cells, fuel cells, and redox flow batteries. Specifically, an oxidation/reduction ion pair, such as Fe^{2+}/Fe^{3+}, or polyoxomatelate, are used as catalyst to oxidize biomass by either absorbing solar light or heat energy. The power density of the solar-induced hybrid fuel cell powered with raw biomass could reach ~50 mW/cm^2. Unlike most cell technologies that are sensitive to impurities, the cell reported in this study is inert to most organic and inorganic contaminants present in the fuels. The fuel cell is completely noble metal free. The similar fuel cell system can also be converted to electrolysis cell for hydrogen production at low temperature using native biomass directly.

504077 Circular Economy: A Path Towards Innovation and Commercialization of Biocomposites for Sustainable Manufacturing

Abstract:

"Circular economy" is an emerging concept that targets waste minimizations through a closed loop economy system, unlike the traditional linear model ("make and dispose" type economy), and therefore supports sustainable development. The underlying principle of this emerging economy is "Nothing is waste; waste is a co-product waiting for another opportunity". In this system, the byproducts and co-products of one industry can find commercially viable applications in another industry, with appropriate design and engineering. There is a significant potential for utilization of biobased coproducts from agro and forest industries, food and beverage industry as well as biofuel and biomass-based pyrolysis industries. Our focus is on biocomposite applications that utilize these materials as reinforcing and filler materials for plastics, including biobased and biodegradable plastics. The developed eco-friendly biocomposites are suitable for use in consumer products, light-weight auto-parts and 100% compostable packaging.

Acknowledgements: Agriculture and Agri-Food Canada and Competitive Green Technologies (CGtech) through AgriInnovation Program (Project #052882 and 051910); the Ontario Ministry of Agriculture, Food, and Rural Affairs - University of Guelph Bioeconomy Industrial uses Research Program (Project # 200425); Natural Sciences and Engineering Research Council, Canada for the Discovery Grants (Project #401111); and the Ontario Research Fund, Research Excellence Program; Round-7 (ORF-RE07) from the Ontario Ministry of Research and Innovation (MRI), currently known as the Ontario Ministry of Research, Innovation and Science (MRIS) (Project # 052665 and #052644).

503670 Digital Alchemy for Assembly Engineeringc

Abstract:

Back in the day, Stuart Churchill was a pioneer in the use of computers for modeling chemical and thermal processes. Since that time, the role of computing in chemical engineering has become indispensable. In this talk, we present new computational methods we call "digital alchemy" that allow for predictive modeling of thermodynamic assembly, and permit the design of building blocks optimized for assembly into target structures with desired properties. In this approach, standard thermodynamic ensembles are generalized to include alchemical variables and their thermodynamic conjugates, allowing for rapid, real-time computational investigation of large, complex design spaces.

503671 Improved Algebraic, Numerical, and Graphical Representations in Fluid Mechanics

Abstract:

The concept of equivalents is applied to fully developed flows through straight channels, and two applications are introduced: the equivalents that produce equal rates of flow through channels of different cross-section, and the diminishment that occurs between the regimes of laminar and turbulent flow. The validity of the eddy diffusivity and the invalidity of the mixing-length as measures of the shear stress due to turbulence have perhaps never before been demonstrated so clearly. The possible choices of primary variables and the consequences are also examined. The objective of these explorations is an improved understanding of the basic fluid mechanics by teachers, students, and industrial practitioners.

503672 The Scaling of Turbulence Near the Wall and the Churchill Turbulent Flux Correlation: Insights with Lagrangian Simulations

Abstract:

The evaluation of mean velocity and mean temperature in wall-bounded turbulent flows has been based on the "classical" theory of turbulence and the law of the wall. The viscous wall parameters, namely the friction velocity, the friction temperature, and the kinematic viscosity of the fluid have been used to scale not only the mean values of velocity, but also other mean turbulent quantities close to the wall, including the Reynolds stresses and the turbulent velocity and temperature fluctuations. Stuart Churchill, as he questioned the validity of the Reynolds analogy between momentum and heat or mass transfer in his works [1], he realized that extending the scaling of momentum to the scaling of scalar quantities in turbulence was not a sound approach. He proposed a scaling of turbulence that depends on the turbulent momentum flux and the turbulent heat flux. In his AIChE Institute Lecture (in Miami, 1998), Churchill presented a summary of his work on this topic, and closed form solutions for the mean velocity and temperature [2]. The turbulent Prandtl number, the most significant parameter for the modeling of turbulent transport not only in theory but also in modern CFD commercial software packages, was also interpreted physically based on these ideas [3]. We have used direct numerical simulation results and results from Lagrangian simulations [4] for the transport of passive scalars in turbulent channel flows to explore Churchill's ideas on turbulent flow scaling and to develop an updated theory for the scaling of heat transport in turbulent flows [5-7]. His passion to remove empiricisms and approximations from the development of predictive correlations in turbulent transport has been an inspiration. In this talk, we will revisit the Churchill scaling in view of our simulation results, and will discuss the counterintuitive finding that wall turbulence can be used to separate rather than mix substances under special circumstances [8].

1. Churchill, S. W., 1996, "A Critique of Predictive and Correlative Models for Turbulent Flow and Convection," Ind. Eng. Chem. Res., 35, pp. 3122–3140.
2. Churchill, S.W., 2000, "Progress in the Thermal Sciences: AICHE Institute Lecture," AIChE Journal, 46(9), pp. 1704-1722.
3. Churchill, S. W., 2002, "A reinterpretation of the Turbulent Prandtl Number," Ind. Eng. Chem. Res., 41, pp. 6393–6401.
4. Papavassiliou, D.V, 2002, "Turbulent transport from continuous sources at the wall of a channel," Int. J. Heat Mass Transfer, 45(17), pp. 3571-3583.
5. Le, P.M., and Papavassiliou, D.V., 2008, "On the scaling of heat transfer using thermal flux gradients for fully developed turbulent channel and Couette flows," Int. Commun. Heat Mass Transfer, 35(4), pp. 404-412.
6. Srinivasan, C., and Papavassiliou, D. V., 2011, "Prediction of Turbulent Prandtl Number in Wall Flows With Lagrangian Simulations," Ind. Eng. Chem. Res., 50(15), pp. 8881–8891.
7. Srinivasan, C., and Papavassiliou, D.V., 2013, "Heat transfer scaling for wall bounded turbulent flows," Applied Mechanics Review 65(3), Art. 031002.

8. Nguyen, Q., Srinivasan, C., and Papavassiliou, D.V., 2015, "Flow induced separation in wall turbulence", Phys Rev E, 91, Art. 033019.

503674 Flow Boiling Using a Piranha Pin Fin Heat Sink

Abstract:

Flow Boiling using a Piranha Pin Fin Heat Sink

Corey Woodcock, Xiangfei Yu, Yoav Peles[1] and Joel Plawsky

Rensselaer Polytechnic Institute; [1]University of Central Florida

The *Piranha Pin Fin* (*PPF*) microdevice is a heat exchanger engineered to dissipate extreme thermal loads using either dielectric fluids such as HFE7000 or water. Fabrication of the device is CMOS compatible and the first generations of the device have been fashioned from silicon substrates. The PPF operates on the principle that heat transfer would be more efficient if heated fluids could be removed nearly as fast as they are created. This process would keep the driving force as high as possible. Thus the heat exchanger is built as two layers; a primary substrate that dissipates the bulk of the heat and shunts the heated fluids to a booster section that continues to heat the fluid and extends the total surface area of the device. Three generations of devices were fabricated. The first two generations were single-layer systems designed to help understand the operation of the primary layer and optimize the shape and location of the pin fins. The third generation was the first, two-layer system to be fabricated and tested.

Figure 1 Representative PPF array

Correlations for the heat transfer coefficient, h_{tp}, based on the Boiling Number and the Weber Number in the PPF design was developed and validated with Gen II. The discrepancy of heat transfer coefficient between predicted and measured values was within 20%. The ratio of two-phase to single-phase pressure drop was correlated with the Boiling Number (Bo) and the Jakob Number (Ja). The discrepancy of the two-phase multiplier between prediction and experimental measurements were within 12%.

The third generation *PPF* heat sinks were the first series to provide both a primary substrate populated with *PPF*s, as well as a booster substrate populated with solid micro pin fins. The substrates are micro-aligned and fusion-bonded to provide a full three dimensional (3D) architecture with bulk silicon properties. Experimental results from the Gen III *PPF* microdevices provide a proof of concept of the true *PPF* heat sink concept with fluid venting into multiple substrates. At low flowrates (G = 670 [kg/s/m^2]) the Gen III *PPF* heat sink has been shown to dissipate heat loads approaching 1.1 [kW/cm^2] at an exit quality of 95%, while at higher flowrates (G = 3,300 [kg/s/m^2]) the thermal limit of the heat sink remains unknown due to exceeding the available power supply at >1.4 [kW/cm^2].

503673 Transport Problems in the Spirit of Stuart Churchill for Teaching and Research at the University of Michigan

Abstract:

Stuart Churchillâ€™s solutions to important heat transport problems have greatly influence both pedagogy and research. The correlation of Churchill and Chu for the natural convection heat transfer to or from a vertical plate, over an astonishing range of Rayleigh numbers is based partly on a similarity solution of a type common in heat, mass, and momentum transfer for problems in chemical engineering. One such problem is asphaltene or wax transport and deposition onto the wall of an oil pipeline, which can be solved using similar methods, as will be discussed. In addition, such problems, and the wealth of their applications make for excellent motivations and examples in the teaching of graduate or undergraduate transport, as will be discussed in the talk.

504347 Design and Redesign of Sustainable Engineered Nanomaterials

Abstract:

Engineered nanoparticles are increasingly being incorporated into devices and products across a variety of commercial sectors â€“ this means that engineered nanoscale materials will either intentionally or unintentionally be released into the ecosystem. The long-term goal of the presented work is to understand the molecular design rules that control nanoparticle toxicity using aspects of materials science (nanoparticle design, fabrication, and modification), analytical chemistry (developing new assays to monitor nanotoxicity), and ecology (monitoring how nanoparticles enter and accumulate in the food web through bacteria and how these nanoparticles influence bacterial function). Taken together, these data suggest that careful consideration of engineered nanoparticle surface chemistry will likely allow design of safe and sustainable nanoscale materials.

504295 DNA Nanotechnology, a Promising Tool to Target Cancer

Abstract:

DNA is a popular material for constructing complex, multi-dimensional nanostructures due to its ability to organize in a precise and predictable manner. Popular approaches to create nanostructures from DNA include DNA origami, DNA tile and DNA brick assembly. An alternate approach to direct the assembly of single-stranded (ssDNA) is to conjugate a hydrophobic moiety (i.e., polymer or lipid-like tail) to the ssDNA to form an amphiphilic molecule that spontaneously self-assembles when added to aqueous solutions. However, this approach has not yet been used to create nanostructures with similar levels of complexity as those achieved by other amphiphiles. To date, the majority of structures created by the assembly of ssDNA-amphiphiles have been only spherical and cylindrical micelles. Our group showed that DNA nanotubes could be created by using molecular self-assembly of ssDNA-amphiphiles. In this presentation I will discuss how we design ssDNA-amphiphiles that self-assemble into supramolecular nanostructures with non-spherical geometries, such as ssDNA nanotubes, and how we use these ssDNA nanotubes to target glioblastoma multiforme (GBM), the most common form of primary brain cancer, in vitro and in vivo in an orthotopic mouse model of GBM.

496483 Functionalization of Iron Oxide Nanoparticles and the Impact on Reactive Oxygen Species Generation for Potential Cancer Treatment

Abstract:

Functionalization of iron oxide nanoparticles and the impact on reactive oxygen species generation for potential cancer treatment

Trang Mai, J. Zach Hilt

Department of Chemical and Materials Engineering

University of Kentucky Lexington, KY 40506

Corresponding Author: zach.hilt@uky.edu

Iron oxide nanoparticles (IONPs) and reactive oxygen species (ROS) generation have attracted much attention in the last several years. It has been shown that IONPs can produce ROS, which play a major role in various cellular pathways, via Fenton and Haber-Weiss reactions. Recent studies also show that alternating magnetic field (AMF) exposure can enhance the production of ROS by IONPs. At increased level, ROS can cause damages to proteins and other biological structures, lead to cellular apoptosis, and contribute to many diseases including cancer. The AMF-enhanced ROS could provide a potential strategy to exploit the elevated ROS stress for cancer therapy.

The aim of this work was to develop iron oxide nanoparticles based platforms with either small molecule or polymer coatings and to study impact of the coating on the surface reactivity of the particles. Iron oxide nanoparticles were synthesized by a co-precipitation process and coated with small molecules (e.g., citric acid, sodium phosphate, aminosilane, dopamine) or polymers, such as poly(ethylene glycol)-based systems. Crystal and functional properties of the synthesized particles were characterized using X-ray diffraction (XRD) and Fourier transform infrared (FTIR) spectroscopy. Hydrodynamic size was determined by dynamic light scattering (DLS) while the amount of coating was investigated using thermogravimetric analysis (TGA). AMF heating ability was also demonstrated. Surface reactivity via ROS generation induced by the particles was evaluated using a methylene blue decolorization assay with hydrogen peroxide and AMF exposure for varying times.

504469 Theranostic Nanovehicles for the Diagnosis and Treatment of Cerebrovascular Diseases

Abstract:

For successful brain delivery, any nanosystem must withstand dominant clearance pressure from the peripheral reticuloendothelial system, marginate from the bulk blood flow to the vascular endothelium, permeate the blood-brain-barrier, and accumulate at the target site. In addition to these common challenges, the nanoparticles intended for cerebrovascular targeting must incorporate appropriate design elements to ensure their retention in the cerebral vasculature. Furthermore, a functionally optimized nanovehicle design demands synergistic amalgamation of the physicochemical properties of various components and their intended physiological effects. Strategies developed in our laboratory to surmount these barriers will be discussed in the talk.

485042 Three-Dimensional Graphene-Based Microbarriers for Controlling Release and Reactivity in Colloidal Liquid Phases

Abstract:

Two-dimensional materials are of great interest as high-performance molecular barriers. Graphene in particular is atomically thin, is impermeable to all molecules, and in some forms can be easily deposited over large areas into planar multilayer films that have been shown to suppress molecular transport. Graphene and graphene oxide sheets are also known to spontaneously self-assemble at liquidâˆ'liquid interfaces on the surfaces of dispersed droplets, but much less is known about the barrier properties of these ultrathin films in 3D curved microgeometries. This article demonstrates that 3D films self-assembled from graphene oxide or reduced graphene oxide sheets can be exploited to control the release of small molecules from dispersed liquid phase droplets by evaporation. The release rate and containment time can be tuned by addition of multivalent cations that recruit additional sheets from the bulk liquid to the interface, which is shown by molecular dynamics to occur by an electrostatic bridging mechanism. 3D graphene-based films on droplet surfaces can also be used to control the release and transport of soluble molecules from the droplet to surrounding bulk solvent phases. In some cases, the release can be effectively stopped to produce unique kinetically trapped emulsion phases consisting of two fully miscible but segregated liquids. Finally, interfacial graphene-based films are also shown to control interfacial chemical reaction processes by serving as transport barriers between the phases or by intercepting reactive cross-phase molecular collisions. This reaction control is demonstrated by using 3D graphene-based microbarriers to protect oxidation-sensitive oils from attack by aqueous-phase reactive oxygen species, which is an undesirable pathway implicated in many chemical product degradation and spoilage processes.

502134 Nanostructured Materials forÂ Sensitive and Selective NH_3 detection

Abstract:

NH3 gas is a prominent air pollutant that is frequently found in industrial and livestock production environments. Exposure to NH3 at an unsafe level can result in pulmonary and lung diseases besides other health issues. Due to the importance in controlling pollution and protecting public health, to develop new sensing platforms for NH3 detection has attracted great attention. In this study, sensitive NH3 gas sensors with enhanced selectivity were developed based on various functional nanomaterials. X-ray diffraction (XRD), scanning electron microscopy (SEM) and X-ray photoelectron spectroscopy (XPS) were used to illustrate the properties of the as-prepared sensing materials and to understand the underlying sensing mechanism. The analytical characteristics of NH3 sensors such as sensitivity, selectivity and LOD will also be discussed.

497359 Enhancing Microfluidic Capture of Circulating Tumor Cells Using Magnetic Nanoparticles

Abstract:

Majority of cancer deaths are attributed to metastasis, the translocation of primary tumor cells to secondary sites. Circulating tumor cells (CTCs), cancer cells found in the blood stream, are believed to be a major contributor to metastasis. Hence, enumeration of CTCs holds promise as a means of monitoring disease progression. Unfortunately, the extreme rarity of CTCs in blood samples (typically ~1 CTC per 10^6 blood cells) makes accurate enumeration challenging. The use of microfluidic devices functionalized with antibodies to capture and enumerate CTCs has been an active area of research for over a decade. In these devices one or more surfaces are typically coated with an antibody against the epithelial cell adhesion molecule (EpCAM), which is believed to be a biomarker for CTCs. Such microfluidic devices can reliably count the number of CTCs in blood samples at low flow rates, but the desire to increase throughput from current values of 2-3 mL/hr leads to loss of specificity towards CTCs. Furthermore, many non-CTC cells are captured in these devices, resulting in poor capture purity. Here we explore the possibility of enhancing the capture efficiency and specificity of microfluidic CTC capture and enumeration systems through the use of magnetic nanoparticles in the presence of a magnetic field gradient to improve cross-stream migration of cells towards the antibody (ab) functionalized surface of the microfluidic channels. Initial studies have focused on demonstrating this principle using cells that have uptaken nanoparticles through non-specific interactions.

First we studied the uptake of dextran coated magnetic nanoparticles by EpCAM-expressing BxPC-3 pancreatic cells. Then capture efficiency for these cells using a microfluidic device with herringbone structure was quantified at 2 uL/s for the cases of bare and antibody-functionalized devices with and without magnetic fields. In the case of microfluidic devices coated with anti-EpCAM, in the presence of an external magnetic field created using a Halbach Array (HA), 62 % capture efficiency was found as opposed to 43 % when the device was not subjected to external magnetic field. We have currently obtained magnetic nanoparticles that can be employed to enhance the capture of CTCs at a throughput of 7.2 mL/hr.

Ongoing studies aim to develop magnetic nanoparticles conjugated with a peptide specific to EpCAM expressing BxPC3 cells. We expect such specific targeting will lead to an even greater improvement in capture efficiency and selectivity, especially in complex whole blood samples.

501656 Actuating Gold Nanoparticle Chemiresistor Sensitivity and Selectivity for Sensing Carbonyl Compounds in Air

Abstract:

Poor sensitivity and selectivity are common challenges faced by gas sensing chemiresistors. This work demonstrates a new approach that uses designed thiol monolayers to functionalize gold nanoparticles (AuNPs) for sensing trace volatile carbonyl compounds. A thiol ligand fitted with a ï·-urea group for molecular recognition of carbonyls dramatically increases both the sensitivity and selectivity of the derived AuNP sensor. The sensitivity for acetone detection surpasses all current chemiresistor limits of detection. The sensor works under ambient temperature and low humidity conditions, and shows a linear relationship between response and a broad carbonyl analyte concentration range of 0.1 ppb to 1000 ppm.

509054 Solve this! Fundamental Approach to Problem Solving in Industrial Processes I (Invited Talks)

Abstract:

Solving complex plant problems is an everyday exercise that affects all projects and budgets, and it is very rarely introduced to students at any level. In fact, by the time young engineers are few years into working in industry, most have lost the engineering skills that they developed in school. We hope to encourage future engineers to retain those skills and learn the importance of plant problem-solving.

492950 Meeting Challenges in a Land Constrained Solar Economy: Co-Production of Energy and Food Using Photovoltaic Systems over Farmland

Abstract:

The main purpose of this presentation is to demonstrate that contrary to the popular belief that ample land area is available to install photovoltaic modules, in an economy mainly supported by solar energy, there will be severe land availability constraints in many regions of the world—including many states in the USA. This will inevitably lead to competition for land between food and energy [1]. *We therefore propose several innovative PV systems that will allow for the co-production of food and energy from the same unit area of farmland thereby allowing local energy needs to be met by local supply of photons*. This allows us to overcome feedstock (land) and transport (long-distance transmission) limitations for electrical generation that would otherwise be difficult to overcome in a solar economy.

A recent study by MacKay showed that the energy consumption per unit area in densely populated countries such as the UK, Germany, Republic of Korea, and Japan is approaching the energy generation per unit area of photovoltaic farms [2]. This suggests that a transition to a solar powered economy in these nations would require deployment of solar on a large fraction of the nation's total land area, which will directly compete with land requirements for other needs (e.g. agriculture). Many other countries, including the United States are also trending toward this scenario [2]. A recent study by National Renewable Energy Laboratories confirms the large land area requirements of solar farms [3].

Using the available data on current U.S. energy consumption and land availability [4]-[5], we have estimated that as many as 31 states may be unable to meet their needs without importing energy long distances from other states or offshore. This scenario arises because much of the open land in the U.S. is devoted to cropland (21.5%), pasture land (32.3%), and forest land (30.4%). Only 3.6% of the land falls in the category of miscellaneous/other use land [5].

As it should remain a priority to preserve our native forest and wildlife areas, one of the greatest opportunities for increasing the availability of land for solar energy generation is over farmland. These land areas are often easily accessed and surround population centers. However, if PV is to be successful deployed over farmland, novel systems will need to be designed that do not negatively impact crop production. As most photosynthetic activity requires only the visible portion of the spectrum (400-750 nm), these wavelengths can be allowed to past through to plants while the IR portion of the spectrum is harvested as electricity. It is also clear that plants can go relatively extended periods of time without solar irradiance (night) and survive extended periods of reduced lighting (cloudy days) without significant impact to crop production so it may be possible to harvest a portion of the visible portion as well.

Figure 1 shows several possible configurations for harvesting solar energy over farmland. Fig. 1 (a)-(b) are systems that could operate optimally using current industrially-proven solar absorbers (e.g. silicon). Fig. 1 (c)-(f) shows systems for which light splitting would result in even higher performance by optimizing the band gap of the absorber for the portion of the spectrum it is absorbing. Fig. 1(a) is the simplest case in which south-facing PV modules are altered such that every other cell is left open or replaced with a light diffuser. In this system crops never experience shading for very long as the shadow will be constantly moving underneath the panel. Fig. 1(b) is the case in which single- or dual-axis tracking panels are used. Again, the shadow will move throughout the day. In the event that crops cannot tolerate extended periods of shade, these panels can also be patterned and be made twice the size to decrease the time intervals of shading, though the total time each plant would spend in the shade over the course of the day would be the same.

Fig. 1(c)-(f) are systems based on industrially available dichroic mirrors that split the spectrum [6]-[7]. These offer the advantage of concentrating the light for higher efficiencies. As a downside, diffuse portions of the spectrum may not be harvested efficiently. In Fig. 1(c), a tracking parabolic trough system is used in which the IR portion of the spectrum is reflected to a PV cell of appropriate band gap while the visible light passes through. As an option, a thin bifacial PV cells may be placed at the back (or a pattern thereof) to harvest a portion of the visible light and light scattered back from the crops. In Fig. 1(d), the system is non-tracking for mechanical simplicity and directs much of the IR spectrum to bifacial PV cells of appropriate band gap. During non-growing seasons, PV panels could be inserted in place of the transparent support. Fig. 1(e) depicts a tracking, long-pass system in which the visible light is concentrated and collected in fiber optic cables and then redistributed to plants underneath (and potentially other uses such as lighting. Fig. 1(f) depicts a non-tracking, long-pass system in which visible light is again collected in fiber optic cables and then redistributed to plants underneath, while the IR passes through to be collected on PV of appropriate band gap underneath. Promising preliminary calculations have been completed and are on-going to determine how much power can be produced and the optimal band gaps for each system.

The choice of which system to use will depend on how the crops respond to different shading characteristics and how much light deprivation and which portions of the spectrum they can tolerate being deprived of. It will also depend on the relative percentage of diffuse to direct light in a geographic location as the systems in Fig. 1(a)-(b) are better suited for areas with more diffuse light. We explore these options and make some general conclusions on system selection.

References

[1] Solar Projects Sow Tension, Wall Street Journal, Wall Str. J. (2017) A3.

[2] D.J.C. MacKay, Solar energy in the context of energy use, energy transportation and energy storage., Philos. Trans. A. Math. Phys. Eng. Sci. 371 (2013) 20110431. doi:10.1098/rsta.2011.0431.

[3] S. Ong, C. Campbell, P. Denholm, R. Margolis, G. Heath, Land-Use Requirements for Solar Power Plants in the United States - Technical Report NREL/TP-6A20-56290, 2013.

[4] State Profiles and Energy Estimates, (n.d.). http://www.eia.gov/state/.

[5] Major Land Uses, (n.d.). https://www.ers.usda.gov/data-products/major-land-uses/.

[6] T. Ulavi, T. Hebrink, J.H. Davidson, Analysis of a hybrid solar window for building integration, Sol. Energy. 105 (2014) 290–302. doi:10.1016/j.solener.2014.03.006.

[7] E. Gençer, C. Miskin, X. Sun, M.R. Khan, P. Bermel, M.A. Alam, R. Agrawal, Directing solar photons to sustainably meet food, energy, and water needs, Sci. Rep. (2017).

498203 Embodied Phosphorus in Interstate U.S. Food Transfers: Sustainability Implications for Food-Energy-Water Nexus

Abstract:

Modern agriculture relies heavily on nitrogen, phosphorus, and potassium based fertilizers to reduce the yield gap and sustain current crop yields. Recently inorganic phosphorus fertilizers, an essential input to agriculture have garnered attention as they are primarily obtained from non-renewable and geographically concentrated rock phosphate sources. Phosphorus content in the soil is heavily skewed across the world with many regions facing water quality issues due to nutrient overloading in surface waters, while some face nutrient deficiency and soil degradation issues. Food trade is essential for supplying diverse and large quantities of food items to feed the growing population at global, national, and regional scale. Food trade also presents a pathway for transferring large volume of virtual nutrients and associated environmental impacts. The spatial heterogeneity of food production systems can increase or decrease the consumption based environmental impact of food trade in a particular region. At domestic level, vast quantities of food is traded between states resulting in a complex interstate food trade network. Although, estimates of United States (U.S.) centric domestic production and international trade of embodied phosphorus are available, we lack an understanding of interstate flux of embodied phosphorus for the U.S. To avoid unintended consequences and ensure optimal outcomes for a resilient food supply system, a holistic systems approach to studying interconnected resources is required. We model the origin and destination of food and embodied phosphorus flows to quantify trends in regional phosphorus use, identify hotspots, and propose interventions for enhancing the fertilizer use efficiency and overall environmental sustainability of the U.S. food system.

Using publicly available disparate datasets, we develop a food trade and associated virtual phosphorus model for the U.S. To this end, we evaluate over 60 crop items covering commodity groups of cereal grains, and fruits, nuts and vegetables. The interstate food trade network is translated to virtual phosphorus network by using crop yields and phosphate fertilizer application rates from estimates provided by the U.S. Department of Agriculture. Preliminary results indicate that over 655 million tons of food commodities were transferred across the U.S. in 2012 along with 61 million tons of embodied phosphorus fertilizer. Additionally, fruits and vegetables category contributed only 30% by mass for food trade but accounted for half of embodied phosphorus trade compared to cereal grains. We also evaluate virtual phosphorus savings for each individual item to identify phosphorus use efficiency in the US food production system. Additionally, we quantify the embodied energy flows associated with embodied phosphorus trade and discuss the inorganic phosphorus impact on water quality. Furthermore, we apply a network theory framework to understand the structure, robustness, and environmental sustainability of the considered trade networks. From a network perspective, our previous work has found that interstate food trade forms a well-connected, robust network with a majority of states participating in a high volume of trade. We characterize interstate virtual phosphorus network through series of higher order network measures to understand robustness and resilience patterns in virtual phosphorus trade. This will provide a necessary framework to evaluate series of interventions in a highly interconnected food trade system.

495017 Comparing Beccs and DAC for Climate Change Mitigation: The Water-Land-Energy Nexus

Abstract:

The Paris Agreement signaled global commitment to limit average global temperature rise to 2°C and to pursue efforts to achieve 1.5°C. The IPCC AR5 cites negative emission technologies, particularly bio-energy with carbon capture and storage (BECCS), as critical to achieving this[1]. BECCS' ability to deliver negative emissions has been questioned with many uncertainties around land-use change emissions. Furthermore, large-scale biomass use poses several biophysical and socioeconomic risks such as exacerbating scarce water resources, biodiversity loss, and large opportunity costs of land-use for biomass. Capturing carbon dioxide (CO_2) directly from the atmosphere, *direct air capture* (DAC), presents an alternative source of negative emissions. Established capture technologies can be used to extract it from the atmosphere, seemingly without the significant natural resource requirements of BECCS. However, the low concentration of CO_2 in air necessitates a high energy cost of capture.

In this work, a comparison of the strengths and weaknesses of each technology based on key performance metrics (land, water and energy requirements) has been done. A negative emissions target of 12 Gt_{CO2}/y was used; this is the central estimate of what is required from RCP2.6[1] and has been cited as the target beyond which BECCS is no longer economically feasible. A *'farm-to-NET plant'* model of BECCS accounting extensively for the biomass supply chain and power plant specifications (coal co-firing proportions, capture rates) has been built to determine the overall GHG emission balance of the technology. Data from DAC processes proposed in the literature and existing pilot projects have been used to quantify the potential land, water and energy requirements of a large-scale system.

For BECCS, high variability in carbon balance outcomes was observed depending on the feedstock type and source, and assumptions on indirect land-use changes. Depending on decisions taken throughout the supply chain, a BECCS facility could take over 50 years to begin removing CO_2 from the atmosphere, or it could start removing CO_2 from day one[2]. Stark contrasts in resource requirements to meet the mitigation target were found for BECCS and DAC. Depending on the conditions of its deployment, a BECCS plant operating at 100% load factor could need 363-2100 Mha of land and 4-16 Tm^3/yr of water dedicated.

DAC's needs were 3-5 orders of magnitude lower at 0.04-3.3 Mha and 180 km^3/yr^3, thus posing little to these resources. It is important to note that DAC's water requirements are very sensitive to relative humidity and hence location of the plant[3]. Furthermore, the land area associated with the electricity used for DAC process was not accounted for, and could be significant when considering low carbon power, such as solar and wind.

BECCS' status as an energy-positive mitigation option was found to be untrue in some scenarios. Its annual energy output in the case of US-sourced Switchgrass could be as high as 22 EJ/yr, however up to 37 EJ/yr could be required if Brazil-sourced Miscanthus is the choice biomass. DAC energy costs, however, are still significantly higher at 81-274 EJ/y, even when the lower-bound from literature is assumed.

Surplus resource availability, particularly land and water, are highly-dependent on geography. BECCS and DAC have been shown to require contrasting amounts of these. Therefore, the preferred technology must be discussed within the context of a specific location, and in the case of BECCS, biomass source if not locally-grown.

References

1. IPCC. *Climate Change 2014: Mitigation of Climate Change. Working Group III Contribution to the Fifth Assessment Report of the Intergovernmental Panel on Climate Change* (2014). doi:10.1017/CBO9781107415416

2. Fajardy, M. & Mac Dowell, N. Can BECCS deliver sustainable and resource efficient negative emissions? *Energy Environ. Sci.* 0–42 (2017). doi:10.1039/C7EE00465F

3. Socolow, R. *et al.* Direct Air Capture of CO 2 with Chemicals Panel on Public Affairs. *Am. Phys. Soc. - Panel Public Aff.* 100 (2011).

499908 Food Manufacturing: The Shift from Centralised to DistributedÂ Production

Abstract:

Projections for global food demand for a growing world population indicates an increase of 70% by 2050 (United Nations, 2017). The competitiveness among the food producers to satisfy production demand and consumer expectations has led to the search of new alternatives and innovations to make food products more attractive for the consumers, thus increasing company profits (Ingram et al., 2013).

With the growing concern about climate change, the consumer expectations are not limited to the relation quality-price, but also they include a low environmental impact of the product (Kemp et al., 2010).

As in other industries, the food manufacturing has benefited from the economy of scale, where the mass production allows the reduction of the manufacturing cost. However, a centralised production leads to a complex distribution chain, where the distances travelled for the product distribution increase and with them the cost and the CO_2emissions related to transportation.

Another production alternative is distributed manufacturing with shorter distances among the facilities (i.e. farm-factory, factory-distribution centre), thus the transportation cost and hence environmental impact are reduced. Nevertheless, other factors can affect the environmental impact of a production-distribution system, e.g. the carbon footprint of the raw materials.

The aim of this work is to analyse from an economic and environmental perspective several manufacturing-distribution scenarios, and identify the factors with greater influence in the choice between a centralised manufacturing and a more distributed one. For this purpose, a novel framework we call the *honeycomb model* is presented for the techno-economic and sustainability assessment of different manufacturing scenarios. This framework can be applied for the estimation of the optimum configuration that minimises the cost and CO_2emissions of the manufacturing-distribution chain, as well as perform studies of sensitivity of the optimum number of processing plants to certain parameters/characteristics of the system. Here, the tomato paste production is used as case study.

The results indicate that increments in the area of the system analysed and the product demand favour a more distributed manufacturing, while the origin of the raw materials and their carbon footprint could lead to the centralisation of the production.

Sensitivity analysis of the optimum manufacturing configuration provides important information about when a more distributed manufacturing is preferred over a centralised one. This information can be useful for further design of strategies to shift a centralised manufacturing-distribution configuration to a more distributed one when the purpose is to boost local economies.

This project is part of the Centre for Sustainable Energy Use in Food Chains (CSEF) from RCUK.

References

K. Kemp, A. Insch, D.K. Holdsworth, J.G. Knight. 2010. Food miles: Do UK consumers actually care? Food Policy 35:504â€'513.

J.S.I. Ingram, H.L. Wright, L. Foster, T. Aldred, D. Barling, T.G. Benton, et al. 2013.Priority research questions for the UK food system. Food Sec. 5:617â€"636.

United Nations. Available in: http://www.un.org/en/development/desa/news/population/2015-report.html. Date accessed: February 18th, 2017.

494133 Understanding Biomass Value Chains and the Environment-Food-Energy-Water Nexus through Whole-Systems Analysis and Optimisation

Abstract:

Biomass has the potential to provide access to a sustainable supply of resources such as energy, chemicals and materials. Since the production of biomass, i.e. plant matter such as forestry, agricultural crops and residues, requires a large amount of land and water, competes with food production and can affect the ecosystem (e.g. deforestation, leaching of fertilisers and pesticides into water bodies), there is a delicate balance between all of these elements. Together, these are known as the environment-food-energy-water nexus and all of the benefits associated with the use of biomass to meet our requirements for energy and other products must be considered along with the potential impact to the nexus as a whole. Therefore, it is important not only to understand how best to exploit the great potential of biomass but also to understand the detailed interactions among all of the elements in the nexus.

The BEFEW (Biomass and the Energy-Food-Environment-Water Nexus) project funded by the Newton Fund and EPSRC (Grant No. EP/P018165/1) is examining the many different possibilities for producing energy and high-value products from biomass that also maintain the balance in the nexus. Detailed maps of existing land use and soil quality, water availability, quality and tolerance to contamination, and climate projections in terms of temperature and rainfall are used in order to understand the potential for growing biomass. Different processing facilities that convert the biomass to energy or other chemicals are researched and a comprehensive database of such technologies, including a site suitability analysis is developed. Thus, a detailed and rich representation of the influence, on the whole of the nexus, of the types of biomass grown, where they are grown, the types of technologies used (and therefore the products that will be made from biomass) and where they are located can be built. A mixed-integer linear programming (MILP) model is developed in order to determine the combinations of crops grown and technologies used to convert biomass to energy and high-value products (thus forming a biomass value chain) that provide the greatest benefit with the lowest impact on the nexus. Using the model, the efficient and robust value chains that are synergistic with the nexus and have the potential for deployment over a wide range of scenarios with different combinations of key socio-economic and environmental performance indicators and constraints are identified. In this conference, the latest progress in the BEFEW project will be presented, with a focus on the MILP model and case studies.

505039 Innovative Approaches to Achieving FEW Nexus Goals that can Influence Near-Term & Long-Term Decision Making

Abstract:

The world population is estimated to reach to 9.7B by 2050, resulting in increased demand for resources and increased production of waste. As a result, today's waste disposal systems and approach to waste management will not be adequate for future applications. Therefore, a systematic approach to sustainable food, energy and water management is needed to meet increased demand for resources and efficient waste utilization systems. Currently, the food, energy and water (FEW) sectors use technologies developed to address the needs of a given sector (*i.e.* energy, water, food or agriculture), and wastes from each sector are usually managed separately. Production systems underlying these sectors have traditionally treated pollution and waste as externalities that are diluted into the ambient environment, and separate infrastructure systems have consequently been developed with unnecessary inefficiencies that are increasingly problematic. Because neither landfilling nor industrial composting provide ultimate reutilization of organic waste, and neither can achieve both energy and nutrient recovery in one process from organic waste, we need to transition from a "Traditional Linear Waste Disposal" approach to a sustainable "Closed Loop Resource Recovery" approach. In addition, finding efficient and affordable solutions to avoid GHG emission leakages from landfills and composting facilities through the use of Anaerobic Digestion infrastructure will be a key element of a sustainable system for future waste management. Thus, exploring how FEW systems can efficiently be optimized to (1) repurpose or recycle organic waste products, (2) internalize traditional externalities, and (3) integrate wastes with resource inputs across systems, by deploying integrative management strategies is essential. In addition, integrating other relevant disciplines that bring the necessary economic, social and educational perspectives to addressing this problem, will achieve broader impacts across the system.

498677 Road Map for Embedding Ethics into ChE Undergraduate Curricula

Abstract:

The author will outline a number of methodologies and a suggested subject sequencing that can be effectively used to further develop an embedded ethics curricula. This curricula will consist of definitions, examples and suggestions on where to embed ethics into any existing ABET-approved, chemical engineering undergraduate course set. The author has lectured extensively at universities on ethics and a number of ethics related subjects.

504018 Views on Ethics in Undergraduate Education

Abstract:

Engineering programs are expected to graduate students who understand their professional ethical responsibilities. A studentâ€™s experience with ethics often begins with academic integrity. Research on student attitudes and behavior regarding academic integrity will be reviewed with suggestions to help faculty nurture a strong ethical foundation in their students.

495849 Ethical Reasoning in the Engineering Curriculum

Abstract:

Although the provision of ethics in engineering curricula has increased in recent years, the teaching of engineering ethics remains patchy and sporadic in a number of countries. Ethics in research, on the other hand, appears to be at a more mature stage. At the same time, the number of ethical codes, codes of practice and research concordats has flourished, both nationally and internationally. The questions remain: are those codes â€œliveâ€? Have those concordats resulted in an improvement of professional integrity? Is the teaching of ethics different and/or disconnected from ethics in research and in the workplace? Is ethics in chemical engineering different from ethics in other engineering disciplines? The paper analyses those questions in relation to an ever changing landscape, influenced by local and international drivers.

RAPID
Transforming Process Industries

Breakthrough Innovation to Transform the Process Industries

Karen Fletcher
CEO - RAPID

Jim Bielenberg
CTO - RAPID

October 30, 2017

ManufacturingUSA

Manufacturing USA Overview

Vision	U.S. Global leadership in Advanced Manufacturing
Mission	Connecting people, ideas, and technology to solve industry-relevant advanced manufacturing challenges, thereby enhancing industrial competitiveness and economic growth, and strengthening our national security.
Goals	1. Increase competitiveness 2. Facilitate technology transition 3. Accelerate the manufacturing workforce 4. Ensure stable and sustainable infrastructure

www.ManufacturingUSA.com

2

RAPID Transforming Process Industries

Manufacturing USA

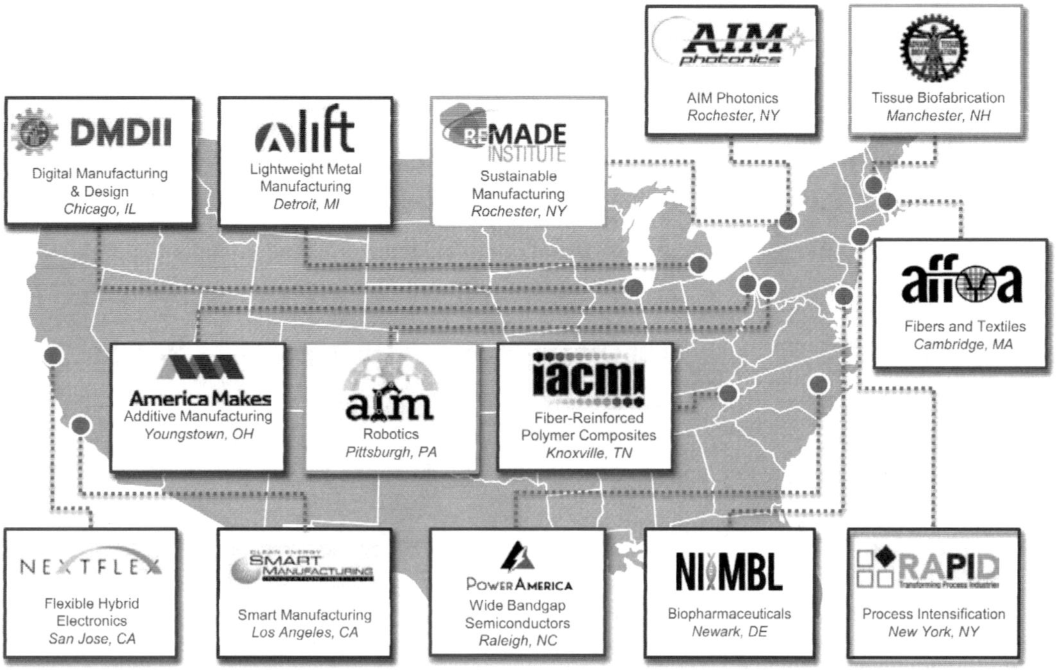

3

RAPID Transforming Process Industries

RAPID's Industry-Led Vision

A dynamic network of partners who collectively build a sustainable ***ecosystem*** *that:*

... researches, develops and advances new technologies for modular chemical process intensification

... builds a strong portfolio of R&D projects & educational tools

... delivers dramatic reductions in energy, environmental footprint, capital and operating cost

... makes U.S. Manufacturing and our workforce more competitive

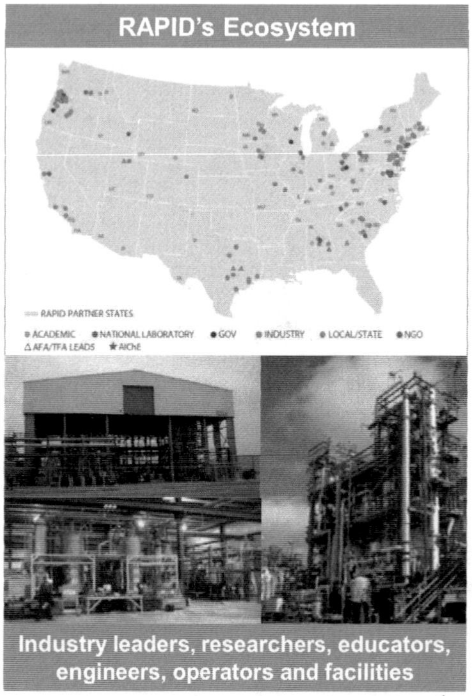

4

Our Mandate

- Lead a national effort to research, develop and demonstrate high-impact modular chemical process intensification solutions for U.S. Manufacturing.
- Actively build RAPID membership through an inclusive and attractive value proposition.
- Leverage $70 million of federal funding with cost share from members.
- Operate the Institute to benefit a wide range of stakeholders.
- Establish an infrastructure that enables access to process intensification resources, tools, expertise, and facilities.
- Bring together private and public entities to co-invest in R&D, commercialization, and deployment of innovative technologies.
- Establish a technical education and workforce development program.

"The goal for these Institutes is to revitalize American manufacturing and support domestic manufacturing competitiveness."
— U.S. DOE

DOE Performance Metrics

- Energy Efficiency
- Energy Productivity improvement
- Intensification in Individual Process Modules
- Cost-Effective Manufacturing of Modules
- Cost Effective Deployment
- Enabling R&D Portfolio
- Industrial Partnerships
- Pathway to Self-Sustainment
- Train the Trainers
- Educate Students
- Annual Planning Process
- Industrial Roadmap
- Emerging Supply Chain
- Diversity of Firms and Individuals in the Eco-System

6

RAPID
Transforming Process Industries

52 Members and growing

RAPID
Transforming Process Industries

Technical Focus Areas

Chemical Commodity Processes

- Develop guidelines for integration of novel reaction and separation modules
- Validate design tools for process intensification

Renewable Bio Products

- Prototype and scale novel bio-conversion processes
- Improve energy & capital efficiency of existing and emerging processes

Natural Gas Upgrading

- Maximize impact through transfer of technologies and learnings across industries
- Mature and demonstrate transformational and enabling technologies for gas utilization

8

RAPID Transforming Process Industries

Technical Focus Areas

Modeling and Simulation

- Develop methods & tools for design, optimization, and intensification across multiple length and time scales
- Establish methods for intensification of dynamic/periodic operations

Intensified Process Fundamentals

- Advance inherently energy efficient separation processes & reaction platforms
- Develop fundamentals for multifunctional modules such as hybrid separation/ reaction schemes

Module Manufacturing

- Standardize modules and components to drive demand and capital investment within the supply chain
- Lower the cost of PI equipment using advanced manufacturing technology

9

RAPID Transforming Process Industries

2017 Accomplishments

- Completed DOE negotiations & all required documentation
- Stood up the Institute, with strong support/services from AIChE
- Established membership base
- Identified, negotiated and launched 4 Jump Start projects
- Formed Education & WFD Committee and completed a roadmap
- Completed focus area roadmaps with 100+ participants
- Executed call for pre-proposals and call for proposals
- Designed and launched a new website www.aiche.org/rapid
- Hired CTO (Mar), Technology Manager (Oct), Comms Leader (Nov)

10

A Public-Private Partnership Between U.S DOE EERE & AIChE®

RAPID Transforming Process Industries

Technology Gaps

RAPID Roadmap

Chemical & Commodity Processing

- **Rxn/Separation**
  Novel reaction/separation schemes that are scalable and drive process efficiencies (e.g., membrane-,or sorption enhanced reactors). Applications include processing light paraffins to olefins, increase p-xylene yield vs conventional processes, hydrogen production, managing oxygen supply to reactions, etc.
- **Non-Thermal Drivers**
  Use of alternative, non-thermal driving forces to activate chemical systems at the appropriate (atomic/molecular) scales.
- **Batch Systems**
  Intensification schemes for batch systems. Transferring concepts largely developed for continuous processes to the batch realm could result in increased productivity/lower cost for specialty/fine chemicals.

Natural Gas Upgrading

- **Selective Conversion**
  Concepts that show dramatic increases in desired product yields via fundamental improvements in catalysis, heat and mass transfer, and process concepts. This could include include alternative energy inputs and/or the use of novel reaction systems.
- **Separations**
  Energy efficient separations technology to purify the reaction product mix, condition the feed in preparation for conversion, and to generate coreactants for participation in natural gas conversion gas.
- **Process Consolidation**
  Process consolidation and modularity to reduce total installed cost by reducing the total number of unit operations and by reducing the amount of field fabrication.

Renewable Bioproducts

- **Primary Separation**
  Technologies to reduce energy demand in primary separation process steps designed to recover organic molecules and biomass components from water.
- **Water management**
  Low capital and energy intensive solutions for dewatering and drying of biomass feedstocks, water removal and drying in pulp and paper process, and drying and removal of low levels of residual water from end products.
- **Couple Rxn/Spn/Hxt**
  Use of novel chemistries and MCPI strategies to couple heat transfer and reaction in thermal processing of biomass and/or novel applications of reactive separation technologies in biological conversion technologies such as fermentation.

Intensified Process Fundamentals

- **Scale-Out Methods**
  Scale out methodologies and models to predict performance of alternative energy input approaches for reactions and mixing and determine the suitable scale for modular manufacturing.
- **Fundamental Data Acquisition/Modeling**
  Approaches to address key issues with lack of data on fluxes, adsorption, and catalyst kinetics for wide classes of materials, enabling model development and experimental testing of novel materials as adsorbents, membranes, catalysts and their integration.
- **Predictive Models**
  Modeling capabilities to screen concepts and configurations of all types and predict optimal structures.

Modeling & Simulation

- **PI Software Tools**
  Software tools for integrated reaction and/or separation processes and/or cyclic process such as pressure swing adsorption (PSA) or temperature swing adsorption (TSA). Such tools must be widely accessible and capable of integrating MCPI solutions with existing unit operations.
- **Data Availability**
  Modeling approaches coupled with data generation and/or analysis to generate databases of physical parameters enabling design with mass separating agents.
- **PI Assessment Tools**
  Tools to asses safety, sustainability, and control in PI and MCPI applications, including tools that address unique issues of uncertainty and reduced control variable options that are present in PI and MCPI applications.

Module Manufacturing

- **Intensified Components**
  Intensified components that drive down the cost of module pre-assembly, transportation, and installation, while driving significant energy savings in chemical processes.
- **Standard Designs**
  Design approaches that limit the amount of non-recurring engineering — during systems integration and installation — needed to support customized modules. This could include standard modules that enable economies of mass production and/or designs that enable incremental capacity additions.
- **Distributed Processing**
  Module design and manufacturing approaches to enable distributed chemical processing. These will provide new paths to capital cost reduction and innovative techniques for maintenance and remote access and monitoring.

www.aiche.org/rapid

A Public-Private Partnership Between U.S DOE EERE & AIChE®

RAPID Transforming Process Industries

RAPID Education & Workforce Development Roadmap

OBJECTIVE: Establish a technical education and workforce development program that will leverage existing resources to train and educate the workforce, who can then research, develop, design and operate processes that incorporate new process intensification (PI) technologies and modular process designs widely within U.S. industry.

Undergraduate Students

GOAL
Create PI & MCPI awareness in future generations of engineers by augmenting the standard (ABET accredited) engineering curriculum with modules and educational content introducing PI & MCPI topics relevant to industry.

WHAT SUCCESS LOOKS LIKE
500 students* trained in PI & MCPI per year by 2019.

METHOD
eLearning course modules, face-to-face courses in conjunction with major meeting, Intern program.

Graduate Students

GOAL
Generate PI & MCPI expertise so that graduate students can connect the relevance of PI & MCPI research to industrial applications and create new opportunities and applications for MCPI.

WHAT SUCCESS LOOKS LIKE
500 students* trained in PI & MCPI per year by 2019.

METHOD
eLearning course modules and face-to-face courses in conjunction with major meeting.

Faculty

GOAL
Update and educate university faculty on PI & MCPI concepts and introduce eLearning modules created by RAPID for integration into their curriculum.

WHAT SUCCESS LOOKS LIKE
50 education/training professionals per year in MCPI by 2019.

METHOD
Face-to-face course in conjunction with major meeting on the Fundamentals of PI & MCPI.

Technicians/Operators

GOAL
Introduce technicians and operators to PI & MCPI, provide awareness of new processes that include PI & MCPI as well as emphasizing process safety.

WHAT SUCCESS LOOKS LIKE
At least 50 professionals** per year in MCPI by 2019.

METHOD
eLearning course modules.

Professional Engineers R&D | Design | Production

GOAL:
Train the current generation of engineers on how to utilize and deploy PI & MCPI technology in order to improve process efficiency.

WHAT SUCCESS LOOKS LIKE:
At least 50 professionals** per year in MCPI by 2019.

METHOD
eLearning course modules, face-to-face courses in conjunction with major meeting, and promotion through Manufacturing Extension Partnerships (MEPs).

Management

GOAL
Create high level awareness of the benefits of PI & MCPI from a techno-economic analysis perspective.

METHOD
Presentation material (includes relevant case studies, sustainability, and safety); short videos aimed at executives.

PI & MCPI Body Knowledge Drives Content Development Topics and Priority

Crosscutting areas identify content topics, levels of understanding for each audience member as well as high-priority topic areas. It includes four sub-sections:

- PI & MCPI Overview
- Modeling & Simulation
- Intensified Process Fundamentals
- Module Manufacturing

Application areas drive relevancy for content and provide industrial applications, examples and case studies within content and include the three sub-sections above.

* Students defined as undergraduate and graduate students.
** Professionals defined as technician, operator or engineer.

www.aiche.org/rapid

RAPID Transforming Process Industries

2017 Call for Proposals

- Overall Results
  - 73 submissions
  - 42 "lead" institutions
  - ~$30M (fed) request for 2018
- Proposals review process completed during Sept/Oct
- Announcement of selected projects November 2nd, 2017

# of Submissions	Submissions*
Chemical Commodity Processes	13
Intensified Process Fundamentals	13
Modeling and Simulation	6
Module Manufacturing	7
Natural Gas Upgrading	18
Renewable Bioproducts	16
	73

Summary Data - RAPID 2017 Proposal Call

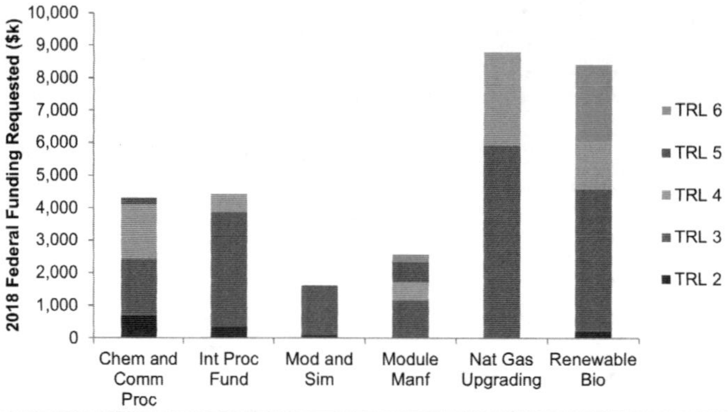

13

RAPID Transforming Process Industries

Dynamic Intensification of the Operation of Dividing Wall Columns

DOW

TEXAS
The University of Texas at Austin

Innovation:

- Demonstrate that a dynamically controlled process (non-steady state) creates opportunity for improved efficiency

Project Outcomes

- Develop theoretical basis for periodic operation of intensified processes
- Experimental proof of concept using dividing wall column (DWC) pilot plant available at UT

Institute Target Addressed

- Develop tools and technologies to reduce the cost of deploying modular chemical process intensification in existing processes

Fall 2017 Update

- Theoretical framework for general value/screening of periodic operation tested
- Initial screening of DWC periodic operation assessed

14

Adsorptive Nitrogen Rejection from Natural Gas

RAPID Transforming Process Industries

PRAXAIR

Georgia Institute of Technology

Innovation

- Utilize breakthrough rate dependent adsorbent to modularize conventional PSA process

Project Outcomes

- Field demonstration of portable, modular PSA system

Institute Target Addressed

- 1,000 hours of pilot scale operating time for a modular and intensified process with reduced capacity , improved energy efficiency, and lower emissions/environmental waste

Fall 2017 Update

- System installation, shakedown, and testing to demonstrate adsorbent performance in a PSA pilot system underway
- Manufacturability studies progressing - potential adsorbent manufacturer identified.
- Early 2018 tasks - additional work on TEA and H2S tolerance testing at Georgia Tech

15

Autothermal pyrolysis of lignocellulose wastes to sugars & other biobased products

RAPID Transforming Process Industries

easy energy SYSTEMS

IOWA STATE

Innovation

- Couple intensified reaction + heat transfer design and novel process chemistry to modularize sugar production from biomass

Project Outcomes

- Establish MCPI benefits of autothermal pyrolysis, leading to design of a commercial scale modular system

Institute Target Addressed

- MCPI RD&D resulting in a doubling of energy productivity ($ Returned/Energy Consumed) by a combination OPEX and CAPEX improvements.

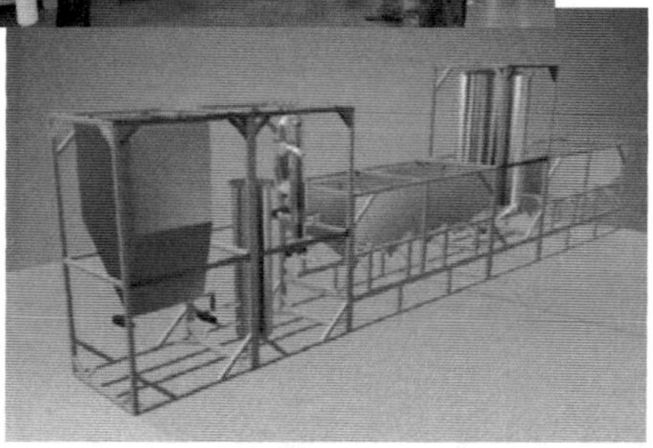

Fall 2017 Update

- Demonstrate sugar yields in excess of BP1 targets on 1 kg/h pyrolysis reactor under auto thermal conditions
- Completed continuous pilot scale pretreatment of multiple feedstocks
- Continuing to work on longer time pilot scale testing

16

RAPID Transforming Process Industries

Mfg Supply Chain Development for the Modular Solar-Thermochemical Conversion Platform

Oregon State University OSU

Pacific Northwest NATIONAL LABORATORY

Innovation

- Coupling material processing approaches and manufacturing techniques that are uniquely well suited to small scales to generate low cost modular components

Project Outcomes

- Develop novel additive manufacturing techniques and materials that enable modular designs

Institute Target Addressed

- Scale-out manufacturing MCPI, with a unit cost reduction of >20% with each doubling in cumulative module manufacturing production

Fall 2017 Update

- OUS has developed unvalidated costs models for key manufacturing steps
- PNNL has established cost targets for key components
- OSU is working in-house capability to test new designs based on selective alloying
- Targeting 1Q2018 for new tool completion in support of novel manufacturing techniques

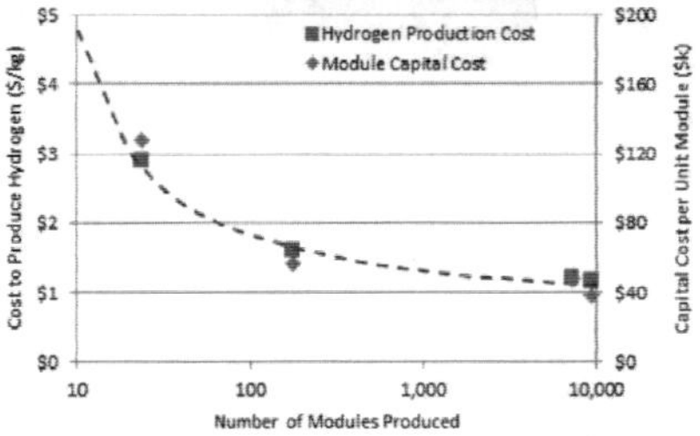

Estimated capital and product costs for individual STARS modules as a function of module mass production.

RAPID Transforming Process Industries

Plans for 2018

- Build the RAPID community
  - Member profiles, webinars, newsletters, website, networking opportunities
  - Conferences / face-to-face meetings (AIChE Spring / Annual)
- $30+M invested in R&D projects (fed funds + cost share)
- DOE (external) peer review
- Updated roadmaps; Gaps and new areas explored
- Project calls
- Training courses and educational webinars
- Increase membership and “reach”
  - New addition to our team: State MEPs*

* MEP = Manufacturing Extension Partnership

RAPID
Transforming Process Industries

To learn more about RAPID:

www.AICHE.org/RAPID

KareF@aiche.org

JameB@aiche.org

19

493251 Modeling and Simulation â€“ a Key Component in Enabling Process Intensification

Abstract:

The RAPID Manufacturing Institute aims to enable the development, de-risking, and implementation of process intensification across the chemical process industry. We will discuss critical needs in modeling and simulation in this area by highlighting the successes and challenges associated with using modeling and simulation to drive RAPIDâ€™s aims.

493349 Modeling and Simulation Challenges for Process Intensification

Abstract:

Modeling and Simulation Challenges for Process Intensification

Efstratios N. Pistikopoulos[1,2], David Sholl[3], M. M. Faruque Hasan[1,2], Salih E. Demirel[1,2] and Yuhe Tian[1,2], (1)Artie McFerrin Department of Chemical Engineering, Texas A&M University, College Station, TX, (2)Texas A&M Energy Institute, Texas A&M University, College Station, TX, (3)School of Chemical & Biomolecular Engineering, Georgia Institute of Technology, Atlanta, GA

The recently established Rapid Advancement in Process Intensification Deployment (RAPID) Manufacturing Institute aims to transform the U.S. industry by developing the next generation of modular, intensified, and innovative manufacturing technologies and processes. The RAPID Modeling and Simulation Focus Area will in particular address the development of strategies and tools for the multi-scale representation, modeling, optimization, and control of intensified systems. In this presentation, we provide an overview of the challenges towards a systematic synthesis, modeling, and simulation approach for process intensification, as well as an exclusive review of process intensification synthesis approach methods in the open literature, such as the ones included in [1-5]. We also introduce an integrated framework for process intensification incorporating operability, control and safety (illustrated in Figure 1).

Figure 1: An Integrated Framework for Process Intensification Incorporating Operability, Control and Safety

[1] Demirel, S. E., Li, J., & Hasan, M. F. (2017). Systematic process intensification using building blocks. *Computers & Chemical Engineering*. http://doi.org/10.1016/j.compchemeng.2017.01.044.

[2] Baldea, M. (2015). From process integration to process intensification. *Computers & Chemical Engineering*, *81*, 104-114.

[3] Tula, A. K., Babi, D. K., Bottlaender, J., Eden, M. R., & Gani, R. (2017). A computer-aided software-tool for sustainable process synthesis-intensification. *Computers & Chemical Engineering*. http://doi.org/10.1016/j.compchemeng.2017.01.001

[4] Papalexandri, K. P., & Pistikopoulos, E. N. (1996). Generalized modular representation framework for process synthesis. *AIChE Journal*, *42*(4), 1010-1032.

[5] Ismail, S. R., Proios, P., & Pistikopoulos, E. N. (2001). Modular synthesis framework for combined separation/reaction systems. *AIChE Journal*, *47*(3), 629-649.

494520 The Sustainable Synthesis-Design-Intensification of Chemical and Biochemical Processes

Abstract:

With the demands for efficient utilization of energy and water, better protection-control of the environment and correct-consistent use of raw materials, it is no longer sufficient to convert a selected raw material to a desired product, making profit and ensuring operability. The synthesized-designed process must now also be sustainable, incorporating not only economic feasibility but also environmental impact constraints, energy-water usage and/or waste constraints, resources availability constraints, product demand-price constraints and, many more. That is, the demands for new, innovative and sustainable processes need to be addressed. Process intensification has been identified as an important means to achieve this demand through integration with process synthesis-design in the early stage development of any chemical and biochemical process.

The presentation will highlight the issues related to defining the measures of sustainability through which new alternatives can be evaluated. Here, emphasis is given to the concept of sustainability, which is relative and therefore, a reference case first needs to be established. That is, compared to the base case, how sustainable are the new and innovative chemical and biochemical process designs?

A systematic multi-stage and multi-dimension computer aided framework which provides the necessary work-flow, the corresponding data-flow, the associated methods and the needed computational tools will be presented.

Through case studies involving chemical and biochemical processes, the application of the framework and the associated methods-tools to determine more sustainable solutions will be illustrated. It will be shown that significant improvements and non-tradeoff sustainable can be obtained through this integrated approach of combining process intensification as an essential element of process synthesis-design.

2017 AIChE Annual Meeting

October 29 - November 3, 2017
Minneapolis Convention Center, Minneapolis, MN

Democratizing Energy Technology

Dane A. Boysen, PhD
Modular Chemical, Inc.

October 30, 2017

Overview

- How did RAPID get established?
- Why do we go big?
- Why is scale-up a problem?
- How do we break the hegemony of scale-up?

2011-2014

arpa·e

Advanced Research Projects Agency • ENERGY

2015-2016

gti®

GAS TECHNOLOGY INSTITUTE

Innovation Challenge

FINANCING ECONOMIES OF SCALE

GTL PLANT

(boe/d)	($/bpd)	plant cost
1	47,000k	$47M
10	10,000k	$100M
100	2,200k	$220M
1k	470k	$470M
10k	100k	$1,000M

AMMONIA PLANT

(ton/d)	($/tpd)	plant cost
1	10,000k	$10M
10	4,500	$25M
100	2,000k	$200M
1k	900k	$900M
10k	400k	$4,000M

calculated from "2/3 scaling law"

Innovation Challenge

FINANCING ECONOMIES OF SCALE

GTL PLANT			AMMONIA PLANT			
(boe/d)	($/bpd)	plant cost	(ton/d)	($/tpd)	plant cost	
1	47,000k	$47M	1	10,000k	$10M	R&D demo
10	10,000k	$100M	10	4,500	$25M	
100	2,200k	$220M	100	2,000k	$200M	pilot demo
1k	470k	$470M	1k	900k	$900M	commercial
10k	100k	$1,000M	10k	400k	$4,000M	

calculated from "2/3 scaling law"

R&D Spending

OFFICE OF FOSSIL ENERGY BUDGET

U.S. DEPARTMENT OF ENERGY

FY 2015	$561M
FY 2014	$570M
FY 2013*	$495M
FY 2012*	$337M
FY 2011	$434M
FY 2010	$660M
FY 2009	$876M
FY 2008	$465M

Pilot chemical and power plants require **more than $200M**

How can we innovate with vastly inadequate funding?

*continuing resolution

FY2016 U.S. Senate Energy and Water Appropriations Bill

"The Committee recognizes that stranded, flared, and vented natural gas is the result of low natural gas prices that make transporting it uneconomic. As topics for additional Clean Energy Manufacturing Institutes are evaluated, the Secretary is encouraged to consider modular chemical processing as a way to address the issue of natural gas flaring and enable advanced manufacturing applications in the oil and gas industry."

https://www.congress.gov/congressional-report/114th-congress/senate-report/54/1

RAPID

U.S. DEPARTMENT OF ENERGY | Energy Efficiency & Renewable Energy

ADVANCED MANUFACTURING OFFICE

RAPID ADVANCEMENT IN PROCESS INTENSIFICATION DEPLOYMENT INSTITUTE

December 2016, the US Department of Energy announced the establishment of the 10th Manufacturing USA Institute, representing a critical step in the federal government's effort to double U.S. energy productivity by 2030. The Rapid Advancement in Process Intensification Deployment (RAPID) Institute is focused on addressing the barriers listed above to enable the development of breakthrough technologies to boost energy productivity and energy efficiency through manufacturing processes in industries such oil and gas, pulp and paper and various domestic chemical manufacturers. RAPID will leverage approaches to modular chemical process intensification (MCPI)—such as combining multiple process steps such as mixing, reaction, and separation into single more complex and intensified processes—with the **goal of improving productivity and efficiency, cutting operating costs, and reducing waste.**

https://www.aiche.org/rapid/about

RAPID Successes

1. Quick to get up and operational
2. Quality R&D projects in the que
3. Good industry engagement

Challenges for RAPID

1. No clear problem statement (goal ≠ problem)
2. No plan for establishing standards (modular = standards)
3. Not clear how progress will be tracked (need baseline)

RAPID Challenges

1. No clear problem statement (goal ≠ problem)
2. No plan for establishing standards (modular = standards)
3. Not clear how progress will be tracked (need baseline)

Why do we go big?

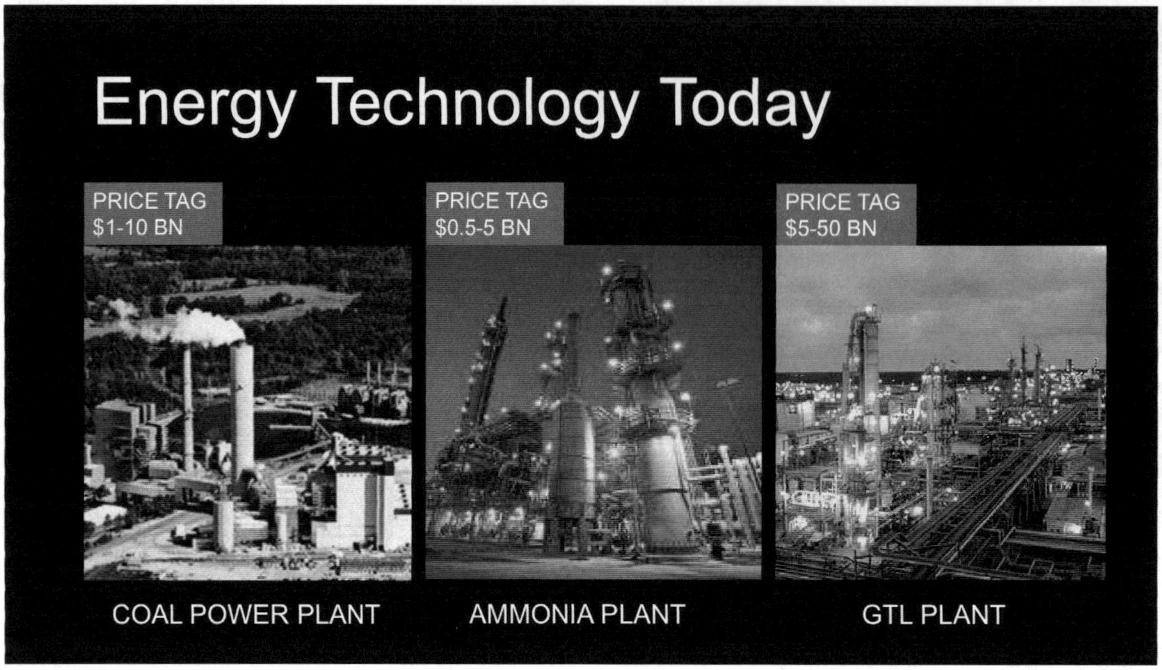
Energy Technology Today
PRICE TAG
$1-10 BN
PRICE TAG
$0.5-5 BN
PRICE TAG
$5-50 BN
COAL POWER PLANT
AMMONIA PLANT
GTL PLANT

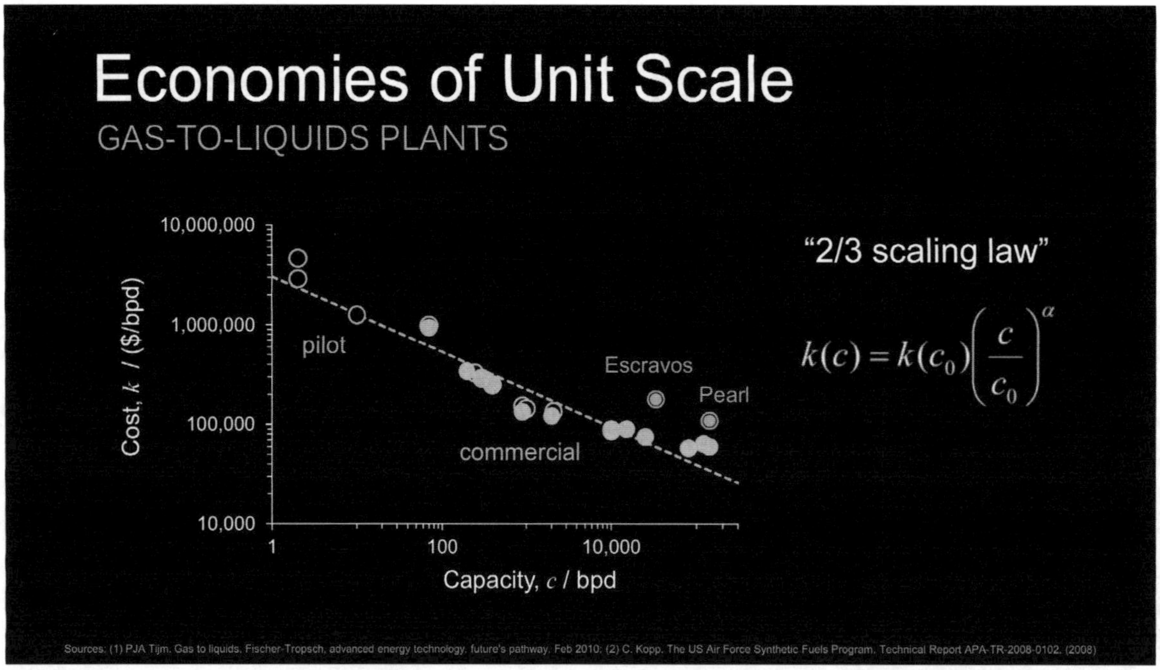
Economies of Unit Scale
GAS-TO-LIQUIDS PLANTS
10,000,000
1,000,000
100,000
10,000
Cost, k / ($/bpd)
pilot
Escravos
Pearl
commercial
1
100
10,000
Capacity, c / bpd
"2/3 scaling law"
$k(c) = k(c_0)\left(\frac{c}{c_0}\right)^{\alpha}$
Sources: (1) PJA Tijm. Gas to liquids. Fischer-Tropsch, advanced energy technology. future's pathway. Feb 2010; (2) C. Kopp. The US Air Force Synthetic Fuels Program. Technical Report APA-TR-2008-0102. (2008)

Why do we scale-up?

ARGUMENT

capital cost $\propto$ area $[L^2]$
capacity $\propto$ volume $[L^3]$
capital cost / capacity $\propto$ $[L^2]/[L^3]$

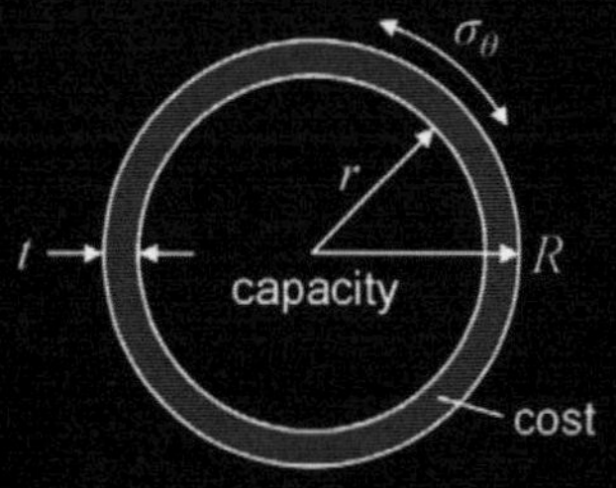

pipe cross-section

FLAW

pipe thickness ignored
hoop stress (constant) $\propto t / r$
capital cost $\propto l \cdot r^2$
capital cost / capacity $\propto$ constant

Why is scale-up a problem?

#1

BARRIER TO INNOVATION

#2

HUGE CAPITAL NEEDS

MEGA-PROJECT ECONOMICS

RAND Study:

- 52 mega-projects
- \$0.5B and \$10B (1984 dollars)
- average over budget = 90%

Source: E.W. Merrow. Understanding the outcomes of megaprojects: a quantitative analysis of very large civilian projects, The RAND Corporation, Santa Monica, CA, 1988.

Capital Resources

U.S. COMPANIES BY REVENUE, 2007

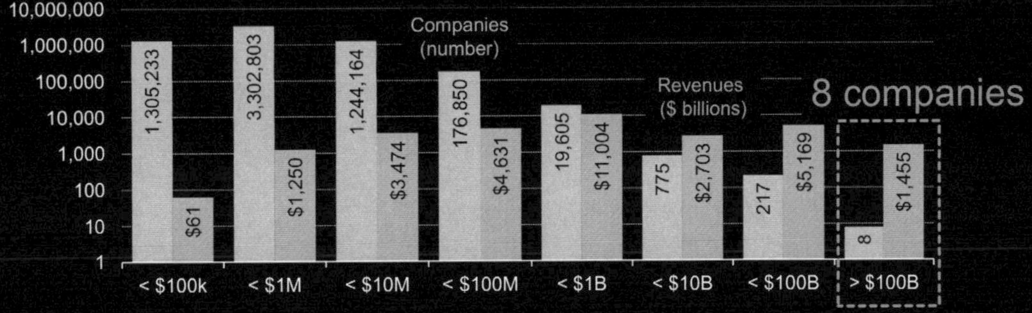

How many U.S. companies can finance a \$1 billion plant?

Capital Resources

GDP OF AFRICAN COUNTRIES, 2015

6 countries

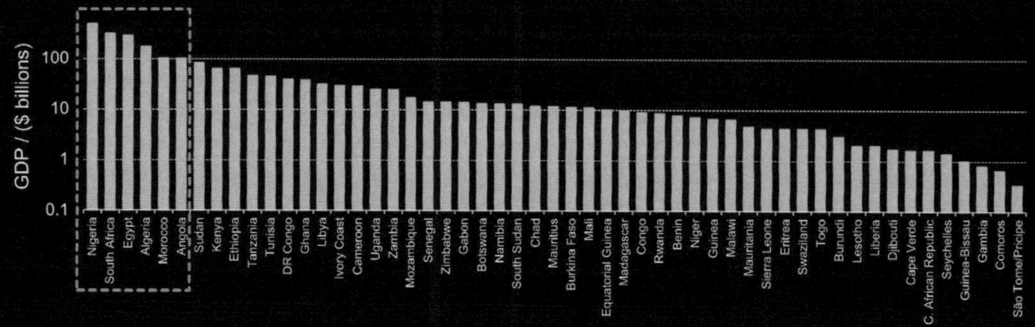

How many African countries can finance a $1 billion plant?

Source: https://en.wikipedia.org/wiki/list_of_african_countries_by_gdp_(nominal)

#3

DISTRIBUTED CHALLENGES

Developing World

DISTRIBUTED MARKETS

AFRICA
17% WORLD POPULATION
1% WORLD FERTILIZER USE

In 1999, Uganda farmers bought urea for $600/ton, global market price was $100/ton, why?

- Market size (< 1% global market)
- Transport cost (>$50/ton, 30% total)
- Finance cost ($300k, 1 kton)

Source: World Bank, 2015

Isaac Mkalia, 20, checks his mobile phone in Kojiado district, near the Tan anian border (Photograph: Sven Torfinn/Oxfam)

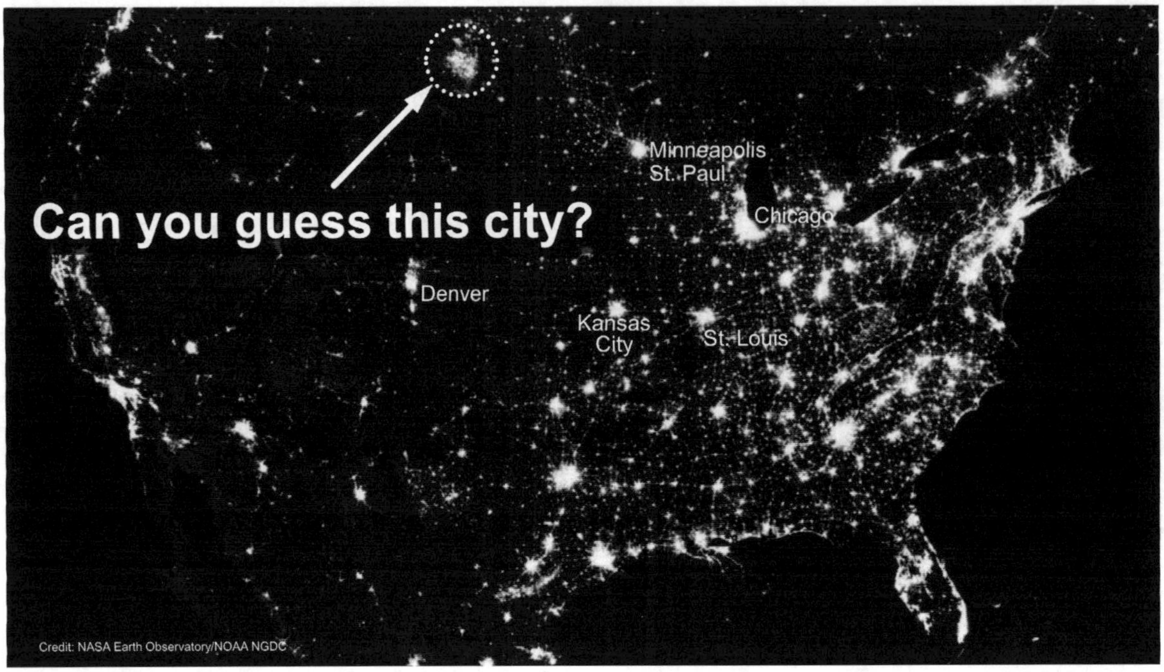
Can you guess this city?
Minneapolis St. Paul
Chicago
Denver
Kansas City
St. Louis
Credit: NASA Earth Observatory/NOAA NGDC

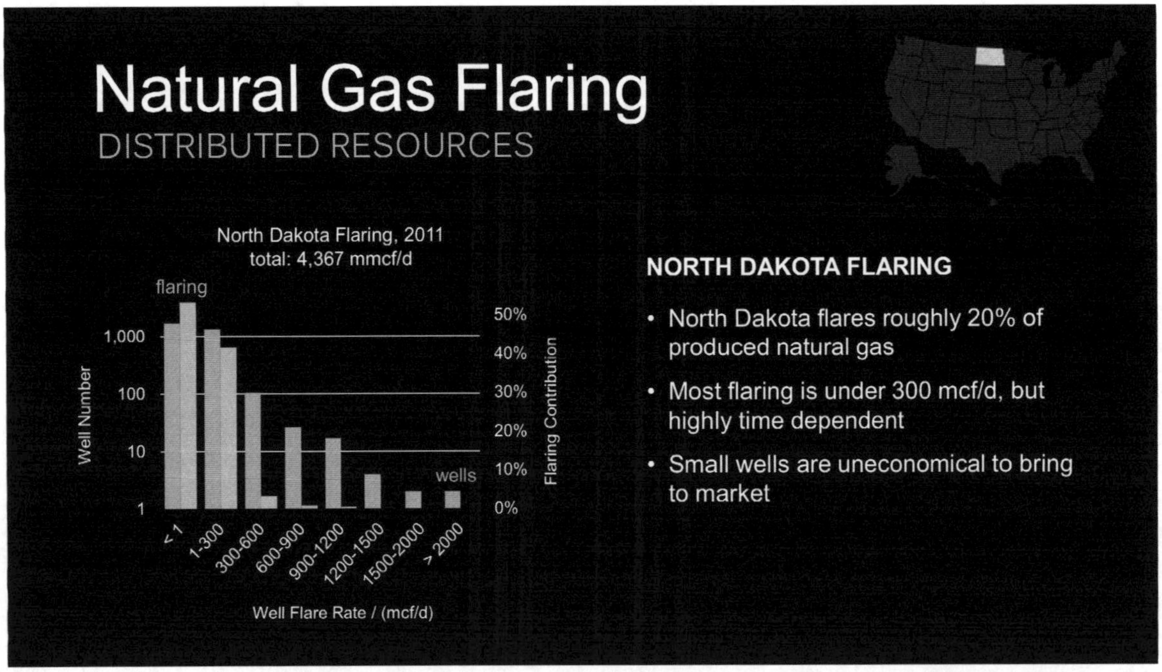
Natural Gas Flaring
DISTRIBUTED RESOURCES
North Dakota Flaring, 2011
total: 4,367 mmcf/d
flaring
wells
Well Number
1,000
100
10
1
50%
40%
30%
20%
10%
0%
Flaring Contribution
< 1
1-300
300-600
600-900
900-1200
1200-1500
1500-2000
> 2000
Well Flare Rate / (mcf/d)
NORTH DAKOTA FLARING
• North Dakota flares roughly 20% of produced natural gas
• Most flaring is under 300 mcf/d, but highly time dependent
• Small wells are uneconomical to bring to market

How do we break the hegemony of scale-up?

Democratizing Technology

HISTORIC EXAMPLES

1450	1908	1950	1973	1977
Gütenburg Press	Ford Model T	ISO Shipping Container	Motorola DynaTAC 8000X	Commodore PET
INFORMATION	TRANSPORTATION	SHIPPING	COMMUNICATION	COMPUTATION

Democratizing Technology

WHAT ARE THE DEFINING FEATURES?

1. Empowers the little guy
2. Never developed by the incumbent
3. Displaces entrenched incumbent
4. Levels the playing field
5. Leverages capital of the many
6. Modular, mass produced, standardized
7. Fast innovation cycles

…and there are many examples

guns	*…displaced*	*swords*
Gutenberg press	*…displaced*	*illuminated manuscripts*
iso-containers	*…displaced*	*bulk shipping*
cellular phones	*…displaced*	*land lines*
personal computers	*…displaced*	*central computing*
automobiles	*…displaced*	*horses*
photovoltaics	*…displaced*	*solar thermal*
electric-arc furnaces	*…displaced*	*blast furnaces*
e-commerce	*…displaced*	*store fronts*
televisions	*…displaced*	*theatres*
record players	*…displaced*	*live musicians*

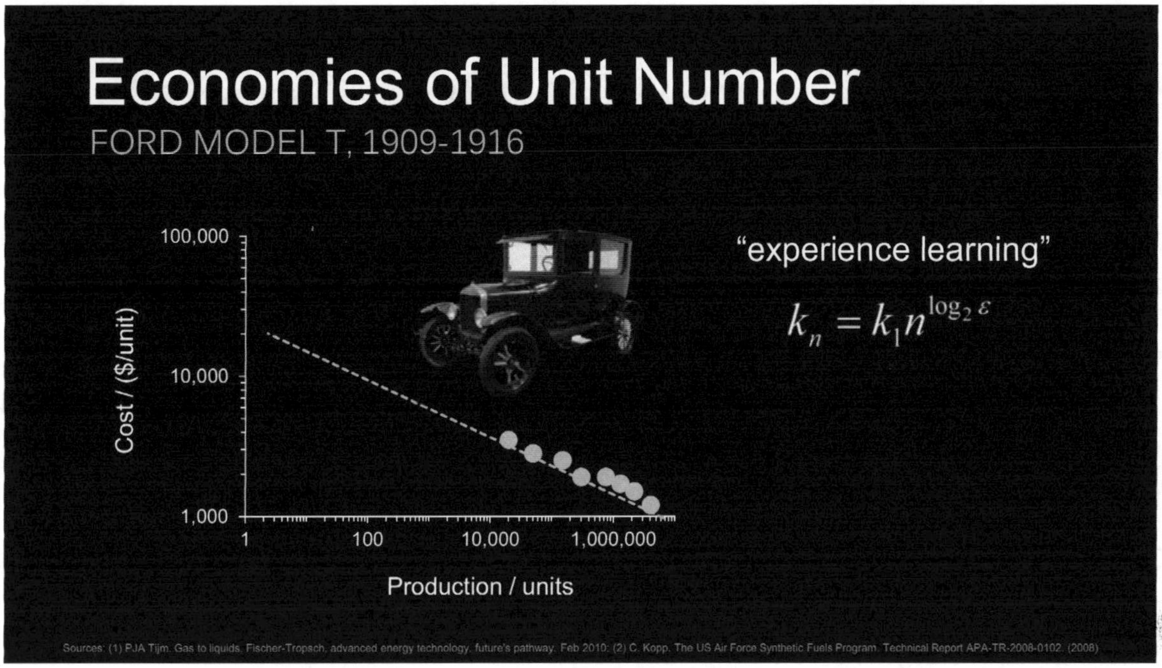
Economies of Unit Number
FORD MODEL T, 1909-1916
100,000
10,000
1,000
Cost / ($/unit)
1
100
10,000
1,000,000
Production / units
"experience learning"
$k_n = k_1 n^{\log_2 \varepsilon}$
Sources: (1) PJA Tijm. Gas to liquids. Fischer-Tropsch, advanced energy technology. future's pathway. Feb 2010; (2) C. Kopp. The US Air Force Synthetic Fuels Program. Technical Report APA-TR-2008-0102. (2008)

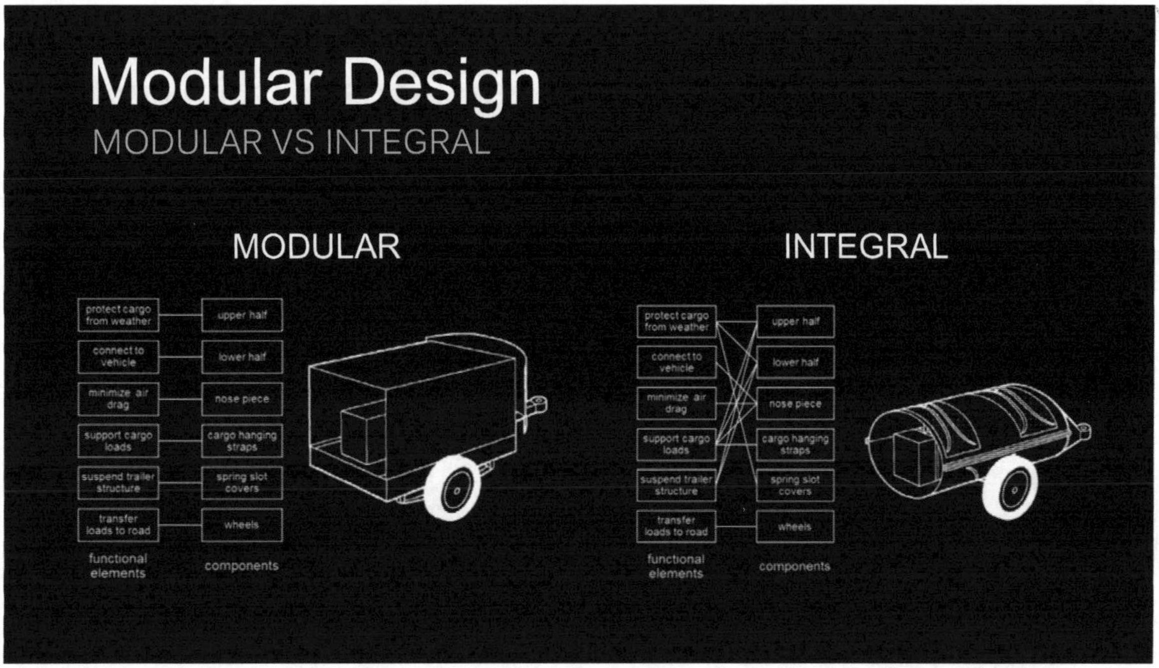
Modular Design
MODULAR VS INTEGRAL
MODULAR
INTEGRAL
protect cargo from weather
connect to vehicle
minimize air drag
support cargo loads
suspend trailer structure
transfer loads to road
functional elements
upper half
lower half
nose piece
cargo hanging straps
spring slot covers
wheels
components

Modular Design

MODULAR VS INTEGRAL

MODULAR

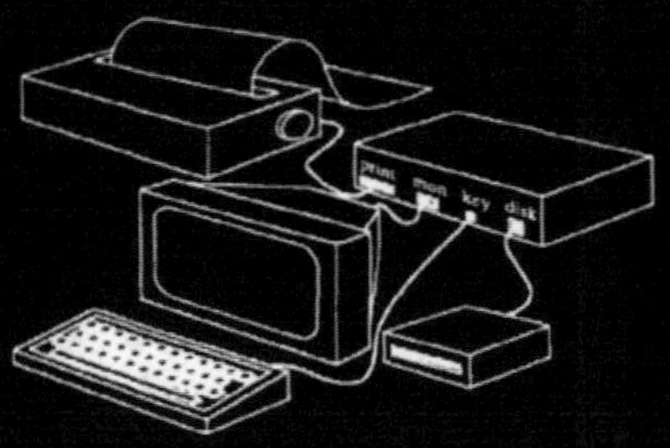

INTEGRAL

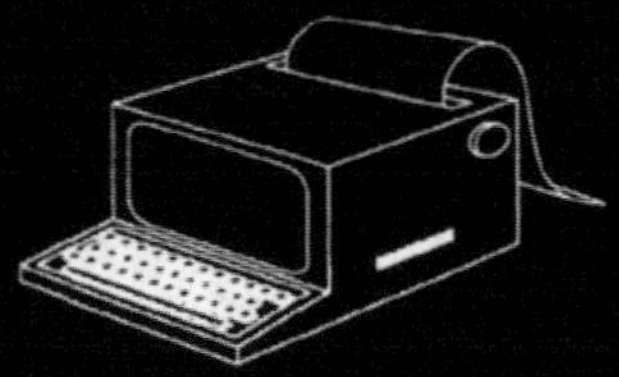

Small Modular

OIL REFINERY (WORLD)

- 7 TW petroleum refining
- 700 plants
- $500/kW capex

AUTO ENGINES (U.S.)

- 1.3 TW motive power
- 250 million engines
- $50/kW capex

Small Modular

GAS TURBINES (U.S.)

- 0.2 TW electricity
- 5000 gas turbine generators
- $1000/kW capex

AUTO ENGINES (U.S.)

- 1.3 TW motive power
- 250 million engines
- $50/kW capex

What about the democratization of chemical processing?

not if, but
when, who, how

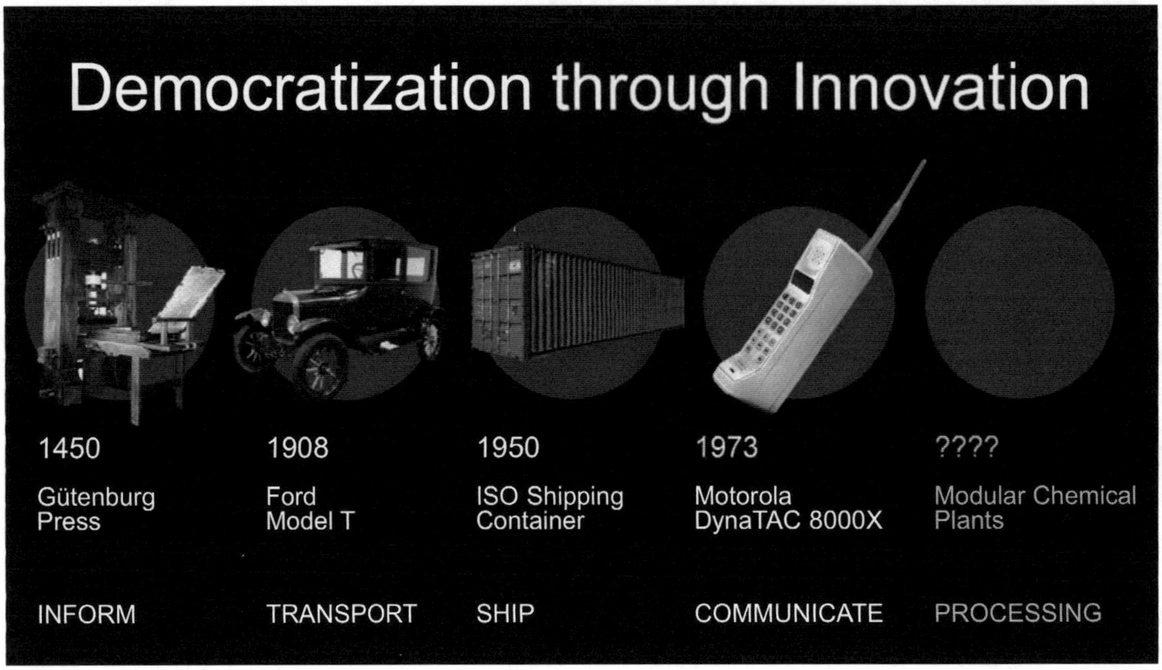
Democratization through Innovation
1450
Gütenburg
Press
INFORM
1908
Ford
Model T
TRANSPORT
1950
ISO Shipping
Container
SHIP
1973
Motorola
DynaTAC 8000X
COMMUNICATE
????
Modular Chemical
Plants
PROCESSING

Isaac Mkalia, 20, checks his mobile phone in Kojiado district, near the Tanzanian border (Photograph: Sven Torfinn/Oxfam)

Thank You

Dane A. Boysen
dane.boysen@gmail.com

498543 Production of Chemicals from the Thermal Cracking of Castor Oil, Castor Oil Methyl Ester and Ricinoleic Acid

Abstract:

This work evaluates the experimental results of three different processes to obtain heptaldehyde, undecylenic acid and methyl undecylenate through castor oil, castor oil methyl ester, and ricinoleic acid. Heptaldehyde and undecylenic acid are important chemicals on food, packaging, ink, solvent and polymers industries. Methyl undecylenate is commonly used as a Flavoring agent and also an important route to obtain undecylenic acid. Methyl undecylenate was obtained by the thermal cracking of castor oil methyl ester. In order to obtain heptaldehyde, methyl undecylenate, and undecylenic acid, the thermal cracking was performed using commercial castor oil, castor oil methyl ester (obtained by castor oil transesterification reaction) and commercial ricinoleic acid, respectively. Thermal cracking temperatures were 530, 545, 560 and 575 °C with 400 g/h mass flow on the feed. All materials were blended with distilled water, considering the same proportion (75 % material, 25 % water). Temperature influence on bio-oil yields was observed, likewise different behaviors to each material. Iodine index, acid index, and density were determined to all bio-oil samples and their content were analyzed by GC-FID. The best condition to obtain higher yields of each chemical were pointed. With regard to the castor oil, higher yields of heptaldehyde and undecylenic acid were found on experiments at 530 °C. The same temperatures resulted in higher yields concerning the ricinoleic acid. The cracking of methyl ricinoleate results on higher yields of heptaldehyde at 530 °C and methyl undecylenate at 560 °C.

NEEM COATED UREA- A FISCAL APPROACH FOR FARMERS

Sara Ahsan, Ali Ayub

Abstract

Soil fertility is determined by three major elements namely Nitrogen, Phosphorus & Potassium (N, P, K) of which nitrogen plays a very important role. For this reason, Urea (Containing 46% of N) consumption is very high all over the world. Generally, the plants are able to recover a fraction of Nitrogen present in Urea and this fraction is known as Nitrogen Use Efficiency (NUE). This NUE stands anywhere between 30 to 50%. Thus, two third of Urea Nitrogen escapes from the soil in the form of nitrogen, ammonia and nitrates etc. and is not used by the plants. Engro Fertilizers, Pakistan, researched and experimented use of "Azadirachta indica" commonly known as neem oil as nitrification inhibiting agent. Based upon the results of extensive laboratory trials, coating urea with neem helps in slow release of nitrogen, nourishing the crops for a longer period. The process reduces pollution of groundwater. There is an increase in crop yield and also serves as an efficient natural insecticide. It also serves as an anti-caking agent and will reduce the amount of urea required by a plant by ~10-20%.

This paper explicates Engro's experience and research on neem coated urea, methods of preparing an effective neem oil emulsion, quantities of raw materials to be used per unit of urea, its uses and results obtained after extensive laboratory and commercial trials.

INTRODUCTION

Snam Progetti divides Urea plant into following major sections i.e.

1. Urea synthesis and NH3, CO2 recovery at high pressure.
2. Urea purification and NH3, CO2 recovery at medium and low pressures.
3. Urea concentration.
4. Urea prilling.
5. Waste water treatment.

Snamprogetti ammonia stripping process is characterized by a urea synthesis loop operating at about 160 Kg/cm2 (abs); with ammonia to carbon dioxide molar ratio at urea reactor inlet of 3.3 to 3.6. This allows a high CO2 conversion into urea in the reactor itself. There are two kinds of chemical reactions at the same time in the urea reactor:

$$2NH_3 + CO_2 \leftrightarrow NH_2\text{-}COO\text{-}NH_4 + 32560 \text{ kcal/kmol of carbamate (at 1 atm; 25°C)}$$

$$NH_2\text{-}COO\text{-}NH_4 \leftrightarrow NH_2\text{-}CO\text{-}NH_2 + H_2O - 4{,}200 \text{ kcal/kmol of urea (at 1 atm; 25°C)}$$

First reaction is very fast & exothermic and the second one is weakly endothermic and occurs in the liquid phase at low speed. Downstream the urea synthesis, the decomposition and relevant recovery of unconverted chemical reagents occurs. It is carried out in three subsequent steps: High Pressure Decomposition in H.P. stripper; Medium pressure Decomposition in M.P. Decomposer and, finally, Low Pressure Decomposition in Low Pressure Decomposer.

The decomposition reaction is the reverse reaction of the first one above showed, viz.:

$$\mathbf{NH_2\text{-}COO\text{-}NH_4 \leftrightarrow 2NH_3 + CO_2 - Heat,}$$

And, as can be inferred from the equation, it is promoted by reducing pressure and/or adding heat.

The urea reactor effluent solution enters the stripper, under slightly lower pressure than the urea reactor, where a fair part of the unconverted carbamate is decomposed in NH3 and CO2. The overall yield of the H.P. synthesis loop referred to CO2 is as high as 83-85% (on molar basis). Downstream of the stripper residual carbamate and ammonia are recovered in two recycle stages operating at 18.5 kg/cm2 (abs) (namely M.P. section) and 4.8 kg/cm2 (abs) (namely L.P. section) respectively.

Ammonia and carbon dioxide vapors from the stripper top, after mixing with the carbonate recycle solution from M.P. section, are condensed, at the same pressure level of the stripper itself, in the 1st and 2nd High Pressure carbamate condensers, thus producing the L.M.P and the L.P. steam which are used in downstream sections. After separating the inert gases which are passed to M.P. section, the carbamate solution is finally recycled to the reactor bottom by means of a liquid/liquid ejector, which exploits H.P. ammonia feed to reactors as motive fluid.

Downstream of the two recovery stages urea is concentrated to 99.75% via two vacuum concentrator stages. The solution leaving the L.P. decomposer bottom with about 70% wt. urea, is sent first to the tube of vacuum pre concentrator and then pumped to first vacuum concentrator both operating at a pressure of 0.34 kg/cm2 (abs). The mixed phase coming out from the process side of the first vacuum concentrator enters the first gas-liquid vacuum separator from where vapors are again extracted by the first vacuum system while the urea melt (~ 95 wt%), by gravity & differential pressure flow, enters the bottom of second vacuum concentrator operating at a pressure of 0.03 kg/cm2 (abs). The extracted urea melt (~99.75% pure) from the second concentrator is sent to prilling tower. Droplets of molten urea from the prilling bucket fall along the natural draught prilling tower and solidify and cool while encountering a countercurrent air flow.

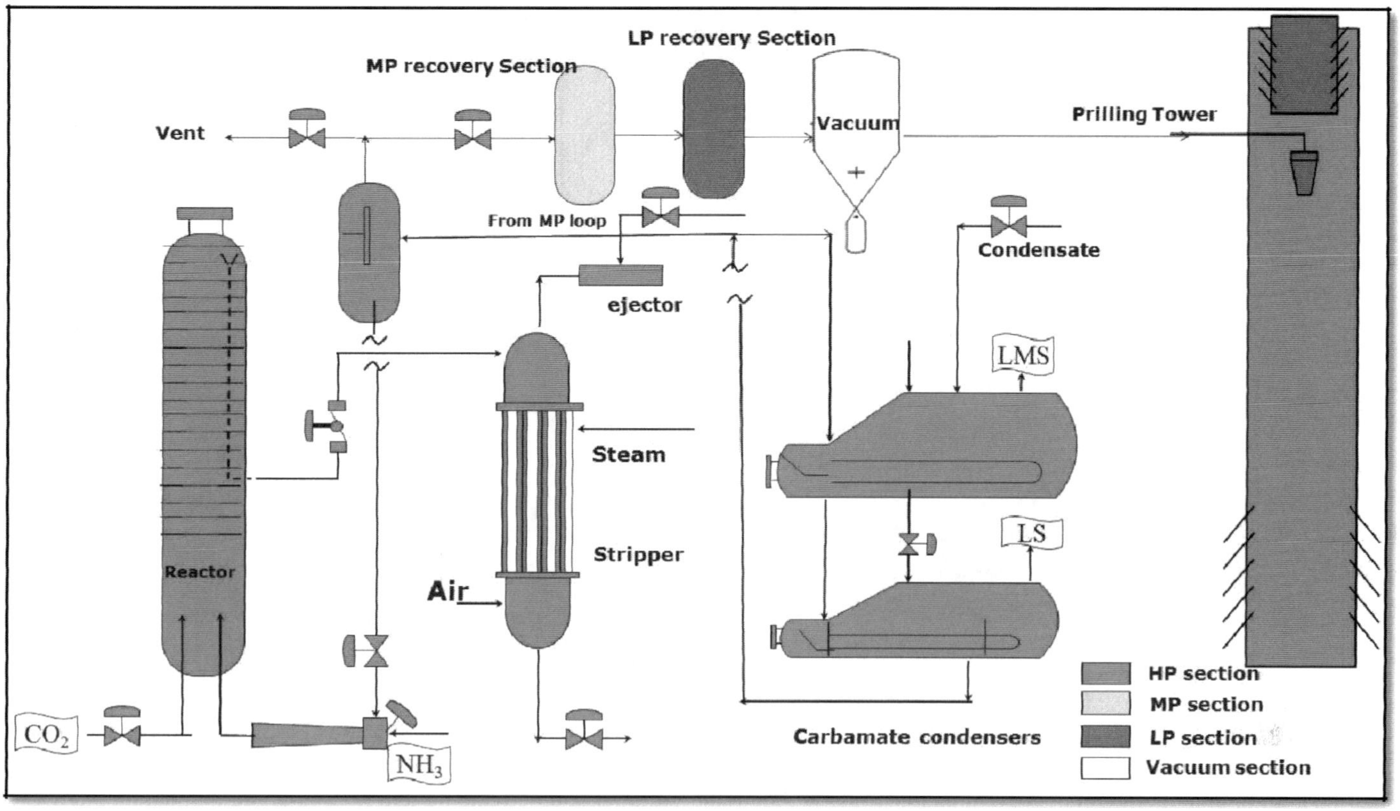

Figure 1. Process Description: Picture elucidating a short process description of Snam's urea plant.

NEEM COATING ON UREA PRODUCT- AN EXPERIENCE AT ENGRO

Purpose of coating Neem on Urea Product:

With 46% N content, Urea is the most popular Nitrogenous fertilizer around the world. When Urea is applied to soil, it is first hydrolyzed into Ammonical form (Ammonium Ion i.e. NH4+) and then to nitrite (NO2-), followed by to nitrate (NO3-) forms by the process called nitrification. Nitrification process is necessary for making nitrogen available to plants. However, if the process of nitrification is too rapid; nitrogen will escape to atmosphere and plants will not be able to recover it from Urea efficiently. Generally, the plants are able to recover a fraction of Nitrogen present in Urea and this fraction is known as Nitrogen Use Efficiency (NUE). This NUE stands anywhere between 30 to 50%. Thus, two third of Urea Nitrogen escapes from the soil and is not used by the plants. Hence this lead us to regulate the urea hydrolysis and nitrification. This regulation is done by some chemical/natural agents called Nitrification inhibitors. However, most of the nitrification inhibitors are costly chemicals (examples nitrapyrin, dicyandiamide and ammonium thiosulphate) which are not economically feasible.

It has been scientifically established that Neem oil serves as an effective nitrification inhibitor whose benefits when coated on urea area mentioned below:

- Neem coating leads to more gradual release of urea, helping plants gain more nutrients and resulting in higher yields.

- Lower underground water contamination due to leaching of urea.
- Neem serves as a natural insecticide and an anticaking agent

Properties of Neem Oil:

The properties of neem oil used are as follows:

1. Neem Oil purity = 98.5%+
2. Color= Yellowish brown clear liquid
3. Moisture = 0.5% Max.
4. Specific gravity = 0.9 to 0.94 at 30 oC
5. Iodine value = 65 to 80
6. Azadirachtin Content = min 150 ppm
7. Saponification Value = 180 to 205
8. Up saponi cable matter = < 2 %
9. Density of neem oil = 0.92g/cm^3
10. Viscosity of neem oil = 35CSt

As per standard limits of PSQA, Azadirachtin content in final product must be greater than 50 ppm corresponding to 250ppm of neem oil coat. Engro started its coating setting a standard in Pakistan of 500ppm of neem oil coat. So, 0.5kg of neem oil is required per ton of urea. Cost of neem oil is 2.21$ per kg.

Process employed for coating of Prilled Urea

Process Description:

Coating of neem oil on urea product requires preparation of solution of neem oil in water. Neem oil is not soluble in water so it requires an emulsification agent. On Laboratory scale, tests were done on the use of Lab grade 25% Ammonical water as an emulsifying agent. The results were favorable for the use of Ammonical water as an agent. It was then decided to use it in our emulsion preparation method.

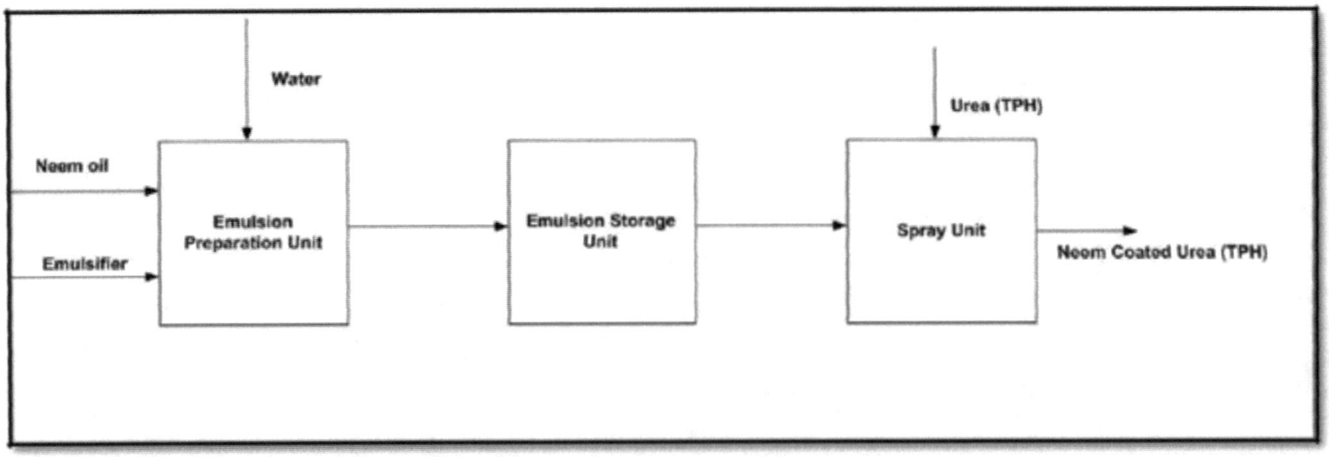

Figure 2: Block Diagram of emulsion preparation process

Below two images depicts neem solution change in color before and after addition of emulsifying agent.

Figure 3: Neem Oil & water (Before addition of Ammonical solution)

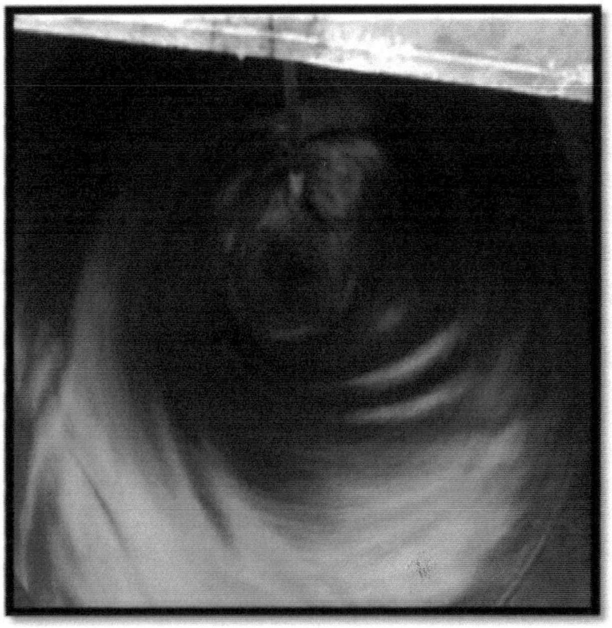

Figure 4: Neem emulsion (After addition of Ammonical solution)

Design capacity of Engro's EnVen plant is 3835 MTPD which is equivalent to 160TPH of product over the belt. For coating 500 ppm of neem oil over this product, first and the foremost step was to add accurate quantities of raw materials and emulsifying agent. For the aforementioned urea production, total neem emulsion required was 3835kg, with pure neem oil amount of 1917.5kg (50% of total emulsion). 0.1grams of 25% Ammonical solution is required per 50grams of pure neem oil used and the remaining being water.

Engro's Laboratory researched and registered its test for determining amount of Azadirachtin in the coated product. Below is the Process flow diagram of the facility used during coating activity

Figure 5: Coated Urea on Conveyer Belt

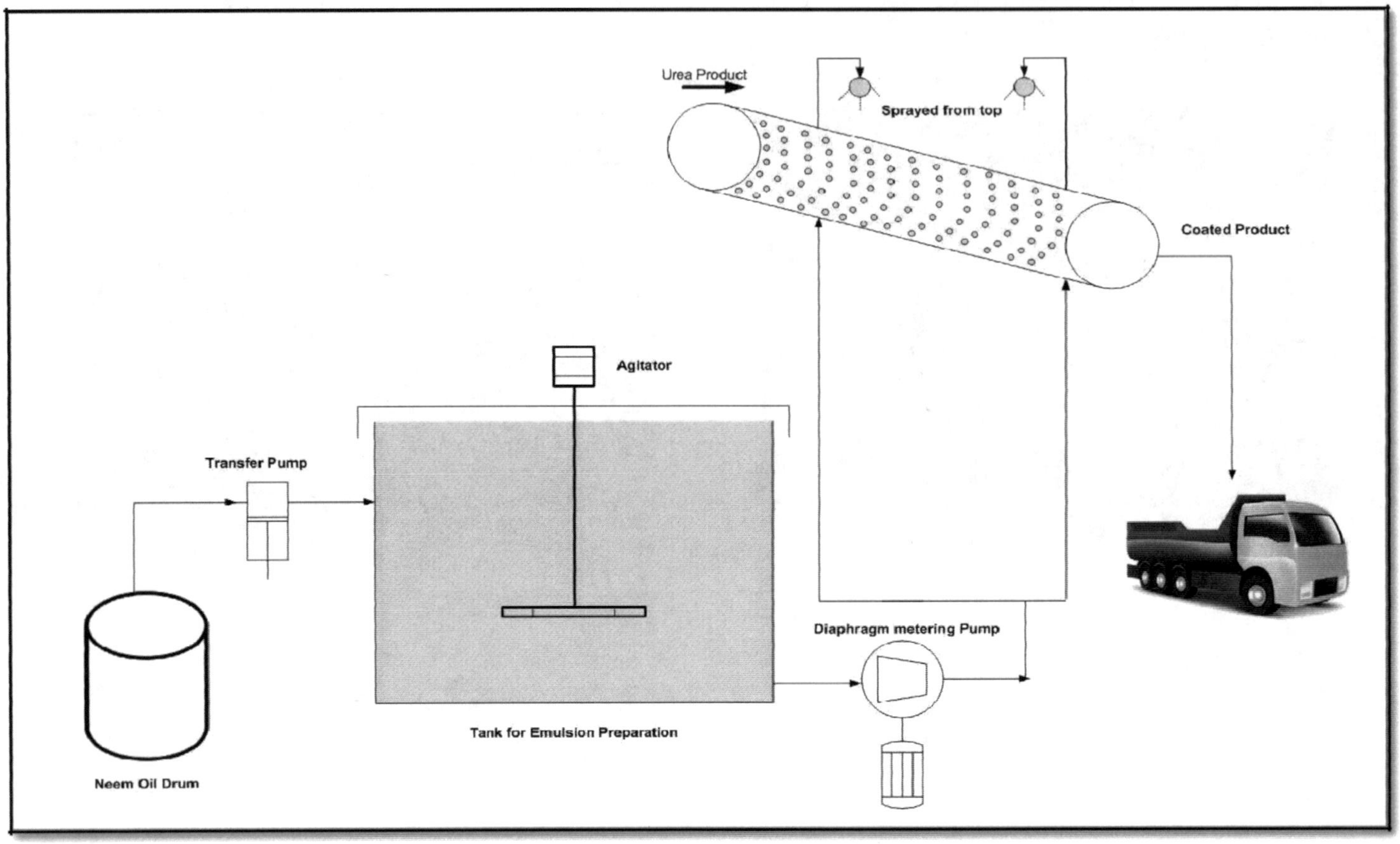

Figure 6: Block Diagram of emulsion preparation process

Before stating coating, stroke testing of the diaphragm metering pump was done. This testing was made basis for selection of the pump strokes during neem coating. The stroke testing results are as mentioned below:

Table 1. Pump stroke testing results

Strokes	Flow (Litre/min)	Flow (Litre/hr)
10	0.5	30
20	0.980	58.8
30	1.37	82.2
40	1.87	112.2
50	2.35	141
60	2.6	156
70	3.15	189
80	3.48	208.8
90	3.9	234
100	4.3	258

For a coat of 500 ppm corresponding strokes needs to be adjusted at 70%. To determine efficaciousness of the spraying unit, we set the strokes primarily at 70% and collected a sample then adjusted them to 50% and collected another sample. These samples reported Azadirachtin content in them to be 99.5ppm and 90ppm. Minimum limit of the active agent was 50ppm, so, strokes were further reduced to 45% and coating as started. Following are the results obtained during coating.

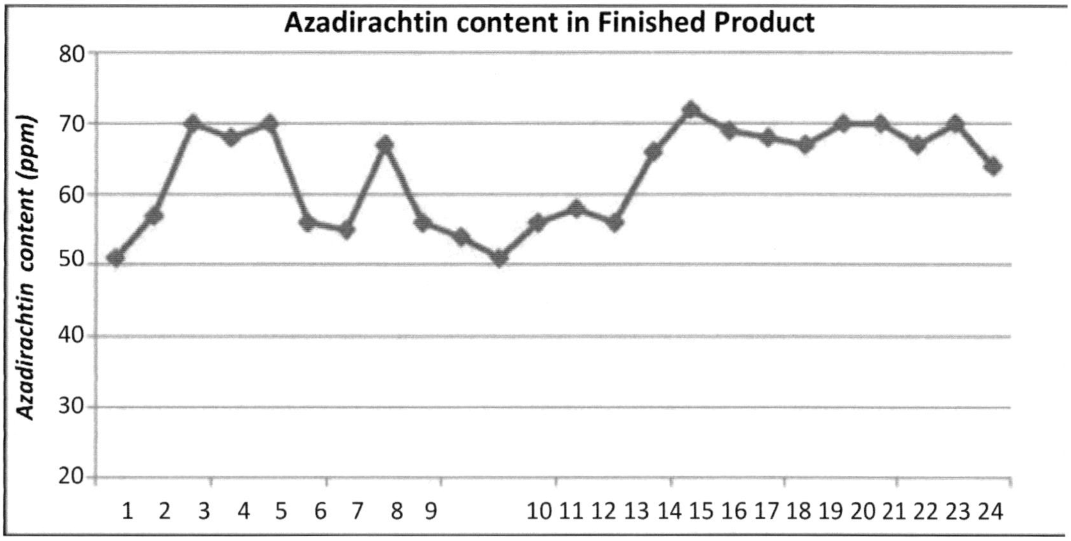

Graphical representation 1: Azadirachtin content in Finished Product

Required Equipment

- One transfer Pump for transferring neem oil into the tank where emulsion is to be prepared (Capacity of used pump: 4m^3/hr)
- A tank of capacity depending on the design capacity of plant (Tons per hour of product passing through belt conveyer) (Capacity of used Tank: 4.82m^3)
- Agitator for mixing and emulsion preparation (Rpm of Used agitator: 70)
- Neem emulsion storage tank (Capacity of used Tank: 4.82 m^3)
- Gear Pump for transferring neem emulsion to nozzles (Capacity of gear pump used: 220LPH)
- Spray nozzles for spraying neem emulsion on urea over the belt conveyer. (Capacity of each nozzle: 1.6LPM)

Neem in Pakistani Market:

Neem story is very different in Pakistan. Its oil extraction is being done in the country but at a small scale. Neem seed and oil are not traded in Pakistani commodity market. As this tree seed and leaves have toxic properties, so it is not fed to the animals, hence each year all its fruit and leaves are wasted. Consequent result of less production of oil is high prices. After lots of struggles, a pharmaceutical company agreed on selling at 9$ per kg of neem oil. But purchasing at this rate was neither economically feasible nor gives a positive message to the market. Hence, if we want to start neem coating in Pakistan permanently, either its market needs to be improved or else oil should be imported from India or any other country where prices of neem oil are low.

In case we are successful in marketing this Neem coated Urea, it will start a total new industry in Pakistan with lots of parties setting up efficient Neem oil extraction plants and lively hood to thousands of people.

Discussion and Benefits of neem coating on Urea

The decrease in the local urea sales whereby increasing the sale of imported urea led local fertilizer producers to slash their prices. Consequently, to deplete our urea inventory price was lowered by PKR 70. It is important to mention here that government, in Fiscal Year 2016-17 (FY17) budget, reduced urea prices by PKR 400/bag to PKR 1,400/bag. But this price cutting didn't result in the expected sales. So, Engro decided of neem coating on Urea Product, hoping to attract the customers with its uses. Indian government decision of coating 100% produced urea further strengthened our decision. For test run purposes, 10,000 MeT of urea was coated and sent to selected warehouses for marketing.

Engro Fertilizers Limited started first ever coating in Pakistan and homogenized the procedures for production of Neem Coated Urea. To have a uniform and consistent coating of Neem emulsion on urea prills and maintaining the contents of Azadirachtin in final product, Laboratory results and tests were performed after every two hours.

Summarizing the benefits of neem coating on urea product as under:

- Increase in price of coated urea per 50kg bag
- Decrease in nitrification and consequent increase in the yield of crops thus strengthening agricultural business
- Reduction in raw material consumption due to decrease in the amount of fertilizer required by plants
- Reduction in environmental pollution of ground water due to leaching of nitrates and gaseous emissions
- Lively hood to thousands of people
- Profit to the farmers due to reduction in the amount of fertilizers required per crop by approximately 30%
- Neem serves as a natural insecticide

496747 Technoeconomic Assessment of Two-Step Process for Upgrading and Utilizing Used Railroad Ties in an Integrated Biorefinery

Abstract:

In the USA, over 21 million wood railroad ties are produced per year to replace the approximately 20 million used ties that have provided excellent primary service in railroad track over 40 years. Due to the presence of carcinogenic preservatives, these used ties will primarily be either landfilled or incinerated as they do not meet U.S. Environmental Protection Agency (U.S. EPA) stringent requirements. Consequently, a two-step pyrolysis process was developed to upgrade used ties by first extracting wood preservatives (Step One) and converting upgraded used ties into biofuels, bioproducts, power, and heat (Step Two). This study evaluates the techno-economic viability of the proposed two-step process embedded in integrated pyrolysis biorefinery with liquid fuel, power and recovered preservatives as product streams.

The technoeconomic analysis assumed 2.33 million metric tons of used railroad ties are available yearly according to statistics by the Railroad tie association. This analysis considers 3 plants spread across the U.S. to minimize transportation cost with 2358 tons/day throughput each. The rail road ties biorefinery consists of eight areas including preprocessing (A100), thermal desorption (A200) and fast pyrolysis and vapor quenching (A300), hydroprocessing (A400), hydrogen generation (A500), power generation (A600), cooling and other utilities (A700) and waste management (A800). The minimum selling prices as well as sensitivity analyses were performed to determine the impact of thermal desorption conditions on the overall economic.

496556 Valorisation of Three-Phase Olive Mill Wastewater with the Addition of High Protein Co-Substrates

Abstract:

Introduction

Nowadays national and international legislation attempt to enforce reductions in greenhouse gas emissions, while minimization of dependency to fossil fuel gathers all-around agreement. At the same time optimization of waste management strategies towards zero waste generation is a major challenge for modern society. Olive Mill effluents (OME) is the effluent generated during the production of olive oil and accounts for the 60% of the overall weight of the olive fruit. OME is rich in carbon and phenols while its nitrogen content together with the pH are very low rendering common treatment methods ineffective to tackle it. A valorization process that can be used for the management of OME is Anaerobic Digestion (AD). A key attribute of energy efficiency schemes has been the utilization of negative value substrates for biogas generation in anaerobic digesters. Anaerobic digestion (AD) is a biological method for the treatment and valorization of organic material. Its product can be recovered and utilized for heat and/or electricity generation. The efficiency of the process is affected by the operational conditions selected for the digesters and it depends highly on the type of the organic substrate used as feedstock.

Common organic substrates for AD systems include municipal solid waste, agroindustial byproducts and waste including slaughterhouse and dairy industry waste. Several types of AD feedstock are rich in fat, oil and grease (FOG). This is a feature that is highly desirable, as fat has the highest biomethane potential of all natural macromolecules. However, high fat content can inhibit the digesters through flotation, accumulation, pH reduction, and a number of intracellular inhibition processes. Furthermore, agroindustial wastewater, including olive mill effluents (OME), is characterized by high concentration of phenols, low nitrogen availability and low pH. All of the above are known to inhibit the digestion process especially in overloaded or unstable systems.

Aim of this work was to investigate the effect of external addition of nitrogen into digesters operating with OME as single substrate. The co-substrates evaluated were low-value byproducts of animal and plant origin, namely bone and meat meal, blood meal, feather meal, mixture of feather and blood meal, fish meal, soy been meal and maize gluten.

Material and methods

OME was collected from a three-phase olive mill. The inoculum used was collected from four, six-liter digesters operating under steady state conditions with cattle manure as substrate and an OLR of approximately 3kg/VSm3-d. Protein-rich substrates were collected from a pet food manufacturer. All substrates were stored in a freezer at -20°C after collection until further utilization.

In this work 250mL Duran bottles were used as batch digesters and for assessing the biomethane potential and biodegradability of the different substrates. In order to evaluate the batch process and generate comparable results two loading rates (10 and 20 $kgVS/m^3$) were evaluated. The I/S ratio was set at 2 and 1 respectively. In the second stage of the experiment, four identical stainless steel reactors were used. The reactors were heated through the recirculation of hot water within the water jacket, feeding and removal of the substrates was taking place manually, while the mixing of the reactors was achieved with the application of immersed propellers and direct current mixing motors.

Results and discussion

Mono digestion of the substrates

The highest biogas production of the substrates under mono-digestion conditions was obtained using bone and meat meal with a value of 511 $mLCH_4/gVS_{added}$. This was approximately 16% higher compared to the production of biogas using blood meal and feather meal mixture with a production of 427 $mLCH_4/gVS_{added}$. Related to the plant-derived substrates the specific methane production of gluten and soy bean meal amounted to 495 $mLCH_4/gVS_{added}$ and 460$mLCH_4/gVS_{added}$ respectively.

Co-digestion of OME with high nitrogen substrates

In the second stage of this study OME was co-digested with high nitrogen substrates. The highest biogas production was attained using a combination of OME and fishmeal with a yield of 482 $mLCH4/gVS_{added}$ corresponding to the 77% of the ultimate production for this mixture. Digestion of the other mixtures gave also very positive results with the mixture of OME & blood meal and of OME & soybean meal offering yields of 454 and 446 $mLCH4/gVS_{added}$ corresponding to the 74 and 72% of the ultimate production for the two mixtures respectively. The lowest biogas production registered was the one obtained from a mixture containing OME & bone and meat meal with a yield of 418 $mLCH4/gVS_{added}$, equal to the 68% of the theoretical value.

Validation of the results in CSTR systems

In order to validate the results of the previous experimental stages, the biogas yield using four substrate mixtures composed of olive mill wastewater (OMWW) and protein-rich substrates were assessed in semi-continuous digesters. Feeding was provided daily throughout the week including the weekends. The four substrates chosen for analysis in the semi continuous systems were:

(a) bone and meat meal as it is the co-substrate with the highest market availability and the lowest price.

(b) blood meal; this substrate was selected due to its negative effects recorded when high loading rate was used in the batch digestion.

(c) fishmeal; this substrate offered the highest bio-methane production of all substrate mixtures assessed in the batch digestion stage.

(d) The last of the substrates assessed was gluten. This was the only plant-derived protein-rich substrate assessed; its selection was based on its higher protein content over the second candidate substrate, namely soybean meal.

In this experimentation stage the highest biogas production was achieved by the mixture containing blood meal with an average production equal to 449 mLCH4/gVS added, followed by biogas production from bone and meat meal with 443 mLCH4/gVS added. The lowest production was achieved when using the mixture containing FM with an average yield of 416 mLCH4/gVS added. Our results indicate that the type of protein added is insignificant; in all cases the yields of the four different mixtures were within 7% of each other ranging between 449 and 416 mLCH4/gVS added. In all cases VFA concentrations were very low and no inhibition was recorded in any of the reactors treating the different substrates.

Conclusion

OMWW, a heavily polluted effluent can be valorized through the anaerobic digestion process into biogas and digestate. The addition of nitrogen-rich substrates offers the opportunity to achieve a steady process with high bioconversion efficiency. Furthermore, the type of the co-substrate seems to be insignificant as no substantial differences in biogas yield were registered when fish, animal or plant derived proteins were used as nitrogen-carrying additives. This work offers a framework for the possibility to valorize the OMWW in treatment plants in the vicinity of the olive mills without the requirement of manure addition as carrier substrate. This development would minimize the risk of bacterial contamination from transportation of manure near olive mills. According to our results one ton of nitrogen-rich substrates can be used for the treatment of more than $50m^3$ of OMWW while at the end of the process the recovered effluent has significantly better characteristics as soil improver when compared to raw OMWW.

492445 Evaluation and Comparison ofÂ the Scale-up Potential for Autotropic andÂ Heterotrophic Algal Oil Production

Abstract:

Significant research has been completed over the past decade to develop technologies to facilitate the production of algal lipids for use as feedstock for next generation renewable fuel facilities. The interest for microalgae is motivated by the substantial increase in yield per land area compared to oilseed crops reducing the pressure to supplant food crops with energy crops. However, the challenges to produce oil from algae at scales approaching crop oil production facilities have been great. One of those barriers is the need to get photons from sunlight or artificial sources to the growing algae as an energy source. It has recently been proposed that growing the algae heterotropically might overcome this scale-up barrier. The transition to heterotrophic halts the photosynthesis process, and requires an organic carbon source to provide energy, as the heterotrophic strain of microalgae is not able to produce energy compared to the autotrophic which produces an energy source though photosynthesis. Additionally, the transition has been proven to increase the lipid content of the microalgae by replacing the chlorophyll cells produced during photosynthesis. If the strain of microalgae used is non-light dependent it will negate the requirement of industrial scale, clear photo bioreactors or open ponds. Additional chemicals and catalysts will be required for the growth and development of the microalgae strains for the growth media.

In this study, a group of senior chemical engineering students developed preliminary designs and then performed an economic analysis for algal oil production utilizing both autotropic and heterotrophic growth strategies. Both proposed systems utilizes the strain of microalgae *Chlorella Vulgaris* as the raw material. *Chlorella Vulgaris* has a lipid content ranging between 15-35 wt%.

The study involved the development of Input/output diagrams, block flow diagrams, and process flow diagrams for compete processes starting from lab algae inoculation through to final lipid oil product purification. Preliminary equipment sizing was performed. The capital costs were estimated using a factored broad cost estimation method based on vendor quotes for the largest equipment and cost chart data for the rest of the equipment. Revenues and operating costs for a 20 year project life were also estimated.

The heterotrophic option will yield a higher lipid content which will generate a larger amount of oil when compared to the autotrophic strain but will require a higher quantity of inputs. The production throughput was comparable to a small scale oilseed production facility. Neither option is currently economically competitive with a commercial soybean oil processing facility. However, the heterotrophic option had significant advantages over the autotropic design. The key barriers to commercialization were identified so that future R&D activities can focus in these areas.

Process Developments in Continuous Silicone-Polyether Copolymers

- Akshay Kundan
- Chauncey Rinard
- Steven Bahr

MOMENTIVE™
inventing possibilities

10/30/2017

BACKGROUND ON SILICONE-POLYETHER COPOLYMERS (SPC)

MOMENTIVE™

Urethane Additives

POTENTIAL APPLICATIONS, KEY PRODUCTS & TYPICAL BENEFITS

- Improve reaction mix compatibility
- Determine reaction rate and productivity
- Provide bulk foam stabilization
- Regulate cells size and openness
- Make smooth surface of molded foam
- Improve properties like fire, comfort
- Improve dimensional stability of low-density PU
- Help CASE systems for easy pouring and faster cure
- Make PU leather smooth and abrasion-resistant
- Enhance foam physical properties and fire behavior

MOLDED FOAMS

- Automotive seating
- Furniture
- Pillows and mattresses

SLAB FOAMS

- Furniture
- Bedding
- Technical foams
- Laminated foams

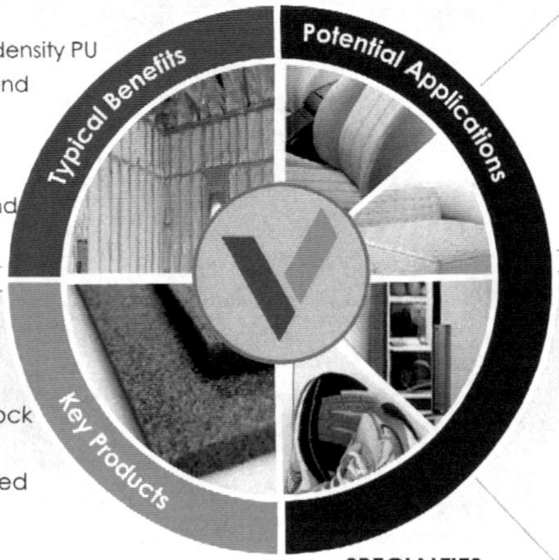

- Niax* silicones and catalysts for slabstock foams (conventional, high resilience and visco-elastic)
- Geolite* process modifiers for slabstock foams
- Niax silicones and catalysts for molded foams based on TDI, MDI and TM systems
- Niax silicones, catalyst and special additives for rigid foams
- Niax additives for specialty PU applications
- Fomrez† special PU catalysts

SPECIALTIES

- Carpet backing
- Electronics
- Footwear
- Artificial turf
- Artificial leather

RIGID FOAMS

- Fridge, cool store, water heater, display units
- Refrigerated trucks, reefer containers
- Construction panels, SIPs
- Insulation panels, boards
- Gap filling
- Insulated pipe & tanks
- Mining/grouting
- Decorative, structural, packaging, floral

3

MOMENTIVE

*Niax and Geolite are trademarks of Momentive Performance Materials Inc.
†Fomrez is a trademark of Chemtura Corporation, used with permission by Galata Chemicals LLC.

Momentive Urethane Additives Portfolio

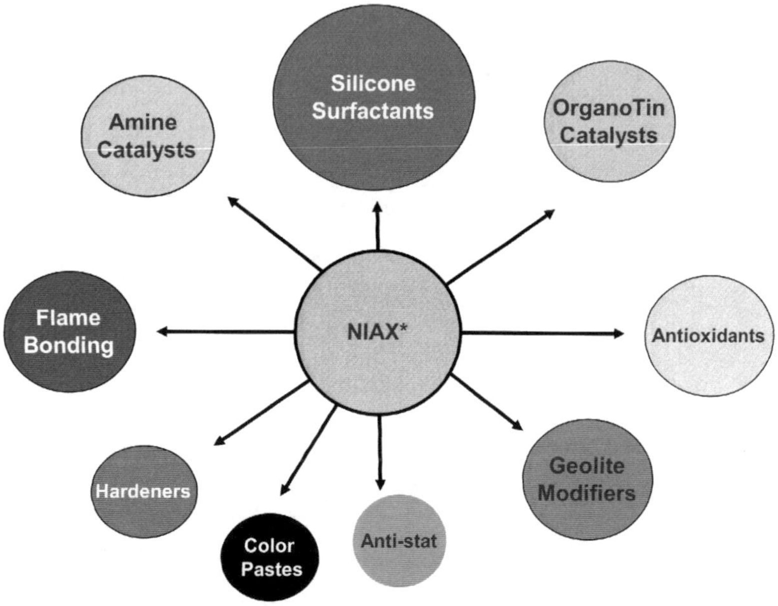

4

MOMENTIVE

Why Silicone Surfactants?
Emulsification and Stabilization

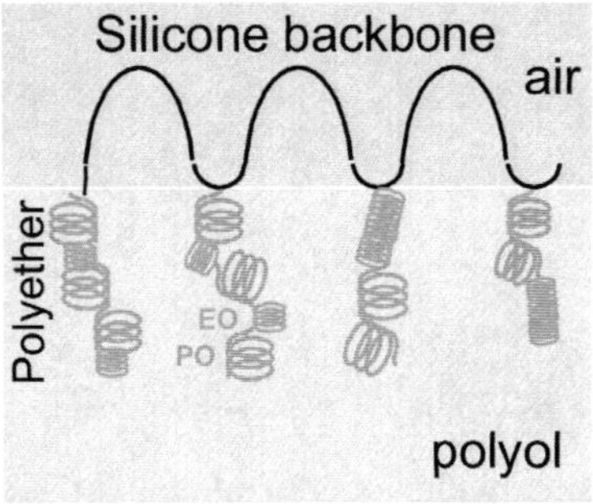

Surfactant 1% Surfactant in a 3000 MW Triol	γ (dynes/cm)
None (pure polyol)	**35**
Nonylphenolethoxylate	**32**
Sulfonated Fatty Ester	**31**
Silicone/Polyether	**25***
Fluorocarbon/Polyether	**24**

*The only product that will stabilize a urethane foam. Surface Tension reduction is a necessary but not sufficient requirement for foam stabilization

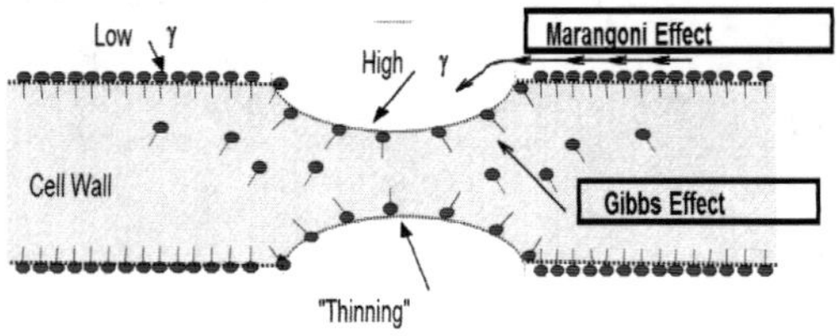

Gibbs-Marangoni: Thinning causes lower local concentration of surfactant which raises γ causing a γ gradient in the film. This gradient is a restorative force drawing more surfactant to lower the γ. The surface is said to become "elastic".

Ewers-Sutherland: Similar concept except this approach suggests that as the surfactant mono-layer moves to reduce the local γ gradient it drags with it sufficient bulk material to help heal the rupture.

MOMENTIVE

Copolymer Chemistry

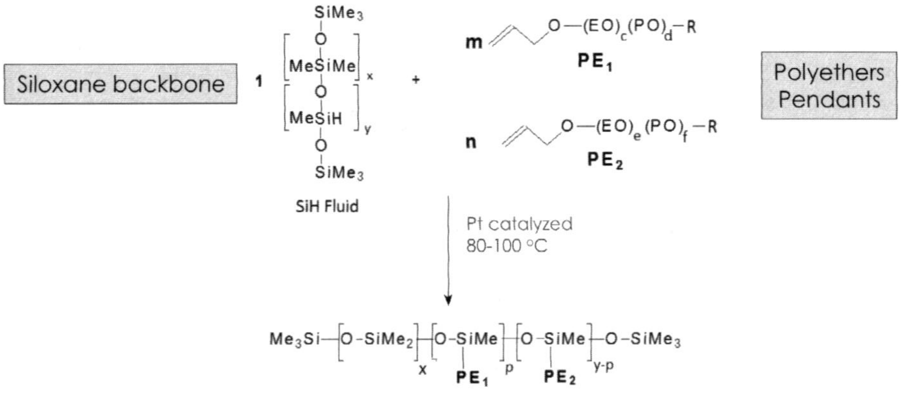

Silicone Polyether Copolymer

- The reaction is governed by compatibility, reaction selectivity and randomization.
- Potential Technologies in MPM:
  - Batch process
  - Continuous Copolymer Unit (CCU)
  - Static mixing plug flow reactors
  - Multistage Mixer (MSM)
  - Patented Technologies: CCU, Static mixing plug flow reactors, MSM.

MOMENTIVE

IP REVIEW

MOMENTIVE

Innovation – Continuous SPC Manufacturing Technology

- Detailed IP search of available patents for producing silicone copolymers continuously.
- Process development boom came in early 2000s.

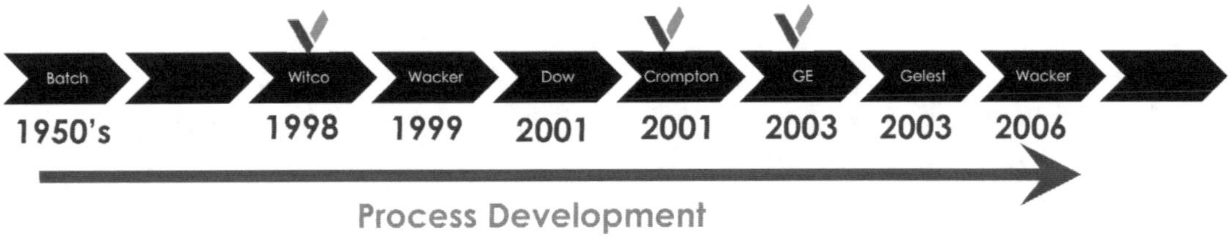

Year filed	Company	Patent #	Process
1998	**Witco Corporation**	**US 5986022**	**CSTR and PFR (CCU)**
1999	Wacker Chemie Gmbh	US6350824 B1	Pre-mixed and fed into the reaction zone
2001	**Dow Corning Toray Silicone**	**US 6,410,772 B2**	**CSTR, tubular reactor and PFR**
2001	Crompton	US20020099137 A1	Static Mixing PFR in series or parallel
2003	**General Electric Company**	**US 6,897,280 B2**	**Multi-stage blade – mixed plug flow reactor (MSM)**
2003	Gelest Inc	US 20040176561 A1	PFR and a tubular reactor packed with a solid, acidic polymerization catalyst
2005	Wacker Chemie Ag	US7145028 B2	Integrated loop tube reactor

MOMENTIVE

CONTINUOUS COPOLYMER UNIT (CCU)

PATENT NO.: 5986022
YEAR APPROVED: 1999
ASSIGNEE: WITCO CORPORATION

MOMENTIVE™

CCU Design

- 2 CSTR's and 1 PFR in series
- Reaction has to complete at least 80-85% before entering PFR to avoid phase separation.
- Properties slightly different from batch products.
- CSTR Limitation: Not all of any of the reactants can be consumed completely.

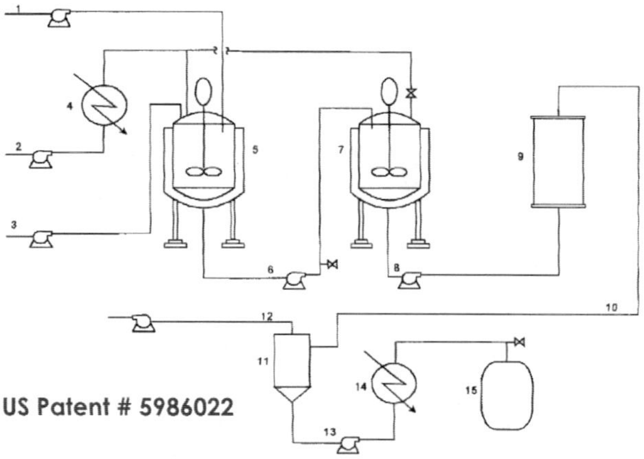

US Patent # 5986022

- PFR Limitation: Immiscible SiH fluid and olefinic compound will phase separate very rapidly and reaction slows down or ceases completely.
- Benefits from a batch: Low Operating Cost (Conversion Cost Equivalent to LCC), Reduced Footprint, Reduced Waste, Solvent Reduction, Capacity.
- Cons: Versatility

Has its advantages over batch but versatility is missing.

10 MOMENTIVE™

MULTI-STAGE MIXER PATENT (MSM)

PATENT NO.: US 6,897,280 B2
YEAR APPROVED: 2005
ASSIGNEE: GENERAL ELECTRIC COMPANY

MOMENTIVE™

MSM Patent Overview

- Evolution of CCU Technology.
- ID 50 mm, L – 0.6 m, 8 stages of 4-8 min residence time, tubular jacketed reactor. Offset channels to prevent back mixing.
- Eddies and vortices of sufficient intensity in multi-stage blades shear biphasic liquid mixture and disperse one phase into another for rapid reaction.
- Last stage of reactor is plug flow. 75-90% of reaction is finished before entering this region.
- Front end equipment required
  Flow meters, buffer, catalyst addition systems, stripping unit
- Back end equipment required
  Product stripper, product cooler, product filters and blend equipment (tanks or mixer)

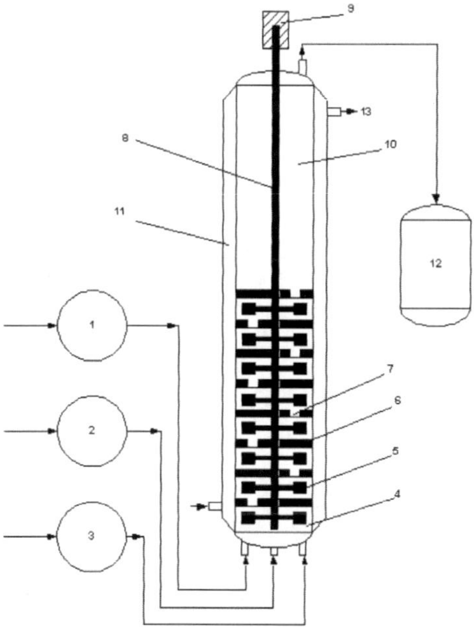

US 6,897,280 B2 (May 24, 2005)

2nd Generation CCU

12

MOMENTIVE™

Batch/CCU/MSM

Comparison of processes for different SPC (US Patent# US 6,897,280)

Raw material	Process	SiH (cc H2/g)	Appearance	Viscosity (cst)	Foam Rise (cm)	Flow Air (porosity)	Cell Structure
Siloxane A + PE_1 + PE_2 + Buffer + Pt catalyzed	Batch	ND	Clear	2349	38.0	5.2	7
	CCU	**0.4**	**Hazy**	2461	**36.8**	4.2	**10**
	MSM	ND	Clear	2482	38.6	6.3	7
Siloxane B + PE_3 + PE_4+ Buffer + Pt catalyzed	Batch	ND	Clear	2771	40.5	2.2	6
	CCU	**0.1**	**Hazy**	2569	**Total Foam Collapse**		
	MSM	ND	Clear	2626	39.2	7.0	7
Siloxane B + PE_5 + PE_4 + Buffer + Pt catalyzed + DPG	Batch	ND	Clear	1003	37.9	6.6	7
	CCU	ND	Clear	942	37.1	6.1	**8**
	MSM	ND	Clear	1036	37.9	5.6	7
Siloxane B + PE_6 + PE_4 + Buffer + Pt catalyzed	Batch	ND	Clear	1794	40.9	5.1	7
	CCU	ND	Clear	1851	39.0	5.9	7
	MSM	ND	Clear	1833	40.7	5.1	7

MSM produced similar to batch grade. CCU is slightly different.

13 MOMENTIVE

DOW PATENT

PATENT NO.: US 6,410,772 B2

YEAR APPROVED: 2002

ASSIGNEE: DOW CORNING TORAY SILICONE CO., LTD.

 MOMENTIVE

Dow Patent Overview

- ID: 50 mm, L = 1 m, 33 disks
- Blade or pin stirrers severely impact plug flow conditions.

- 2 or more disks maintain plug flow conditions.
- Rotation of disks facilitates mixing of various components & improves efficiency while maintaining plug flow conditions.

- Reactor capacity significantly improved by adding a pre-mixer.
- Static reactor downstream (25 mm ID, L - 3m, 1.5L) solves the feed rate & elongation of tubular reactor issues.

- Loading of SiH fluid to PE is about 1.
- Residence time of 4 min for feed rate of 50kg/hr achieved.
- **Limitations:** Single PE product.

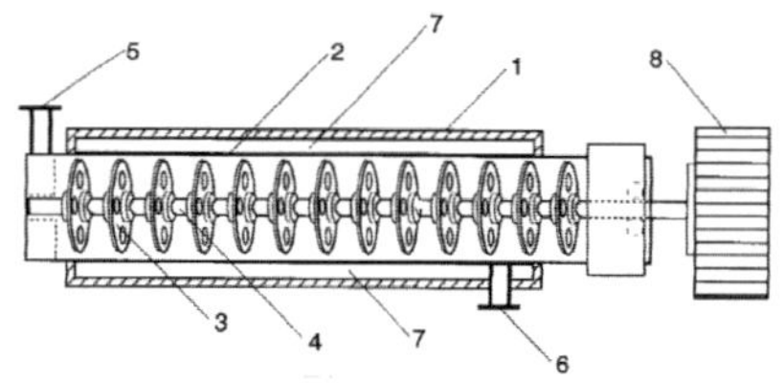

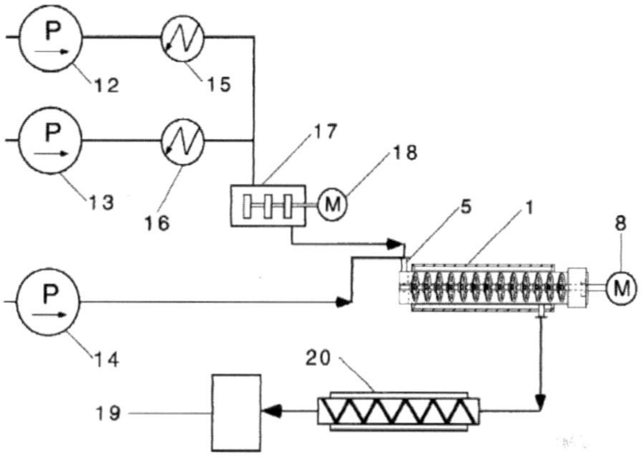

US Patent # 6,410,772 B2

  MOMENTIVE

Summary of Comparisons

	Batch	CCU	MSM
Reactor	Single tank	At Least One 100-300 Gallon Tank Plug Flow Reactor	Tubular reactor with multiple stages of stirring blades
Pros/Cons	Versatility	Low Operating Cost (Conversion Cost Equivalent to LCC) Reduced Footprint Reduced Waste Solvent Reduction Capacity	Low Operating Cost (Conversion Cost Lower than LCC) Versatility Small Footprint Reduced Waste Solvent Reduction Capacity
Capability	Terminal/Pendant Copolymers Single/Multiple Polyethers Wide Range of MW's/Viscosities	Terminal/Pendant Copolymers Single Polyether Lower End MW's/ Viscosities	Terminal/Pendant Copolymers Single/Multiple Polyethers Wide Range of MW's/Viscosities
Product grade		Different than batch	Similar to batch

Process Development key to achieving good quality SPC

  MOMENTIVE

Thank You.

DISCLAIMER: THE MATERIALS, PRODUCTS AND SERVICES OF Momentive Performance Materials Inc. and its subsidiaries and affiliates DOING BUSINESS IN LOCAL JURISDICTIONS (collectively "SUPPLIERS") ARE SOLD BY THE RESPECTIVE LEGAL ENTITY OF THE SUPPLIER SUBJECT TO SUPPLIERS' STANDARD CONDITIONS OF SALE, WHICH ARE INCLUDED IN THE APPLICABLE DISTRIBUTOR OR OTHER SALES AGREEMENT, PRINTED ON THE BACK OF ORDER ACKNOWLEDGMENTS AND INVOICES, AND AVAILABLE UPON REQUEST. ALTHOUGH ANY INFORMATION, RECOMMENDATIONS, OR ADVICE CONTAINED HEREIN IS GIVEN IN GOOD FAITH, SUPPLIERS MAKE NO WARRANTY OR GUARANTEE, EXPRESS OR IMPLIED, (i) THAT THE RESULTS DESCRIBED HEREIN WILL BE OBTAINED UNDER END-USE CONDITIONS, OR (ii) AS TO THE EFFECTIVENESS OR SAFETY OF ANY DESIGN INCORPORATING SUPPLIERS' PRODUCTS, MATERIALS, SERVICES, RECOMMENDATIONS OR ADVICE. AFOREMENTIONED EXCLUSIONS OR LIMITATION OF LIABILITY ARE NOT APPLICABLE TO THE EXTENT THAT THE END-USE CONDITIONS AND/OR INCORPORATION CONDITIONS CORRESPOND TO THE RECOMMENDED CONDITIONS OF USE AND/OR OF INCORPORATION AS DESCRIBED BY SUPPLIER IN ITS PRODUCT DATA SHEET AND/OR PRODUCT SPECIFICATIONS. EXCEPT AS PROVIDED IN SUPPLIERS' STANDARD CONDITIONS OF SALE, SUPPLIERS AND THEIR REPRESENTATIVES SHALL IN NO EVENT BE RESPONSIBLE FOR ANY LOSS RESULTING FROM ANY USE OF ITS MATERIALS, PRODUCTS OR SERVICES DESCRIBED HEREIN. Each user bears full responsibility for making its own determination as to the suitability of Suppliers' materials, services, recommendations, or advice for its own particular use. Each user must identify and perform all tests and analyses necessary to assure that its finished parts incorporating Suppliers' products, materials, or services will be safe and suitable for use under end-use conditions. Nothing in this or any other document, nor any oral recommendation or advice, shall be deemed to alter, vary, supersede, or waive any provision of Suppliers' Standard Conditions of Sale or this Disclaimer, unless any such modification is specifically agreed to in a writing signed by Suppliers. No statement contained herein concerning a possible or suggested use of any material, product, service or design is intended, or should be construed, to grant any license under any patent or other intellectual property right of Suppliers or any of its subsidiaries or affiliates covering such use or design, or as a recommendation for the use of such material, product, service or design in the infringement of any patent or other intellectual property right.

Momentive and the Momentive logo are trademarks of Momentive Performance Materials Inc.

17

MOMENTIVE™

497659 Esterification of Carboxylic Acids Present in the BIO-OIL Produced By Thermal Cracking of Triacylglycerols

Abstract:

The production of alternative fuels from renewable sources has been object of several studies in recent years. A promising alternative is the conversion of biomass to biofuels by thermal cracking. Through this process, it is possible to obtain a liquid product called bio-oil, with physicochemical properties similar to the petroleum crude oil. However, bio-oil might contain a high amount of carboxylic acids that results on a high acid index ranging from 110 mg KOH/g of sample to 207 mg KOH/g of sample, whereas the maximum required by the current legislation (EN 14104) is 0.5 mg KOH/g of sample. In this work, the bio-oil was produced from commercial soybean oil at 525°C and a mass flow rate of 350 g/h. The objective of this study is to perform esterification reactions in order to reduce the acidity index of the bio-oil. These reactions present a methodology of easy operation, low cost and may reduce the bio-oil corrosivity. For this, alcoholic reagents, ethanol and methanol, were used for the esterification reactions. The influence of different temperatures, reaction times, catalysts and, stoichiometric ratios between the reagents on the reduction of the acid number and improvement of the bio-oil quality were evaluated. Good results were obtained with the esterification reaction using ethanol, sulfuric acid, bio-oil to ethanol ratio of 10:5, temperature of 60°C and reaction time of 90 minutes. The acid index of the bio-oil sample decreased from 111.33 mg KOH/g of sample to 34.64 mg KOH/g of sample, resulting in a reduction of approximately 70%.

506291 Flex-to-Stretch Electronics

Abstract:

Stretchable electronic devices maintain electrical connections when subjected to stress or strain and are therefore useful for enabling electronics with new mechanical properties. However, these devices often require power or electrical signals from non-stretchable components. Connecting flexible and stretchable substrates is a possible solution that maintains the electrical connection within a medium that can withstand deformation and prolonged use. The ability to create a robust electrical connection between these mechanically disparate components may enable new types of hybrid devices. In this study we present a basic method to fabricate such new hybrid devices, utilizing liquid metals that form fantastic stretchable interconnects because they are both conductive and fluidic. We characterized the mechanical and electrical properties of these hybrid devices. While we focus on silicon elastomers and liquid metals, there are numerous other methods of fabricating stretchable devices that share many properties with the devices in this work.

506292 Systematic Analysis of Cloud Point and Crystallization in Fatty Acid Ethyl Ester Biodiesel Mixtures

Abstract:

Saturated biodiesel components, fatty acid methyl esters (FAMEs) or fatty acid ethyl esters (FAEEs), crystallize at high temperatures, which limits their commercial application. FAEEs are particularly advantageous biodiesel components because they can be renewably sourced and have lower cloud points (CP) than FAMEs with the same carbon-chain length. This research systematically evaluates FAEE cloud point and crystallization data to aid in the development of FAEE biodiesel blends with practical cold flow properties. The effects of cold-flow improving additives dimethyl azelate (DMA) and triacetin (TA) on FAMEs and FAEEs are also analyzed. Neat FAEE CP increases with carbon-chain length and decreases with greater degree of unsaturation, and saturated/saturated FAEE mixtures show partial cocrystallization. Additives lowered the CP, but did so via different mechanisms. Thermodynamic ideal solution modeling accurately predicts binary FAEE CP with and without additives within 2% of experimental values.

Feasibility Study of Ionic Liquid Desalination Design

Material Physics and Applications Division, Los Alamos National Laboratory
PI: Zachary Cosenza RLM: George Scott Goff

LA-UR 17-26682

Introduction

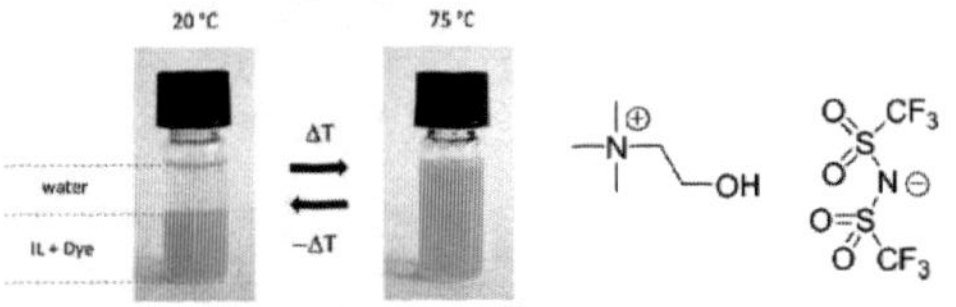

Courtesy of Nockemann et al. [1]

- Ionic liquids (ILs) are a class of room temperature salts in liquid form.
- Water has been shown to be soluble in ILs with T dependence, while the IL remains insoluble.
- Temperature may be used to drive water from a saline into the IL and back out at modest T.
- Current desalination methods (reverse osmosis and multiple-effect evaporation) run at high T or P
- Proposal is to **develop liquid-liquid extraction system using ILs** and saline.

Baseline System

Continuous Process uses Octanoic Acid as solvent

Batch Process (N = 1) uses [Choline][TF2N] as solvent

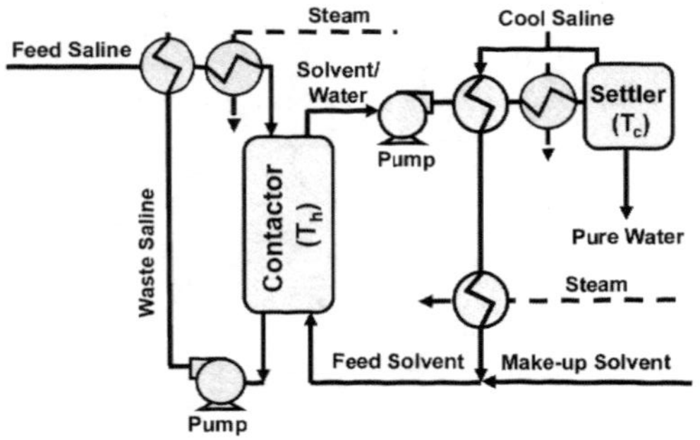

Major Process Parameters

Temperature – Separation efficiency is driven by the difference in contactor (T_h) and settler (T_c) temperatures.

Solvent – The process efficiency is limited by equilibrium, and in turn by the solvent used.

Saline Input – It has been shown [3] that equilibrium is altered by saline concentrations, with higher salt concentrations reducing the driving force.

Methods

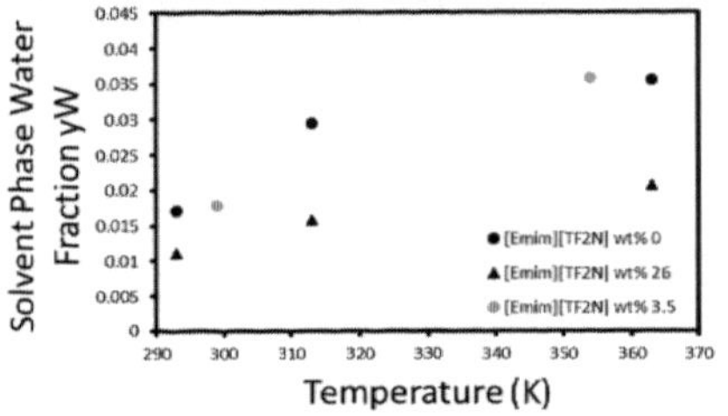

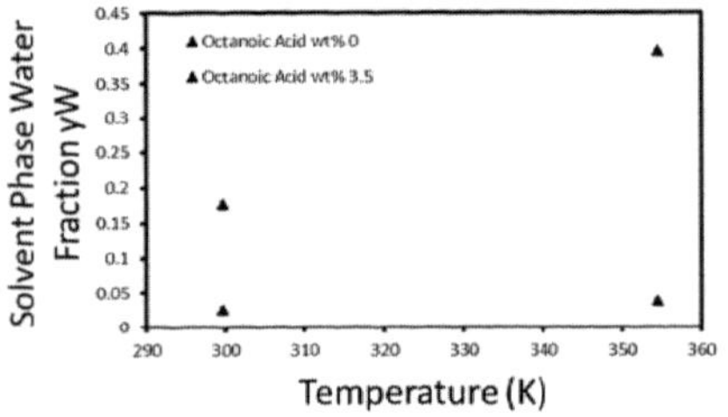

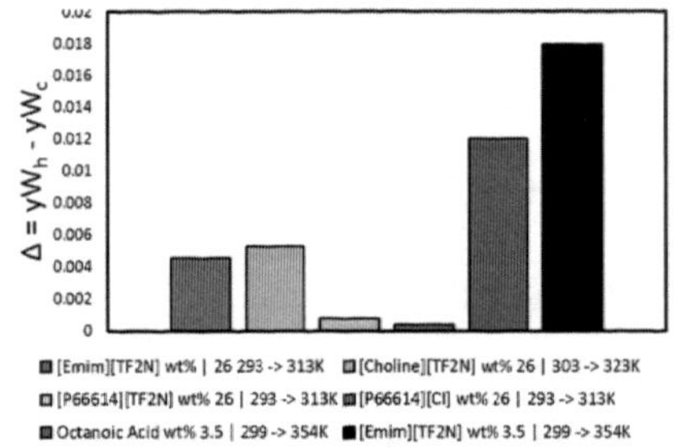

Analysis

Operating Line (OL):

$$\frac{V_m}{F} = \mathcal{N} * \iota_m + \frac{V_1}{F} - \mathcal{N} * \iota_1$$

Boundary Conditions:

(i) Set $\mathcal{N}$ (slope)
(ii) Set inlet water conc.
(iii) Pinch Point solution
(iv) Set inlet solvent conc.

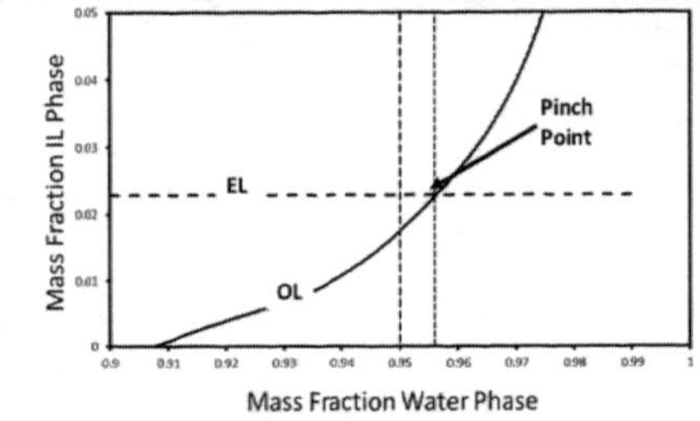

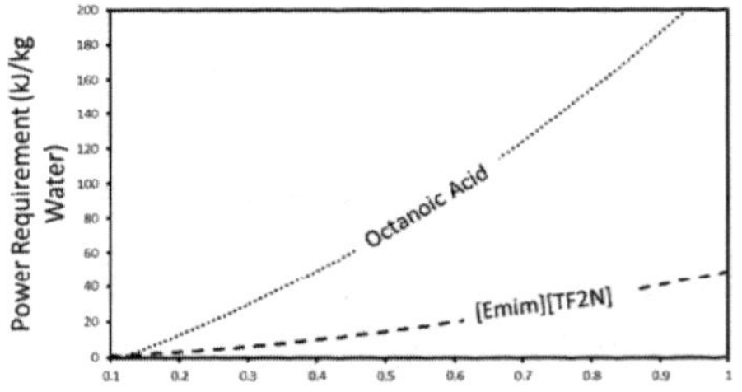

Energy Requirements OA and Emim, 353K with L = 3.0 kg/s 3.5 wt% water baseline

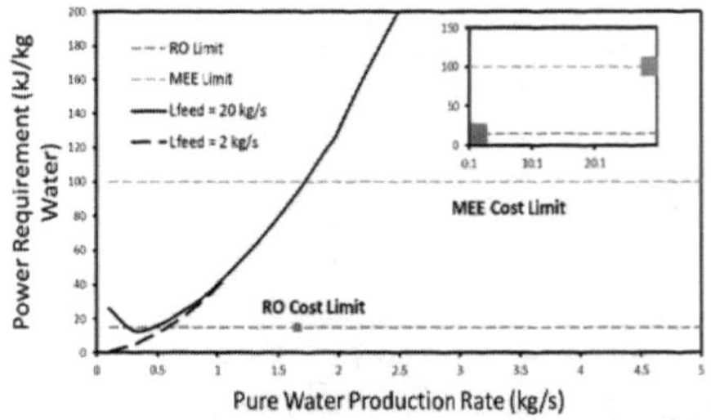

Cost Curve w/ Parameter Variation, RO ■ [14, 23] and MEE ■ [14, 24] based on pure water production (353K) and 3.5 wt% water

Conclusions

- Like the conclusion reached by Sanap et al [2], the continuous process is highly pinched at N = 1 at all upper process temperatures.
- Compared to the current literature work (Octanoic Acid as a solvent), [Emim][TF2N] has higher extractive and cost efficiency.
- If coupled with waste heat the marginal energy cost is close to that of RO, but far from MEE.
- Data on droplet size, mixing times, mass and heat transfer coefficients, and fixed costs would help scale-up process beyond basic thermodynamics.

References

[1] Nockmann, *J. Phys. Chem.* Ed 113. 1429-1437 (2009)
[2] D.B. Sanap et al, *Desalination* Ed. 357 150-162 (2015)
[3] Bajpayee, Anurag Directional Solvent Extraction Desalination, MIT 2012, *MIT Libraries*
[4] Zhong Yujiang, *Environ. Sci. Tech.* Ed. 50 1039-1045 (2016)

Acknowledgements

Los Alamos NATIONAL LABORATORY

506320 Preparation and Characterization of Shape Memory Assisted Self-Healing Coatings

Abstract:

Shape memory assisted self-healing (SMASH) coatings have the ability heal nearly completely after damage with the application of heat, making them useful for corrosion resistance in steel structures (1). This summerâ€™s project sought to expand upon existing research to develop an efficiently healing and transparent SMASH coating that could be used in technological applications, such as on cell phone or computer screens (2) . The materials tested were comprised of electrospun poly(Îµ-caprolactone) (PCL) fiber mats coated in epoxy with differing glass transition temperatures (Tg). The healing efficiency of the samples were tested by damaging each one with a razorblade and healing them over varying times and heats in an oven. The transparency of each sample was observed visually before and after heating, a technique which will be refined later in the research process. Once the most ideal combination is found, further experimentation will be conducted to determine whether the ideal epoxy/PCL combination is also one that can withstand multiple damage-healing cycles to the same spot without becoming opaque. References (1) Nejad, H. B. Thermally/Mechanically Responsive Polymeric Composites with Shape Memory/Self-Healing Properties. Syracuse University 2015, 79-95. (2) Luo, X. Thermally Responsive Polymer Systems for Self-Healing, Reversible Adhesion and Shape Memory Applications. Syracuse University, ProQuest Dissertations Publishing 2010. 3437600.

506372 Defluoridation of Ground Water Using Impregnated Aluminum

Abstract:

Groundwater is an important source of drinking water and it is estimated that more than 200 million people worldwide are drinking groundwater with fluoride concentrations greater than the WHO guideline value of 1.5 mg/L. The majority of these cases occur in the developing world. fluoride has long been found to have a beneficial effect on dental health, although when present in drinking water at concentrations much above the WHO guideline value (and national drinking water limit for most countries) of 1.5 mg/L, long term use can result in development of dental fluorosis or at its worst, crippling skeletal fluorosis. The effects on health depend on a number of factors including dose from other sources of fluoride (toothpastes, food), age, gender and general health status.

506536 Ordering pH-Responsive Polymer-Grafted Nanoparticles in Flow Coating Process

Abstract:

Poly(acrylic acid) (PAA)-grafted silica nanoparticles with different grafting densities and brush lengths were synthesized via surface-initiated ATRP, and their deposition patterns on silicon substrates were investigated by systematically varying the speed of moving substrates, pH values, and free polymer poly(vinylpyrrolidone) (PVP) concentrations in aqueous solution. Brush-brush interactions are tuned by varying the PAA chain conformations with pH and also with the addition of free PVP. Different deposition patterns were observed for nanoparticles at low grafting density and long brush length at a velocity of 0.4 mm/min and a PVP concentration of 0.24 mg/ml. At pH 11, nanoparticles assembled into well-regulated lines parallel to the moving direction of the blade. However, at pH 9, nanoparticles showed stronger repulsion, which led to Rayleigh-instability, and a dot pattern formation. At pH 3, particle attractions get strong due the hydrogen bonding between PAA and PVP, hence Bénard cell patterns were observed. The mechanisms of hydrogen bonding between two different polymers and also physical bridging within particles are demonstrated to organize the grafted nanoparticles in a flow-coating process. Our findings will enable the easy fabrication of patterned particle domains for functional polymer films.

494879 A DEM Modeling of Biomass Fast Pyrolysis in a Double Screw Reactor

Abstract:

Biomass is an abundant renewable resource that has been used for thousands of years to provide heat and energy in rural areas. More efficient thermochemical conversion ways of utilizing biomass such as fast pyrolysis are introduced in recent studies [1, 2, 3]. Screw reactors are one of the commonly used reactors for biomass fast pyrolysis [3, 4, 5]. In a screw reactor, mechanical forces are provided to the bed to enhance particle mixing and heat transfer, and achieve particulate matter transport for continuous operation. The pyrolysis process in the screw reactor configuration is a reactive granular flow system which involves particle flow, heat transfer and biomass devolatilization reactions at the same time. However, reactor-scale evaluation of heat and mass transfer effects on the biomass fast pyrolysis is very limited. Efforts are needed to discover the underlying physical phenomena and the interplay between the physical transport phenomena and chemical reactions for reactor design and optimization. Numerical simulation is capable of predicting the granular flow and heat transfer behavior in the reactor and providing useful information of particle flow such as particle residence time, particle mixing degree, and heat transfer coefficient [6].

In this research, we proposed an extended DEM method for modeling the reactive granular flow in a double screw reactor. A semi-detailed biomass devolatilization kinetics is adopted in the simulation and the decomposition dynamics of the major biomass components and pyrolysis products are analyzed at different operating conditions. The heat transfer coefficient information is extracted from the DEM simulation and provided as heat transfer boundary conditions for a single particle model. The intra-particle heat and mass transport phenomena are modeled and the effects on the biomass pyrolysis process are evaluated. We report that the heat transfer coefficient in the double screw reactor varies within a wide range of 20 to 90 $W/(m^2K)$ for different operating conditions. Results also indicate the particle-fluid-particle conductive heat transfer pathways are the dominant contributors to the total heat flux, which accounts for 70%-80% in the total heat flux. Radiation heat transfer contributes 14%-26% to the total heat flux. The reported result is helpful for reactor performance evaluation and reactor design optimization.

Reference

[1] C. Stevens, R. C. Brown, Thermochemical processing of biomass: conversion into fuels, chemicals and power, John Wiley & Sons, 2011.

[2] H. Goyal, D. Seal, R. Saxena, Bio-fuels from thermochemical conversion of renewable resources: a review, Renewable and sustainable energy reviews 12 (2) (2008) 504–517.

[3] R.Venderbosch,W.Prins,Fastpyrolysistechnologydevelopment,Biofuels,bioproductsandbiorefining 4 (2) (2010) 178–208.

[4] E. Butler, G. Devlin, D. Meier, K. McDonnell, A review of recent laboratory research and commercial developments in fast pyrolysis and upgrading, Renewable and Sustainable Energy Reviews 15 (8) (2011) 4171–4186.

[5] A. V. Bridgwater, Review of fast pyrolysis of biomass and product upgrading, Biomass and bioenergy 38 (2012) 68–94.

[6] F. Qi, T. J. Heindel, M. M. Wright, Numerical study of particle mixing in a lab-scale screw mixer using the discrete element method, Powder Technology 308 (2017) 334–345.

484227 Hydrocarbon Produced from Upgrading Rich Phenolic Compound Bio-Oil with Low Catalyst Coking

Abstract:

Catalytic upgrading of raw bio-oil and liquidâ€"liquid extracted bio-oil (high concentrated phenolic with trace acid and acetaldehyde) with methanol over ZSM-5 catalyst had been studied in this work. Temperature played vital function and leaded to increasing gas yield but less catalyst coking. It also changed both chemical distribution and selectivity on both gas and liquid products, with aromatic concentration increasing by 34.42%. Temperature of 400 _C was selected as the optimized reaction conditions with liquid yield of 10.47 wt.% and 75.00% aromatic hydrocarbon in liquid product from 100 g biomass, only with coke yield of 1.42 wt.%. Phenolic-rich extracted bio-oil obtained higher aromatic hydrocarbon yield (7.3 wt.% increased from 1.1 wt.%) and lower coke yield (1.42 wt.% decrease from 15.79 wt.%) than raw bio-oil. Catalyst regenerated from this type of feedstock also achieved higher activity and longer useable running times on aromatic yield compared to fresh catalyst. This result suggested that lignin derived phenolic and guaiacol compounds were not the only reason caused catalyst coking on ZSM-5 catalyst. Small high active molecules of acetic acid and acetaldehyde also acted as important precursors of catalyst coke formation.

501928 Renewable Hydrocarbon Production from Catalytic Upgrading of Fast Pyrolysis Bio-Oil from a Feedstock Blend

Abstract:

Production of renewable fuel blendstock from catalytic hydroprocessing of fast pyrolysis oil has been considered as one of the promising thermochemical conversion pathways of biomass to renewable fuel. However, this technology requires overcoming technical and economic barriers for it to be considered competitive with petroleum-derived fuels. The use of low-cost, blended feedstocks in place of a single source of biomass is one way to reduce the overall process costs. In this presentation, we report the catalytic upgrading of a fast pyrolysis bio-oil derived from a mix of clean pine, forest residue and demolition waste. Due to the thermal instability of the bio-oil, it was pretreated in the presence of Ru/C catalyst using a continuous flow reactor (400 ml) with nominal bio-oil space velocity of 0.25-0.5 h^{-1} and 500-1000 L H_2/L bio-oil and operated at 140°C and 1200 psig. Further catalytic hydrodeoxygenation and hydrocracking reactions were accomplished in a two-bed continuous flow reactor consisting of a sulfided Ru/C catalyst held at 170°C, followed by a sulfided commercial hydrotreating/hydrocracking catalyst held nominally at 400°C. The operating pressure was set at 2000 psig, with 0.1 h^{-1} bio-oil space velocity and 2500 L H_2/L bio-oil. The two-phase product was subsequently separated, with the organic phase (less than 3 wt% oxygen) distilled in a batch set-up into gasoline (20-184°C), diesel (184°C-338°C), jet (150°-250°C) and heavy (>338°C) fractions. Characterization of these fractions will be presented.

496589 CFD-DEM Model for Predicting Producer Gas Contaminants during Biomass Gasification

Abstract:

Biomass gasification is valuable method for extracting energy in the form of producer gas from biomass. The presence of contaminants in producer gas however creates minor and severe downstream problems. We developed a quasi-3D CFD simulation of biomass gasification using Eulerian-Lagrangian concept in an open source C++ code (OpenFOAM). The particle-particle collisions were described using Hertz-Mindlin particle contact model. Navier-Stokes equation was utilized to resolved the gas flow field variables and turbulence was included through a standard Îº-Îµ turbulence model. The pyrolysis, homogeneous and heterogeneous reactions in the gasifier were considered. Six case evaluating the effects of gasification temperature (790, 934, and 1078 Â°C) and equivalence ratio (0.15, 0.25, and 0.35) were simulated. The gas and particle flow field variables such as velocity, pressure, and void fractions inside the gasifier were obtained. Also, the concentration of permanent gases (CO, CO_2, H_2O, CH_4, and H_2O) and contaminant gases (NH_3, HCN, and H_2S) inside the gasifier and across the gasifier outlet were obtained. The simulation results were then compared with experimental data from a bench-scale bubbling fluidized bed gasifier.

498686 Site-Directed Mutagenesis of Family 64 CBM Provides Insights into the Anomalous Binding Interactions with Pretreated Cellulose during Biomass Saccharification

Abstract:

Lignocellulosic biomass, composed of cellulose, hemicellulose and lignin, is broken down into simple fermentable sugars in nature by cellulolytic enzyme cocktails secreted by various microbes. Enzymatically-catalyzed breakdown of cellulose into sugars like glucose has critical significance towards cost-effective production of cellulosic biofuels. Lignocellulosic biomass pretreatment methods can increase the enzymatic accessibility of cellulose microfibrils and/or change the crystalline ultrastructure of cellulose that can also impact enzyme activity (1). Cellulose III, an unnatural allomorph of cellulose, is formed during anhydrous liquid ammonia based pretreatment of the native crystalline cellulose I allomorphic state (*1-2*). Cellulose III has already been shown to have anomalous binding interactions with carbohydrate-binding modules and cellulolytic enzymes (e.g., cellulases) in our previous studies *(3)*. Here, we report results from our site-directed mutagenesis study of a novel carbohydrate-binding module (CBM) belonging to family 64, followed by detailed binding characterization of the mutant CBM library with both cellulose I and cellulose III. These studies shed light into the structure-function relationships guiding the binding of native and engineered CBMs to unnatural cellulose III, as opposed to native cellulose I. This work has direct implications on the cost-effective enzymatic hydrolysis of biomass into fermentable sugars in lignocellulosic biorefineries employing pretreatments that can modify the cellulose ultrastructure.

1. Chundawat, S. P. S. et al. Restructuring the crystalline cellulose hydrogen bond network enhances its depolymerization rate. J. Am. Chem. Soc. 133, 11163â€"11174 (2011).

2. da Costa Sousa, L. et al. Next-generation ammonia pretreatment enhances cellulosic biofuel production. Energy Environ. Sci. 9, 1215â€"1223 (2016).

3. Gao, D. et al. Increased enzyme binding to substrate is not necessary for more efficient cellulose hydrolysis. Proc. Natl. Acad. Sci. 110, 10922â€"10927 (2013).

Direct Succinic Acid Production from Non-Hydrolyzed Biomass Using Sequential Solid-State and Slurry Fermentation with Mixed Fungal Cultures

Jerico Alcantara [a,b], Andro Mondala [a], Logan Hughey [a], Shaun Shields [a]
jerico.alcantra@wmich.edu

[a] Western Michigan University
Department of Chemical and Paper Engineering

[b] University of the Philippines Los Baños
Department of Chemical Engineering

Succinic Acid

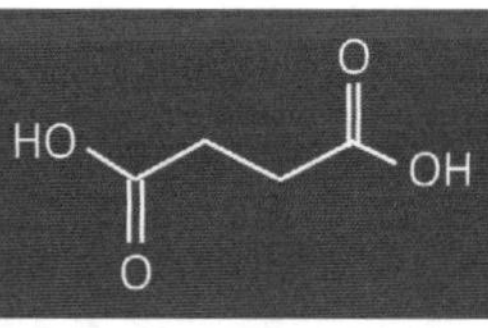

- a four-carbon 1,4-dicarboxylic acid
- one of US Department of Energy's top 12 value-added chemicals from biomass

Uses

Platform chemical	Specialty chemical
1,4-butanediol	Detergent and surfactants
Adipic acid	Ion chelators
2-pyrrolidone	Food ingredients
Tetrahydrofuran	Pharmaceuticals

Zeikus et al. (1999)

2013 Succinic acid market shares

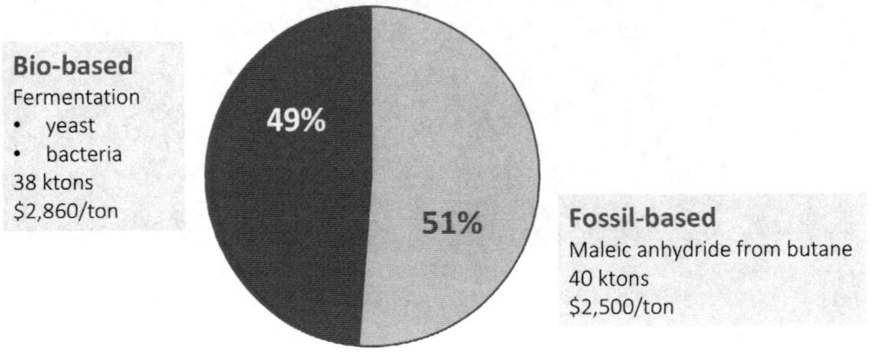

E4tech, RE-CORD and WUR (2015)

Metabolic Pathway for Bacterial Succinic Acid Fermentation

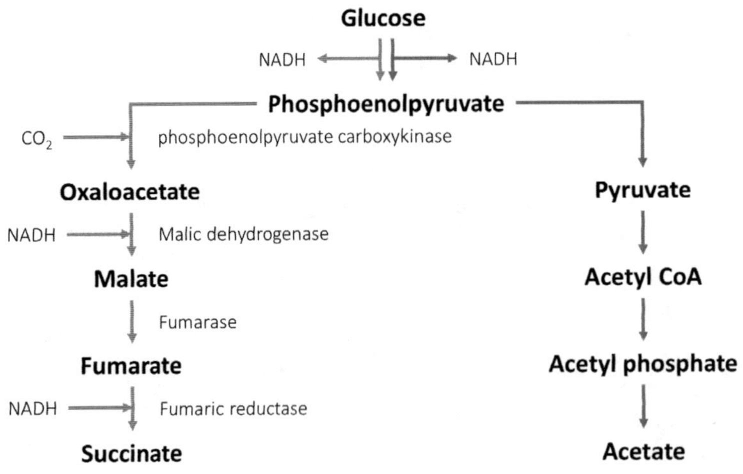

van Heerden and Nicol (2013)

Bacteria vs Fungi

	Bacteria	Fungi
Feedstock	Corn syrup Starch Sugar from energy rich crops	Non-pretreated plant biomass
Operational Requirement	Carbon dioxide/carbonate salt	not costly
Nutrient Requirement	Yeast extract Biotin	not costly

Jansen and van Gulik (2014)

Werpy and Petersen (2004)

Two-step fermentation process

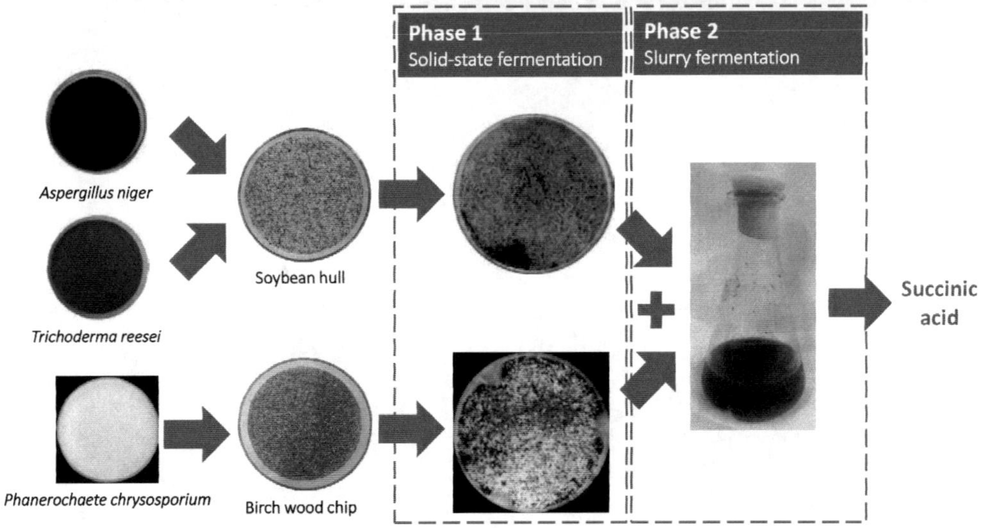

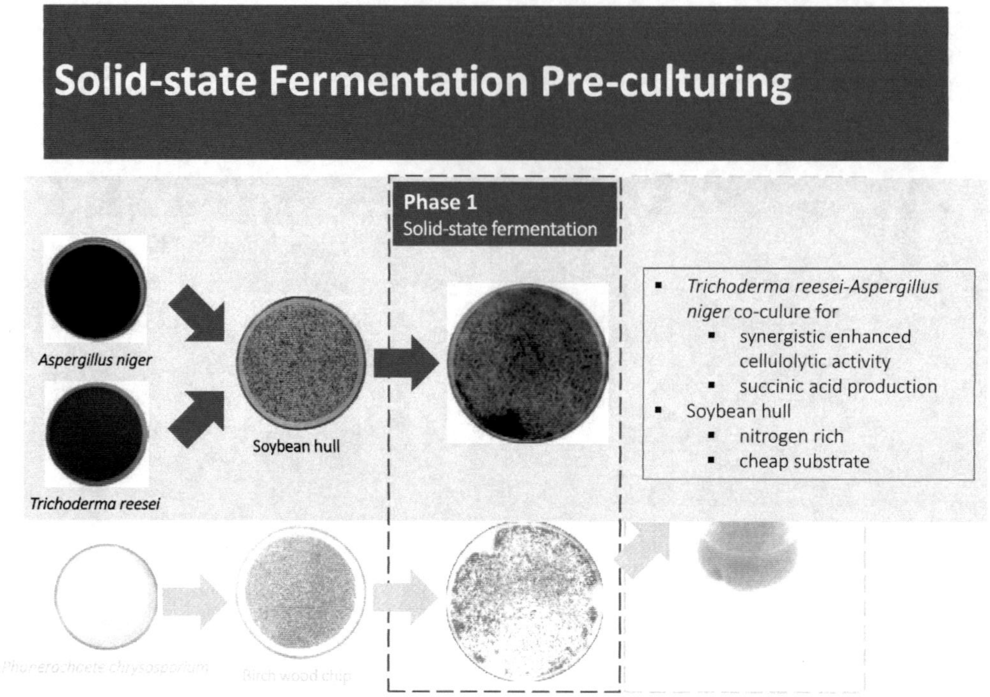
Solid-state Fermentation Pre-culturing
Phase 1
Solid-state fermentation
Aspergillus niger
Trichoderma reesei
Soybean hull
Trichoderma reesei-Aspergillus niger co-culure for
synergistic enhanced cellulolytic activity
succinic acid production
Soybean hull
nitrogen rich
cheap substrate
Phanerochaete chrysosporium
Birch wood chip

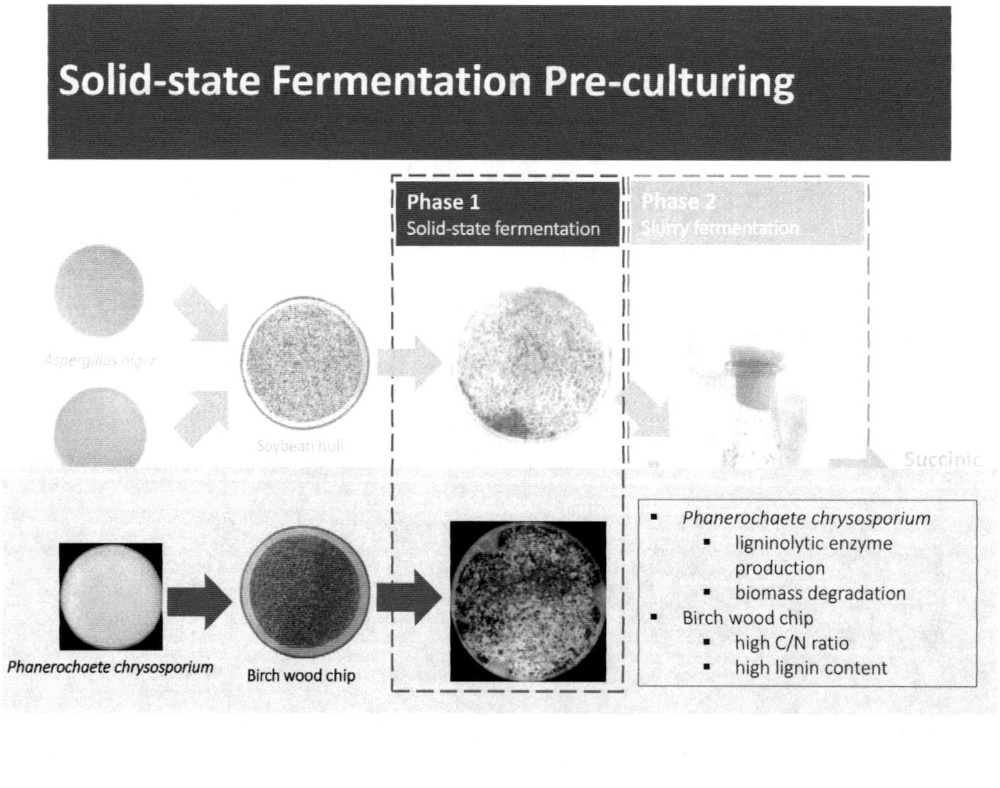
Solid-state Fermentation Pre-culturing
Phase 1
Solid-state fermentation
Phase 2
Slurry fermentation
Aspergillus niger
Soybean hull
Phanerochaete chrysosporium
Birch wood chip
Phanerochaete chrysosporium
ligninolytic enzyme production
biomass degradation
Birch wood chip
high C/N ratio
high lignin content

Slurry Fermentation

- Solid-state fermentation setups were combined and submerged in buffer with constant mixing
- Mixed fungal culture for simultaneous
  - lignocellulolytic enzyme production
  - biomass degradation and enzymatic saccharification
  - succinic acid production

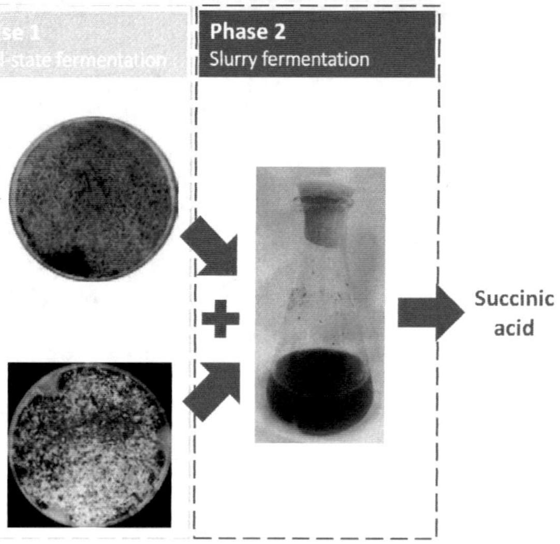

Methodology

Biomass preparation
(0.50 – 1.00 mm)

Inoculum preparation
(2 days)

Solid-state fermentation
(30 °C, 7 days, 95% RH, 70% moisture)

Slurry fermentation 1 (SF1)
(30 °C, 150 rpm, pH 4.8 0.05 M acetate buffer 20% solids)

3:1 BWC to SH ratio

Slurry fermentation 2 (SF2)
(30 °C, 150 rpm, pH 4.8 0.05 M acetate buffer 20% solids)

4:1 BWC to SH ratio

non-pretreated wood chip
(< 0.5 mm)

Slurry Fermentation 1 (SF1)

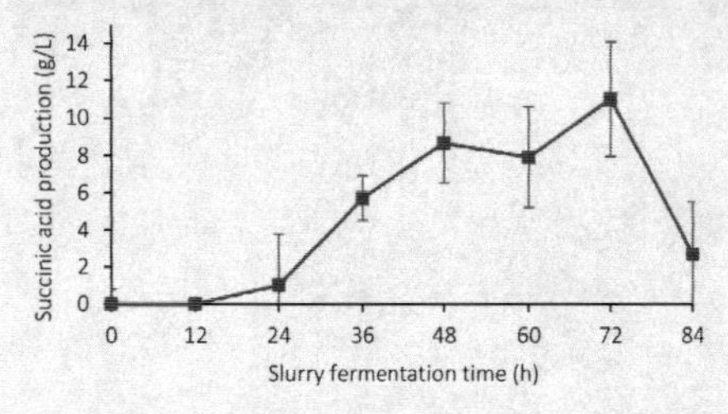

Succinic acid production

- Succinic acid was the main product of slurry fermentation
- Oxalic acid and citric acid were detected in small quantities (< 1 g/L)
- Little to none of succinic acid produced within 24 h
- Maximum succinic acid production of 11.01 g/L after 72 h
- Volumetric productivity of 0.15 g/L·h

Slurry Fermentation 1 (SF1)

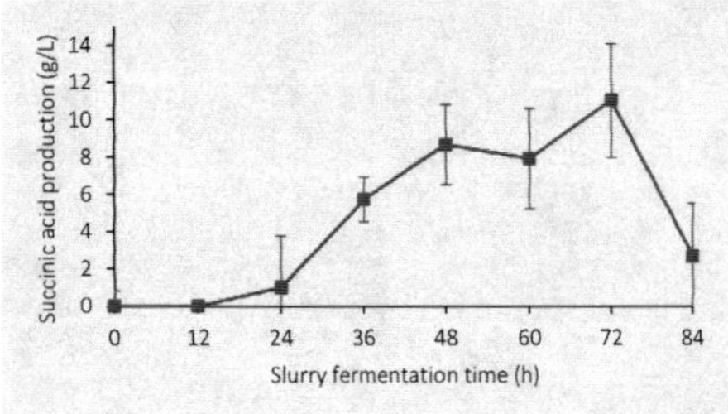

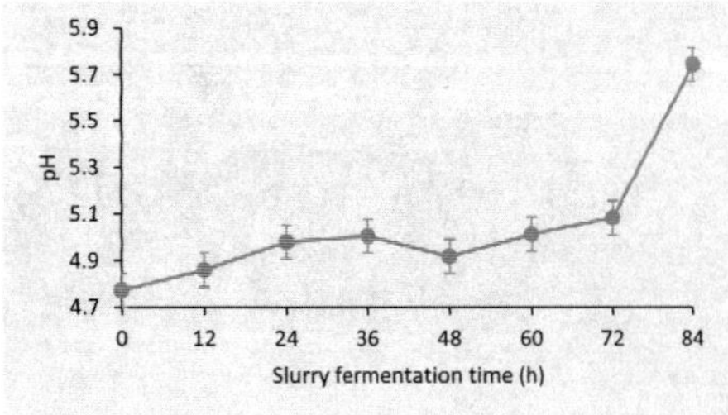

pH

- Initial pH was set at 4.8 for optimal cellulolysis
- Succinic acid production occurred within pH range of 4.77 to 5.08
- High succinic acid present in the slurry mixture resulted to low pH
- After 72 h, succinic acid production dropped resulted to drastic increase in pH

Slurry Fermentation 1 (SF1)

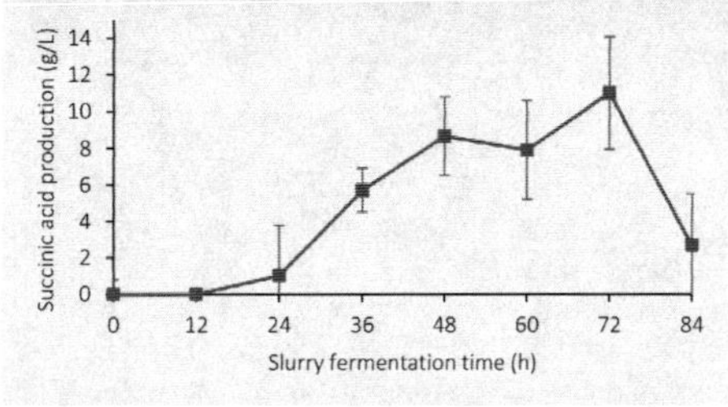

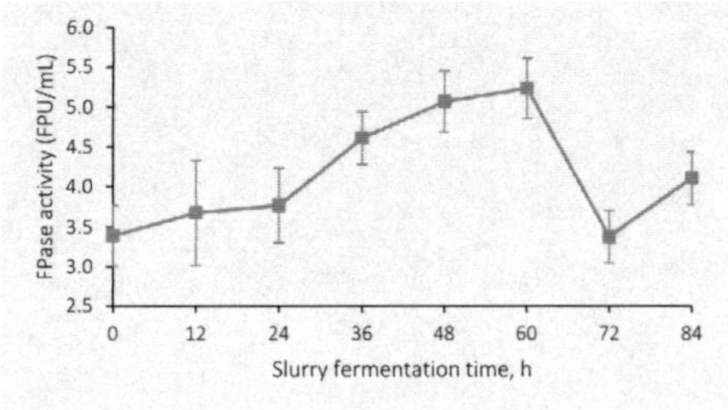

Cellulolytic enzyme activity

- Succinic acid concentration increases with increasing cellulose activity
- Sugars were released from the lignocellulosic matrix resulting to increase in succinic acid production
- After 60 h, cellulase activity decreased, high succinic acid concentration may inhibited cellulolytic enzyme activity
- After 72 h, cellulase activity started to increase again as succinic acid concentration decreased.

Slurry Fermentation 1 (SF1)

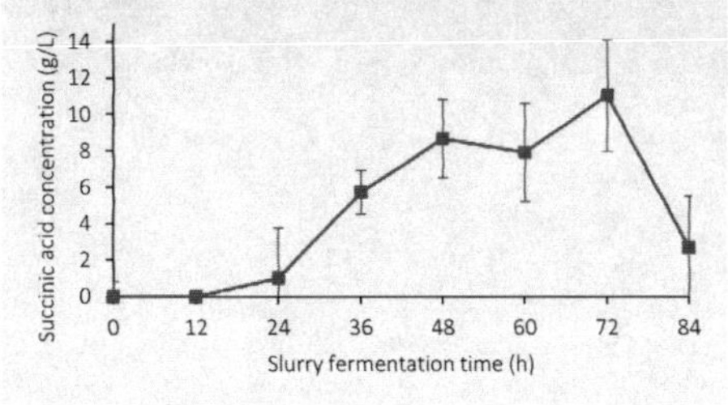

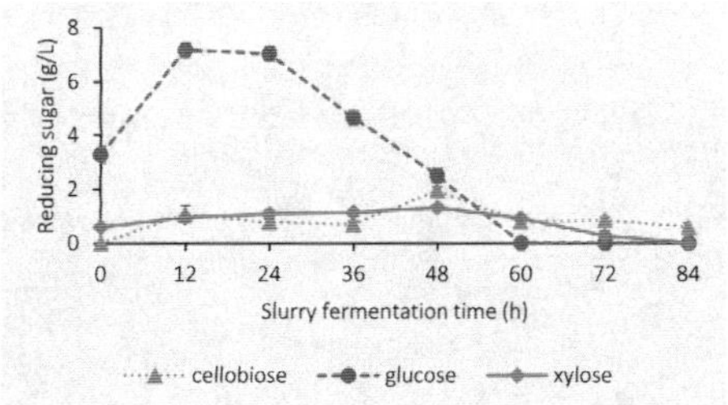

Reducing sugar

- Residual glucose was doubled within 12 h
- After 24 h, glucose started to decline
- Succinic acid increases as residual sugar decreases
- Residual sugars decreased with decrease in cellulase activity

Slurry Fermentation 2 (SF2)

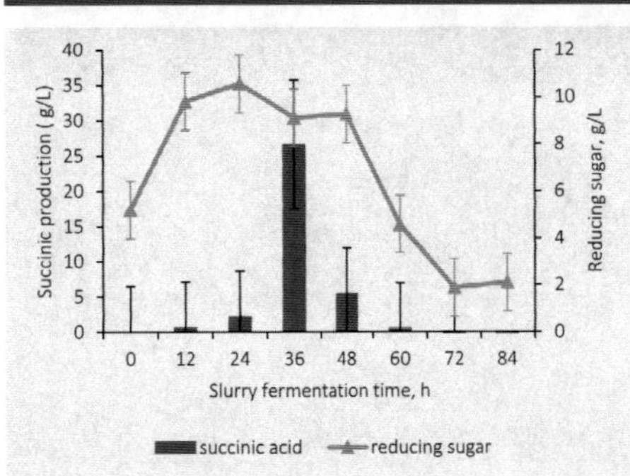

Addition of non-pretreated wood chip

- Non-pretreated birch wood chip was added at the start of slurry fermentation
- Additional carbon rich substrate resulted to higher reducing sugars liberated during cellulolysis
- Maximum succinic acid production of 26.62 g/L succinic acid after 36 h
- Time to reach maximum production shifted
- Volumetric productivity increased to 0.74 g/L·h
- Productivity increased 5 folds

Batch fermentation for succinic acid production

	Microorganism	Substrate	Time (h)	SA ($g\ L^{-1}$)	Productivity ($g\ L^{-1}\ h^{-1}$)	References
Bacteria	*A. succinogenes*	glucose	78	105.8	1.34	Guettler et al. (1996)
	A.succiniciproducens	glucose	19	34.2	1.8	Lee et al. (1999)
	A.succiniciproducens	glucose + corn steep liquor	13.5	17.8	1.32	Lee et al. (2000)
	A. succinogenes	wheat bran	54	64	1.19	Du et al. (2008)
	A. Succinogenes	sugarcane bagasse	121	52	0.43	Borges and Pereira (2011)
	A.Succiniciproducens	wood hydrolysate + corn steep syrup	32	23.8	0.74	Lee et al. (2003)
Fungi	*A. Niger*	molassses			0.14	Ikram-ul et al. (2004)
	Fungal mixed culture	birch wood chip + soybean hull (3:1)	72	11.01	0.15	this study
	Fungal mixed culture	birch wood chip + soybean hull (4:1)	36	26.62	0.74	this study

Summary

Highlights

- Maximum production of 26.62 g/L after 36 h of slurry fermentation
- Maximum volumetric productivity of 0.74 g/L·h
- Productivity increased 5 folds when fresh carbon rich substrate was added

Advantage

- Much cheaper substrate
- No chemical pretreatment
- Minimal operational and nutrient requirement

Disadvantage

- Biological pretreatment (i.e., solid-state fermentation) is relatively slow

Engineering implication

- Results of this study can be used for future scale up of this consolidated bioprocessing technology

References

Borges, Pereira Jr., E. R. N., 2011. Succinic acid production from sugarcane bagasse hemicellulose hydrolysate *by Actinobacillus succinogenes*. J. Ind. Microbiol. Biotechnol. 38, 1001–1011.

Du, C., Lin, S. K. C., Koutinas, A., Wang, R., Dorado, P., Webb, C., 2008. A wheat biorefining strategy based on solid-state fermentation for fermentative production of succinic acid. Bioresour. Technol. 99, 8310-8315.

E4tech, RE-CORD and WUR, 2015. From the Sugar Platform to biofuels and biochemicals. Final report for the European Commission, contract No. ENER/C2/423-2012/SI2.673791

Guettler, M.V., Jain, M.K., Rumler, D., 1996. Method for making succinic acid, bacterial variants for use in the process, and methods for obtaining variants. US patent 5,573,931

Ikram-ul, H., Ali, S., Qadeer, M. A., Iqbal, J., 2004. Citric acid production by selected mutants of *Aspergillus niger* from cane molasses. Bioresour. Technol. 93 (2), 125-130.

Jansen, M. L. A., van Gulik, W. M., 2014. Towards large scale fermentative production of succinic acid. Curr. Opin. Biotechnol. 30, 190-197.

Lee, P. C., Lee, W. G., Lee, S. Y., Chang, H. N., 1999. Effects of medium components on the growth of Anaerobiospirillum succiniciproducens and succinic acid production. Process Biochem. 35, 49-55.

Lee, P. C., Lee, W. G., Lee, S. Y., Chang, H. N., Chang, Y. K., 2000. Fermentative production of succinic acid from glucose and corn steep liquor by Anaerobiospirillum succiniciproducens. Biotechnol. Bioprocess Eng. 5, 379-381.

Lee, P. C., Lee, S. Y., Hong, S. H., Chang, H. N., Park, S. C., 2003. Biological conversion of wood hydrolysate to succinic acid by *Anaerobiospirillum succiniciproducens*. Biotechnol. Lett. 25, 111–114.

van Heerden, C. D., Nicol, W., 2013. Continuous and batch cultures of Escherichia coli KJ134 for succinic acid fermentation metabolic flux distributions and production characteristics. Microbial Cell Factories, 12(80).

Werpy, P., Petersen, G., 2004. Top value added chemicals from biomass: Volume I - Results of screening potential candidates from sugars and synthesis gas. US Department of Energy.

Zeikus, J. G., Jain, M. K., Elankovan, P., 1999. Biotechnology of succinic acid production and markets for derived industrial products. Appl. Microbiol. Biotechnol. 51, 545-552.

494849 Enhanced Alcohols, Ketones and Organic Acids Production Via Co-Fermentation of Sugars and Gases

Abstract:

Butanol is a better fuel than ethanol due to its compatibility with gasoline and existing fuel infrastructure. Butanol can also be converted into drop-in diesel and jet fuel by a simple hydrogenation step. Butanol has been produced biologically by the traditional acetone-butanol-ethanol (ABE) fermentation using molasses, starches and, recently, lignocellulosic biomass. The ABE process suffers from low conversion yields that hinder its commercialization. More than 50% of the carbon in sugars is wasted in producing H_2 and CO_2. Increased conversion efficiency of renewable raw materials to butanol could make this butanol production viable. The present study reports on a novel process for biological production of alcohols, organic acids and ketone. This process consists of a co-fermentation of sugars and gaseous substrates for ABE production in a single fermentation system while converting the off gases containing H_2 and CO_2 to produce ethanol and acetic acid. Depending on the microorganism and reactor configuration used, the novel process can enhance alcohol yield by more than 25%. Media containing pure sugar and redcedar hydrolyzate were used in this study. The use of two-stage reactors resulted in a 19% improvement in ABE yield from sugar by *Clostridium acetobutylicum* ATCC 824 in the first stage with 60 g/L glucose and the conversion of generated H_2 and CO_2 by *Clostridium ragsdalei* to ethanol and acetic acid in the second reactor. The efficiency of H_2 and CO_2 conversion by *C. ragsdalei* was 72% and 16%, respectively. The total organic acid production in the co-fermentation process in the two-stage reactors was also increased by 141%. In addition to reducing CO_2 emissions, the co-fermentation of sugars and gases has the potential to enhance the feasibility of ABE fermentation by producing more alcohols from an otherwise waste gas stream. The results demonstrate that co-fermentation of sugars and gases to various products is feasible, and can be exploited to enhance a bioprocessing economy.

502541 Attainable Region Analysis for Production of Butyl Butyrate Via Biochemical Route

Abstract:

EcoLodge aims at providing a proof-of-concept for a new, integrated biotechnological production process for C8 ester butyl butyrate, a promising supplement/substitute for diesel and jet engine fuels, from lignocellulose derived C5 and C6 sugars. Previous studies suggest that butyl butyrate has excellent properties both as gasoline and as a diesel component [1]. Butanol and butyric acid are the process intermediates produced via anaerobic fermentation, which undergo enzymatic esterification yielding the product, butyl butyrate.

The attainable region concept employs a computational/geometrical procedure for determining the boundaries of the region that include all possible reactor products for a known feed condition [2]. In this paper, we applied the attainable region concept to study different process configurations for biochemical production of butyl butyrate. These are i) butyl butyrate reactor with the feed from the butanol and butyric acid reactors, ii) combined butanol and butyl butyrate production with the feed from butyric acid reactor, and iii) combined butyric acid and butyl butyrate production with the feed from butanol reactor. We used mathematical models and model coefficients from the literature for cell growth, butyric acid and butanol fermentations, and their esterification for butyl butyrate production. Models include cell growth inhibition due to butanol and butyric acid, and enzyme inhibition due to butanol, butyric acid and glucose [3-5].

In this paper, we demonstrated the capabilities of the attainable region analysis as a tool to assess different process configurations for butyl butyrate production. Process configuration ii) showed the highest capability for butyl butyrate production, since the butyric acid enhances the butanol production, and butanol esterification decreases butanol inhibition on the cell growth. The results of this study will be employed in the experimental phase of the project.

1. Lange, J. P., Price, R., Ayoub, P. M., Louis, J., Petrus, L., Clarke, L., & Gosselink, H. (2010). Valeric biofuels: a platform of cellulosic transportation fuels. Angewandte Chemie International Edition, 49(26), 4479-4483.
2. Milne, D., Glasser, D., Hildebrandt, D., & Hausberger, B. (2004). Application of the attainable region concept to the oxidative dehydrogenation of 1-Butene in inert porous membrane reactors. Industrial & engineering chemistry research, 43(8), 1827-1831.
3. Yang, X., & Tsao, G. T. (1994). Mathematical modeling of inhibition kinetics in acetone-butanol fermentation by Clostridium acetobutylicum. Biotechnology progress, 10(5), 532-538.
4. Varma, M. N., & Madras, G. (2008). Kinetics of synthesis of butyl butyrate by esterification and transesterification in supercritical carbon dioxide. Journal of chemical technology and biotechnology, 83(8), 1135-1144.
5. Song, H., Eom, M. H., Lee, S., Lee, J., Cho, J. H., & Seung, D. (2010). Modeling of batch experimental kinetics and application to fed-batch fermentation of Clostridium tyrobutyricum for enhanced butyric acid production. Biochemical Engineering Journal, 53(1), 71-76.

500314 Biomass Conversion to Functionalized Carbon

Abstract:

Biomass Conversion to Functionalized Carbon

Masoudeh Ahmadi, Jacek Jasinski, Jagannadh Stayavolu, Mahendra Sunkara*

Conn Center for Renewable Energy Research, University of Louisville, Louisville KY 40292, United States

*Corresponding Author: mksunk01@exchange.louisville.edu

In this research emphasis is placed on the applied methodology to produce functionalized carbon from hemp agriculture by products. We have studied an effective, economic, and novel technology to transform lignocellulosic biomass into the functionalized carbon. The effects of various parameters on the preparation such as carbonization methodology and impregnation ratio were studied and their effects on the textural properties such as surface area, pore volume and crystalline structure were discussed. There has been a large amount of research and interest in the conversion of biomass to carbons for various applications. However, the challenges remain both in terms of processing time scale and also with the ability to tune the resulting carbonÕs properties. In addition, the functionalization schemes for using the resulting carbons toward catalyst schemes require additional development. The challenges in utilizing carbon supports for catalysts include uniform dispersion of active catalyst metals and metal oxide clusters and improved catalyst and support interactions.

In this study, Hemp fibers were mixed with KOH solution with different impregnation ratio. Then carbonization was done in a horizontal tube furnace under nitrogen flow at 700ûC for one hour. Textural and chemical properties for the samples were investigated by means of XRD, SEM, BET, FTIR and XPS. The textural properties of carbon were found to be strongly dependent on the carbonization method and impregnation ratio. Reduction technique was used to load Pt particles on to the resulting carbon structures using tetraammine platinum nitrate as Pt precursor. The carbon supports were first impregnated with tetraammine platinum nitrate to produce 5%wt Pt/C. The mixture was further dried overnight at 100 ¡C followed by reduction under flowing hydrogen at 450¡C for 2 h using tube furnace to remove the ligands of the precursor and to reduce the metal to its active elemental state. In this presentation, various mechanistic models will be presented on the carbonization of biomass using different means.

501921 Synthesis of Water-Soluble Oligosaccharides As Potential Prebiotics Via Non-Enzymatic Sugar Glycosylation

Abstract:

Bioactive oligosaccharides have found promising applications as functional food, prebiotic compounds, immunostimulating agents, and pharmaceuticals. Oligosaccharides synthesis prevalently relies on biological methods such as enzymatic transglycosylation and controlled hydrolysis of polysaccharides. In this study, we demonstrated a novel non-enzymatic glycosylation method to produce water-soluble oligosaccharides in acidic lithium bromide trihydrate from monomeric sugars, in which new glycosidic bonds were formed between the C1 oxocarbenium on a protonated sugar and the hydroxyl groups (C2, C3 or C6) of another sugar. The oligosaccharide products were analyzed using HPIC, MALDI-TOF/MS, and NMR. The results indicated that up to 70% of the glucose could be converted to gluco-oligosaccharides. The resultant gluco-oligosaccharides were linked predominantly through Î±-1,6-glycosidic bonds with small qantity ofÂ Î²-1,6-, Î±-1,2-, and Î±-1,3-glycosidic bonds. Preliminary evaluation indicated that the oligosaccharides could be metabolized by gut microbiota; therefore, the oligosaccharides have the potential as prebiotics.

498564 A General Data-Driven Method for Analysis of Energy Efficiency in Manufacturing Zones

Abstract:

Industrial energy use is about 30% of the total energy consumption in the U.S. In manufacturing plants, significant energy loss occurs in process systems, and energy generation, conversion, and distribution steps. Hence, how to improve energy efficiency has been continuously a main focus. Numerous energy consumption/GHG emission analysis methods have been developed for in-depth analysis for industrial systems. However, further methodological study is needed for systematic analysis of process and non-process energy consumption and loss. In this work, we introduce a simple, general data-driven analysis method to study energy consumption and loss in industrial zones that many have a variety of manufacturing sectors. The database from the US Census Bureau and US Energy Information Administration is used to perform the analysis of geographical regions. In the presentation, we will also provide analysis of the effectiveness of known strategies for energy efficiency improvement. A case study on the energy efficiency of the automotive-centered manufacturing sector, where various chemical supply companies exist, will be presented to demonstrate methodological attractiveness.

499757 Optimal Design of Multi-Enterprise Industrial Waste-to-Energy Networks Under Uncertain Conditions

Abstract:

Energy-intensive industrial activities, as many chemical processes, are facing increasing pressures to decrease their dependence on non-renewable energy sources. This represents a challenge but also an opportunity to explore alternative and more reliable energy sources that were previously considered of secondary importance, such as industrial waste incineration. In fact, chemical and pharmaceutical industrial sectors generate significant amounts of high-calorific liquid waste residues that are often incinerated for simultaneous treatment and energy recovery. However, in case of energy deficits due to low energy content of residuals, additional fuels, such as natural gas, are required to sustain the combustion process. Recent studies have shown that a systematic management of waste solvent incineration in integrated chemical sites can lead to an important decrease of the siteâ€™s primary energy consumption, reducing both utility costs and environmental impact (Abaecherli et al., 2016; Wassick, 2009). Nonetheless, potential opportunities arising from a coordinated multi-site management have only been partially investigated, especially when considering their contribution to a more sustainable energy system. This is of particular interest for countries where the chemical and pharmaceutical industry consists in several smaller enterprises with a centralized treatment capacity localized in few independent incineration sites, and a resulting waste-to-energy network that can be considered as a multi-enterprise supply chain. In fact, such small businesses often have neither the resources nor the necessary waste amounts to operate own incineration facilities and have then to ship the generated waste to third companies for treatment. These are usually the aforementioned chemical sites with incineration facilities, with the addition of some cement plants that can also use some residues as substitution fuel. Hence, waste shipments play an important role concerning costs and energy consumption, in particular when the supply chain network consists of sparse enterprises.

As typical industrial waste-to-energy networks are made of a large number of enterprises, the generation of hazardous waste can vary significantly in time, both in terms of amounts and composition. This is especially true in presence of multi-purpose plants, where type and quantity of residuals completely depend on ongoing production campaigns. Such variation also affects management and operation of incineration facilities, as it is common practice to use combustible waste solvents to sustain the incineration of non-combustible residues, in order to minimize the consumption of auxiliary fuels, such as natural gas. Although former studies have already considered the simultaneous optimization of process operations and financial decisions in the case of chemical supply chains (Guillén et al., 2006), possible decisions are usually constrained to allocating and sequencing tasks to different process units. In this sense, a formulation combining management of waste incineration sites and supply chain logistics with a more detailed description of process conditions, such as upper and lower temperature bounds, has only been recently proposed (Bolis et al., 2016). Nevertheless, such methodology still lacks uncertainty tools to tackle variations in waste amounts and composition, which can critically affect both shipments and incineration processes.

This study proposes an optimization framework for the design of multi-enterprise industrial waste-to-energy networks, focusing on long-term investment strategies and synergies between multiple incineration sites and cement plants. The considered system is formulated as a mixed-integer linear problem (MILP) with discrete time representation and consists of: i) a set of incineration and cement plants with site-specific features; ii) a set of various types of hazardous liquid waste with different physicochemical properties; as well as, iii) a transportation network connecting waste producers and treatment facilities. The overall objective is to optimize the costs of the whole network by managing: i) the planning of waste shipments between enterprises; ii) the expansion of both incineration and temporary storage capacities; as well as iii) the management of incineration sites, comprising temporary storage, waste mixing and combustion processes. In order to obtain robust solutions to the problem, a combination between a scenario planning approach (Tsiakis et al., 2001) and random perturbations is implemented in the model to address parameter uncertainty. In practice, a set of scenarios is first defined for covering a broad range of plausible future megatrends concerning energy systems and the chemical industry. Whereas energy-related parameters define gas, electricity and shipment costs, together with the share of rail and road transportation, future levels for quantity and composition of the hazardous waste can be roughly estimated as a function of possible changes in the chemical and process industry, exploring highly different conditions from business-as-usual to strong focus in pharmaceutical and life-science branches. Concerning very long optimization horizons, investment are probably the most important decisions, since they affect the system for a long period, while limited adjustments are still possible for shipments in case of unforeseen events. For this reason, it is imposed that investment variables have to be equal for all the scenarios, whereas all other variables can assume different values in each scenario. In practice, this can be achieved by inserting an additional set for the scenarios and solving the problem simultaneously for all of them, with common investment decisions. Although it is in principle possible to insert weights related to the probability of each scenario to occur, in this case no clear priority could be defined because of the length of the considered horizon and the consequent high uncertain conditions. Hence, the new objective function has simply been defined as the sum of the single scenariosâ €™ objectives. Next, random perturbations concerning waste amounts and composition are used to represent unforeseen events and the varying waste generation typical of multi-purpose plants. By comparing the results obtained under several random perturbations, it is possible to further investigated the robustness of the solution, which is not anymore limited to the given scenarios.

The methodology has been applied to the whole Swiss industrial waste-to-energy network, which comprises 5 independent incineration sites, 6 cement plants and several sparse waste-generating companies, grouped in 7 regions. A set of five different scenarios describing a broad range of plausible future conditions, divided in two consequent time horizons of 15 years, starting in years 2020 and 2035, has been investigated with the discussed uncertainty tools. Moreover, the influence of future uncertainty on the optimization results has been studied by applying three simulation approaches. First, a long-term strategy has been obtained with a single optimization over the whole horizon, including the two time points. Secondly, weights have been introduced for each time horizon, to consider the higher importance of near future in present decision-making. Finally, a stepwise procedure, solving the problem for the three horizons in series, has been applied to define a time priority for investments. The comparison of the different approaches indicates a significant difference in the timing of investments, and provides a complete overview for the design of industrial waste-to-energy networks.

References

M.L. Abaecherli, D. Santos González, E. Capón-García, K. Hungerbühler, 2016, Mathematical Optimization of Real-time Waste Incineration Scheduling in the Industry, 26th European Symposium on Computer Aided Process Engineering

V. Bolis, E. Capón-García, K. Hungerbühler, 2016, Optimal Design of Industrial Waste-to-Energy Networks, 26th European Symposium on Computer Aided Process Engineering

G. Guillén, M. Badell, A. Espuña and L. Puigjaner, 2006, Simultaneous optimization of process operations and financial decisions to enhance the integrated planning/scheduling of chemical supply chain, Computers & Chemical Engineering, 30, 421-436

P. Tsiakis, N. Shah and C. C. Pantelides, 2001, Design of Multi-echelon Supply Chain Networks under Demand Uncertainty, Industrial & Engineering Chemistry Research, 40, 3585â€“3604

J.M. Wassick, 2009, Enterprise-wide optimization in an integrated chemical complex, Computers & Chemical Engineering, 33, 1950-1963

501480 Survey-Driven Sustainability Assessment Using Global Optimization Techniques

Abstract:

In this work, we pursue an extension of our previously-developed Sustainability Index Interval (SII) method [1] that directly incorporates the opinions of the public into the scheme.

A case study involving the sustainability assessment and ranking of leading aluminum-producing enterprises will be pursued to demonstrate the proposed method. Appropriate sustainability assessment metrics and basic indicators (e.g. greenhouse gas emissions) are selected from Phillis et al. [2] and our previous work [1]. Values (or ranges of values) for each of the metrics are obtained for all enterprises via publicly-available documentation. Once this data is obtained, surveys listing this data are compiled and then presented to three groups of people: the general public, university students, and university professors. Each group's opinion is sought on each enterprise's sustainability efforts regarding the aforementioned metrics. The results of these surveys are then incorporated into an SII for each enterprise. Subsequently, recommendations for each enterprise are developed on areas in which sustainability improvements will yield the greatest SII improvement.

References:

[1] Conner, J.A., Phillis, Y.A., Manousiouthakis, V.I. (2012), "On a Sustainability Interval Index and Its Computation Through Global Optimization", AIChE Journal, 58(9), 2743-2757.

[2] Phillis, Y.A, Davis, B.J. (2009), "Assessment of Corporate Sustainability via Fuzzy Logic", J Intell Robot Syst., 55 3-20.

500526 On the Likely Competition for Land between Food and Energy in a Solar Economy

Abstract:

The main purpose of this presentation is to demonstrate that contrary to the popular belief that ample land area is available to install photovoltaic modules, in an economy mainly supported by solar energy, there will be severe land availability constraints in many regions of the world—including many states in the USA. This will inevitably lead to competition for land between food and energy [1]. Our analysis will highlight the urgency with which solutions must be sought to avoid competition between these basic needs of humanity to enable an earth with more than 10 billion people in a few decades.

The vast potential for solar energy cannot be disputed. However, the land area requirements often stated to meet our energy needs from solar energy are often much lower than will actually be required in practice. A recent study by MacKay showed that the energy consumption per unit area in densely populated countries such as the UK, Germany, Republic of Korea, and Japan is approaching the energy generation per unit area of photovoltaic farms [2]. This suggests that a transition to a solar powered economy in these nations would require deployment of solar on a large fraction of the nation's total land area, which will directly compete with land requirements for other needs (e.g. agriculture). Many other countries, including the United States are also trending toward this scenario [2]. A recent study by National Renewable Energy Laboratories confirms the large land area requirements of solar farms [3].

In this study, we estimate the United States' land area requirements for an economy fully powered by solar energy by using actual production data of PV farms, which have an average production throughout the year of 4-11 W/m^2. Using available data [4] we have calculated the energy use per square meter for the U.S. (excluding Alaska and Hawaii) to be 0.43 W/m^2. We can thus estimate that the actual land requirement will likely be 1.9-5.2% of the U.S. land area, when reasonable allowances are made for nighttime storage losses. (Note that in performing these calculations we have accounted for the fact that a kWh of coal does not have the same intrinsic value as a kWh of electricity and have appropriately converted between various energy sources and end use sectors based on average conversion efficiencies of each energy transformation.) Unfortunately, much of the available land needed for deployment of PV is often located great distances from the population centers where the energy is needed. When the same analysis is performed on a state by state basis we find that the land area needed to meet individual states' needs ranges from 0.2-0.5% (Montana) to 16.4-45.0% (New Jersey). While a solar economy is feasible for many of the large, sparsely populated states of the Plains States and Mountain West, a transition to a solar economy will be challenging for many other states.

A remaining important question is which of these states have sufficient available land not dedicated to other uses to meet these energy needs. *We answer this using data from the United States Department of Agriculture on current land usage and estimate that as many as 31 states would be unable to internally meet their energy needs given the available land in those states not already dedicated to other uses.* This calculation is based on what is categorized as "miscellaneous land" in each state—that land not currently devoted to cropland, pasture/range land, forest use, urban areas, and special uses (parks and wildlife areas). Note that miscellaneous land includes deserts, swamps, bare rock areas, rural residential, etc.—some of which will be unsuitable for PV, so land constraints may be more severe than estimated. The reason for the lack of available space is that much of the open land in the U.S. is devoted to cropland (21.5%), pasture land (32.3%), and forest land (30.4%). Only 3.6% of the land falls in the category of miscellaneous/other use land [5]. While some land currently devoted to other purposes (e.g. rooftops in urban areas) could be used for PV as well, many states would have to rely on long distance transmission to meet their energy needs under current strategies. Rooftop PV also has inherently higher cost compared with larger scale PV installations.

Assuming it will remain a priority to preserve the wildlife habitat embodied by forests and other special uses and in the absence of radical improvements in agricultural output that allow us to greatly reduce the percentage of land devoted to agriculture, innovation will be required to find ways of sharing PV with current land uses as the available space is filled up. One way this could be accomplished is by incorporating solar energy harvesting with agricultural activities, as has been done successfully with wind farms in many areas. While this seems counterintuitive, PV systems can be designed that allow large fractions of the visible portion of the solar spectrum needed for photosynthesis to pass through, while capturing other photons for electricity or heat. We will briefly describe some of possible configurations and material research needs to accomplish these objectives.

References

[1] Solar Projects Sow Tension, Wall Street Journal, (March 9, 2017) A3.

[2] D.J.C. MacKay, Solar energy in the context of energy use, energy transportation and energy storage., Philos. Trans. A. Math. Phys. Eng. Sci. 371 (2013) 20110431. doi:10.1098/rsta.2011.0431.

[3] S. Ong, C. Campbell, P. Denholm, R. Margolis, G. Heath, Land-Use Requirements for Solar Power Plants in the United States - Technical Report NREL/TP-6A20-56290, 2013.

[4] State Profiles and Energy Estimates, (n.d.). http://www.eia.gov/state/.

[5] Major Land Uses, (n.d.). https://www.ers.usda.gov/data-products/major-land-uses/.

488856 Energy Efficient Design of Ionic Liquid Based Gas Separation Processes

Abstract:

Gas separation processes have been one of the most important operations in the oil and gas related industries, where gases of interest are present in significant amounts to justify their separation for use as raw gas in chemical production utilization. Currently, the gas separation technologies include energy intensive distillation and solvent based absorption operations or flux limited adsorption and membrane-based operations. For most gas mixtures, however, the most common separation technology applied is distillation, which consumes large amounts of energy to give the high purity products. These distillation columns operate at low temperatures and high pressures and therefore require high energy consumption, leading to negative environmental impacts. In certain regions of their operation, all these technologies work very efficiently. Therefore, an alternative scheme, taking advantage of the regions where each individual technology operate best, a hybrid gas separation scheme of combining distillation with absorption separation processes is considered. In this novel gas separation process design strategy, an absorber is employed followed by flash-evaporation operations with a low energy consumption distillation operation used in a pre-separation step only to reduce flux (if necessary). Since no energy is required in an absorber operation, this hybrid gas separation process strategy can significantly reduce the energy consumption, where distillation is only employed for a partial separation where significantly less energy is required.

Because of non-volatility, good stability, tunable viscosity and designable properties, ionic liquids (ILs) are considered as novel potential solvents and alternative media for gas absorption. Therefore, a strategy for hybrid gas separation process synthesis where distillation and IL-based absorption are employed for energy efficient gas processing has been developed. However, the potentially thousands of ILs that may be applicable, makes it a challenging task to search for the best one for specific gas absorptions in different raw gas systems. Therefore a selection-screening method for ILs is also necessary for the development of novel hybrid gas separation process can be designed.

In this presentation, a three-stage methodology proposed for hybrid gas separation process design and evaluation will be highlighted. The first stage involves IL screening, where a systematic screening method together with a database tool is established to identify suitable ILs based on a collection of gas solubility data, Henryâ€™s constant data as well as data estimated through reliable predictive models (for example, COSMOS-RS). The second stage is process design, where the important design issues (amount of solvent needed, operating temperatures and pressure, evaporation conditions, etc.) are determined. A hybrid gas separation scheme is designed to replace the conventional distillation process. Since the only energy requiring step in the hybrid process is the flash-evaporation step (and the low energy consuming pre-distillation step, if employed), potentially a large reduction of energy consumption is possible by switching from distillation to the hybrid-absorption scheme the selective gas separation tasks. For example, replace distillation by absorption to remove only the gases present in smaller amounts in the gas mixture, thereby letting the larger amounts free to go out as the exit (raffinate) gas. This small amount absorbed gas is then easily separated through evaporation or distillation, which only consumes a small fraction of the total energy of the conventional

distillation based process. The third stage involves verification and sustainability analysis based on rigorous process simulation of the generated hybrid gas separation process strategy. The gas separation problem for a model shale gas mixture is selected as a case study to highlight the application of this hybrid separation process design method.

The presentation will highlight the method, the data and applications of the method to generate novel, innovative and sustainable gas separation processes requiring significantly less energy than known processes.

494975 Maximising the Mitigation Potential of Curtailed Wind: A Comparison between Carbon Capture and Utilisation, and Direct Air Capture Processes for the UK

Abstract:

Carbon capture and storage (CCS) with fossil fuel or biomass plants (BECCS) is considered a critical technology to meet mitigation targets set by the Paris Agreement[1]. However, several drawbacks including high upfront investment costs, significant energy penalty and long-term permanent storage challenges have limited the uptake of CCS on the required scale. Carbon capture and utilisation (CCU) provides an alternative route to recycle CO_2 into chemical feedstock and/or synthetic transport fuels (e.g. methanol, DME) that can displace fossil-derived fuels. As the carbon is only *transformed*, CCU must be integrated with capture/storage to actually offset subsequent emissions from the vehicles consuming them. The mitigation of decentralised emissions poses significant challenges and necessitates the use of carbon dioxide removal technologies (CDR), one of which is direct capture of CO_2 from the atmosphere (DAC).

The last decade has seen increasing penetration of wind power in the UK electricity system to meet mitigation targets. Because of this, periods of surplus wind generation and low demand or limited/full storage capacity arise. Constraint payments then have to be made to wind farms to curtail generation. This work investigates two possible options to achieve mitigation with this curtailed electricity. In **Process A**, curtailed electricity is used to run electrolysis and methanol synthesis plants. It is then integrated with a direct air capture (DAC) plant to recapture and recycle emissions from the vehicles. **Process B** assumes curtailed electricity is used to run a DAC plant directly in order to capture decentralised carbon emissions and provide CO_2 feedstock for CCU processes.

The UK was used as a case study and the methanol synthesis process described by Rihko-Struckmann et al.[2] was used as the reference. A range of energy requirements for DAC are cited in literature; the lower and upper bounds of 6.7 GJ/tCO_2 and 12.6 GJ/tCO_2[3], respectively, were used. This work has taken a base case curtailment level of 2.5% of the UK total electricity demand, which is equivalent to 390 GWh/y[4]. Both processes have been compared on the basis of mitigation potential, defined by the proportion of CO_2 emissions from gasoline vehicles that are avoided, and mitigation costs per tonne of CO_2 captured.

Process A resulted in avoiding 0.12% of gasoline emissions (~0.05 MtCO_2/y). Surplus energy (~64% of the curtailed electricity) was required to run the DAC plant and an associated air separation unit. The mitigation of potential of Process B was 0.10% or 0.18%, depending on energy requirement used. Therefore, the process that maximises mitigation potential depends on the DAC process considered; using the lower-bound energy requirement, surplus electricity for DAC only is preferable. Neither process is economically viable. CCU costs ($905/t$CO_2$) were found to

be double the DAC-only costs ($449/t$CO_2$), mainly due to high H_2 costs. It will remain financially-unattractive unless the methanol production becomes profitable. This is unlikely as it requires methanol price to almost double, a carbon price of $313/t to be in effect, or H_2 price to reduce to a third of todayâ€™s price to $1800/t.

References

1. IPCC. *Climate Change 2014: Mitigation of Climate Change. Working Group III Contribution to the Fifth Assessment Report of the Intergovernmental Panel on Climate Change* (2014). doi:10.1017/CBO9781107415416

2. Rihko-Struckmann, L. K., Peschel, A., Hanke-Rauschenbach, R. & Sundmacher, K. Assessment of methanol synthesis utilizing exhaust CO2 for chemical storage of electrical energy. *Ind. Eng. Chem. Res.* **49,** 11073â€“11078 (2010).

3. Socolow, R. *et al.* Direct Air Capture of CO 2 with Chemicals Panel on Public Affairs. *Am. Phys. Soc. - Panel Public Aff.* 100 (2011).

4. Messiou, A. Centre for Environmental Policy Investigating the role of power storage in accommodating the future wind. (2012).

498759 Thermodynamic Model-Based Synthesis Methodology to Design Optimal Heat-Integrated Work Exchanger Network

Abstract:

Heat integration has been widely and successfully practiced for recovering thermal energy, mostly through synthesizing heat exchanger networks (HENs) in the process industry over the past three decades. It was recognized another form of energy, mechanical energy, should be also recovered systematically. In 1996, Huang and Fan (1996) introduced a notion of work integration and a new type of process synthesis called work exchanger network (WEN) synthesis. Over the past few years, various process analysis and design methodologies have been proposed to develop WEN systems, some of which also involve thermal energy recovery (Liu et al., 2014; Onishi et al., 2014; Razib et al., 2012; Fu and Gundersen, 2016; Huang and Karimi, 2016). Recently, Amini-Rankouhi and Huang (2017) presented a mathematical modeling and analysis method to predict the maximum amount of mechanical energy that can be feasibly recovered using direct work exchangers prior to WEN configuration development. In this work, we will introduce a thermodynamic model-based synthesis approach to develop a heat-integrated work exchanger network (HIWEN), in which direct work exchangers may work under different operating conditions. Case studies will demonstrate that the resulting HIWENs can recover the maximum amount of mechanical and thermal energy simultaneously at the lowest cost. In this presentation, network implementation and operational challenges will be also discussed.

References:

Amini-Rankouhi, A. and Y. Huang," Prediction of Maximum Recoverable Mechanical Energy via Work Integration: A Thermodynamic Modeling and Analysis Approach," submitted, *AIChE J.*, 2017.

Fu C, Gundersen T. Heat and work integration: Fundamental insights and applications to carbon dioxide capture processes. *Energy Convers Manag*. 2016; 121:36â€“48.

Huang K, Karimi I. Work-heat exchanger network synthesis (WHENS). *Energy*. 2016; 113:1006-1017.

Huang YL, Fan LT. Analysis of a work exchanger network. *Ind Eng Chem Res*. 1996; 35:3528-3538.

Liu GL, Zhou H, Shen RJ, Feng X. A graphical method for integrating work exchange network. *Appl Energy*. 2014; 114:588-599.

Onishi VC, Ravagnani M, Caballero JA. Simultaneous synthesis of work exchange networks with heat integration. *Chem Eng Sci*. 2014; 112:87-107.

Razib MS, Hasan MMF, Karimi IA. Preliminary synthesis of work exchange networks. *Comput Chem Eng*. 2012; 37:262-277.

494713 Optimal Use of LNG Cold Energy in Air Separation Units

Abstract:

Liquefied natural gas (LNG) is a major source of natural gas supply for long-distance transport. After a long transportation, the LNG is vaporized in a receiving terminal in order to supply natural gas at desired conditions through domestic pipeline networks. During regasification, the cold energy of the LNG (ca. -163 ^{o}C) is just released to the environment by heat exchange with seawater or air, wasting a large amount of exergy. Even a part of the boil-off gas, produced during unloading and storing LNG, is burnt to vaporize LNG [1]. Therefore, there have been various attempts to recover such cold exergy in the import terminal, for example, power generation by the cold Rankine cycle, desalination of seawater using LNG cold energy, food processing (freezing food using the cold LNG as refrigerant) and air separation units (ASU) integrated with the cold LNG stream [2]. Due to the low operating temperature of air separation units (from -170 ^{o}C to -190 ^{o}C), which is closer to the LNG temperature than other options, supplying parts of the cold duty of an ASU is regarded as a promising alternative for utilizing the cold energy of LNG. This integration has already been deployed in several LNG import terminals [3]. Unlike ASUs, other applications have higher operating temperatures and they cover smaller temperature ranges, which will result in the sub-optimal use of LNG cold energy.

However, air separation has complex process schemes with a single column, double column, or even triple column distillation, including sophisticated internal heat integration. This increases the complexity of design in LNG regasification systems when integrated with an ASU. Moreover, the structure and performance of the ASU process integrated with LNG regasification will vary depending on the way LNG cold energy is used. Thus, there have been several suggestions for integration of LNG regasification processes with different ASU designs, showing distinctive characteristics. Therefore, these integration solutions were simulated and assessed to understand their strengths and weaknesses in depth under fair conditions.

In this work, ASU processes integrated with LNG regasification are categorized based on the type of ASU system and the method for utilization of LNG cold energy. A comparison between the processes is conducted by considering exergy as a performance measure. The fact that the each system brings different products makes exergy efficiency a reasonable objective function [4]. In addition, a novel method of integrating one column ASU with LNG regasification is developed for optimal utilization of LNG cold energy, based on the result of the exergy analysis. A discussion is also made on the optimal use of LNG cold energy, depending on the type of ASU.

[1] Patel, D., Mak, J., Rivera, D. & Angtuaco, J. 2013. LNG vaporizer selection based on site ambient conditions. The 17th International Conference & Exhibition on Liquefied Natural Gas. Houston.
[2] Kim, H. & Hong, S. 2006. Review on economical efficiency of lng cold energy use in South Korea. 23rd World Gas Conference. Amsterdam.
[3] Xu, W., Duan, J. & Mao, W. 2014. Process study and exergy analysis of a novel air separation process cooled by LNG cold energy. Journal of Thermal Science, 23(1), 77-84.
[4] Marmolejo Correa, D. & Gundersen, T. 2015. A new efficiency parameter for exergy analysis in low temperature

processes. International Journal of Exergy, 17(2), 135-170.

494356 IGCC Process Intensification with the Reforming Technologies for Enhanced Power Generation with CCS Technology

Abstract:

IGCC (Integrated gasification and combined cycle) is a pre-combustion technique which represents higher thermal efficiency with large scale implementation of CO_2 capture. On the other hand, NGCC (Natural gas combined cycle) power plants offers higher process performance compared to IGCC process with CCS technologies but the fluctuating and comparatively higher costs of natural gas limits its extensive implementation over coal based power generation systems. Therefore, process intensification of coal based gasification processes with the natural gas reforming technology has been proposed in this study to enhance the overall process performance while limiting the natural gas consumption. In this study, two IGCC based process models have been developed and evaluated in terms of both the process performance and economics with CO_2 capture. Case 1 is taken as the conventional IGCC process, whereas, case 2 presents an idea of integrating methane reforming process with an IGCC technology. Coal is used as primary fuel in both the cases for the coal gasification process, however, small amount of natural gas is used as secondary fuel only in the case 2 design. The high enthalpy steam generated during coal slurry gasification process in case 2 is used to assist the methane reforming to generate additional H_2. The integration of IGCC with methane reforming process not only supplies the heat required for the endothermic reforming process but also increases the heating value of the resulting syngas. Moreover, the H_2O generated in the gasification process in case 2 can be consequently utilized in the natural gas reforming process without utilizing any extra steam. This concept also provides an opportunity for the process economics since shared water gas shift reactors and CO_2 capture units will suffice the process needs. The net electricity generation capacity and efficiency for case 1 and case 2 is calculated as (375.08 MW_e, 472.92 MW_e) and (35.93%, 40.70%), respectively. While comparing the results, it has been seen that case 2 design offers nearly 4.77% higher efficiency compared case 1 design with CO_2 capture. The process economics analysis showed that the case 2 design requires a higher CAPEX and OPEX throughout the project life compared to the case 1 design. However, due to the higher power generation capacity of case 2 design, it has a potential to reduce the LCOE (levelized cost of electricity) by 5.81% compared to the case 1 design. Moreover, case 2 design not only requires the less CO_2 capture and avoidance costs compared to the case 1 but also exhibits higher rate of return on the investment with the less payback time. With the higher efficiency and least case 2 has been considered as the most feasible option for electricity production at an economical price compared to the case 1 design.

500239 Vegetable Oil Process Intensification with Sterically Hindered Alcohols to Biofuels and Biochemicals

Abstract:

Vegetable oils, refined to different degrees, are the raw material to manufacture biodiesel. Commercial batch times can exceed 8 h due to poor mass transfer and miscibility between oil and methanol. Moreover, free fatty acids methyl esters (FAME) have a cloud point (CP) and pour point (PP) ranging from −3 °C to 12 °C and −15 °C to 10 °C, respectively [1], i.e. poor cold flow properties unsuitable for most climates unless blended with petro-diesel. Substituting methanol with branched alcohols to transesterify or esterify vegetable oil matrices improves the cold flow properties of the fatty alkyl esters. For example, the CP and PP of isopropyl esters range from −14 °C to 0 °C and −20 °C to −5 °C, respectively [1]. We propose a fluidized catalytic bed and ultrasound-based processes to either transesterify vegetable oils with alcohols other than methanol or esterify free fatty acids with a trimethylolppropane to produce biolubricants.

A fluidized catalytic bed concurrently transesterified and cracked canola oil with iso-propanol (iPrOH) to fatty acid isopropyl esters in the vapour phase. We alternated the injection of vegetable oil + iPrOH (reaction stage) and air (regeneration stage) to remove the coke from the CaO/Al_2O_3 catalyst. The reaction stages and regeneration stages lasted 2 min. We varied the iPrOH:oil ratio from 35 to 51 (mol:mol) and the air flow for during the regeneration step. The coke yield ranged from 3.5% to 34% and the isopropyl esters yield from 2% to 31%. The highest hydrocarbons yield was 42% (GC analysis). Fluidized beds are excellent candidates for biodiesel process intensification because as little as 10 t of catalyst may produce 80 kt/y of biodiesel, which corresponds to a reactor size at least 10X lower than the current commercial processes [2-4].

We propose a new ultrasonic-assisted mixing device to transesterify vegetable and waste cooking oil with ethanol and iso-propanol in a continuous process. We adapted a 13 mm diameter tip ultrasound probe to a media mill, where a rotating basket circulated the oil: alcohol mixture at 1200 rpm in the vicinity of the tip [5]. The alcohol:oil ratio was stoichiometric and we adopted KOH as basic catalyst. Ultrasound operated at 20 KHz and 500 W nominal power and with pulses 1 s on and 1 s off. After 1 min reaction, 40 %, 55 % and 90 % oil converted to isopropyl, ethyl and methyl esters, respectively. In absence of ultrasound, the same conversions were reached after 2 hours. Ultrasound creates a very fine oil: alcohol emulsion that increases the contact among the reagents and the catalyst.

Vegetable oils as biolubricants suffer from poor oxidative stability because of the hydrogen in b position to the ester groups, which have the tendency to give b-elimination [6]. Trimethylolpropane (TMP) esterify free fatty acids to biodegradable and stable tri-esters whose application is as biolubricants. We apply ultrasound to esterify oleic acid with TMP with heterogeneous catalysts including Al_2O_3-supported $H_3PW_{12}O_{40}$ and sulphated ZrO_2 systems. We vary the power amplitude (20 to 60 % of 500 W) and adopt different ultrasound pulses lengths. Sulphated zirconia systems convert over 80 % of substrate in 2 hours, while Al_2O_3-supported $H_3PW_{12}O_{40}$ is mostly inactive.

References

[1] Joshi R.M., Pegg M.J., Flow properties of biodiesel fuel blends at low temperatures, *Fuel*, 2007, 86:143–151

[2] Boffito D.C., Galli F., Pirola C., Patience G.S., CaO and isopropanol transesterify and crack triglycerides to isopropyl esters and green diesel, *Energ. Conv. Mangement.*, 2017, 139, 71-78

[3] Boffito D.C., Neagoe C., Edake M., Pastor-Ramirez B., Patience G.S., Biofuel synthesis in a capillary fluidized bed, *Cat. Today*, 2014, 237, 13–17

[4] Boffito D.C., Blanco-Manrique G., Patience G.S., One step cracking/transesterification of vegetable oil: Reaction–regeneration cycles in a capillary fluidized bed, *Energ. Conv. Manage.*, 2015, 103, 958–96

[5] Boffito D.C., Galli F., Martinez P.R., Pirola C., Bianchi C.L., Patience G.S., Transesterification of triglycerides in a new ultrasonic-assisted mixing device, *Chem. Eng. Trans.*, 2015, 43, 427-432.

[6] Zaccheria F., Mariani M., Psaro R., Bondioli P., Ravasio N., Environmentally friendly lubricants through a zero waste process, Appl. Catal. B., 2016, 181, 581–586

487349 Targeting Maximum Energy and Water Efficiencies for the Sustainable Total Textile Waste Refinery

Abstract:

The global textile industry is among the top ten manufacturing economic activities attaining a market value of €0.82 trillion. In 2008, 24 million tonnes of cellulosic and 39 million tonnes of synthetic fibres were sold globally (Oerlikon report, 2009). Despite industry's great economic value, the textile industry is among the most polluting ones. To only cultivate 1 kg of cotton; 16 g of pesticides, 460 g of fertilizer and 22 $m^3$of water is consumed (Saleem et al., 2010). In EU alone, 10 million tonnes of textile waste are produced each year (EASME, 2015). Specifically, in central Europe, Nordic countries, USA and Japan, the total post-consumer textile waste going to landfill is estimated 0.6, 1.9, 0.6, 0.1, 0.4, 0.9 million tonnes respectively. The present work coordinates established and original Process Integration technologies to the solution of a new problem; the design of the sustainable Total Textile Waste Refinery (TTWR); a new biorefinery type, that it is being currently developed in the course of the European Horizon2020 research project (RESYNTEX, 2015). It uses the complete spectrum of waste textiles as feedstock, aiming to provide a holistic response to the largely overlooked waste textile management problem. TTWR is now just before the phase of pilot construction and process integration technologies are currently implemented to accelerate the design solution.

TTWR consists of gate links between textile materials with commodity and specialty products and chemicals. Feedstock includes the entire range of textile fibres, consisting mostly of cotton, polyethylene terephthalate (PET), polyamides (PAs), wool but also other material in smaller quantities like acrylics and elastane. Glucose, amino-acid solutions, and PET and PA monomers and oligomers are the gate chemicals produced by the decomposition of the feedstock before the process eventually produces bio-ethanol, adhesives used in the wood-based panel production, PET bottles and PA-derived value added chemicals (caprolactam, amylamine, hexanoic acid etc.). The decomposition of natural and synthetic textile fibres is realised using patented chemistries with emphasis on chemical and enzymatic hydrolysis. The overall process consists of 20 sections dedicated to pretreatment for the reduction of the feedstock size, discolorations for the removal of dyes and impurities, depolymerizations for the decomposition of material and the production of gate chemicals, separations and purification of gate chemicals, reactions and separations for the production of end products and finally solid and liquid waste treatment processes. At particular stages (discolorations, separations and purifications), the TTWR process can be considered as the reverse process of textile production (pretreatment, dying, polishing etc.) and therefore, it has similar characteristics regarding water consumption. Without any water integration attempts the water consumption is as high as 52 tonnes per tonne of feedstock, and it is shared among reaction and separation stages (mostly washing and filtration steps) at a 2:1 ratio. Contrary to the textile fabrication processes which uses an average of 30 MJ/kg of fabric (Kalliala and Nousianinen, 1999), TTWR is more energy intensive, requiring more than 53 MJ/kg of feedstock. The majority of the energy requirements (~40%) is used in reactions, which are mainly neutral thermochemical hydrolyses at high temperatures for PAs and PET, and biochemical hydrolyses at mild temperatures. Considering that water is used at high ratios compared to the feedstock material in all reactions, it is also the main energy carrier in the reactors. This evidence indicates clear scope for a combined energy and water integration study, since water is used both as a cleaning agent for removing dyes and impurities as well as the main heating medium.

The TTWR process design procedure benefits from the rich technical know-how that has been accumulated so far for the design of conventional chemical processes. The present work takes advantage and improves on the significant contributions in process synthesis and integration (Mountraki et al., 2011a, 2011b; Tsakalova et al., 2012) made in the course of the European research project BIOCORE (Biocore, 2014) dedicated to the evolution of 2ndgeneration biorefineries.

Novel process integration tools have been developed for assessing alternative process integration scenarios and achieving maximum efficiencies in energy and water use in modern biorefineries (Koufolioulios et al., 2014; Nikolakopoulos and Kokossis, 2016). Also, total waste management and recycle design technology has been produced to enable the development of designs for processes with minimum environmental impact (Mountraki et al., 2016; Nikolakopoulos et al., 2016a).

Water and energy Integration is applied with a combined use of process modelling targeting, process synthesis and mathematical optimization. Targeting sets the scope to match and improve efficiencies for energy and water (Pinch, Water Pinch), explaining efficiency bottlenecks, modifications required to improve targets, means to re-use and regenerate water and solvents, as well as means and technologies appropriate to treat waste.

A new tool is proposed In this paper for simultaneously targeting maximum energy and water efficiencies in the TTWR. It is developed by integrating a mathematical programming heat cascade sub-model for targeting and optimally allocating heating and cooling utilities (Koufolioulios et al., 2014) with a two-step iterative procedure to coordinate the targeting of minimum fresh water requirements, the calculation of recycle and treatment flows and the selection of water treatment processes (Nikolakopoulos et al., 2016b).

For the first model, combinations of utilities are built by matching pseudo-cooling utilities with steam levels so as to permit alternation of the purpose of the utility from cooling to heating and thus exploit excess of heat. This actually defines a dual function for a subset of cold utilities. The method calculates for each combination of utilities the minimum utility requirements through an expanded transhipment model of (Papoulias and Grossmann, 1983). The model incorporates the utilities as process streams with the objective of minimizing total utility cost. Heat receivers (i.e. cool streams matched against steam at the same level) are being maximized by being assigned a negative sign. Hot and cold utilities at the same level are assigned different costs and prices respectively. The model calculates the utility combination that minimizes the annualized total cost and maximizes profit.

The second model, consists of two interlinked transhipment models, where successively increasing recycle flows are introduced until convergence to the optimal recycle flow value based on a total cost criterion. The first is a transhipment model used to identify the flowrates and concentrations of the wastewater streams at the effluent water mains. The second model, originally intended for targeting treatment flowrates (Nikolakopoulos et al., 2014) is extended in screening treatment technologies and suggesting processes to integrate. The models demonstrated cascade characteristics by allowing the transfer of contaminant between concentrations intervals.

The application of the proposed tool in the design of the TTWR is able to produce significant energy and water saving targets, that may reach up to 50% for energy and 70% for water. An additional advantage of the procedure is the possibility to detect critical parameters of the system, where retrofitting changes improve on total cost. These are the number and order of washing and filtering stages and their concurrent or countercurrent mode of operation. The results of total water network suggest the use of centralized treatment using ultimately ultra and nano-filtration preceded by simple pretreatment filtering stages.

Acknowledgement

This project has received funding from the *European Union's Horizon 2020 research and innovation program*under grant agreement No 641942.

References

Biocore, 2014. EU research project FP7-241566, http://www.biocore-europe.org

Kalliala, E.M., Nousianinen, P., 1999. Life Cycle Assessment Environmental Profile of Cotton and Polyestercotton Fabrics. AUTREX Res. J. 1, 8–20.

Koufolioulios, D., Nikolakopoulos, A., Pyrgakis, K., Kokossis, A., 2014. A mathematical decomposition for the synthesis and the application of total site analysis on multi-product biorefineries, Computer Aided Chemical Engineering.

Mountraki, A., Tsakalova, M., Panteli, A., Papoutsi, A.I., Kokossis, A.C., 2016. Integrated Waste Management in Multiproduct Biorefineries: Systems Optimization and Analysis of a Real-Life Industrial Plant. Ind. Eng. Chem. Res. 55, 3478–3492.

Mountraki, A.D., Nikolakopoulos, A., Mlayah, B.B., Kokossis, A.C., 2011a. BIOCORE- A systems integration paradigm in the real-life development of a lignocellulosic biorefinery. Comput. Aided Chem. Eng. 29, 1381–1385.

Mountraki, A.D., Nikolakopoulos, A., Pyrgakis, K.A., Mlayah, B.B., Kokossis, A.C., 2011b. BIOCORE - On a paradigm for the design and synthesis of real-life biorefineries, in: 11AIChE - 2011 AIChE Annual Meeting, Conference Proceedings.

Nikolakopoulos, A., Faskiotis, D., Kokossis, A., 2016a. Integrated Transhipment Models for Synchronous Screening of Treatment Technologies and Targeting of Fresh Water and Recycle Flows, Computer Aided Chemical Engineering.

Nikolakopoulos, A., Faskiotis, D., Kokossis, A., 2016b. Cascade Models for Targeting and Synthesis of Total Water Networks. Chem. Eng. Trans. 52, 943–948.

Nikolakopoulos, A., Kokossis, A., 2016. A Problem Decomposition Approach for Developing Total Water Networks in Lignocellulosic Biorefineries. Process Saf. Environ. Prot. in press.

Nikolakopoulos, A., Thomaidis, L., Kokossis, A., 2014. Mathematical Programming Shortcut Screening Models for the Design of Integrated Waste Treatment Systems. Comput. Aided Chem. Eng., Computer Aided Chemical Engineering 34, 381–386.

Papoulias, S.A., Grossmann, I.E., 1983. A structural optimization approach in process synthesis—II. Comput. Chem. Eng. 7, 707–721.

Tsakalova, M., Nikolakopoulos, A., Kokossis, A., 2012. On the systematic synthesis screening and integration of real-life biorefineries, in: AIChE 2012 - 2012 AIChE Annual Meeting, Conference Proceedings.

495001 Advances in Process Intensification: Using Reactive Distillation for the Conversion of Crude Glycerol

Abstract:

Advances in Process Intensification: Using Reactive Distillation for the Conversion of Crude Glycerol

Obakore Agbroko and Tracy J Benson (Lamar University)

With the rise in biodiesel production over the last couple of decades has come the abundance of glycerol, a byproduct from biodiesel manufacturing. A typical biodiesel process includes the transesterification of triglycerides to produce fatty acid methyl esters, followed by fuel clean up steps: product separation, water washing, and acidulation. In short, the glycerol stream also includes water and methanol and is typically considered â€œcrude glycerol.â€ In this study, crude glycerol was converted to propanediols using isopropanol as a hydrogen donor molecule in a reactive distillation column.

Combining reaction engineering and distillation into a single unit operation presents some interesting challenges, yet, reduces the operational footprint as well as the operating and capital costs of manufacturing. Operational and design challenges include mismatched temperatures between the separating components and the reacting components, unfavorable volatilities, and slow reaction rates generally due to poor reactant/catalyst/product mass transfer rates. Reactive distillation is a likely pathway towards process intensification to minimize overall energy consumption, particularly since distillation accounts for 20 % of total energy used within the U.S. manufacturing sector. Increased reflux ratios typically yield higher product purities but at the cost of reboiler heat duties. In comparison with conventional approaches of reaction followed by distillation, the proposed scheme holds great potential due to its ability for lowering both operating and capital costs.

In this study, Aspen Plus was used to model the conversion of glycerol to 1,3-propranediol and 1,2-propanediol while being separated from methanol, water, and other inorganic components that are commonly found in waste glycerol from biodiesel production processes. Using the Non-Random Two Liquid activity coefficient model, the number of trays, conversion, and product selectivities were estimated. To verify simulation results, laboratory experiments were conducted using a 1-inch ID, packed distillation column. Column temperature profiles as well as reboiler and condenser heat duties were determined. The complete reactive distillation analysis will be presented as well as the extension of this technology into other areas of the chemical process industry.

495999 Process Design and Techno-Economic Assessment of a Seaweed-Based Biorefinery: Integration of Thermochemical and Biochemical Conversion Processes

Abstract:

Biofuel production using carbohydrate rich brown seaweed is being intensely researched in the last few years as a viable replacement for 1st and 2nd generation biomass. In this study, for the first time, integrated pyrolysis and anaerobic digestion is simulated on an industrial scale using Aspen Plus software and experimental data, and technoeconomical assessment is made. Main points of integration include the pyrolytic conversion of the digestate, instead of raw seaweed and further digestion of the carbohydrate rich aqueous phase, which is a byproduct of both pyrolysis and catalytic upgrading. The synergistic effect of this integrated process should yield a more economically and environmentally viable biofuel production, compared to standalone pathways.

491067 Process Intensification and Modularization for Sustainability

Abstract:

We will describe various concepts for intensifying catalytic reactions and reactors, including bifunctional materials and cascade reactions, fairly mature by now microtechnology, hybrids of reactors and separations, and multifunctional reactors including static micromixers and microheat exchangers (spatial domain), periodic operation (temporal domain), and use of alternate energy forms such as microwaves. These concepts will be elucidated with examples taken from biomass, shale gas, and ammonia chemistry as means of sustainable chemical processing.

499573 Hydrothermal Carbonization of Wastes for Simultaneous Nutrient Recovery and Energy Capture

Abstract:

Hydrothermal carbonization (HTC) is a process that converts wet biomass into carbon-rich, energy dense solid fuel, called hydrochar, and an aqueous product. The process is done in hot, compressed water, typically in a temperature range of 180 °C - 260 °C. The thermodynamic properties of water change greatly in the subcritical region, and as a result, subcritical water behaves as a non-polar solvent with high ionic activity. When subjected to HTC, biomass undergoes rapid hydrolysis and produces solid hydrochar, as well as an aqueous product containing organic acids, sugars, and sugar derivatives [1].

Industrialized societies produce large waste streams with high moisture content, such as wastewater solids, animal manure, food processing wastes, and some industrial wastes. HTC can readily produce a carbon-rich, hydrophobic solid hydrochar from these waste streams. Most nitrogen compounds (derived from protein) are found in the aqueous stream, while much of the phosphorous is tightly bound in the hydrochar matrix [2]. Ideally, the nutrients can be captured in the aqueous stream, while the carbon (energy) is retained in the hydrochar. Recently, it has been shown that the presence of Ca, Mg, and Fe, as well as pH and reaction temperature, all affect the extraction of P and N from swine manures [3].

In this study, we evaluate the effects of reaction temperature, reaction time, pH, and especially addition of specific metal oxides on HTC of cow manure. We will show that the energy can be most effectively concentrated in the hydrochar at high reaction temperature with short reaction time. Under most reaction conditions, the nitrogen is readily partitioned to the aqueous phase. However, phosphorous capture requires longer residence times and addition of MgO, CaO, or Fe_2O_3.

References

[1] J. G. Lynam, M. T. Reza, W. Yan, V. R. Vásquez, and C. J. Coronella, "Hydrothermal carbonization of various lignocellulosic biomass," *Biomass Convers. Biorefinery*, vol. 5, no. 2, pp. 173–181, 2015.

[2] M. T. Reza, A. Freitas, X. Yang, S. Hiibel, H. Lin, and C. J. Coronella, "Hydrothermal carbonization (HTC) of cow manure: Carbon and nitrogen distributions in HTC products," *Environ. Prog. Sustain. Energy*, vol. 35, no. 4, pp. 1002–1011, Jul. 2016.

[3] U. Ekpo, A. B. Ross, M. A. Camargo-Valero, and L. A. Fletcher, "Influence of pH on hydrothermal treatment of swine manure: Impact on extraction of nitrogen and phosphorus in process water," *Bioresour. Technol.*, vol. 214, pp. 637–644, Aug. 2016.

496722 Using Fly Ash As pH Adjustment for Efficient Phosphorus Immobilization and Reutilization from Swine Manure in Hydrothermal Treatment

Abstract:

Concentrated animal feeding operations (CAFOs) are an important part of the U.S. economy, providing the country with a substantial low-cost source of consumable meat and dairy products such as beef, pork, poultry, milk, and eggs. However, CAFOs are becoming a greater public/government concern due to the amount of animal manure (waste) generated by such facilities, resulting in both air and water pollution. The United States Department of Agriculture (USDA) estimates that 335 million tons of dry manure waste is annually generated by CAFOs. This waste stream contains significant quantities of nutrients, accounting for nitrogen and potassium totals of approximately 6.17 and 1.85 million metric tons, respectively.

Application of manure to pasture and crop land is the most prevalent CAFO waste management practice. This practice can be beneficial as it improves soil quality by replenishing important nutrients for plant health. However, the amount of applied manure along with nutrient mobility leads to runoff into local watersheds, causing a host of local/regional public health and economic issues. Nutrient runoff (especially P) leads to growth of harmful algal blooms, which has led to the loss of water supply to U.S. cities (due to elevated microcystin levels), loss of aquatic life, and over $100 million in economic damage since 2008.

Due to its distributed nature, nutrient runoff is one of the most prevalent and challenging environmental issues in the U.S. To address this issue, Ohio University (OHIO) is developing a supercritical water (SCW)-based modular process to be operated at CAFO sites in order to recover the nutrients as a solid product. This nutrient-rich solid product may be used as a fertilizer supplement for field application, with significantly reduced potential for runoff into local watersheds. In addition, the process produces a methane-rich gas product, which may be utilized at the CAFO site or upgraded into pipeline-quality natural gas. SCW provides an advantageous media for recovery of nutrients from animal manure; its lower fluid density and hydrogen bonding strength promotes precipitation of solid P components, and provides a gasification medium for undigested carbonaceous species.

In this research, basic environment was evaluated at various temperatures to determine whether it has impacts on nutrient immobilization and phosphorus mobility of biochars. For a more controlled basic environment, 0.1 M NaOH was used to replace DI water; and for a more economical approach, fly ash was used for adjusting pH. After hydrothermal treatment, nutrient elements were found to be immobilized into the solid phase more efficiently with addition of basic materials, especially with addition of fly ash. In addition to being more efficient in immobilizing nutrient elements, fly ash also altered crystal phases formed at different temperature treatments which thus results in different profile of phosphorus mobility of the biochars. Results to be presented in this presentation include experimental results such as nutrient immobilization as a function of process conditions, post-trial analyses of biochar products (including XRD and SEM/EDS), and biochar phosphorous mobility.

494564 Recovery of High-Value Chemicals from Organic Waste: Economic Potential and Logistical Issues

Abstract:

Hexanoic (C6) and caprylic Acid (C8) are chemicals of high value due to their versatility, high demand, and relatively high energy density [1]. Studies have shown that such chemicals can be recovered from different types of organic waste streams such as biomass waste, food waste, and municipal solid waste; but such studies have not evaluated economic effectiveness [1-4]. In this work, we present supply chain models to study the economic potential and logistical issues associated to the recovery of C6 and C8 from different kinds of organic waste. We conduct a case study using real data from the State of Wisconsin.

In our supply chain framework, we consider four different sources of organic waste (urban centers, wastewater treatment plants, large dairy farms, and landfills), three different types of waste products (cow manure, sludge residues from wastewater treatment, and food waste) and five different types of waste processing technologies. We assume that 2-4% of the chemical oxygen demand (COD) in the waste streams can be converted as C6 or C8 [5,6]. The constraints in the model involve product balances and conversion constraints, capacity constraints, and logical constraints associated to technology sizing and placement. The objective function is to maximize total profit which includes capital and transportation costs and revenue from the recovered product. We conduct the following studies: a) trade-off analysis of profit against total amount of treated waste in the state, b) effect of uncertainty on yield factors and on prices of waste, C6, and C8, c) effect of waste mixing, and d) comparison of profitability from C6/C8 against the popular biogas (C4) recovery alternative.

Preliminary results indicate that, under current market conditions, it is profitable to recover C6 and C8 from organic waste and that much higher mobility can be achieved compared to C4. The results can be refined by conducting experimental studies that resolve uncertainties in key yield factors and technology costs.

References:

[1] Angenent L T, Richter H, Buckel W, et al. Chain elongation with reactor microbiomes: open-culture biotechnology to produce biochemicals. Environmental science & technology, 2016, 50(6): 2796-2810.

[2] Grootscholten T I M, Strik D, Steinbusch K J J, et al. Two-stage medium chain fatty acid (MCFA) production from municipal solid waste and ethanol. Applied energy, 2014, 116: 223-229.

[3] Kucek L A, Xu J, Nguyen M, et al. Waste conversion into n-caprylate and n-caproate: resource recovery from wine lees using anaerobic reactor microbiomes and in-line extraction. Frontiers in Microbiology, 2016, 7.

[4] Grootscholten T I M, dal Borgo F K, Hamelers H V M, et al. Promoting chain elongation in mixed culture acidification reactors by addition of ethanol[J]. Biomass and Bioenergy, 2013, 48: 10-16.

[5] Neves L, Oliveira R, Alves M M. Co-digestion of cow manure, food waste and intermittent input of fat[J]. Bioresource Technology, 2009, 100(6): 1957-1962.

[6] Ge S, Usack J G, Spirito C M, et al. Long-term n-caproic acid production from yeast-fermentation beer in an anaerobic bioreactor with continuous product extraction[J]. Environmental science & technology, 2015, 49(13): 8012-8021.

498471 Optimization of Biorefinery Production Chains and Decision-Making through Sustainability Evaluation: A Biojet Fuel Case Study

Abstract:

The aviation industry is a large contributor to the environmental pollution. For example, during the course of a roundtrip flight from Paris to New York, 2.2 Tons of CO2 are released per passenger into the atmosphere; a number that is roughly equivalent to sixty five 100km trips in a medium sized car (2). Clearly, there is a need for solutions that help mitigate these emissions, and advocacy groups such as the Air Transport Action Group (ATAG) push for actions towards more sustainable operations. According to the International Air Transport Association (IATA) sustainable aviation fuels derived from biomass (i.e. biojet fuels) "can reduce the overall carbon footprint by around 80% over their full lifecycle" (3), and actively supports their development.

Previous contributions have addressed the production of biojet fuel; as examples Crawford et al. (4) and Klein-Marcuschamer et al. (5) have analyzed the economic viability of biojet fuel production under different geographical and technological contexts, while Klein-Marcuschamer et al. (5) have studied the air emissions for its production. Going a step forward, Agusdinata et al. (6) have recognized that multiple actors/stakeholders (e.g. national/international policymakers, biorefineries, farmers) are involved in the development of biojet-fuels, each of which has differents objectives to consider in their decision-making.

In this work, we combine mathematical programming tools (optimization) and the GREENSCOPE approach developed by the U.S. Environmental Protection Agency (7) to assess the sustainability of production of biojet fuels from the perspective of the different actors that participate in the overall production chain. The GREENSCOPE approach allows for the quantification of process sustainability in four main areas: Efficiency, Energy, Economics, and Environment; in here, each E is characterized by a series of metrics that indicates the performance of a particular process. Optimization is applied to find the processing pathways that provide the best performance for each E, and for clusters of the indicators that represent the preferences of the different stakeholders. Multi-objective optimization is then used to find the Pareto frontier that describes the trade-offs among their optimal solutions.

As a case study, we consider the production of biojet fuels from three different organic material sources: eucalyptus, pine, and macauba, through three main technological pathways: fast pyrolysis, hydrothermal liquefaction, and gasification followed by Fischer-Tropsch synthesis. In all cases, the benefits of adding a follow-up cracking step of long-chain (hydrocarbon) residues, is also considered.

References

1. Brasseur, G.P., M. Gupta, B.E. Anderson, S. Balasubramanian, S. Barrett, D. Duda, G. Fleming, P.M. Forster, J. Fuglestvedt, A. Gettelman, R.N. Halthore, S. Jacob, M.Z. Jacobson, A. Khodayari, K. Liou, M.T. Lund, R.C. Miake-Lye, P. Minnis, S. Olsen, J.E. Penner, R. Prinn, U. Schumann, H.B. Selkirk, A. Sokolov, N. Unger, P. Wolfe, H. Wong, D.W. Wuebbles, B. Yi, P. Yang, and C. Zhou "Impact of Aviation on Climate: FAA's Aviation Climate Change Research Initiative (ACCRI) Phase II.", Bull. Amer. Meteor. Soc., 2016, 97, 561–583.

2. Simulations done on the basis of the tool available at https://co2.myclimate.org/. Accessed on April 16, 2017.

3. IATA website: http://www.iata.org/whatwedo/environment/Pages/sustainable-alternative-jet-fuels.aspx. Accessed on April 16, 2017.

4. Crawford, J. T., Shan, C. W., Budsberg, E., Morgan, H., Bura, R. and Gustafson, R., "Hydrocarbon bio-jet fuel from bioconversion of poplar biomass: techno-economic assessment", Biotech. for Biofuels, 2016, 9 (1), 141.

5. Klein-Marcuschamer, D., Turner, C., Allen, M., Gray, P., Dietzgen, R. G., Gresshoff, P. M., Hankamer, B., Heimann, K., Scott, P. T., Stephens, E., Speight, R. and Nielsen, L. K., "Technoeconomic analysis of renewable aviation fuel from microalgae, Pongamia pinnata, and sugarcane". Biofuels, Bioprod. Bioref., 2013, 7: 416–428.

6. Agusdinata, D. B., Zhao, F. and DeLaurentis, D. A., "Sustainability of Biojet Fuels: A Multiactor Life Cycle Assessment Approach", IEEE Potentials, 2012, 31(1), 27-33.

7. Ruiz-Mercado, G. J., Gonzalez, M. A. and Smith, R. L., "Sustainability Indicators for Chemical Processes: III. Biodiesel Case Study", Ind. & Eng. Chem. Res., 2013, 52, (5), 6747-6760.

502474 Eco-LCA Methodolgy for Sustainability Assessment of Bio-Wwts Focused on Energy Recovery

Abstract:

Introduction. The energy potential of agro-industrial effluents receiving biological treatment is often not much exploited, particularly in emergent regions because this alternative energy is forced to compete disadvantageously with conventional energy sources from an economic point of view. Consequently, smart strategies are required to prove the environmental/economic potential of biological wastewater treatment systems (Bio-WWTs) depolluting agro-industrial effluents committed with improved energy recovery. In this context, the following contribution introduces a methodological framework for refined sustainability assessment of Bio-WWTs treating agro-industrial effluents. It combines life cycle assessment (LCA) and ecological economics methods, specifically the emergy accounting approach. Results of a first proof of concept of this methodology are presented here using a Bio-WWTs of the poultry sector and a Bio-WWTs treating beer production effluents as case studies.

Methods. The methodology is developed as follows:

(i) Obtaining of Net Environmental Benefits Indicators-(NEB) by means LCA. The analyses includes avoided or shifted conventional energy in the borders of Bio-WWTs (Godin et al. 2012.) by the recovery of biogas from the Bio-WWTs operation. For each case study the NEB indicators were calculated for two life cycle scenarios considering different energy recovery routes in order to transform these NEBs in Environmental Influence Indicators (IAM). Application of the analytical identification method enables obtaining a reduced set ofIAM (Grunwald-Rosch, 2011).

(ii) Identification of potential co-benefits or "add-values" related with changes in the ecosystem services offer, which are promoted by differences between life cycle scenarios. These "add-values" are specifically appraised by emergy methods (Zhang Y, 2010). Then a set of normalized add-value indicators (NAV) are proposed to assess the "eco-economic" dimension of sustainbility.

Results. A set of Sustainable-Development Indicators (SDIs) has been obtained and defined as NAV/IAM ratios, similarly to ecoefficiency indicators. These SDIs are related to: CO_2-capture capacity, resources utilization, water pollutants and airborne emissions dilution. An overall sustainability indicator NAV_T/IAM_P has been also tailored from the contribution of single SDIs. Charts as presented in figure 1 were obtained and used to identify sustainability conditions.

Discussion and conclusions.A methodology unique for sustainability assessment of Bio-WWTs focused on energy recovery has been developed. This is based on integration of LCA/emergy methods (Eco-LCA) and allows to build representative SDIs for Bio-WWTs. These SDIs are dimensionless quantities because emergy (NAV) and environmental (IAM) terms are normalised values ranging from -1 to 1. Positive values of SDIs indicate "additionality", potential "add-value" or a co-benefit obtained, which is promoted by reducing a direct environmental impact (positive IAM). A strong non-additional condition as far as an

individual SDI results from both emergy and environmental terms minus numbers. A weak non-additional condition is produced when the emergy term is negative but the respective environmental is not. Results show that sustainability requires synergy among different SDIs and is not only dependent of climate change control/mitigation, which is the more generalized and accepted sustainbility criteria for wastewater treatment systems. Bio-WWTs focused on energy recovery increase resources utilization, this is an "emergy cost" which is "not additional" but it is required to promote additionality through other SDIs. The study shows that sustainability of Bio-WWTs with higher energy availability or production (beer case) is more dependent on the contribution to climate change control than Bio-WWTs with lower energy potential (poultry case). The baseline choice and the application of the NEB approach are key aspects to prove additionality of Bio-WWTs by means the outlined methodology.

References.

1. Godin, D., Bouchard, C., Vanrolleghem, P.A. 2012. Net environmental benefit: introducing a new LCA approach on wastewater treatment systems. *Water Science and Technology*, **65**(9), 1624-1631.
2. Grunwald, A., Rosch, C. 2011. Sustainability assessment of energy technologies: towards an integrative framework. *Energy, Sustainability and Society*, **1**(1), 3.
3. Meneses-Jacome, A., Osorio-Molina, A., Parra-Saldivar, R., Gallego-Suarez, D., Velasquez-Arredondo, H.I., Ruiz-Colorado, A.A. 2015. LCA applied to elucidate opportunities for biogas from wastewaters in Colombia. *Water Science and Technology*, **71**(2), 211-219.
4. Wang, X., Liu, J., Ren, N.-Q., Yu, H.-Q., Lee, D.-J., Guo, X. 2012. Assessment of Multiple Sustainability Demands for Wastewater Treatment Alternatives: A Refined Evaluation Scheme and Case Study. *Environmental Science & Technology*, **46**(10), 5542-5549.
5. Zhang, Y., Singh, S., Bakshi, B.R. 2010. Accounting for Ecosystem Services in Life Cycle Assessment, Part I: A Critical Review. *Environmental Science & Technology*, **44**(7), 2232-2242.

501481 Sustainability Assessment Water Resources Use Mining in Colombia, By the Emergy Analysis

Abstract:

Due of its great climate diversity, Colombia has an abundance of natural resources and possesses a wide range of mining products, including coal, gold, platinum, nickel, emerald, limestone, among others that are extracted at a smaller scale. This extraction is carried out in two ways: a formal large-scale extraction and other smaller-scale, traditional and artisanal; the latter lacks adequate and defined technology, usually in some cases it is considered an illegal activity, making it unsafe, unprofitable, uncompetitive and environmentally unsustainable.

In the last decade, there has been a significant increase in both the volume and value of production of primary goods, particularly in the mining and hydrocarbons sector, which has positively and significantly affected the countryâ€™s gross domestic product (GDP) (FEDESARROLLO, 2012), currently being one of the most dynamic of the Colombian economy. The GDP from this sector has grown from 5.2 to 10.3 trillion pesos between 2000 and 2011 (Colombian Ombudsman, 2010), although in the last two years (2013 - 2014) this activity has gone into recess.

Thus, according to the large-scale mining sector (LSM), in 2011 mining represented 24.2% of exports; 2.4% of GDP; 20% of the total foreign direct investment; 650 billion pesos in construction of infrastructure; 2.6 trillion pesos in purchases from domestic suppliers, 65 billion pesos of investment in social responsibility and 178 billion pesos in environmental responsibility (LSM, 2012). This constitutes evidence for the mining sector in Colombia to be considered one of the main economic engines, since it does not only contribute with employment creation (there were 836000 in 2011, between direct and indirect jobs), but also with investments in infrastructure, public utilities, and social and environmental management (LSM, 2012). In royalties, the mining sector contribution to the country is 16.5%; 13% of it comes from coal companies (1.2 trillion pesos a year).

This constitutes evidence for the mining sector in Colombia to be considered one of the main economic engines, since it does not only contribute with employment creation (there were 836000 in 2011, between direct and indirect jobs), but also with investments in infrastructure, public utilities, and social and environmental management (LSM, 2012).

However, it is impossible to avoid discrepancies and political, economic, environmental and social polemics around this productive activity in the country, since mishandled mining practices lead to: social impacts, since many of these unformalized activities allow child labor, make use of intensive techniques in unskilled labor, generating poor quality jobs with low levels of industrial, social, and health security for miners, it is usually developed in remote and deprived areas; furthermore, it generates conflicts between large mining industries due to concession contracts awarding. Regarding political and economic impacts, a minimum working capital and scarce financial resources for investment are evidenced, and therefore an improper handling of inputs and royalties from this activity is presented. In a greater relevance, there are the irreversible environmental impacts such as destruction of ecosystems, damage to protected areas, air pollution, and ***pollution and impact on the availability of water resources***. Globally, water demand is driven primarily by the population and economic growth, and it is anticipated that in the coming decades, it will increase considerably, while supplies will remain almost constant or decreasing by factors such as climate change, overpumping and pollution of aquifers (Drelich, 2012). It is not unknown that the primary input in gold extraction is water; the water that is used in this sector comes from a variety of sources, and sources and quality are the main areas of controversy and debate in this productive sector. ***The main problem of the mining industry is to generate confidence in the development of a responsible, sustainable and transparent strategy in the use of water*** (Drelich, 2012).

Undoubtedly, the flow of resources from this activity is increasing and projections show that the growing trend will continue, since the international scene shows an important dynamism of the mining market. This situation could be exploited by Colombia, given its geographical location and its geological potential (FEDESARROLLO, 2012); it is estimated that US$250 billion will be invested in mining projects in Latin America between now and 2020. Thus, the mining sector gains importance among public policies, not only due to the high revenues generated by its operation, but also because of the social and environmental impact of the sector activity. Hence the importance of seeking for solutions, methodologies and sustainable alternatives to assimilate the challenge of turning the mineral wealth in an opportunity for development, and thus respond to future generations about how resources from non-renewable assets were invested without affecting renewable resources.

This highlights the fact that, despite the importance that the mining sector has taken, the debate about the true benefits and costs, or the environmental, economic, energy and social problems of the sector's activity has intensified. It is misleading to decide whether mining offers a significant potential or a threat to society without a methodological tool that allows predicting how sustainable is this productive activity, which represents a profitable business for many, which provides social and economic welfare (ultimate goal of the implementation of development projects of this magnitude), and for other sectors of the society symbolizes the opposite, considering as innefective and undervalued all corrective and preventive measures that have been implemented by the government agencies or those regulatory entities that have some kind of control in the field of sustainable management of the processes of extraction and use of non-renewable resources, including products that are generated in such activity, in response to the environmental awareness that has been growing in recent years.

Based on the concept of sustainability as "the compatibility between energy, economics (maximum performance) and environmental aspects" (Redclif, 1987), all development projects especially those that threaten the environmental integrity such as the exploitation of natural resources and mining processes, should be focused on being an economic use alternative that provides an energy yield with acceptable environmental burden. That is, this highly demanding production activity in the country (including byproducts generated in it) should be given a particular management by means of methodological tools to harmonize the economic with the energy and environmental aspects; since these three factors can not be assessed independently, but they must be analyzed together and in an interrelated way, in order to give management on a sustainable basis, which avoids the discrepancies noted above. In addition to these tools, it is possible to identify where the weaknesses lie, which apparently make this activity a threat to some sectors of the society.

It is then necessary to establish, among the many valuation methodologies available, those that allow a sustainable assessment of the current management and mining process in Colombia in both artisanal and large-scale mining; likewise identify the gaps in it, whose purpose is to make mining a lever for development not only with good labor, industrial and environmental safety practices, but also going further in its commitment to the country and the city that hosts it. That is, for the management of mining resources, a valuation method should be used thereof, which is bearable and balanced from the economic, energy and environmental levels, as in the current regulations governing this sector a large imbalance is evident between these aspects. That is to say, in many situations mining is presented with a great economic benefit, but it entails irreversible environmental damage or major social problems that trigger collective chaos; or on the contrary, many mining licenses are denied impeding economic growth because of an environmental impact that may be acceptable, discrepancies caused by non-uniformity of a regulatory framework for the mining sector and the lack of methodological tools to assess the sustainability of different productive sectors of society, which and/or balance economic, energy and environmental aspects.

Globally, the possibility of quantifying the sustainability of different production sectors (agriculture, energy, legislative, sanitation) has been subject of study using the method of emergy which allow not only to quantify parameters, but also to analyze the implications and interrelationships of economic and environmental aspects of these long-term processes.

Thus, this research aims to develop a methodological tool; emergy analysis to assess and compare the sustainability of the current small and large scale mining systems in Colombia, including byproducts generated in this process, emphasizing on the **use of water resources**, gaps in these systems which may not make them sustainable, and a uniform methodology that can be used as a road map to regulate that productive sector in a sustainable way, because without doubt, regulation constitutes the source of many of the limitations of the mining sector

497207 Thermodynamic Analysis of an Ion-Exchange Based Waste Water Treatment for Phosphorus Recovery

Abstract:

Waste water treatment and recovery of nutrients from waste water is a strategy that will be used for long term sustainability of both water and nutrients. Waste water generally contains large quantities of useful nutrients such as Nitrogen and Phosphorus (P) that has detrimental impact on environment such as eutrophication of water bodies when lost to environment, further reducing the quality of water resources. Therefore, containing and treating waste water is an important activity to maintain the supply of water. Further, P is a critical element for agricultural production and reserves are limited, hence it is crucial to recover P from waste for long term availability for food production. Currently, there are several industrial scale existing technologies for P recovery from waste water such as AirPrex, Phosnix and, Crystalactor with recovery efficiencies 98%, 90%, and 75% respectively. All these established technologies recover P as struvite and/or calcium phosphate and depend extensively on large quantities of chemicals for processes like acid leaching and precipitation. These upstream chemical requirement imply high energy usage, thus resulting in an energy-water nexus problem for sustainable supply of both energy and water. Hence, it is necessary to study the thermodynamic performance of these recovery technologies at large scale that can inform the future expansion of these waste water treatment and recovery technologies. In this work, we focus on using thermodynamic metrics for performance of a novel Ion-Exchange based P recovery technology. The proposed Ion-Exchange based technology is independent of large quantities of additional chemicals; potentially being more sustainable relative to current technologies. However, the energy requirements for resin regeneration and maintaining the ideal Temperature and Pressure may result into higher energy requirements. Thus, before these technologies can be scaled up for adoption in expanding waste water treatment facility the energy requirement along with water and P fertilizer productions needs to be analyzed. To address this necessity, we use Entropy generation or S_{gen} as a metric that can provide information on the theoretical minimum energy requirement for the successful operation of this ion-exchange process.

Our approach involves rigorous systems modeling to quantify the total S_{gen} where process modeling tools are used for each of the sub-processes in the ion exchange based P recovery technology. The proposed model relies on two unit operations: Ion-exchange and anaerobic-digestion. ASPEN Plus process modeling software package for process modeling of all the sub-processes has been utilized. Results show that both these unit operations involve an entropy change of 34.5 kCal/mole-K and -478 kCal/mole-K respectively. We will also perform an energy analysis of the system to demonstrate the difference between insights obtained from entropy vs energy analysis along with performing the reaction level entropy change in both cases. Availability of theoretical minimum energy value from S_{gen} will potentially inform the energy implications of treating waste water for clean water generation on large scale.

497737 Shale Gas Wastewater Management Using Membrane Distillation: An Optimization Based Approach

Abstract:

Unconventional shale gas industry offers tremendous economic benefits but is also accompanied by a number of environmental challenges. One such critical challenge is sustainable management of high salinity wastewater that returns to the surface after hydraulic fracturing and poses potential negative impacts on the surrounding ecosystems. Currently, about 90% of shale gas wastewater is being recycled for fracking purposes in Marcellus shale play. Recycling shale gas wastewater is a practical short-term solution to minimize total water use in the fracturing process, however, it may not be a viable strategy from a long-term management perspective especially when a well-pad becomes a net water producer. Another common strategy for shale gas wastewater management is direct disposal of the high salinity wastewaters into Class II disposal wells. However, this strategy is not cost effective and has come under scrutiny because of increased seismicity in regions close to class II disposal wells. Shale gas wastewater could also be sent to commercial centralized wastewater treatment (CWT) facilities. However, this strategy is not viable in the long term as the volumetric rate of shale gas wastewater to CWTs is limited to 1%. In order to choose the best strategy for integrated shale gas wastewater management for a specific shale gas extraction site, a holistic systems approach is required that takes into account associate cost of wastewater treatment, transportation, feasible treatment technology, and the potential beneficiary uses for the treated wastewater. A handful of recent studies have focused on shale gas water management using optimization techniques. However, the majority of existing work is focused on short term planning where shale gas wastewater could be recycled for future hydraulic fracturing requirements with little emphasis on long term planning for wastewater management. Furthermore, most conventional desalination technologies are not suitable for handling the high total dissolved solids (TDS) content of shale gas wastewater which could be as high as 350,000 mg/Liter.

We present a comprehensive optimization-based decision-making framework for guiding environmentally and economically conscious management of high salinity wastewater with application to the Marcellus shale play. We evaluate water management alternatives ranging from direct disposal in Class II injection wells to advanced centralized, decentralized, and onsite treatment options. We propose to use membrane distillation (MD) technology which holds great promise for desalination of high salinity wastewaters from shale production. MD is a thermally driven membrane-based process with relatively higher energy requirements compared to conventional desalination techniques such as reverse osmosis (RO) and forward osmosis (FO). However, the application of RO and FO is limited to wastewaters with TDS levels up to 40,000 and 70,000 mg/Liter, respectively. We performed techno-economic assessment (TEA) for shale gas wastewater treatment using MD technology with and without availability of waste heat. The results of TEA show that the major cost driver for a MD plant is the cost of thermal energy required for distillation process. We specifically focus on natural gas compressor stations (NG CS) in the U.S. where two-thirds of the energy input is currently rejected as waste heat. Preliminary results of our work reveal that only 56% of the available waste heat at NG CS in Pennsylvania is sufficient for desalination of shale gas wastewater generated in this state using MD technology. A mixed-integer non-linear programming (MINLP) model will be presented that optimizes the shale gas wastewater management. The model investigates optimal management strategies for wastewater by incorporating detailed treatment cost data obtained from TEA of MD as the treatment technology as well as detailed transportation cost data from shale gas sites to treatment or disposal facilities. We address the problem of finding the optimal design for shale gas wastewater management as well as determining the size and location of treatment facilities in order to minimize the total cost of handling wastewater. The developed optimization framework is applied to a case-study in Greene County in southwest PA where major shale gas development activities takes place. Preliminary results of our optimization model reveal that treating shale gas wastewater at onsite MD units at shale gas extraction sites is the most economically promising wastewater management option. Current work is conducting detailed scenario analysis to investigate how changes in production scenarios and regulatory environments may affect the choice of wastewater management alternatives. The implications of these findings for long-term management of shale gas wastewater will be described.

Systematic Analysis and Optimization of Water-Energy Nexus

Spyridon D. Tsolas, M. Nazmul Karim, M. M. Faruque Hasan

Artie McFerrin Department of Chemical Engineering
Texas A&M University
College Station, Texas, USA.

ARTIE MCFERRIN DEPARTMENT OF CHEMICAL ENGINEERING TEXAS A&M UNIVERSITY | TEXAS A&M UNIVERSITY Texas A&M Energy Institute | TEES TEXAS A&M ENGINEERING EXPERIMENT STATION

2017 Annual AIChE Meeting, Minneapolis, MN
Session: The Energy-Water Nexus
October 31, 2017

Systems Optimization and mULti-scale analySis (SOULS) Lab: http://chenhasan.wpengine.com

Outline

- Introduction
  - Motivation
  - Challenges and objectives
- A graph-theoretic approach for nexus analysis
  - Source-sink representation
  - Results from Graph-theoretic nexus analysis
- Systematic methods for nexus design and optimization
  - Graphical approach
  - Mathematical programming-based optimization approach
- Case study: A country-wide nexus
- Conclusions

2

Introduction

- **Increasing global energy and water demands**
  - Population growth
    - Increased consumption of electricity and water
    - Increased withdrawal of fossil fuels and freshwater
- **Energy needs water**
  - 15% of global water
- **Water needs energy**
  - 8% of global energy

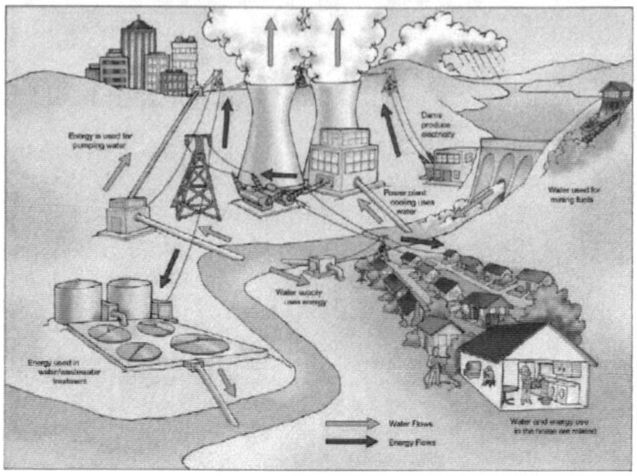

Source: U.S. Department of Energy, 2006

3

Key Challenges

- **Complex Interactions**
  - Increased water demands
    - → Increased energy consumption
- **Additional constraints**
  - Shortage in water availability
    - → Power plant capacity reduction

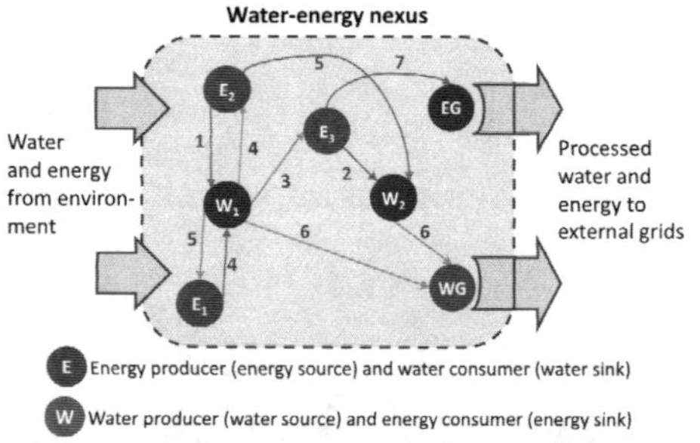

- **Critical Unresolved Questions:**
  - How can we systematically:
    - elucidate the complex interactions and trade-offs?
    - quantify the overall performance of a water-energy nexus?
    - identify redundant entities/facilities to increase overall efficiency?
  - How can we eventually:
    - identify and eliminate excess use of water and energy for sustainability?
    - redesign existing systems to meet increasing water and energy demands?

4

A Graph-Theoretic Approach

Definition: A nexus is a system of two interconnected networks of energy and water

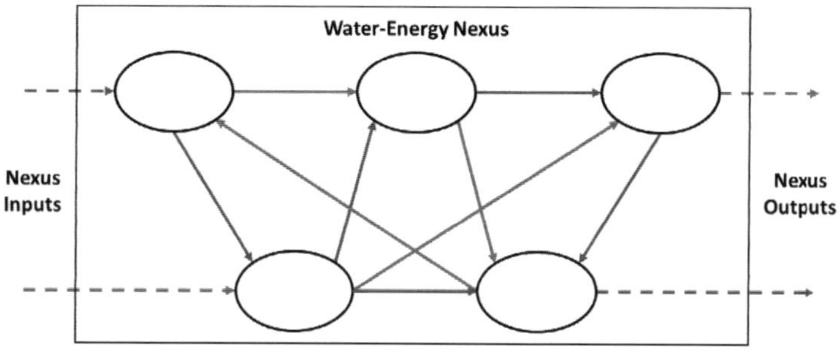

- **Nexus Input: Natural Resources**
- **Nexus Output: External Grids**
- **Shared Processing Nodes:**
  - Energy sources – water sinks
  - Water sources – energy sinks

5

A Graph-Theoretic Approach

- **Multi-scale nexus identification**
  - Cell → Equipment → Plant → Regional
- **Regional Level application:**
  - Energy sources – water sinks
    - Fuel mining and processing facilities
    - Power generating plants, solar parks, etc.
  - Water sources – energy sinks
    - Pumping, treatment and distribution facilities

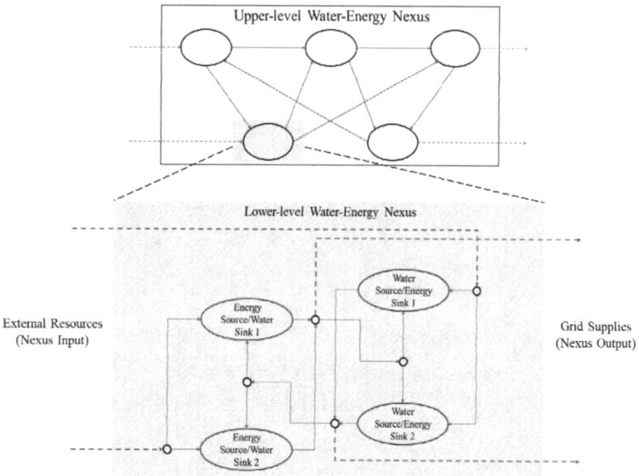

6

A Motivating Example

- **Assumptions:**
  - Energy input to an energy source is implied
  - Energy sources only provide to energy sinks, and vice versa
  - No distance variation between the sources
  - Incoming water streams correspond to withdrawal and not consumption

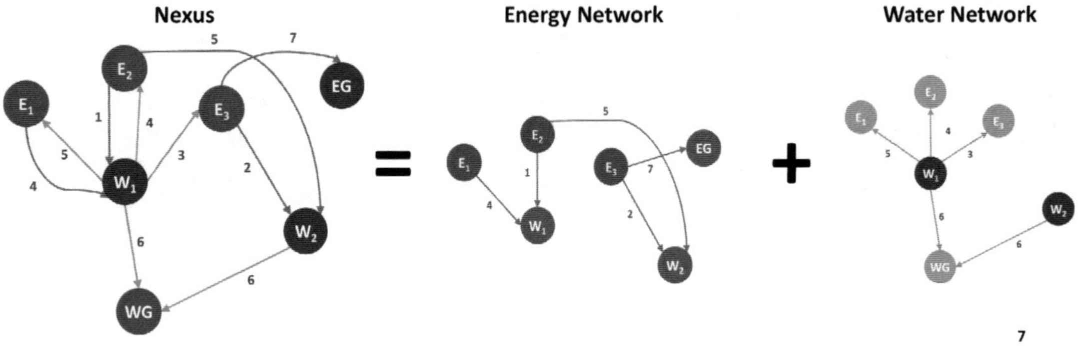

7

Graph Theory Definitions

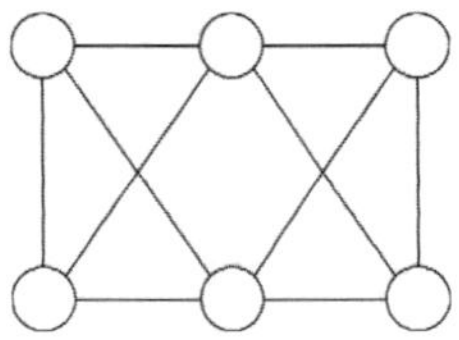

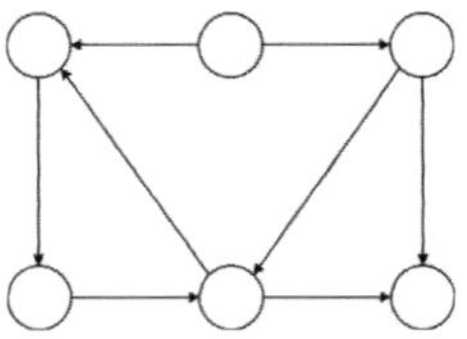

- **Definition: Graph – G=(V, E)**
  - Set of vertices: V
  - Set of edges: E
  - Vertices are connected via edges
- **Definition: Directed graph – G=V(V,E)**
  - Set of vertices: V
  - Set of arcs (oriented edges): E
  - Vertices are connected via edges
- **Bipartite Graph**
  - Vertices belong to subsets X, and Y
  - Adjacent vertices only from different subsets

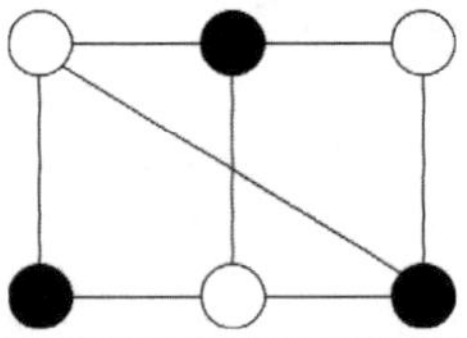

[1]J. A. Bondy and U. S. R. Murty, Graph theory with applications. Citeseer, 1976.
[2]D. Jungnickel, Graphs, networks and algorithms. Springer, 2008.

8

Graph Theory Definitions

- **Walk:**
  - Sequence of edges and vertices
  - **Closed walk:** starting node – end node
- **Trail:**
  - A walk with distinct edges
- **Path:**
  - A walk with both edges and vertices distinct
  - **Cycle:** a closed path

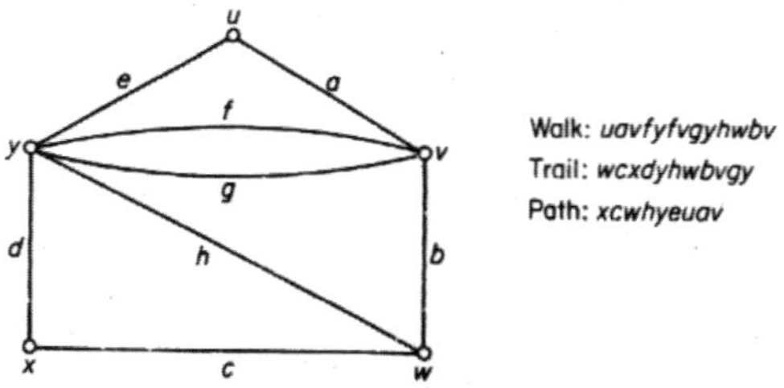

Source: J. A. Bondy and U. S. R. Murty, Graph theory with applications. Citeseer, 1976.

> **Theorem[1,2].** *A graph is bipartite if and only if it contains no odd cycles.*

[1]J. A. Bondy and U. S. R. Murty, Graph theory with applications. Citeseer, 1976.
[2]D. Jungnickel, Graphs, networks and algorithms. Springer, 2008.

9

Water-Energy Nexus (WEN) Graph

- **Bipartite Directed Graph**
  - Red arcs → Energy streams
  - Blue arcs → Water streams
- **Product transformation:**
  - **Red edges**
    → Energy sources/water sinks
  - **Blue edges**
    →Water sources/energy sinks

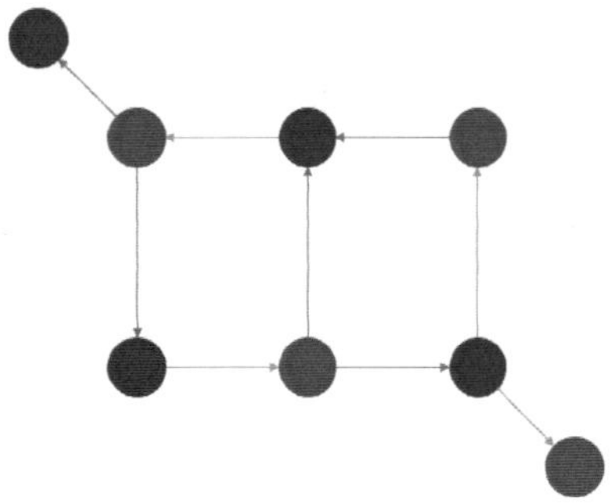

- **Intensity factor (inverse efficiency):**
  - Water source/energy sink: $\phi_w = \frac{energy\ consumption}{water\ production}$
  - Energy source/water sink: $\phi_e = \frac{water\ consumption}{energy\ production}$

10

Types of Redundancies in a WEN Graph

1) Redundant Graph – Zero output

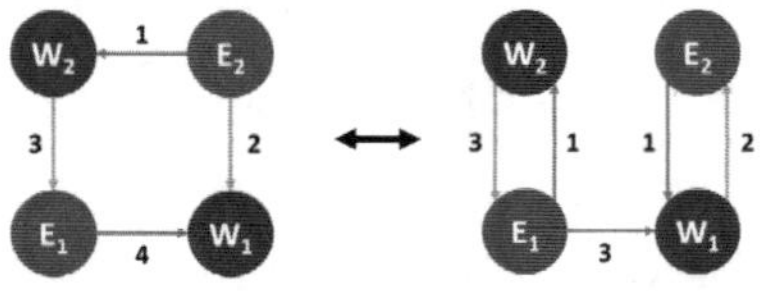

2) Essential Graph – Non-zero output

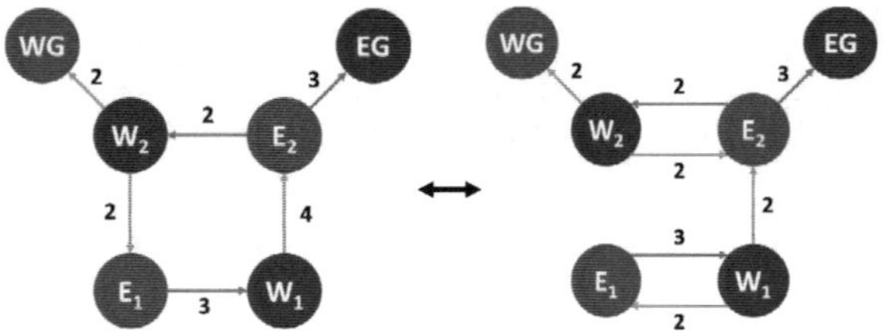

3) Partially Redundant Graph

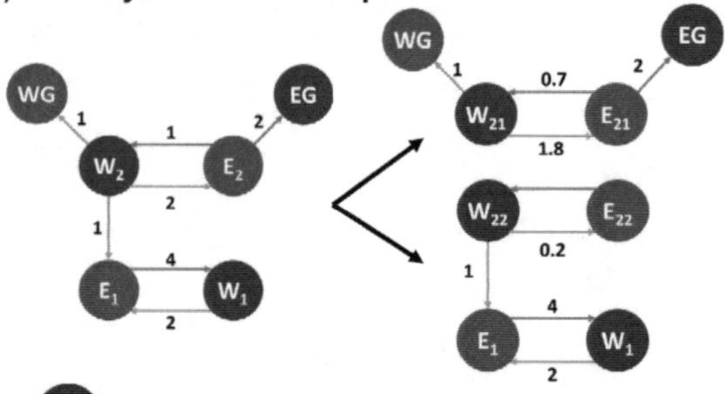

Can be split into:
- **Redundant subgraph**
- **Essential subgraph**

11

Results from Graph-theoretic Nexus analysis

Lemma 1: *No cycle can have vertices more than 2*min(N_W, N_E)*

Lemma 2: *A graph with no-zero output with φ_w>0 and φ_e>0, must contain at least one cycle.*

Lemma 3: *For specified grid demands, the optimal nexus configuration with minimum generation is the one that has no redundant sub-graphs.*

Lemma 4: *For specified external inputs/generation, the optimal nexus configuration with maximum outputs to external grids is the one that has no redundant sub-graphs.*

12

New Challenges

- How can we identify and isolate redundant sub-graphs that are hidden in a complex nexus configuration?
- How can we optimally break/combine sources and/or alter the connectivity among the sources and sinks?

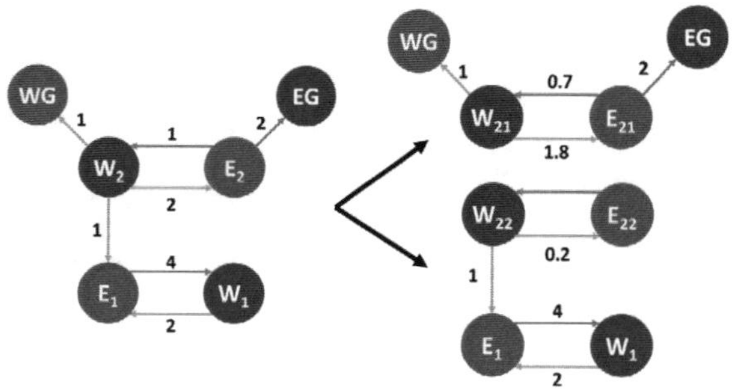

13

Outline

- Introduction
  - Motivation
  - Challenges and objectives
- A graph-theoretic approach for nexus analysis
  - Source-sink representation
  - Results from graph-theoretic nexus analysis
- Systematic methods for nexus design and optimization
  - Graphical approach
  - Mathematical programming-based optimization approach
- Case study: A country-wide nexus
- Conclusions

14

Water-Energy Nexus Diagram

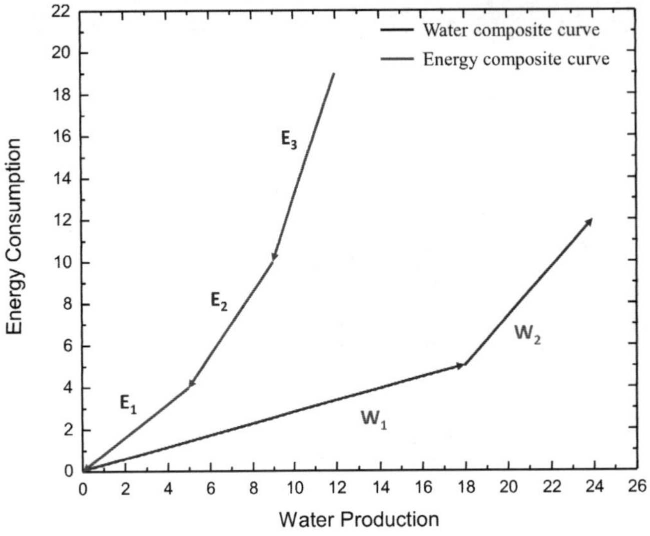

- **From graph to diagram:**
  - Y-axis: Energy consumption
  - X-axis: Water Production

Energy Sources	Water Consumed	Energy Generated
E_1	5	4
E_2	4	6
E_3	3	9
Water Sources	**Water Produced**	**Energy Consumed**
W_1	18	5
W_2	6	7

15

Water-Energy Nexus Diagram

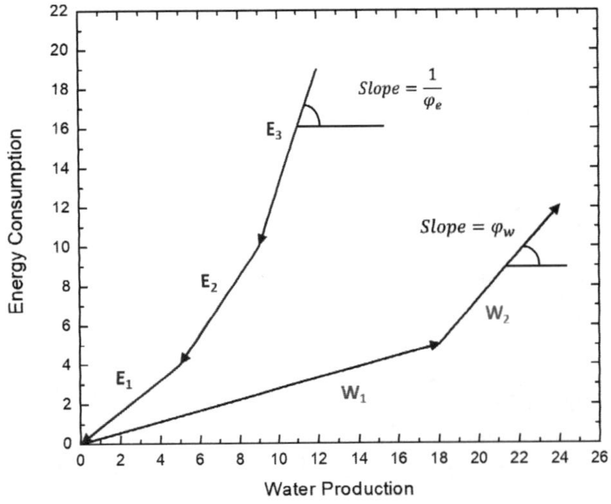

- **Slope of linear segments:**
  - Energy sources: Inverse intensity factor
  - Water sources: Intensity factor
- **Rearrangeable source segments**
  - No distance variation

16

Water-Energy Nexus Diagram

- **Illustration of:**
  - Stream supplies
  - Source-sink matching

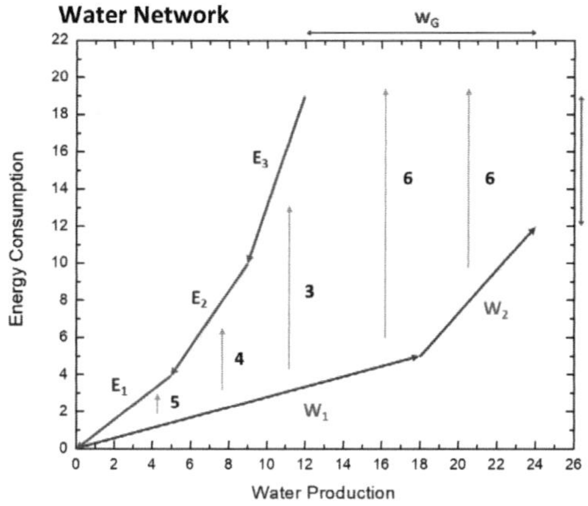

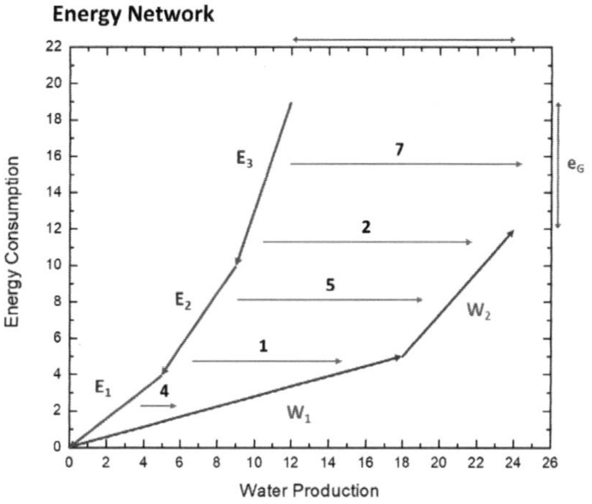

17

Water-Energy Nexus Minimum Generation

- **Rearrange segments:**
  - Decreasing Intensity
  - Energy sources: Increasing slope
  - Water sources: Decreasing slope
- **Check for overlapping**
- **Isolate and separate:**
  - Essential subgraph
  - Redundant subgraph

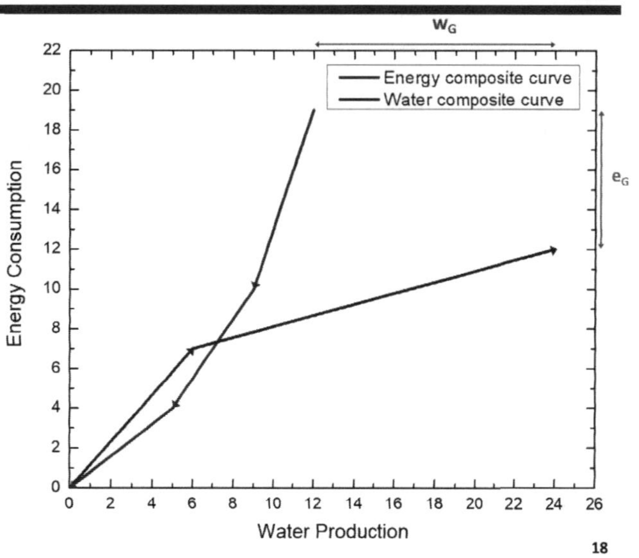

18

Water-Energy Nexus Minimum Generation

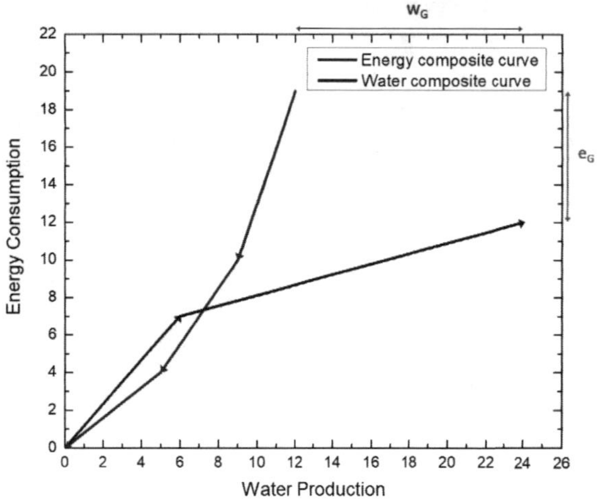

Corresponding Graph

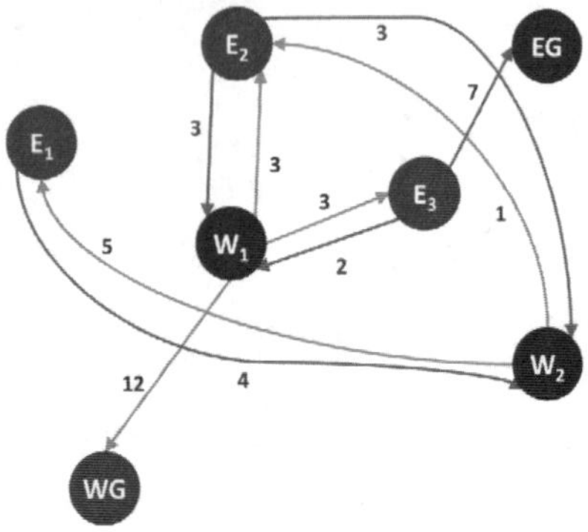

Water-Energy Nexus Minimum Generation

- **Subsystem under crossing point**

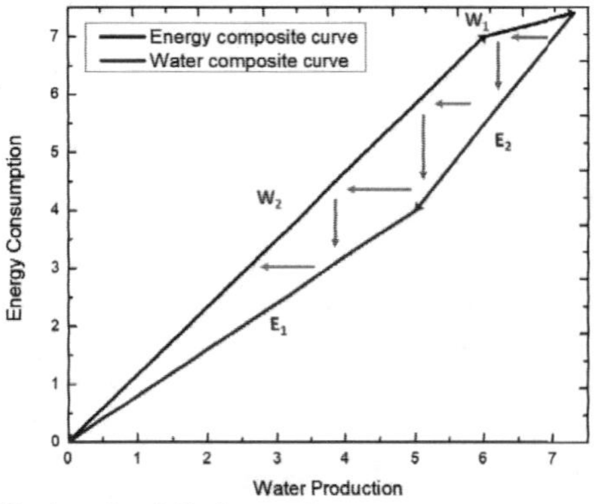

Corresponding Graph

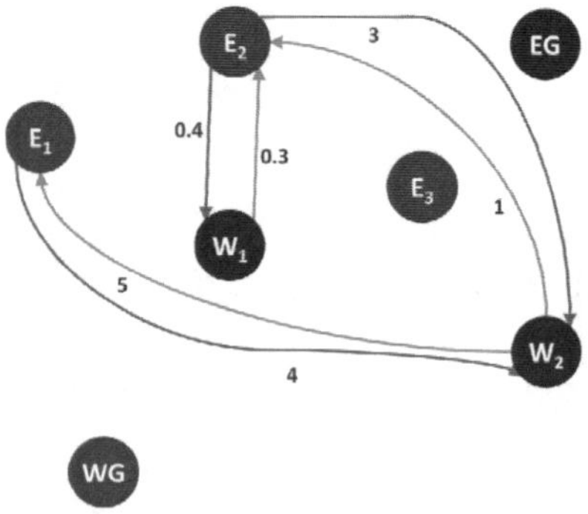

- **Redundant Subgraph**
  - Zero output to the grids

Water-Energy Nexus Minimum Generation

- **Subsystem over crossing point**

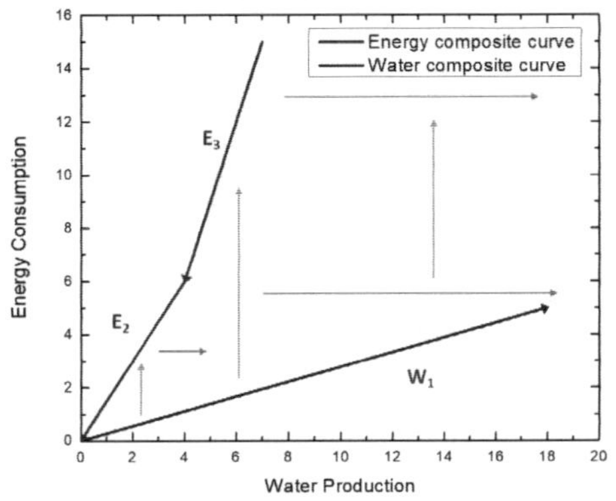

Corresponding Graph

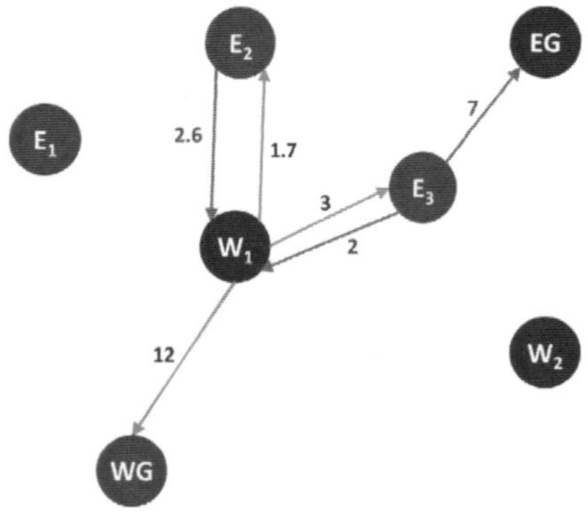

- **Essential Subgraph**
  - Minimum nexus to satisfy the same grid demands

21

Energy Grid Maximization

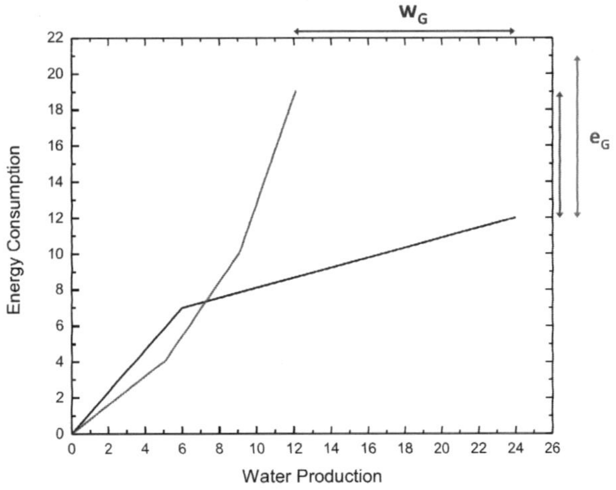

- **Start from essential subgraph**
- **Vertical shift: EG Increases**
- **"Pinch-point": Last position where composite curves intersect → EG_{MAX}**
- **Further shift:** No additional output

Corresponding Graph

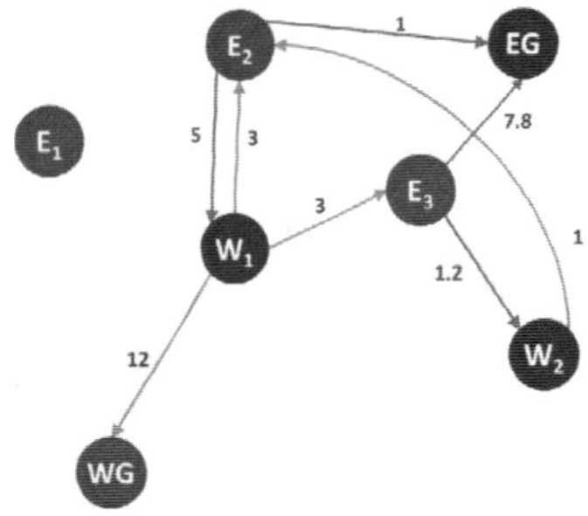

22

Water Grid Maximization

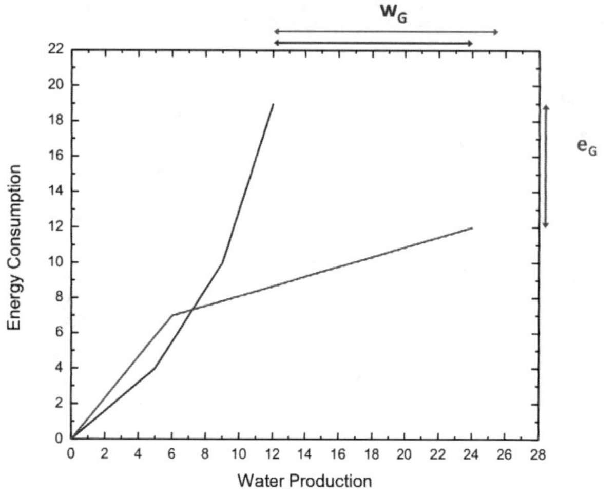

- Start from essential subgraph
- Horizontal shift: WG Increases
- "Pinch-point": Last position where composite curves intersect → WG_{MAX}
- Further shift: No additional output

Corresponding Graph

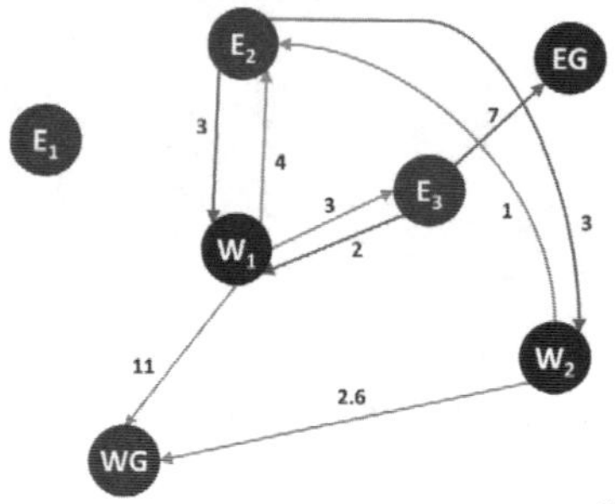

23

Outline

- Introduction
  - Motivation
  - Challenges and objectives
- A graph-theoretic approach for nexus analysis
  - Source-sink representation
  - Results from graph-theoretic nexus analysis
- Systematic methods for nexus design and optimization
  - Graphical approach
  - Mathematical programming-based optimization approach
- Case study: A country-wide nexus
- Conclusions

24

A Linear Programming (LP) Formulation

- **Transshipment-based[1] modeling**
- **General Equations:**

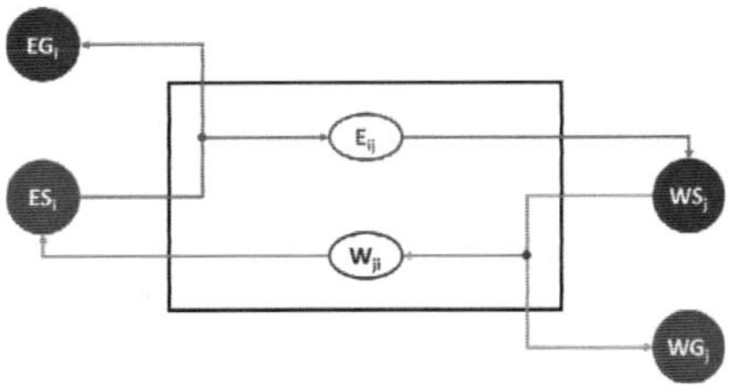

Balance of energy streams: $ES_i = EG_i + \sum_j E_{ij}, \quad \forall\, i\ stream$

Balance of water streams: $WS_j = WG_j + \sum_i W_{ji}, \quad \forall\, j\ stream$

Satisfaction of water needs for ES_i: $\sum_i E_{ij} = \varphi_{w,j} \cdot WS_j, \quad \forall\, j\ stream$

Satisfaction of energy needs for WS_j: $\sum_j W_{ji} = \varphi_{e,j} \cdot ES_i, \quad \forall\, i\ stream$

Current Energy Grid Supply: $EG_0 = \sum_i ES_{max,i} - \sum_j \varphi_{w,j} \cdot WS_{max,j}$

Current Water Grid Supply: $WG_0 = \sum_j WS_{max,j} - \sum_i \varphi_{e,i} \cdot ES_{max,i}$

ES_i	Energy source i
WS_j	Water source j
$\varphi_{E,i}$	intensity factor
$\varphi_{W,j}$	intensity factor
E_{ij}	energy flow from ES_i to WS_j
W_{ji}	water flow from WS_j to ES_i

[1]S. A. Papoulias and I. E. Grossmann, "A structural optimization approach in process synthesis—II: Heat recovery networks," Computers & Chemical Engineering, vol. 7, no. 6, pp. 707-721, 1983.

25

Model Formulation

Minimization of total generation: $\min z = \sum_i ES_i + \sum_j WS_j$

st.

General equations

Satisfaction of current grid demands:

$$EG = \sum_i EG_i = EG_0$$
$$WG = \sum_j WG_j = WG_0$$

Decision variables bounds:

$$ES_i, WS_j, E_{ij}, W_{ji} \geq 0$$
$$ES_i, \leq ES_{i,0}$$
$$WS_j \leq WS_{j,0}$$

Maximization of grid supplies: $\max z = EG\ (WG)$

st.

General equations

Satisfaction of current grid demands:

$$WG = \sum_j WG_j = WG_0 \left(EG = \sum_j EG_j = EG_0 \right)$$

Decision variables bounds:

$$ES_i, WS_j, E_{ij}, W_{ji} \geq 0$$
$$ES_i, \leq ES_{i,0}$$
$$WS_j \leq WS_{j,0}$$

26

Outline

- **Introduction**
  - Motivation
  - Challenges and objectives
- **A graph-theoretic approach for nexus analysis**
  - Source-sink representation
  - Results from graph theoretic nexus analysis
- **Systematic methods for nexus design and optimization**
  - Graphical approach
  - Mathematical programming-based optimization approach
- Case study: A country-wide nexus
- **Conclusions**

27

Case Study: A Country-wide Nexus (Spain)

- **Data Collected for 2013**
  - Energy/Electricity generation by source
  - Water withdrawal by source

- **Requirements Calculated:**
  - Intensity factors from bibliography
  - Differentiation between:
    - open-loop and
    - closed-loop cooling

Water sources:

- Water produced: 38,105 million m^3
- Electricity Required: 25,355 GWh_e

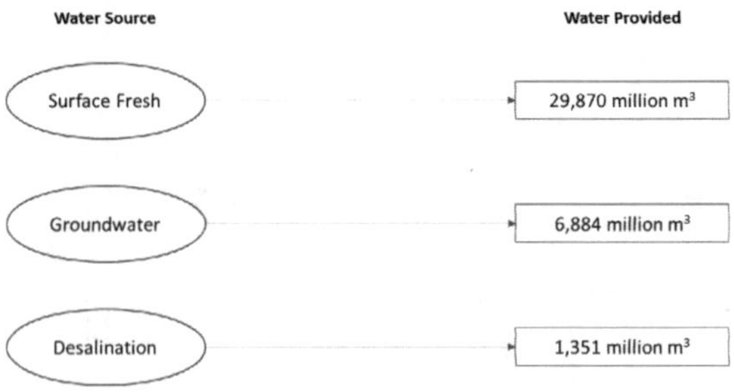

28

Case Study: A Country-wide Nexus (Spain)

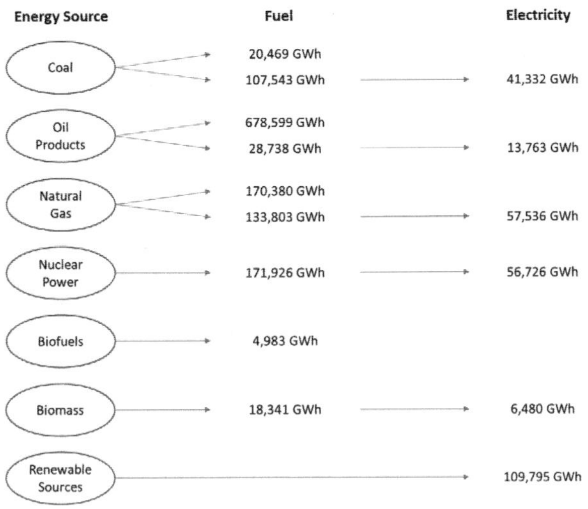

Energy sources:

- Total Fuel: 1,162,854 TJ
- Fuel Grid : 3,147,949 TJ
- Water for Fuel Grid: 7,872 million m^3

Electricity sources:

- Electricity generated: 285,632 GWh_e
- Fuel needed: 3,311,845 GWh
- Water for fuel: 391 million m^3
- Water for electricity: 18,466 million m^3
- Total water: 18,877 million m^3

29

Case Study: A Country-wide Nexus (Spain)

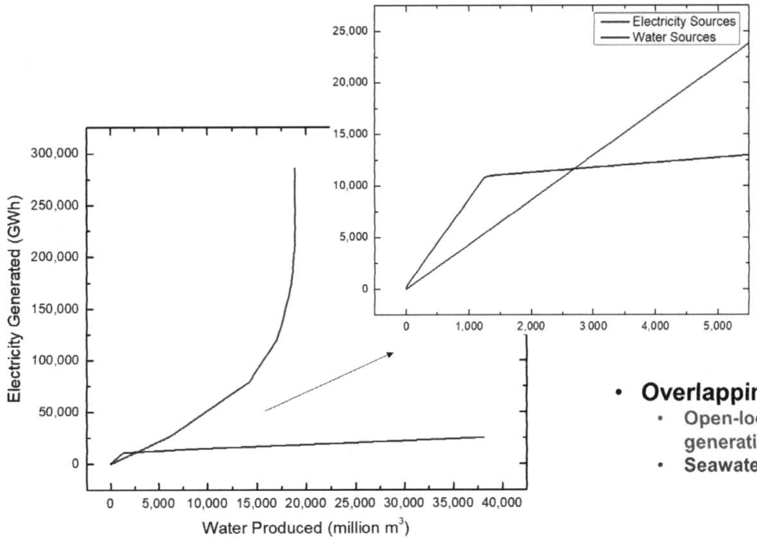

- **Overlapping between:**
  - Open-loop Nuclear power generation
  - Seawater Desalination

30

Case Study: A Country-wide Nexus (Spain)

Energy Source	Current	min GEN	max EGrid
Nuclear (OL)	25,527	13,874	20,147
Biomass (OL)	2,916	2,916	2,916
Coal (OL)	18,599	18,599	18,599
Oil (OL)	6,193	6,193	6,193
NG (OL)	25,891	25,891	25,891
Hydro	41,052	41,052	41,052
Coal (CL)	22,733	22,733	22,733
Nuclear (CL)	31,199	31,199	31,199
Oil (CL)	7,570	7,570	7,570
NG (CL)	31,645	31,645	31,645
Biomass (CL)	3,564	3,564	3,564
Solar Thermal	4,770	4,770	4,770
PV	8,327	8,327	8,327
Wind	55,646	55,646	55,646
Total Electricity	**285,632**	**273,980**	**280,253**
Improvement		-11,652	-5,379
Energy Grid	**260,297**	**260,297**	**265,749**
Improvement			5,453

Water Source	Current	min GEN	max EGrid
MSF	14	-	-
MED	41	-	-
RO	1,188	-	-
ED	108	-	108
Ground Fresh	6,884	5,543	6,884
Surface Fresh	29,870	29,870	29,870
Total Water	**38,105**	**35,413**	**36,862**
Improvement		-2,691	-1,242
Water Grid	**19,227**	**19,227**	**19,227**
Improvement			**0**

- Nexus reduction:
  - 11,652 GWh (4.1%)
  - 2,691 million m^3 (7.6%)

  can be omitted while satisfying existing grid supplies
- Electricity grid maximization:
  - 5,453 GWh → $545.3 million profit
  - Potential profit if exported
  - Potential usage for future increased societal demands

31

Conclusions

- **Source-sink representation of Water-energy nexus established**
- **Water-energy nexus graph utilized for WEN illustration**
- **Water-energy diagram utilized for systematic optimization of WEN**
  - **Isolation of essential and redundant subgraphs**
  - **→Minimization of generation or maximization of grid supplies**
- **Nexus existing infrastructure retrofit design:**
  - **Screening of network's sources and sinks**
  - **Intensity factors → technology efficiency**
- **Potential improvement of Spain's WEN:**
  - **4.1% electricity generation reduction**
  - **7.6% water withdrawal reduction**
- **Future Challenges:**
  - **Distance between process nodes inclusion**
  - **Consider capital and operating cost of sources**

32

Systematic Analysis and Optimization of Water-Energy Nexus

Spyridon D. Tsolas, M. Nazmul Karim, M. M. Faruque Hasan

Artie McFerrin Department of Chemical Engineering
Texas A&M University
College Station, Texas, USA.

ARTIE MCFERRIN DEPARTMENT OF CHEMICAL ENGINEERING TEXAS A&M UNIVERSITY

TEXAS A&M UNIVERSITY Texas A&M Energy Institute

TEXAS A&M ENGINEERING EXPERIMENT STATION

2017 Annual AIChE Meeting, Minneapolis, MN
Session: The Energy-Water Nexus
October 31, 2017

Systems Optimization and mULti-scale analySis (SOULS) Lab: http://chenhasan.wpengine.com

482698 Biologically Active Filters: A Sustainable Treatment Process for Emerging Contaminants

Abstract:

With widespread occurrence and increasing concern of emerging contaminants in source water, the hypothesis that existing filters and adsorbents in water treatment plants can be converted to biologically active filters (BAFs) to treat these compounds was tested. The media included granular activated carbon (GAC) and anthracite/sand dual media, which were collected from two treatment plants. In a bench-scale study with duplicate columns, BAF performance was monitored studying DOC removal, UV_{254} reduction, and adenosine triphosphate (ATP) concentration as function of depth. For GAC BAFs, greater oxygen consumption, increased pH drop, and greater DOC removal normalized to ATP were observed suggesting increased microbial activity as compared to dual-media BAFs. Sixteen emerging contaminants were spiked in the source water. At an 18 min EBCT, GAC BAFs were highly effective with overall removals greater than 80% without pre-ozonation; exceptions included tri(2-chloroethyl) phosphate (TCEP) and iopromide. With a 10 min EBCT, the degree of removal reduced with less than half removed at greater than 80%. The dual-media BAFs showed limited removals with only four compounds removed greater than 80% (acetaminophen, ibuprofen, trimethoprim, and 17Î²-estradiol), and 10 compounds reduced by less than 50% with either EBCT. Pre-ozonation (3 mg/L) improved the removal of PPCPs in GAC and to a greater extent in dual media at both EBCTs. The microbial community on BAF media is believed to play a crucial role in removal of natural organic matter and emerging contaminants. Next generation sequencing (Illumina MiSeq) was used to characterize the microbial communities in the influent, effluent, and media. *Proteobacteria* was observed to be most dominant bacteria in BAF influents and BAF media serving as a source for the latter. *Planctomycetes* phyla were dominant as well in both GAC and dual-media BAF. Based on a factorial analysis, media type was the most significant factor affecting the abundance of five bacterial phyla and 10 bacterial classes. EBCT impacted the abundance of the dominant bacteria phylum *Proteobacteria* at the phyla level, while the effect of pre-ozonation was observed to be significant at the class level for the two dominant bacteria, *Proteobacteria* and *Planctomycetes*.

498192 Optimal Design of Macroscopic Water-Energy Networks Under Uncertainty

Abstract:

Optimal Design of Macroscopic Water-Energy Networks under Uncertainty

Â

Rajib Mukherjee[1], Nasreen A. Elsayed[1], Ramon Gonzalez-Bravo[2], Jose Maria Ponce-Ortega[2], Patrick Linke[3], Mahmoud El-Halwagi[1]

1. Gas and Fuels Research Center, Texas A&M Engineering Experiment Station, College Station, TX 77843, USA

2. Chemical Engineering Department, Universidad Michoacana de San Nicolaì s de Hidalgo, Morelia, Michoacan 58060, Mexico

3. Department of Chemical Engineering, Texas A&M University, Qatar, Education City, Doha

Abstract

The Design of water distribution network should recognize uncertainties in nodal demands, reservoir/tank levels, availability of system components and so on. The inclusion of uncertainty in the model will make the solution robust taking into consideration future projected developments. In this work, optimization of water network that involves residential use, desalination units, cooling systems and industrial processes leading to water-energy nexus, has been performed taking into consideration uncertain demands at different nodes. The model also accounts for sustainable goals through environmental objectives. The network is optimized with both centralized as well as decentralized options. Multiple water sources are also used for optimal selection including desalinated sea water. In this work, we are comparing the deterministic solution with the proposed stochastic solution where uncertain water and energy demands at different time have been used to create the probabilistic model. Excess heat from industries is used to drive water treatment technologies that include thermal membrane desalination. In solving the problem, it is assumed that in future, the demand uncertainty is known but the demand curve is unknown. The stochastic MINLP problem has been formulated as a two stage stochastic integer programming using dual decomposition with Lagrangian relaxation. The designed network is optimal over different scenarios under various conditions. The results are compared with deterministic solutions where the exact demand at different time is known a-priori. Results from the deterministic as well as a probabilistic model applied to an eco-industrial park in Qatar will be presented in this paper.

491116 Water Expulsion from Carbon Rods at High Humidity

Abstract:

Water Expulsion from Carbon Rods at High Humidity

Satish K. Nune, David J. Heldebrant, David Lao, Matthew J. Olszta, Yongsoon Shin, Xiao-Ying Yu and Juan Yao

Porous materials including carbons, silica gels and metalÐorganic frameworks (MOFs) are used for water management in various applications including catalysis, dehumidification and heating, ventilation and air conditioning (HVAC). Materials performance has been enhanced through chemical modifications, surface treatments or impregnation with hygroscopic salts. Regardless of the sorption behaviour/ mechanism, all porous materials exhibit an isotherm in which the quantity of water adsorbed increases with an increase in relative humidity. Recently, we have produced new class of carbon based materials the exhibited unique water adsorption phenomenon. The carbon-based rods can adsorb water at low humidity and spontaneously expel about half the adsorbed water at high relative humidity. Rods synthesised under different reaction conditions (eg., reaction temperature, and the amount of $FeCl_3$ catalyst used) have different morphology, size and exhibited different RH for the onset of evaporation and amount of water expelled. We will present the physical and spectroscopic evidence for this behavior, carbon-based rod growth mechanism and its potential applications.

504631 Development and Demonstration of Advanced Supply Chain Equipment for Harvest, Delivery, and Processing of Herbaceous Biomass

Abstract:

Results to date and future plans from several multi-year cooperatively-funded industry/government development and demonstration projects will be presented. These projects have been co-funded by the U.S. Department of Energyâ€™s Bioenergy Technologies Office to develop and demonstrate innovative new equipment and methods for round- and square-baled herbaceous biomass, and to develop and demonstrate new methods for sustainably building and harvesting future herbaceous biomass supplies. A primary project objective is to demonstrate an advanced biomass supply chain for delivering high impact, high quality feedstocks from the field to the throat of a biorefinery. Photos, video, and measured performance results from the projectâ€™s demonstrations will be provided. Among the developments discussed will be: a new header that will enable efficient variable-rate sustainable harvest of corn stover; a new header enabling a 2x to 3x reduction of field passes for baling equipment; a self-propelled baler operating with crop-adjustable headers and a 6-pack bale accumulator; a bale gathering vehicle capable of collecting 36 bales per trip to the field, self-loading and unloading bale trailers that can reduce load/unload times to 5 minutes per truckload; new automated receiving, handling, pre-processing equipment for round and square bales for biorefinery or biomass depot operations; a near-infrared bale probe that enables instantaneous determination of moisture, ash, glucan, and xylan content in the field or a process line; and harvest and processing operations to help reduce dirt content in biomass delivered to the process. Dirt content in delivered corn stover has been a significant challenge for initial commercial-scale biorefineries. Harvest operations and results from corn stover harvests from one to over two tons per acre, and from switchgrass harvests of up to eight tons per acre will be presented and discussed. Preliminary results of pre-processing testing at Idaho National Labâ€™s Process Demonstration Unit will also be presented and discussed.

504343 Feedstock-Driven Operational Challenges Facing Pioneer Biorefineries in the Emerging Biofuel Industry

Abstract:

Recent evidence indicates that the operational performance of DOE's 1st-plant Integrated Biorefineries (IBRs) are significantly impacted by feedstock variability. This is a high impact problem because several have already failed due in part to these unexpected operational issues. The remaining IBRs are working to solve the issues, however, solving one invariably creates other issues because of interactions across the system.

Conventional feedstock supply systems (CFSS) provide biomass to existing markets that have less stringent quality requirements than do biorefineries. They are vertically integrated, designed around limited markets, single biomass types, single biorefineries and a limited supply radius. In contrast, an Advanced Feedstock Supply System (AFSS) dynamically would provide a "conversion-ready" engineered feedstock from a myriad of diverse raw biomass sources, decoupling the supply of this engineered feedstock from the biorefinery infeed. "Conversion-ready" implies that the engineered feedstock meets all requirements for the conversion process, including physical requirements such as flowability and particle size distribution, as well as more conventional specifications such as convertible composition, convertibility and ash content.

Typically CFSS and AFSS are compared on an nth-plant delivered feedstock cost, which ignores the operational impacts of constant variation in moisture, ash and composition of the biomass entering preprocessing for introduction to the conversion process. A discrete event simulation model was constructed to examine the impacts of reduced throughput and equipment downtime on feed handling and preprocessing systems, replicating the operations and flow of material to the reactor throat of the pretreatment reactor of an integrated corn stover-to-ethanol biorefinery. Results indicate that neither CFSS nor AFSS were capable of achieving design capacity, due primarily to moisture and fines generated during processing. A significant result was the identification of the plug screw feeder to the pretreatment reactor as the primary cause of down time in the modeled pioneer biorefinery scenario, specifically as a result of fines (60-70% of total down time). Moisture-caused failure of grinders far exceeded any other failure, but accounted for only 20-30% of total down time. Additionally, decoupling the feedstock supply from the reactor throat with a five day feedstock buffer only marginally improved minimum scale throughputs for both CFSS and AFSS.

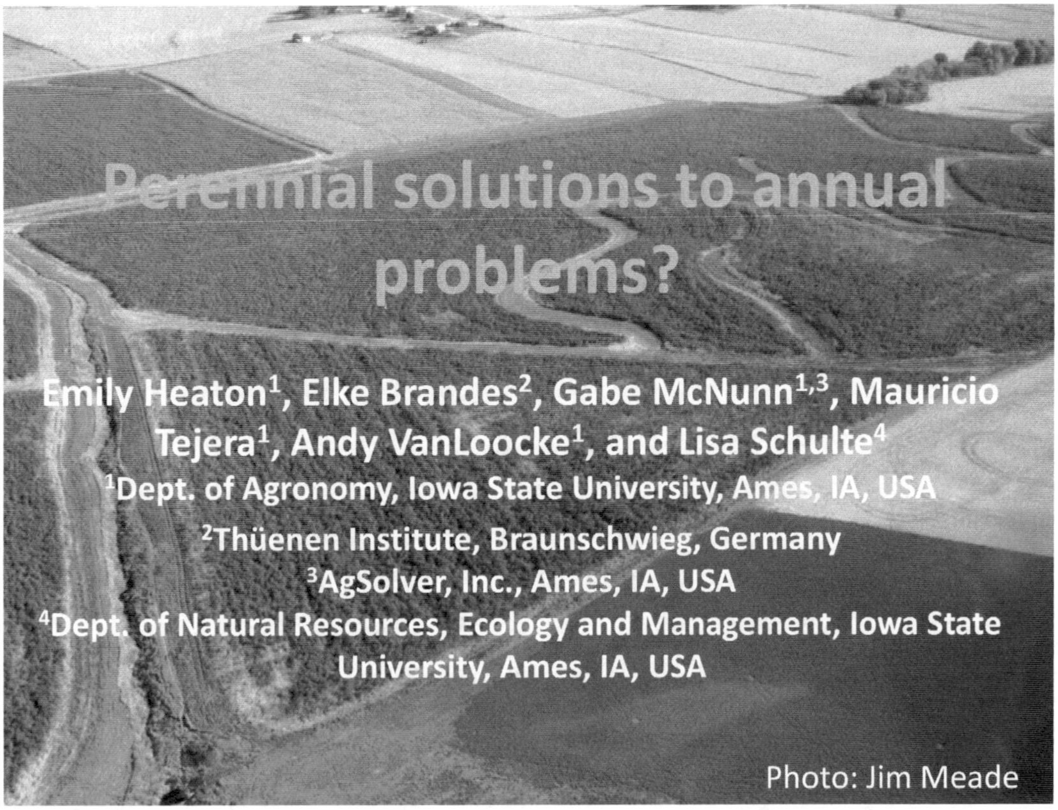
Perennial solutions to annual problems?
Emily Heaton[1], Elke Brandes[2], Gabe McNunn[1,3], Mauricio Tejera[1], Andy VanLoocke[1], and Lisa Schulte[4]
[1]Dept. of Agronomy, Iowa State University, Ames, IA, USA
[2]Thüenen Institute, Braunschwieg, Germany
[3]AgSolver, Inc., Ames, IA, USA
[4]Dept. of Natural Resources, Ecology and Management, Iowa State University, Ames, IA, USA
Photo: Jim Meade

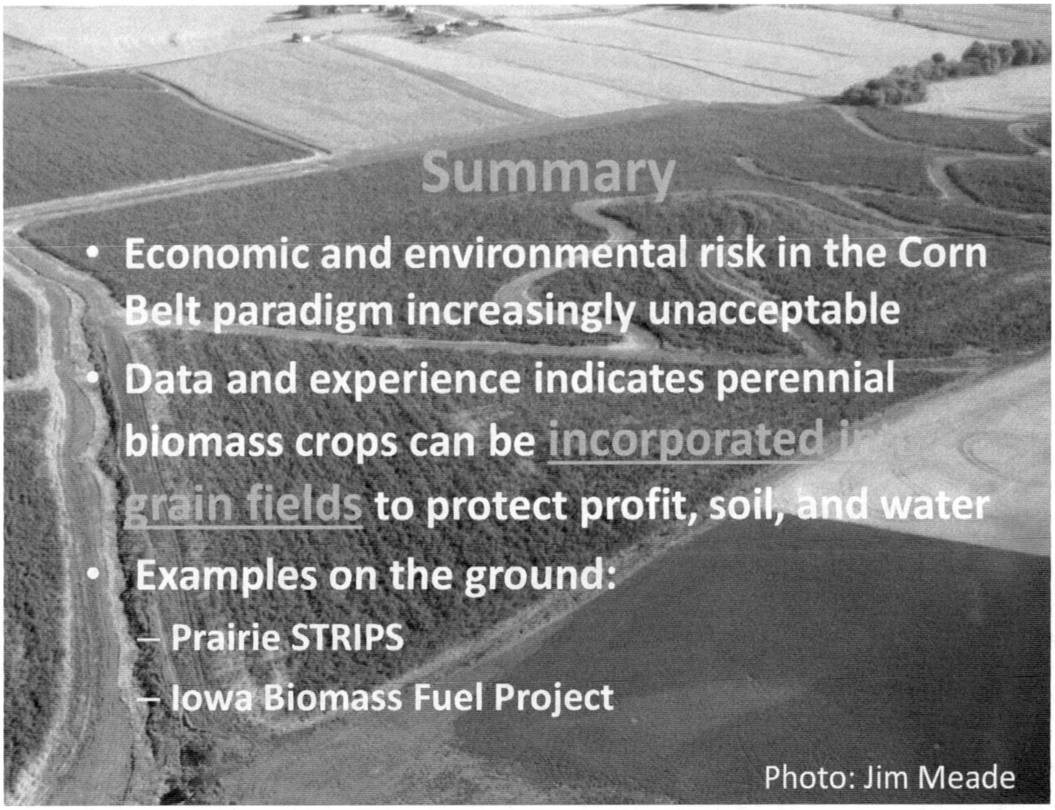
Summary
• Economic and environmental risk in the Corn Belt paradigm increasingly unacceptable
• Data and experience indicates perennial biomass crops can be incorporated into grain fields to protect profit, soil, and water
• Examples on the ground:
– Prairie STRIPS
– Iowa Biomass Fuel Project
Photo: Jim Meade

HARVEST PUBLIC MEDIA

Covering Our Food System From The Heartland

KCUR 89.3 | KBIA | IOWA PUBLIC RADIO

Donate

Home | Food | Energy | In The Field | In Depth | Video | About | Search

Farm Economy Downturn Prompts Fears Of A 'Crisis'

By PEGGY LOWE • JUL 23, 2017

Twitter
Facebook
Google+
Email

Image: Sac County, Tim Youngquist

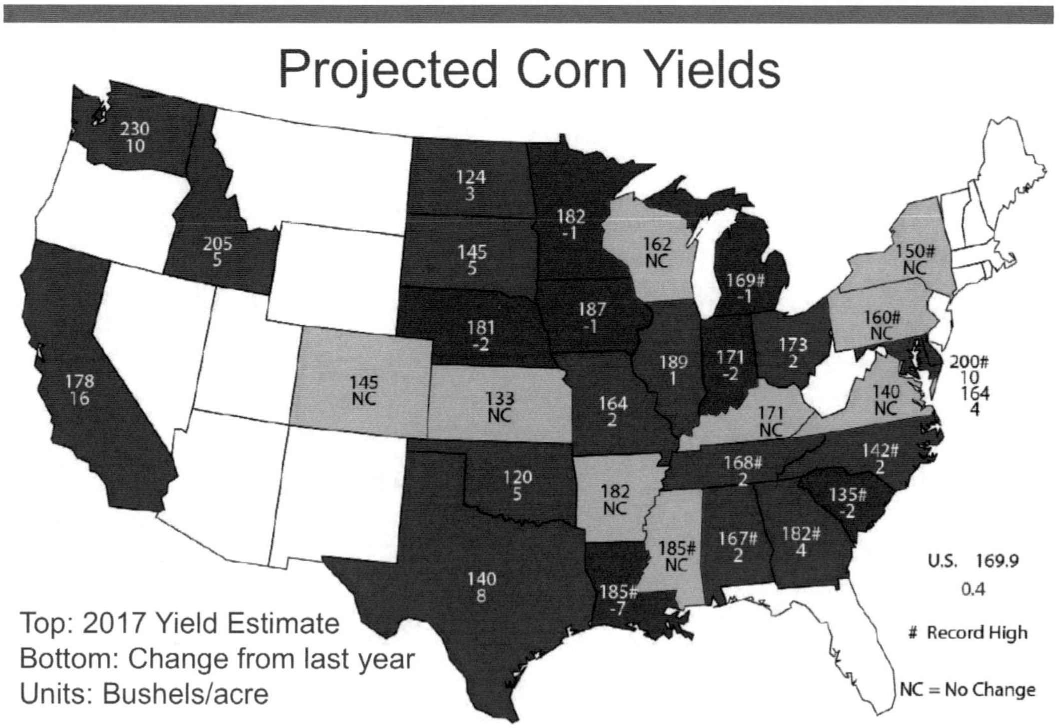

Source: USDA-NASS; slide courtesy of Chad Hart

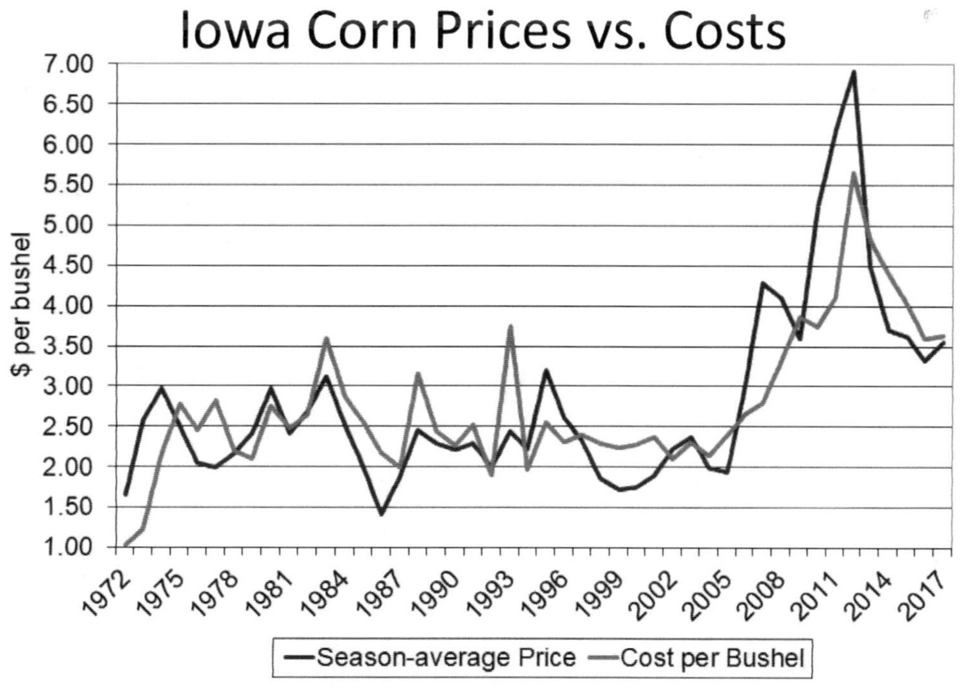

Source: ISU Ag Decision Maker; slide courtesy of Chad Hart

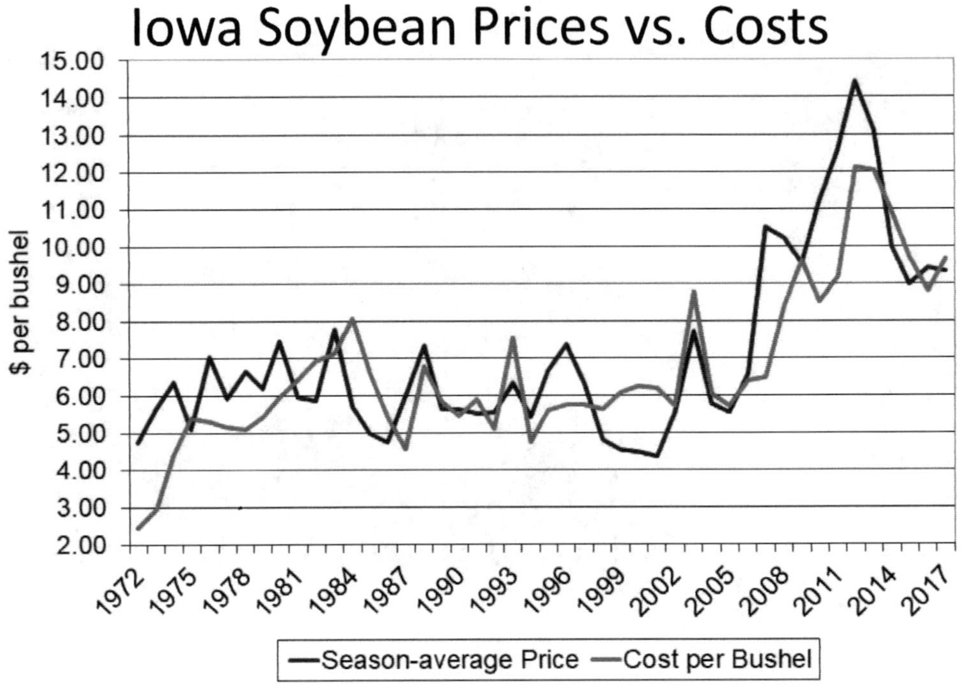

Source: ISU Ag Decision Maker; slide courtesy of Chad Hart

Digging Into the U.S. Farm Balance Sheet

created by: The ERS Farm Income Team

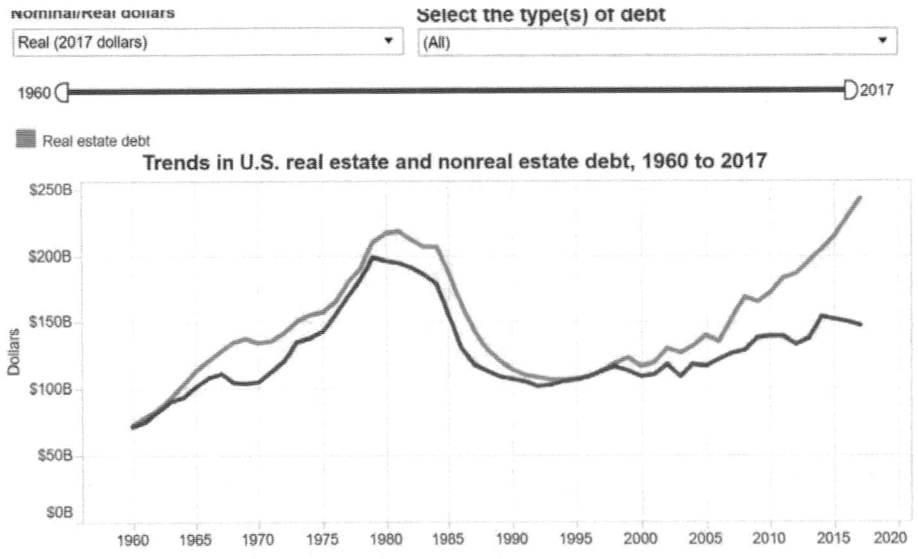

Note: Data for 2017 are forecasts. Values are rounded to the nearest thousand dollars. Inflation-adjusted values use the chain-type GDP deflator. 2017=100 from BEA. DOC. K=thousand. M=million. B=billion. T=trillion.

https://www.ers.usda.gov/data-products/farm-income-and-wealth-statistics

Image: Boone Co., Todd Ontl

Image: Story Co., Drake Larsen

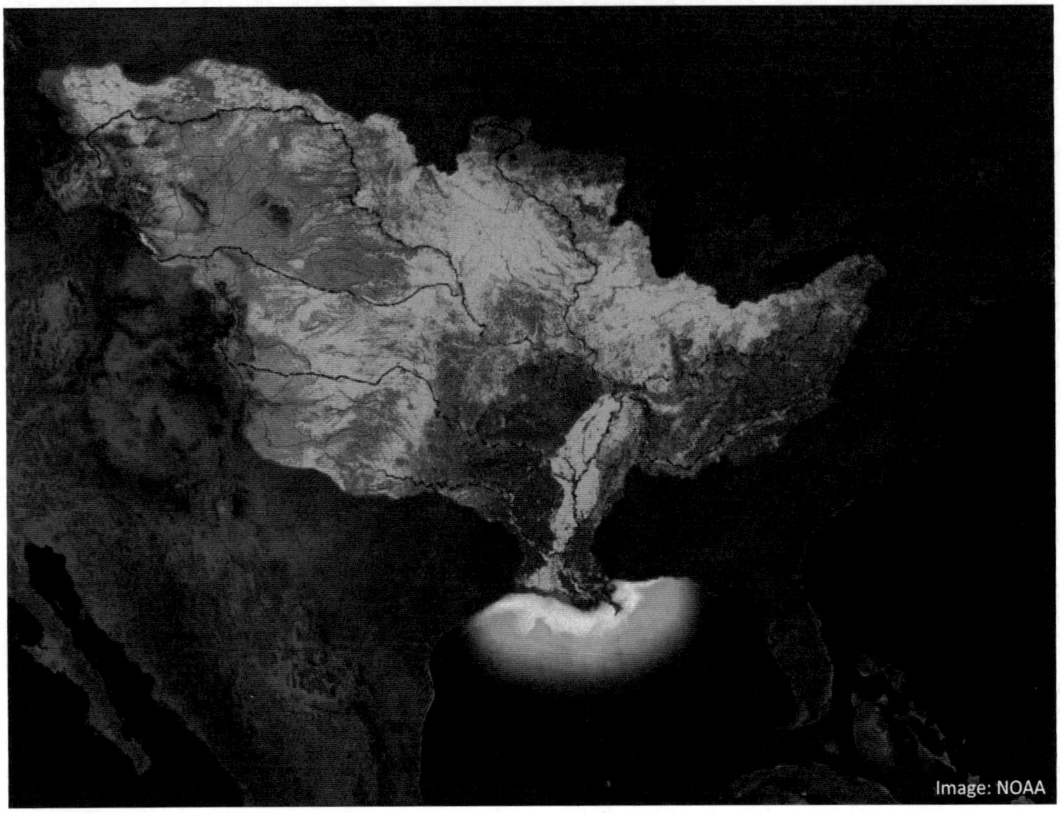
Image: NOAA

Image: Mark Licht

Precision Business Planning Workflow – Alternative Systems as a Business Decision

AGSOLVER

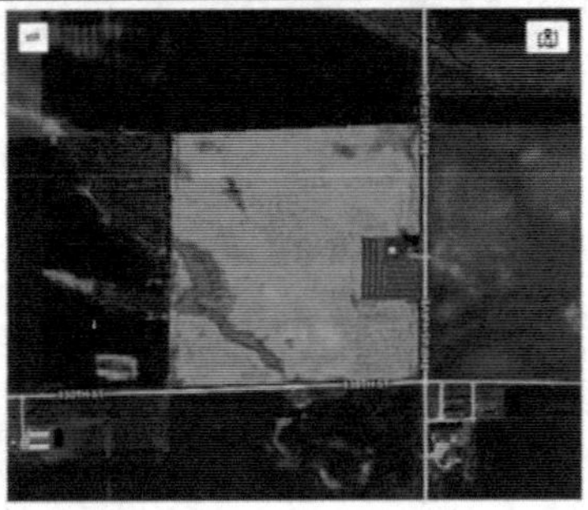

Scenario: Actual Production

Parameter	Value
Field Acreage	143.3 ac
Average Yield	170.2 bu/ac
Profit	$49.63/acre
ROI	6.2 %
Production Efficiency	212.4 bu/$1000
Acreage Opportunity Ratio	23 %
Working Capital Opportunity	$25,973.83
Total Field Expenses	$114,800.50
Total Field Revenue	$121,912.06
Total Field Profit	$7,111.56

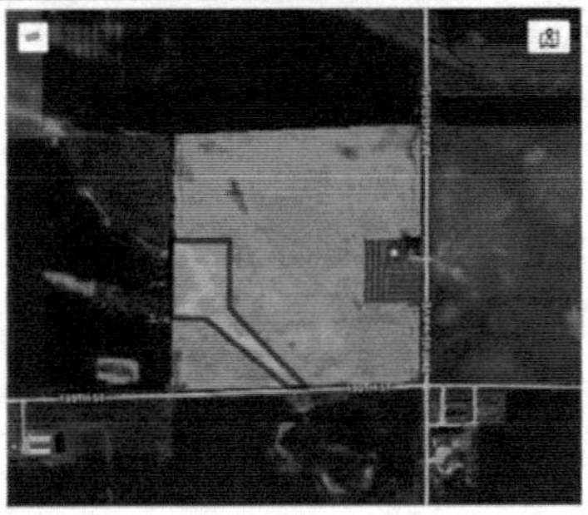

Scenario: Conservation-Final

Parameter	Value
Field Acreage	143.3 ac
Average Yield	179.2 bu/ac
Profit	$93.85/acre
ROI	12.6 %
Production Efficiency	239.7 bu/$1000
Acreage Opportunity Ratio	22 %
Working Capital Opportunity	$19,494.23
Total Field Expenses	$107,085.95
Total Field Revenue	$120,534.99
Total Field Profit	$13,449.04

turning data into decisions for agriculture™

Large Scale Project: What is the economic story?

AGSOLVER

Parameter	Total
number of fields	3,796
total acres	207,937
total acre-years analyzed	641,998
average years of data per field	3.1
average acres per field	54.8
total profit	$ 5,703,472
per field total profit (all years)	$ 1,502
average per acre profit (all years)	$ 27.43
average max annual per acre profit	$ 128.23
average min annual per acre profit	$ (147.78)
total revenue	$ 405,192,182
total expenses	$ 399,488,710
total annualized ROI	0.46%

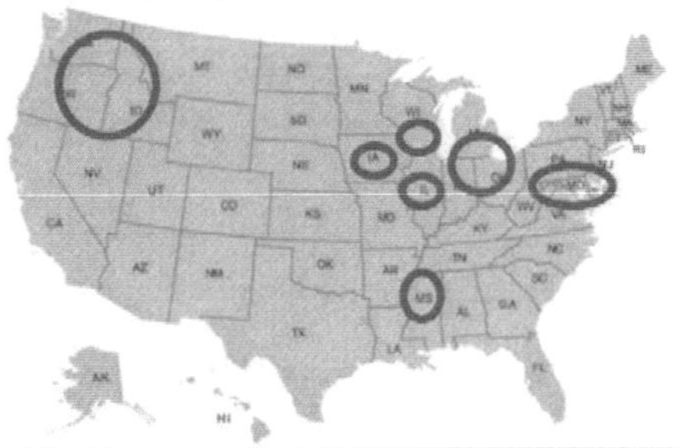

turning data into decisions for agriculture™

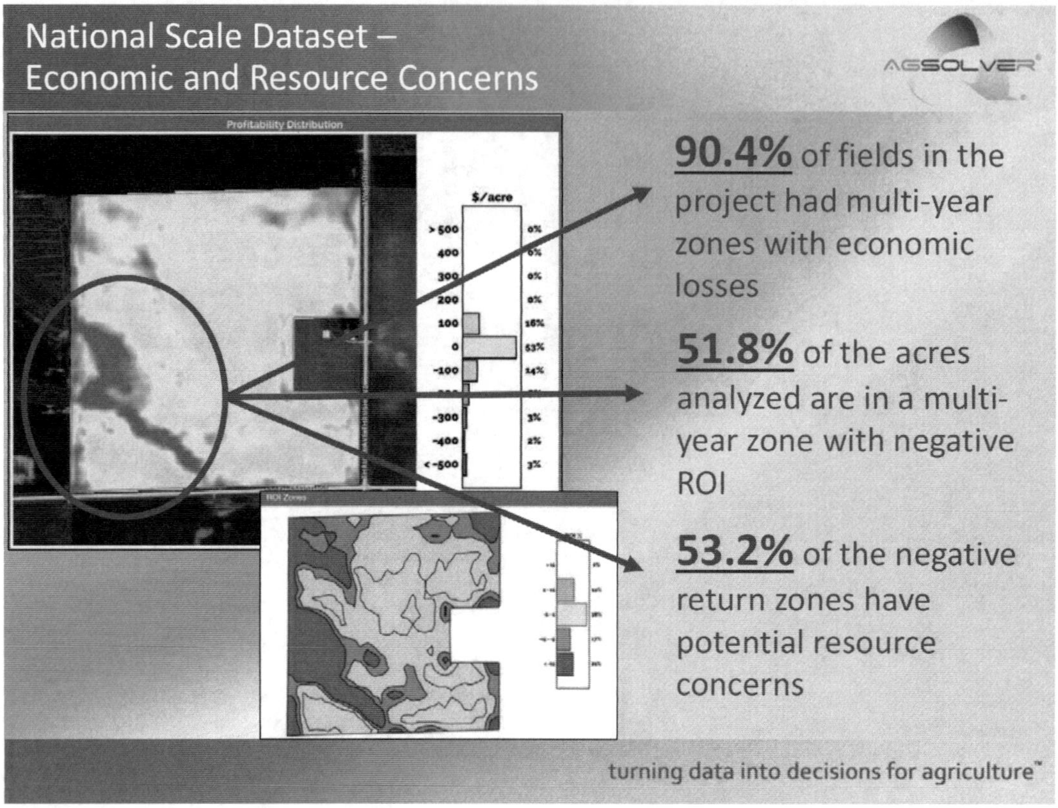

Is there an economic rational to grow

something other

than

corn or soybean in Iowa?

Modeling Framework

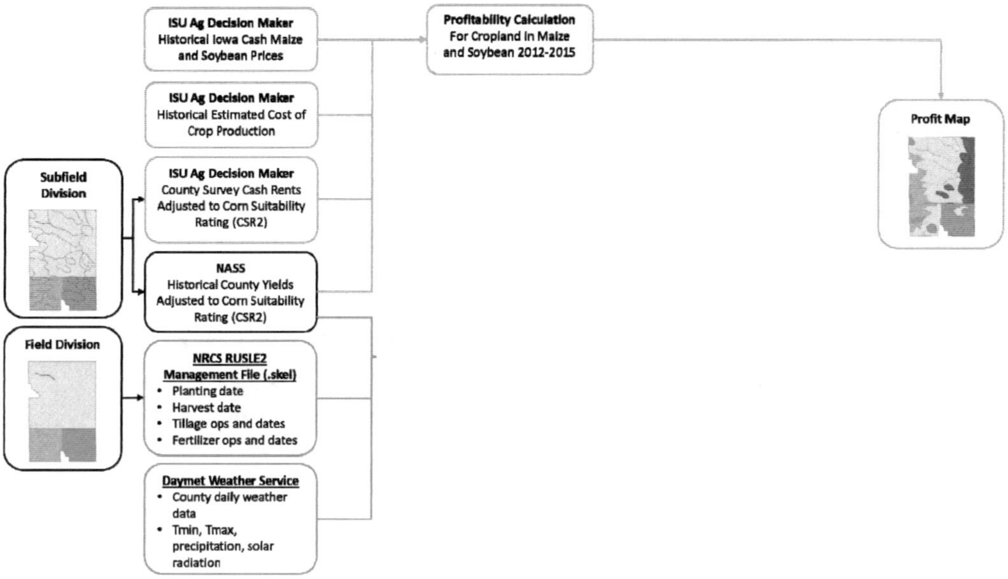

Source: Brandes et al. 2017 Global Change Biology Bioenergy

Corn and soy profitability is down in Iowa.

Some parts of fields are less profitable than others.

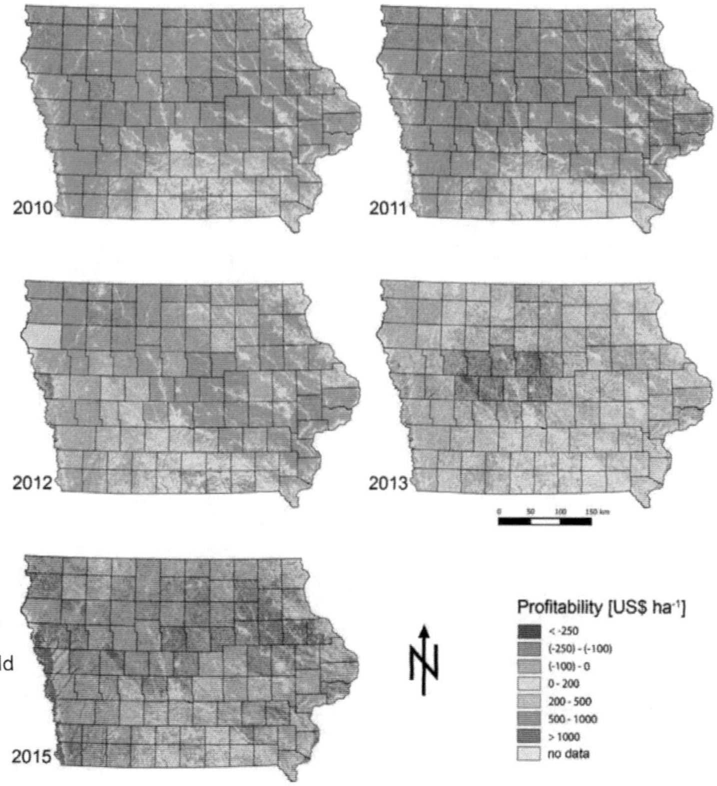

Brandes E, McNunn GS, Schulte LA, Bonner IJ, Muth DJ, Babcock BA, Sharma B, Heaton EA (2016) Subfield profitability analysis reveals an economic case for cropland diversification. *Environmental Research Letters,* **11**, 014009.

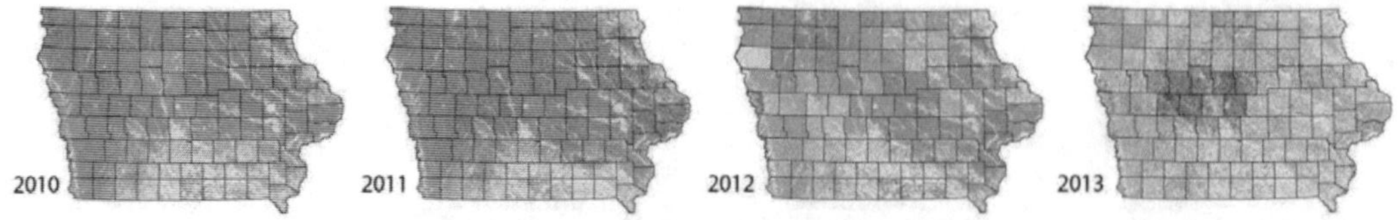

Zoom in to learn more at:
http://mesonet.agron.iastate.edu/GIS/apps/profit/

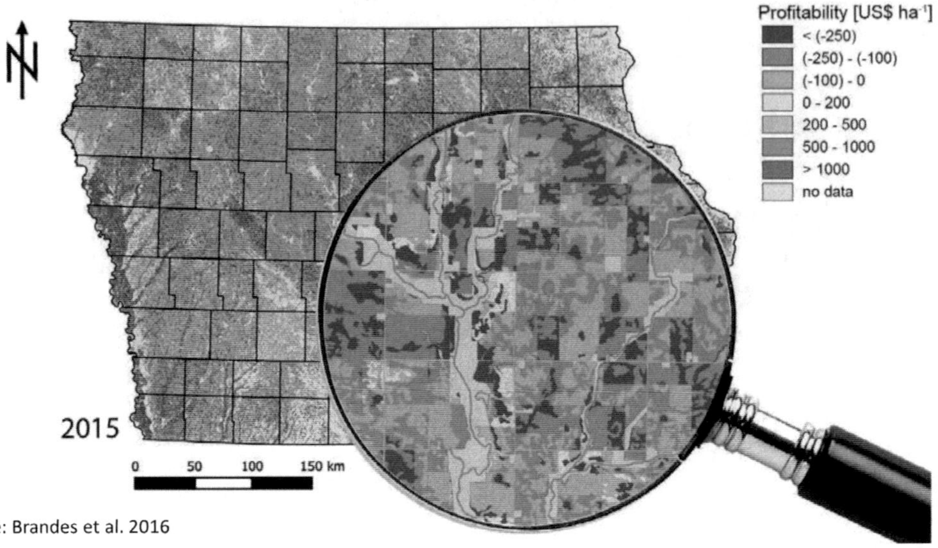

Source: Brandes et al. 2016

Corn/Soy Profitability (avg. 2012-2015)

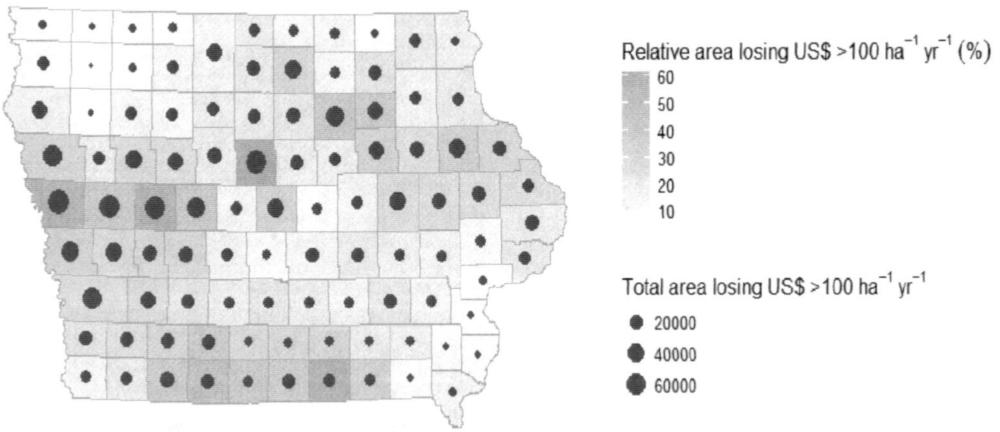

Source: Brandes et al. 2017, in press

What is the impact of replacing
annual crops...

with a high yielding
perennial crop?

Why switchgrass?

1. Year-long activity & potentially cover
2. Deeper, abundant roots
3. 'Liberty' variety trialed nationally by USDA ARS and NIFA CenUSA CAP

Modeling Framework

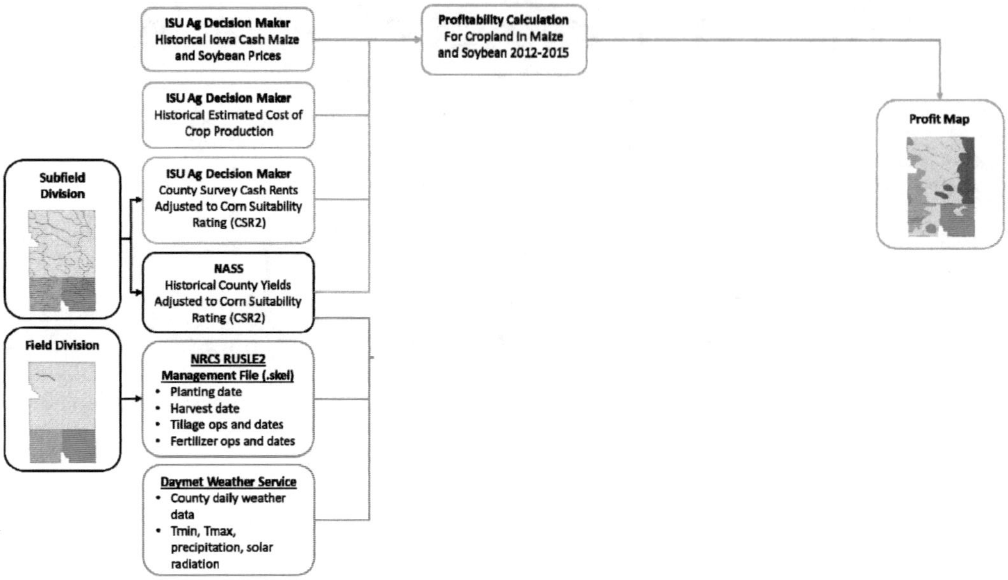

Source: Brandes et al. 2016

3 scenarios

Baseline:

Cropland in corn and/or soybeans between 2012 and 2015

Conservative:

Converting the most unprofitable cropland leaching the most nitrate to switchgrass

Nutrient reduction:

Converting unprofitable cropland with moderate nitrate leaching potential to switchgrass

Baseline nitrogen and profit loss in corn/soy fields

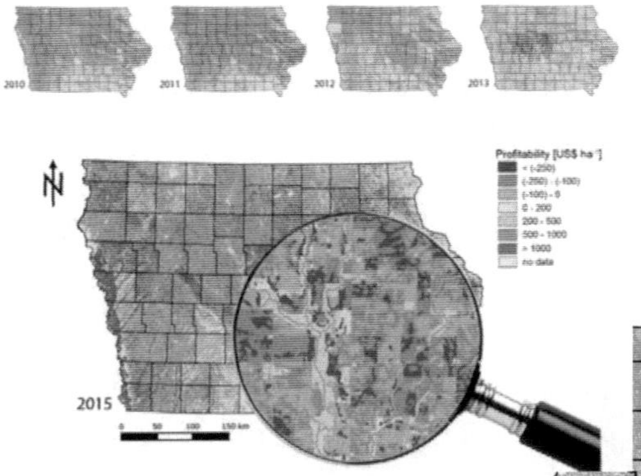

Subfield-scale profitability on land continuously in corn/soy during 2012-2015.

Brandes et al., (2016) *Env. Res. Let.*

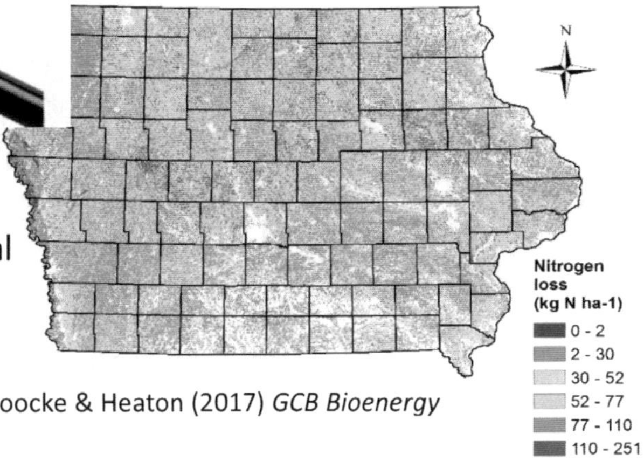

Subfield-scale NO_3-N leaching on corn/soy cropland. Values are annual rates averaged over the years 2012-2015.

Brandes, McNunn, Schulte, Muth, VanLoocke & Heaton (2017) *GCB Bioenergy*

3 scenarios

Baseline:

Cropland in corn and/or soybeans between 2012 and 2015

Conservative:

Converting the most unprofitable cropland leaching the most nitrate to switchgrass

Nutrient reduction:

Converting unprofitable cropland with moderate nitrate leaching potential to switchgrass

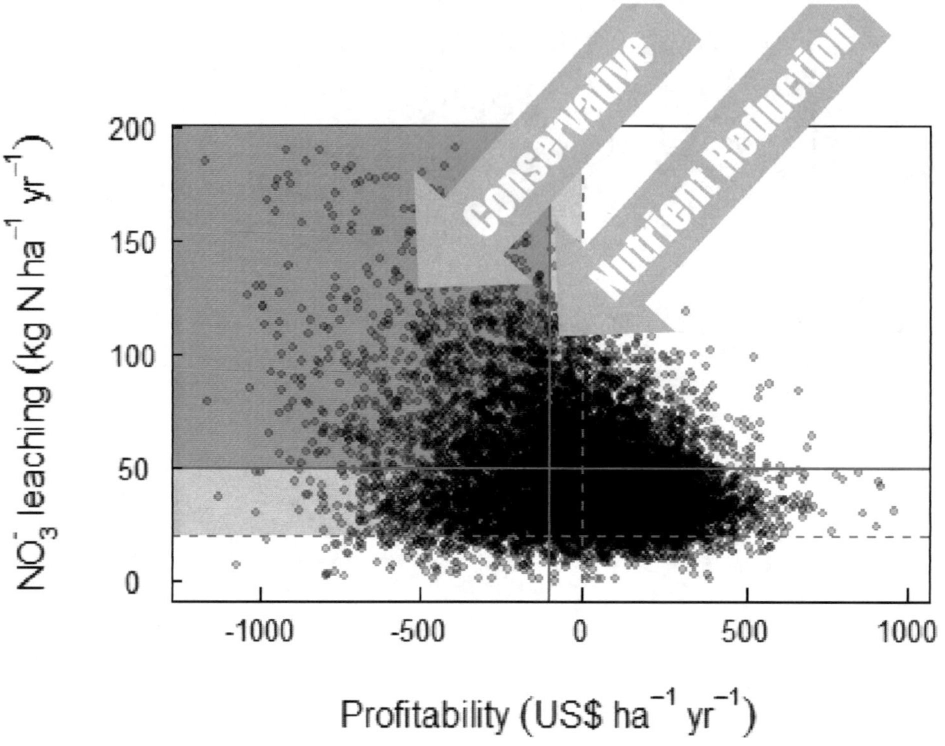

Brandes, McNunn, Schulte, Muth, VanLoocke & Heaton (*in press, GCB Bioenergy*)

Integrating perennials on unprofitable parts of fields can meaningfully increase profitability *and* retain N

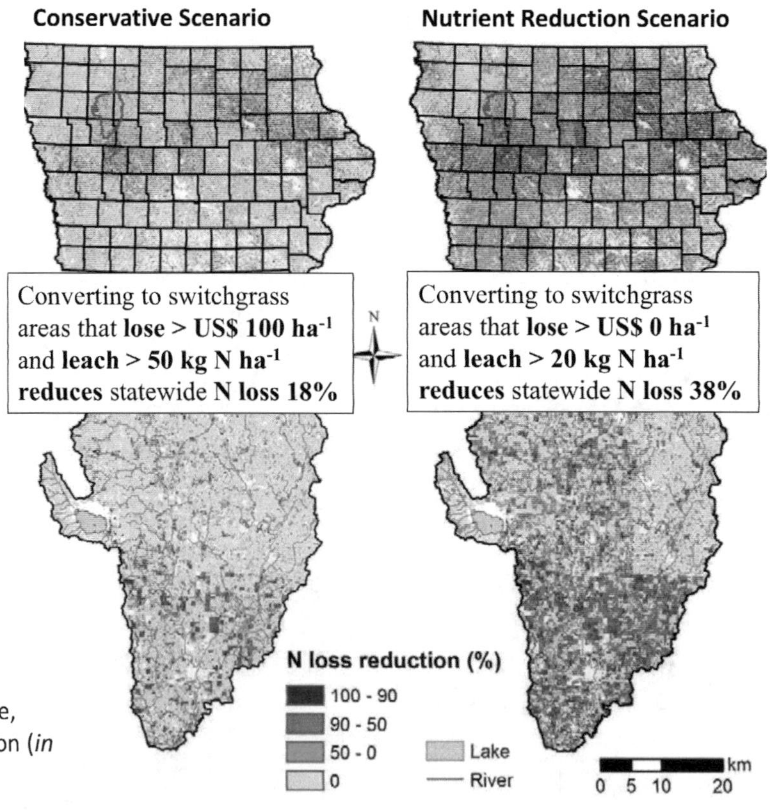

Brandes, McNunn, Schulte, Muth, VanLoocke & Heaton (*in press, GCB Bioenergy*)

Tradeoffs and Biomass

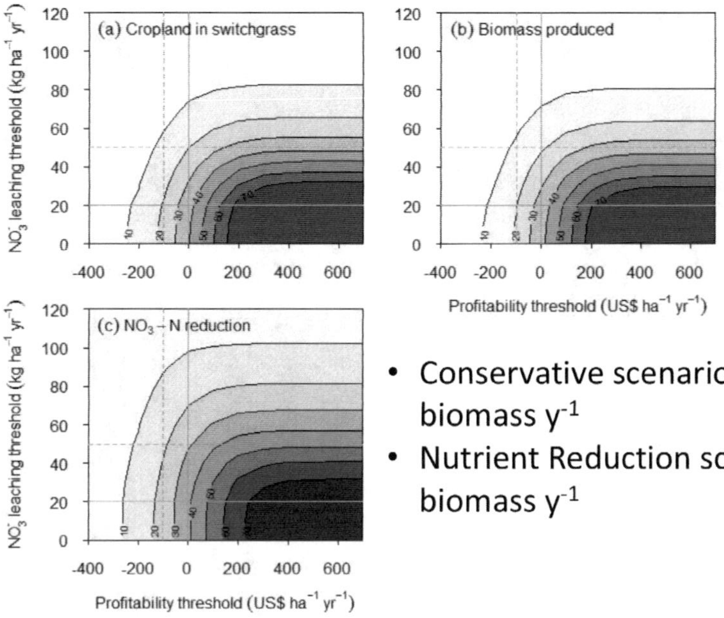

- Conservative scenario: ~10 M Mg biomass y^{-1}
- Nutrient Reduction scenario: ~30 M Mg biomass y^{-1}

Brandes, McNunn, Schulte, Muth, VanLoocke & Heaton (*in press, GCB Bioenergy*)

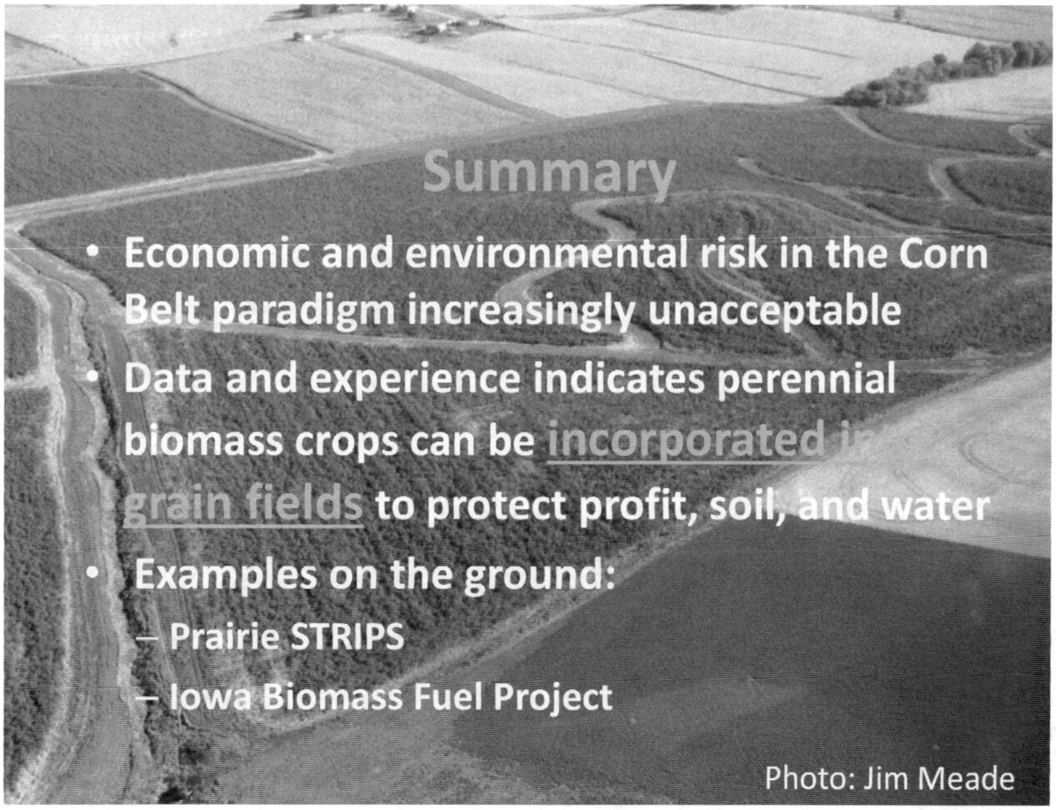

STRIPS
Prairie Strips Collaborator
www.prairiestrips.org
Slide courtesy Lisa Schulte Moore
Photo: Tama Co., Tim Youngquist

Slide courtesy Dr. Lisa Schulte Moore
Photo: Wright Co., Lynn Betts

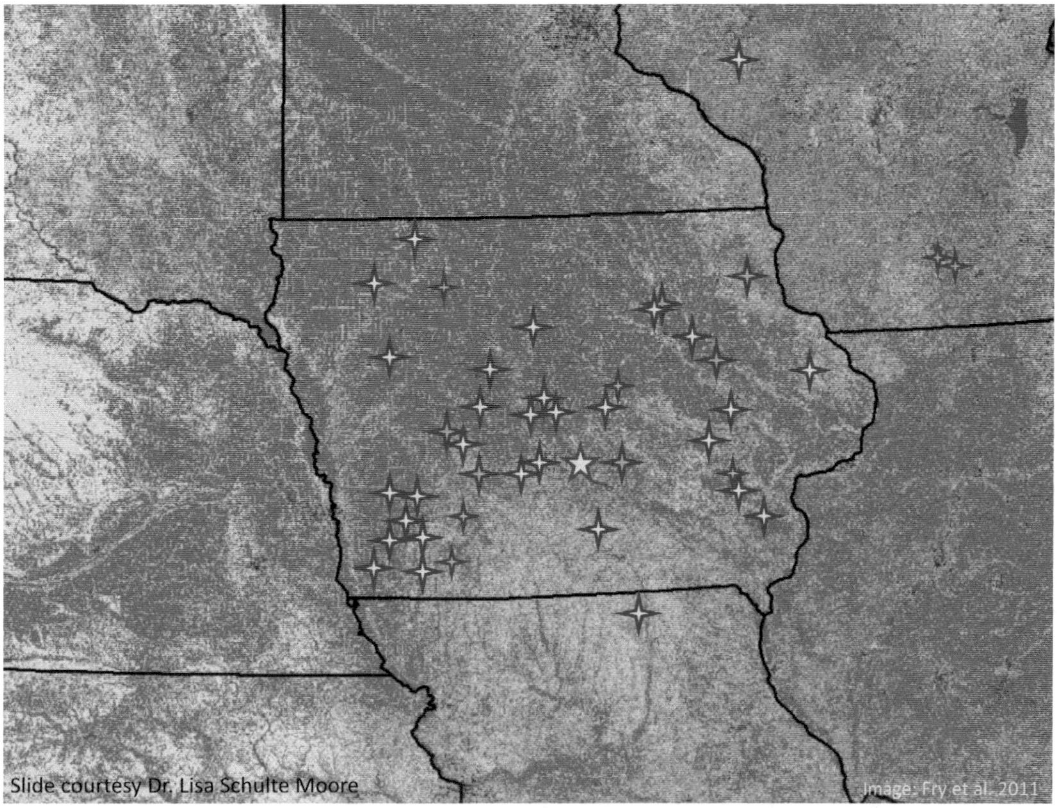

Sources: Zhou et al. 2012, Helmers et al. 2012, Hernandez-Santana et al. 2013, Iqbal et al. 2014, Mitchell et al. 2014, Zhou et al. 2014
Slide courtesy Dr. Lisa Schulte Moore

Strategically adding ~10% prairie to annual row crop fields...

1. 44% reduction in water runoff
2. 95% reduction in soil loss through runoff
3. 90% reduction in phosphorus runoff
4. 84% reduction in nitrogen runoff and 70% reduction in subsurface nitrate loss (not tiled)
5. 2-3 times more beneficial insects and birds
6. No reduction in per acre yields
7. Costs less than terraces; comparable to cover crops

Source: Schulte et al. (2017) PNAS. Data collected by STRIPS team, 2007-2014 at Neal Smith National Wildlife Refuge

Slide courtesy Dr. Lisa Schulte Moore

Pigs & Prairie Goal

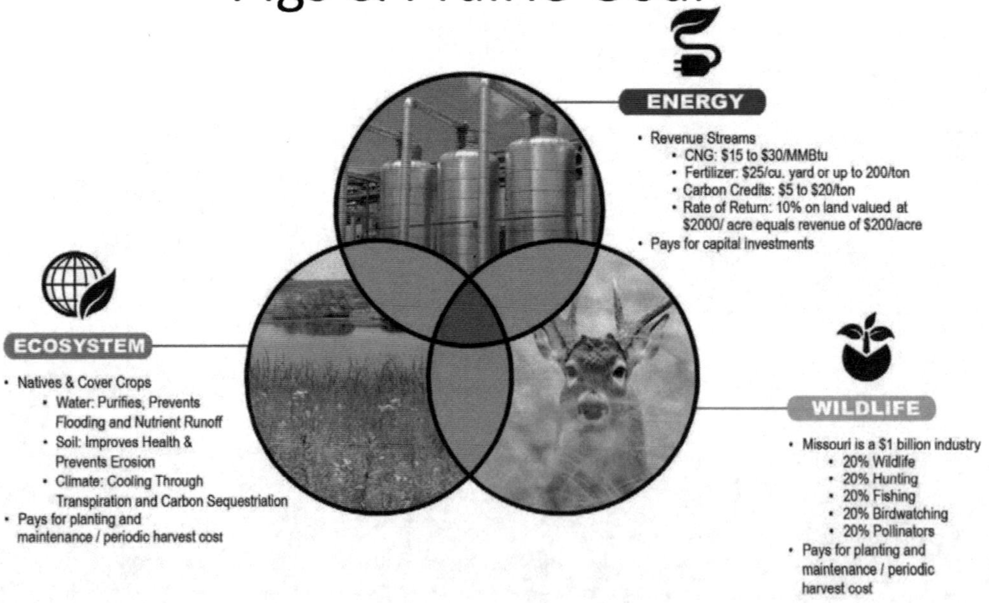

...producing pipeline-quality natural gas ultimately with the energy equivalent of 17 million gallons of diesel fuel each year.

http://sustainability.uiowa.edu/biomass

USDA
United States Department of Agriculture
National Institute of Food and Agriculture

Facilities Management
The University of Iowa

Conversion
- (processing)
- combustion
- heat & power

AGgrow TECH

Agricultural Services
- planting
- management
- harvest

Iowa State University
Department of Agronomy

Agronomy
- research
- extension
  - nitrogen, water, temperature

iowa energy center

Leopold Center for Sustainable Agriculture

Iowa EPSCoR

New Holland Agriculture

Current Biomass Fuels at U Iowa

- Total fuel budget = \$14.6M

- Biomass demand by 2025: 110,000 t y[1]

BOILER	FUEL	CAPACITY (LB/hour)
7	Gas	140,000
8	Gas	145,000
10	Coal, Biomass, & Gas	170,000
11	Coal & Biomass	170,000

Images courtesy Ingrid Gronstal Anderson and Erin Hazen

Giant miscanthus (*Miscanthus × giganteus*)

Looking ahead:
- Densification - blend into Energy Pellets
- Optimize Storage & Logistics
- Focus on minimum time from field-to-boiler

- Crop selected & managed in close partnership with Iowa State University and AgGrow Tech (formerly Repreve Renewables)
- Perennial, grows well in marginal soil
- Displaces 4 tons coal per planted acre
- Co-fired in stoker boiler at 5% by heat (37.5% Volume)
- Over 800 acres planted since 2013
- Target 2500 acres (22,500t) = 10% of fuel mix

University of Iowa Miscanthus Business Plan Review

		Manageable Risk Profile
TECH	Miscanthus Combustibility	low
	Compatible Spec	mid
	Technology to Produce Spec	high
	Technology Development	high
OPERATIONAL	Land	
	Participation in Project Area	low
	Land Suitability	low
	Crop Establishment	
	Planting Stock / Material Source	low
	Planting Schedule	mid
	Planting Method	low
	Successful Crop Establishment	mid
	Annual Yield Expectations	mid
	Winterkill / Crop Loss	mid
	Invasive potential	low
	Supply Chain	
	Harvesting	mid
	Storage	low
	Collection / Transportation	low

Slide courtesy Ferman Milster

Long Term Assessment of Miscanthus Productivity and Sustainability (LAMPS)

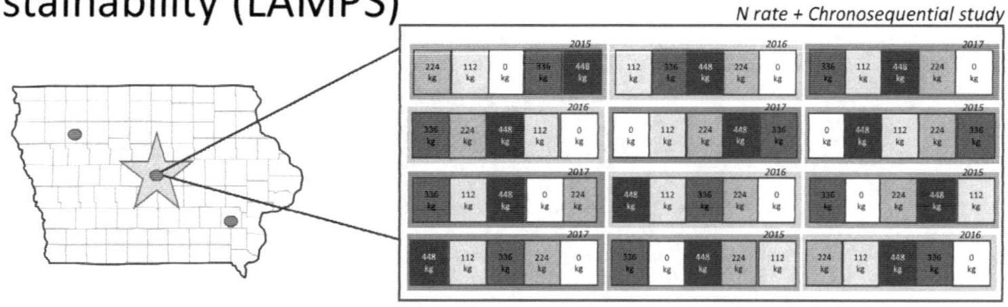

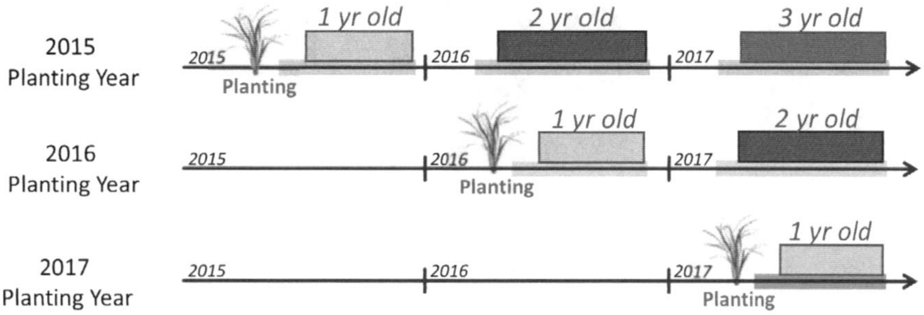

Slide courtesy Mauri Tejera

IMPACT

By burning Miscanthus instead of coal the UI will annually:

- keep **\$10 million y^{-1} in Iowa** by buying biomass instead of out-of-state coal
- Reduce **nitrate leaching and soil loss by ~90%** while building soil and providing wildlife habitat
- Provide farmers **~\$200/acre**, similar to Conservation Reserve Program (CRP payments)
- directly displace **>50,000 tons of fossil CO_2**, (>20 tons of CO_2 per acre from >4 tons of coal per acre)

References and more information at:

- http://faculty.agron.iastate.edu/heaton/
- Facebook and Twitter: @ISUBiomass
- www.prairiestrips.org
- http://sustainability.uiowa.edu/biomass
- www.agsolver.com
- www.newbio.psu.edu

Questions?
WORKMASTER 55
LAND

504321 Methanol to Propylene Process: An Integration of Catalyst Development and Reactor Engineering

Abstract:

Methanol, which can be produced from coal or natural gas, is one of the most important C1 chemicals. The utilization of methanol to make light olefins including ethylene and propylene has been recently commercialized. However, in many applications, it is desired to maximize the yield of propylene, and thus methanol to propylene (MTP) process becomes increasingly interesting. It is normally hard to achieve high selectivity to propylene in methanol conversion. In this work, we have developed a multi-functional catalyst, over which three reactions, i.e. methanol conversion, ethylene alkylation, and C4+ cracking, can be catalyzed. By integrating these three reactions, we attempted to develop a MTP process with the selectivity to propylene higher than 75%. Evaluation of this catalyst has been carried out in laboratory-scale micro-fluidized bed reactor. It shows that the selectivity to propylene of 50% has been achieved in the methanol conversion reaction. We also studied the ethylene alkylation reaction and C4+ cracking reaction, and identified the optimal operation conditions. Based on the results obtained in the micro-fluidized bed reactor, we proposed an integrated process to maximize the selectivity of propylene. In achieving this goal, a turbulent fluidized bed reactor, in which the optimal operation conditions suitable for both methanol conversion and ethylene alkylation have been considered, has been designed and constructed in our pilot plant. The reactor can be operated with 300t/a methanol feed rate. Pilot plant experiments have been conducted for validating the process concept and scaling up the reactor. It shows the selectivity to propylene of 75%, in addition to a selectivity of ethylene of 10%, can be achieved in the pilot plant. This MTP process can be seen as a successful example of the integration of catalyst development and reactor engineering.

504320 Direct Production of Value-Added Chemicals Via Fischer-Tropsch Synthesis

Abstract:

Fischer-Tropsch synthesis (FTS) is a heterogeneous catalytic process for sustainable production of fuel and value-added chemicals via syngas derived from coal, biomass and natural gas. It is considered as an attractive alternative non-petroleum-based production route. FTS is a very complicated catalytic reaction process with co-production of paraffin, olefin, alcohol and other products. The direct synthesis of high value-added olefins and higher alcohols from syngas has drawn particular attention due to due to its process simplicity, low energy consumption and clean utilization of carbon resource. For the purpose of high selectivity to specific olefins or higher alcohols, it is very desirable to regulate the reaction network for certain reactions to take place preferentially. Obviously, it is necessary to control CO activation and then both carbon chain growth and termination for high selectivity to target product by development new catalyst systems and optimization of reaction conditions.

Reaction network during syngas conversion was firstly investigated by probe molecules techniques [1]. Alcohol, aldehyde, olefin as probe molecules were in-situ introduced into the reaction system. Several parallel reactions and second reactions were clarified. The existence of surface acyl species indicated that the insertion of CO rather than hydroxyl was the underlying mechanism for higher alcohols synthesis over modified FT catalysts along with the support from surface simulation. In addition, the dual active sites for higher alcohols formation was further investigated [2]. Several kinds of dual active sites (Cu/FeC_x, CuCo, and Co/Co_2C) were controllably prepared for higher alcohol synthesis. Both Cu/FeC_x and Co/Co_2C could afford high selectivity for long chain alcohols, while only short chain alcohols were produced for CuCo. The difference of the detailed surface structure and the corresponding catalytic performance was discussed in detail.

Fischer-Tropsch to olefins (FTO) by using new active phase was also investigated. Co_2C nanoprisms with exposed facet of {101} and {020} were prepared and tested for syngas conversion [3]. It is widely accepted the formation of Co_2C is considered to be one of the main reasons for deactivation of the Co-based FT reaction because it possesses very low activity for CO hydrogenation, with methane and CO_2 as the main products. However, due to the strong facet effect of Co_2C, Co_2C nanoprisms afforded very high selectivity for the production of lower olefins (60.8 C%), while little methane (about 5.0 C%) was generated with the ratio of olefin/paraffin amongst the $C_{2\text{-}4}$ products being as high 30. The product distribution deviates markedly from the classical ASF distribution with the highest selectivity to propylene. The effect of several promoters and the model for modified ASF distribution were discussed [4]. In addition, the nanoeffects of Co_2C were studied in details to reveal the structure-performance relationship.

References

[1] *Catalysis Science & Technology*ï¼Œ2015, 5, 4224

[2] *Catalysis Science & Technology*ï¼Œ2013, 3, 1591

[3] *Nature*, 2016, 538, 84..

[4] *ACS Catalysis*, 2017, 7, 3622.

504319 An Investigation into the Effect of Li and Mn Promotions on the Activity and Selecivities to Olefins and Alcohols of Co@Co_2C/Activated Carbon (AC) for Fischer-Tropsch Reaction

Abstract:

Abstract:

Utilization of non-precious transition metals for high alcohols and olefins synthesis is of a great importance in heterogeneous catalysis. We synthesized successfully Li and Mn promoted AC supported cobalt metal-carbide (Co@Co_2C/AC) catalysts, which present remarkable activity of more than 200g/kg-cat.h and selectivities of ca. 65 C% both for alpha-alcohols and alpha-olefins via the Fischer-Tropsch reaction at the mild conditions. The formation of the stable cobalt carbide and the Co@Co_2C interface are found essential for the observed reactivity. The results of characterization from In-situ EXAFS, in-situ XRD, HRTEM, XSP, and chemo-adsorption of H_2 and CO experiments showed that Mn promotes the formation of Co_2C species during Fischer-Tropsch reaction and Li increased the CO uptake, while both promoters have little impact on the amount of H_2 uptakes. Density functional theory calculations show that Co_2C is highly efficient for CO non-dissociative adsorption, behaving as noble-metal-like, whereas the Co metal is highly active for CO dissociative adsorption and the subsequent carbon-chain growth. The interface between the cobalt metal and its carbide phase as well as the dual sites available at the interface for facile CO insertion to C_nH_m-groups formed on the Co sites. The remarkably high alpha-alcohol and alpha-olefins selectivities from the Li and Mn promotion effects, namely the formation of more interfacial active sites between Co and Co_2C and enhancement of the ratio for chemo-adsorption CO to H_2 uptake on the surface of Co@Co_2C/AC catalysts, paves a new way for the design of the new catalysts for synthesizing various high value-products in syngas applications.

504154 Direct Production of Gasoline-Range Hydrocarbons from Carbon Dioxide over Iron-Based Multifunctional Catalysts

Abstract:

Converting CO_2 from a detrimental greenhouse gas into value-added liquid fuels not only contributes to mitigating CO_2 emissions, but also reduces dependence on petrochemicals. Unfortunately, the activation of CO_2 and its hydrogenation to hydrocarbons or other alcohols are challenging tasks because CO_2 is a fully oxidized, thermodynamically stable and chemically inert molecule. Most research to date, not surprisingly, is focusing on selective hydrogenation of CO_2 to short-chain products, while few studies on long-chain hydrocarbons, such as gasoline-range (C_5–C_{11}) hydrocarbons. The key to this process is to search for a high efficient catalyst.

Herein, we have succeeded in preparing high efficient, stable, and multifunctional Na–Fe_3O_4/Zeolite catalysts for the direct production of gasoline-range (C_5–C_{11}) hydrocarbons from CO_2 hydrogenation. This catalyst displayed record selectivity towards C_5–C_{11} hydrocarbons (78%) as well as low CH_4 and CO selectivity under industrial relevant conditions, and gasoline fraction are mainly isoparaffins and aromatics thus favouring the octane number. Moreover, the composition of C_5–C_{11} can be tuned by the choice of zeolite type and the integration manner of multifunctional catalyst. When Na–Fe_3O_4 nanocatalyst was matched with HZSM-5 zeolite and integrated by granule-mixing manner, up to 63% of aromatics in gasoline fraction were produced. While Na–Fe_3O_4 was matched with HMCM-22 zeolite, up to 60% of isoparaffins in gasoline fraction were obtained under the dual-bed configuration. Not only that, the multifunctional catalyst exhibited a remarkable stability for 1,000 h on stream, which definitely has the potential to be a promising industrial catalyst for CO_2 utilization to liquid fuels.

In-depth characterizations indicate that, during CO_2 hydrogenation reaction, this catalyst enables RWGS over Fe_3O_4 sites, olefin synthesis over Fe_5C_2 sites, and oligomerization/aromatization/isomerization over zeolite acid sites. The concerted action of the active sites calls for precise control of their structures and proximity. This study paves a new path for the synthesis of liquid fuels by utilizing CO_2 and H_2. Furthermore, it provides an important approach for dealing with the intermittency of renewable sources (sun, wind and so on) by storing energy in liquid fuels.

References

1. G. A. Olah, A. Goeppert and G. S. Prakash, *J. Org. Chem.*, 2008, **74**, 487-498.

2. R. W. Dorner, D. R. Hardy, et al., *Energy Environ. Sci.*, 2010, **3**, 884-890.

3. W. Wang, S. Wang, X. Ma and J. Gong, *Chem. Soc. Rev.*, 2011, **40**, 3703-3727.

4. F. Jiao, J. Li, X. Pan, et al., *Science*, 2016, **351**, 1065-1068.

5. P. Zhai, C. Xu, R. Gao, et al., *Angew. Chem. Int. Ed.*, 2016, **55**, 9902-9907.

6. J. Wei, Q. Ge, R.Yao, et al., *Nat. Commun.*, 2017, **8**, 15174.

Innovative Syngas Production Catalyst for Utilization of CO_2 (CT-CO_2AR™)

Takenori Kanda, Osamu Hirohata, Tomoyuki Mikuriya and Fuyuki Yagi
Chiyoda Corporation, Kanagawa, Japan

1. Introduction

Production of synthesis gas from natural gas and conversion of synthesis gas to fuels and chemicals have become increasingly attractive from the viewpoint of energy resource diversification. The technology of synthesis gas production by the reforming of natural gas with CO_2 (hereafter referred to as "CO_2-reforming") is an effective solution for the issues of efficient energy resource utilization and environmental preservation.

Chiyoda has developed and commercialized a new noble metal based reforming catalyst named CT-CO_2AR™ which has a higher resistance to carbon formation and produces synthesis gas more effectively compared with commercial reforming catalyst.

In this paper, we will elaborate on the characteristics and the competitiveness of the CT-CO_2AR™ catalyst and the results of the operation of the commercial plant.

2. Characteristics of CT-CO_2AR™ catalyst

(1) The characteristics of CO_2/steam reforming

Synthesis gas is used for hydrogen production and as the feed stock of chemicals, and the H_2/CO ratio of synthesis gas depends on the purpose.

For example, hydrogen production requires an H_2/CO ratio of more than 3.0. For ethylene glycol production, the required H_2/CO ratio is 2.0. For chemical production through oxo synthesis, the required H_2/CO ratio is 1.0. For the feedstock of acetic acid or MMA requires only CO or as low as possible H_2/CO ratio.

When hydrocarbons such as natural gas or naphtha and steam are used as reformer feedstocks, the steam reforming reaction shown in Eq. 1 is normal and the H_2/CO ratio of produced synthesis gas becomes more than 3.0.

$$CH_4 + H_2O \Leftrightarrow 3\,H_2 + CO \qquad (1)$$

However, when CO_2 and steam are used as reforming agents, the CO_2 reforming reaction shown in Eq. 2 occurs.

$$CH_4 + CO_2 \Leftrightarrow 2\,H_2 + 2\,CO \qquad (2)$$

Therefore, by changing the ratio of feed CO_2 and steam, various H_2/CO ratios of synthesis gas can be produced.

However, during the reforming reactions, the carbon formation reaction shown in Eq. 3 occurs on the catalyst surface. Deposited carbon causes the deactivation of the catalyst and the occlusion of the catalyst tube[1] leading to an unstable operation.

$$CH_4 \Leftrightarrow C\,(s) + 2\,H_2 \qquad (3)$$

To prevent carbon formation for a stable operation, commercial Ni based reforming catalysts require steam and CO_2 greater than the stoichiometric ratio.

Fig. 1 shows the relation between the operation condition and risk of carbon formation[2]. The bold line in the center represents carbon limit line at 600 ℃. In the black area to the left of this line, the process gas has enough high thermodynamic potential for carbon formation at 600 ℃. Operation around this line cannot be carried out with commercial Ni based reforming catalysts. They are usually operated within the excess steam carbon ratio (S/C) zone (blue area).

On the other hand, our new commercialized noble metal based reforming catalyst (CT-CO_2AR™) enables a stable operation under a lower S/C condition (red area) because it has a higher resistance to carbon formation.

In next section, the advantages of a lower S/C operation will be shown.

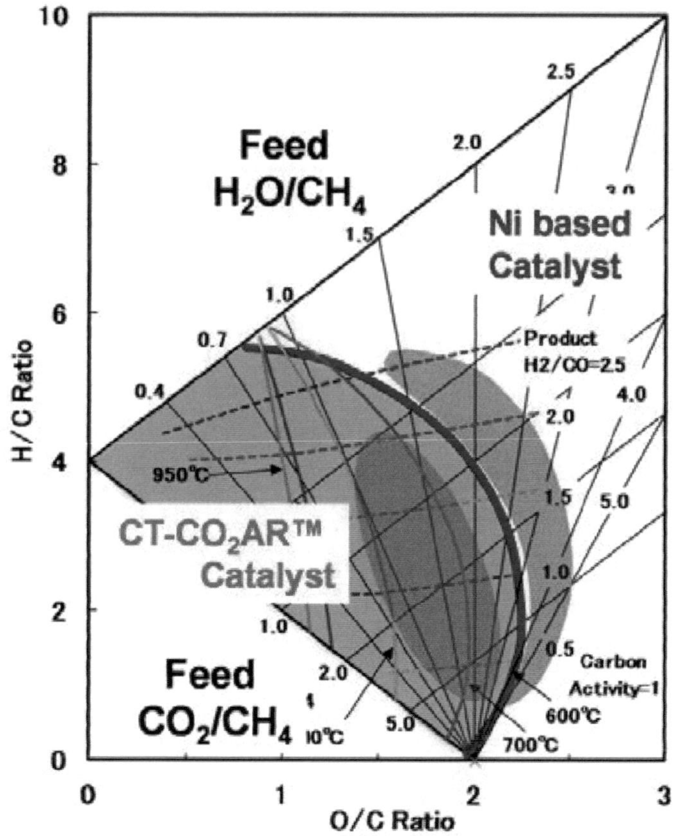

Fig. 1. Carbon Limit Diagram

(2) The advantages of a lower S/C ratio operation

In the case of the synthesis gas production process shown in Fig. 2, the advantages of a lower S/C operation are as follows:

1. Since the amount of steam and CO_2 is reduced, the heat duty to heat steam and CO_2 can be reduced. Therefore, the fuel consumption of the reformer can be reduced.
2. Fig. 3 shows the comparison of synthesis gas concentration at the reformer outlet between a high S/C (=2.2) operation and a low S/C (=1.0) operation in the case of $H_2/CO=1.0$
   In the case of a low S/C operation, the amount of CO_2 can be reduced and as a result, the system volume of CO_2 removal is downsized.
   In addition, because of lower CO_2/C, the capacity of the CO_2 compressor is downsized.
   Therefore, both the CAPEX and the OPEX of CO_2 removal and CO_2 compressor are reduced.
3. Fig. 3 shows that a lower S/C operation can reduce the amount of H_2O at the reformer outlet and it is possible to reduce the amount of process condensate.
   In the case of a low S/C operation, the total amount of produced steam is less than that of a high S/C operation. However, the amount of steam uses as the reforming agent in a low S/C operation is significantly less than that in a high S/C operation.
   As a result, a low S/C operation can increase the amount of export steam.

From these reasons, the CT-CO_2AR™ catalyst is preferable to a commercial Ni based catalyst.

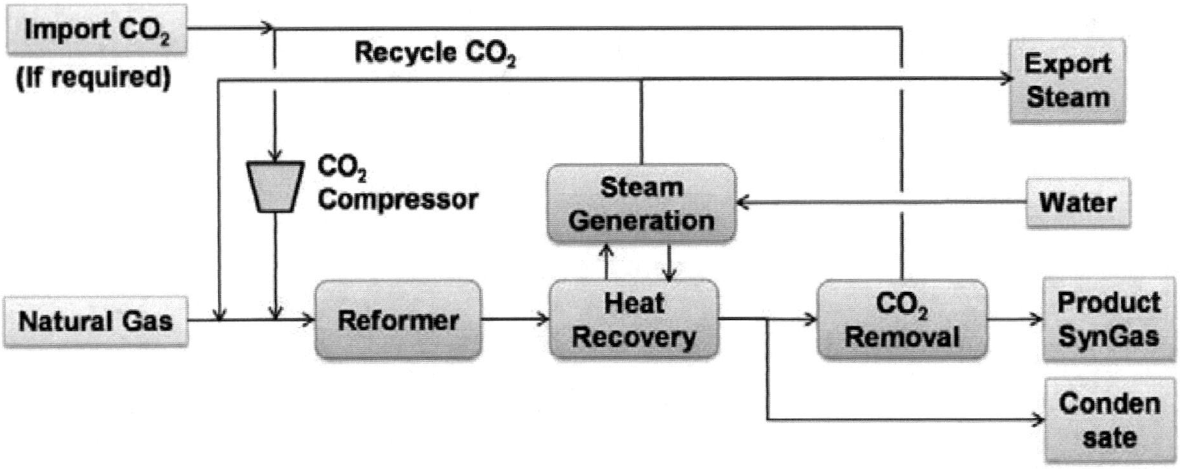

Fig. 2. Block Flow Diagram of Syngas production

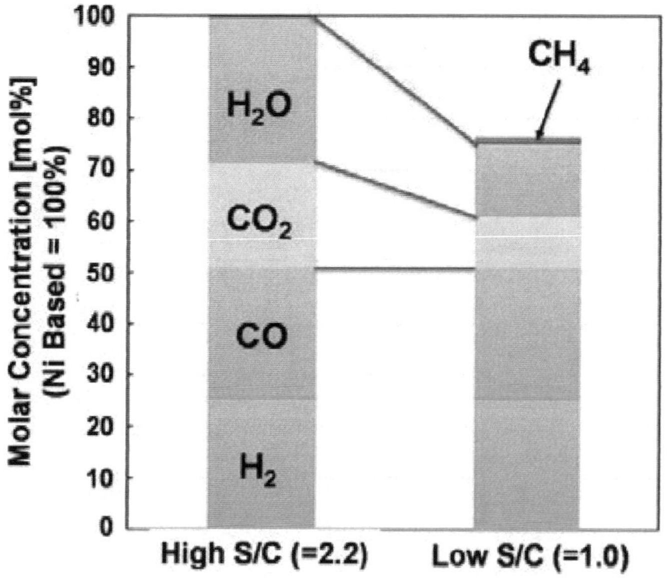

Fig. 3. Typical Example of synthesis gas composition in H_2/CO=1.0

(3) The characteristics of the CT-CO_2AR™ catalyst

Fig. 4 shows the result of EPMA (Electron Probe Micro Analyzer) of the CT-CO_2AR™ catalyst. Generally, the surface reaction on the reforming catalyst is the rate-determining step. Therefore, to reduce the amount of active metal, the active metal is supported on only the outer surface of the carrier like egg-shell.

Fig. 5 shows the accelerated experiment result. The CT-CO_2AR™ catalyst keeps its activity due to its high resistance to carbon formation.

We built the reaction kinetics model shown in Fig. 6 to estimate the behavior of the CT-CO_2AR™ catalyst.

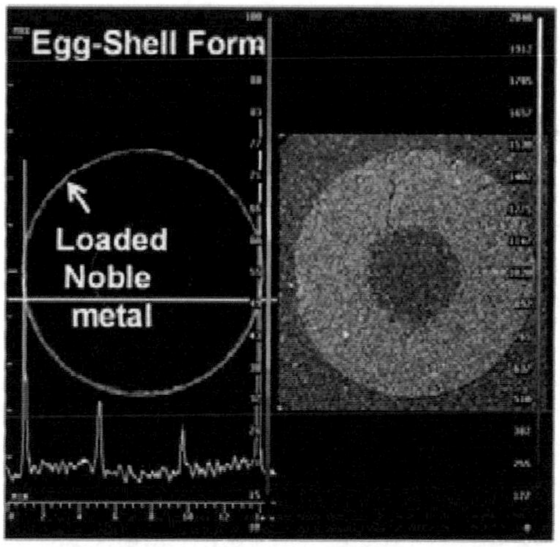

Fig. 4. EPMA result of CT-CO_2AR™ Catalyst

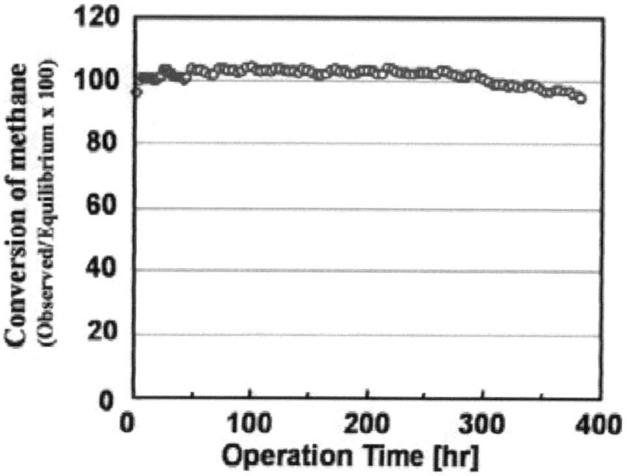

Fig. 5. Result of accelerated experiment

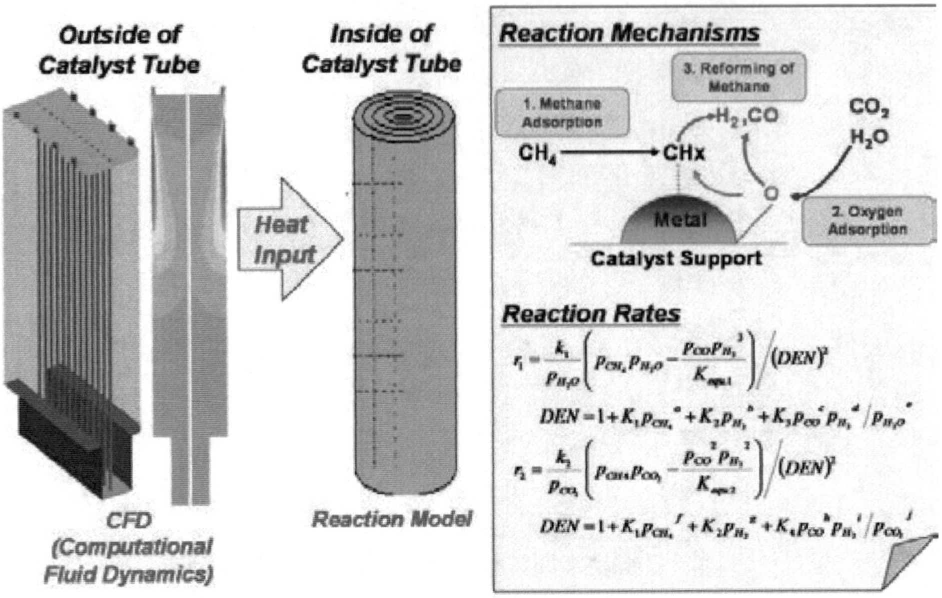

Fig. 6. Reaction model of CT-CO_2AR™ catalyst

3. Advantages of CT-CO_2AR™ catalyst

(1) Study Basis

Regarding the competitiveness of the CT-CO_2AR™ catalyst compared to a commercial Ni based reforming catalyst, we have compared the CT-CO_2AR™ catalyst and Ni based catalysts from the following three points of view based on the block flow shown in Fig. 2.

1. Total energy input (feed + fuel hydrocarbon + utility consumption)
2. Manufacturing costs of synthesis gas
3. CO_2 Emission per final products

[CO_2 Emission] = [CO_2 in Flue and Utility] - [Import CO_2] + [CO_2 in downstream process to produce the final products]

In this study we considered cases with H_2/CO ratios of 1.0 and 0.5. The conditions and estimated final product of each case are listed in Table 1.

(2) Study Result

From Fig.7-9 show the results of the process study. In the case of H_2/CO=1.0, compared to Ni based catalysts, the CT-CO_2AR™ catalyst can cut total energy input by 11% (Fig. 7) manufacturing cost by 12% (Fig. 8).Regarding CO_2 emission, the CT-CO_2AR™ catalyst can achieve a reduction of 22% compared with a Ni based catalyst (Fig. 9). In the case of H_2/CO=0.5, CT-CO_2AR™ catalyst is more competitive. Approximately 30% of both total energy input and manufacturing cost can be reduced and 76% of the amount of CO_2 emission can be reduced. From this study, the CT-CO_2AR™ catalyst is competitive compared with Ni based catalysts both in terms of CAPEX and OPEX.

Table 1. Process study conditions

H_2/CO Ratio	Catalyst	Feed S/C	Amount of (H_2+CO)	Final Products	Capacity
[-]	[-]	[-]	[Nm3/h]	[-]	[ton/hr]
CO (0.5)	Ni Based	3.3	37,500 (25,000 as CO)	Acetic Acid	500,000
	CT-CO_2AR™	0.4			
1.0	Ni Based	2.2	25,000	n-butanol (oxo synthesis)	270,000
	CT-CO_2AR™	1.0			

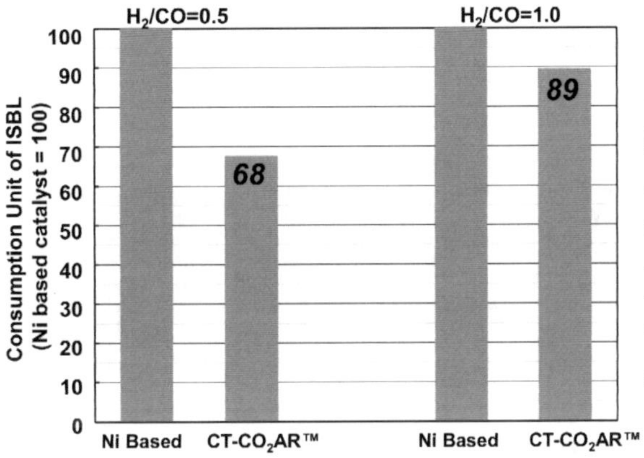

Fig. 7. Total Energy Input

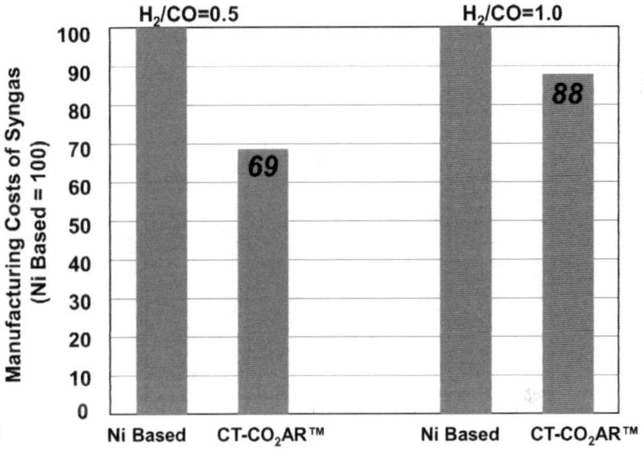

Fig. 8. Manufacturing Costs of Syngas

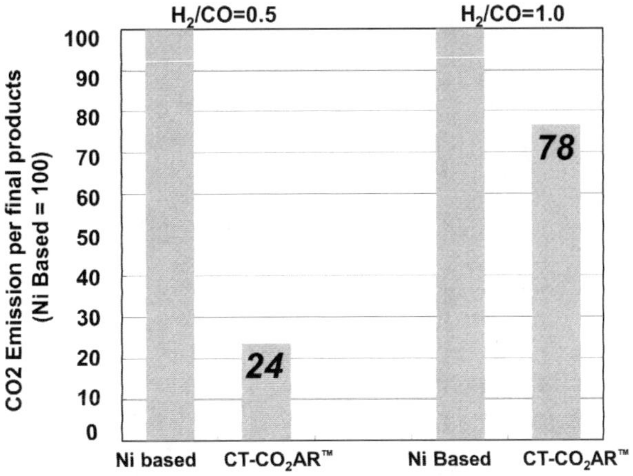

Fig. 9. CO_2 Emission per final products

4. Application to Commercial Plant

The CT-CO_2AR™ catalyst has been operated in a commercial plant in Japan since 2014. The information of the commercial plant is listed in Table 2. In this project, the reforming catalyst has been changed to the CT-CO_2AR™ catalyst and the S/C ratio at the inlet of the prereformer has reduced from 2.2 to 1.4 without changing the amount and specification of the produced synthesis gas.
Regarding the comparison between analysis result using the reaction model and the operation result, a pressure drop through the reformer tube is shown in Fig. 10 and the methane concentration of the synthesis gas is shown in Fig.11.
Since a significant increase in the pressure drop and methane concentration was not observed, a stable operation was achieved during the whole operation time.
In addition, our reaction model can accurately predict the performance of the CT-CO_2AR™ catalyst.

Table 2. Information about the commercial plant

Client's Name :	KH Neochem Co., Ltd
Plant Location :	Yokkaichi, Japan
Product :	Synthesis gas H_2/CO ratio=1.0
Product capacity :	5,000 Nm^3/hr (as H_2+CO)
Project Scope :	Catalyst supply to the existing plant
Feed Hydrocarbon :	Natural gas or Naphtha

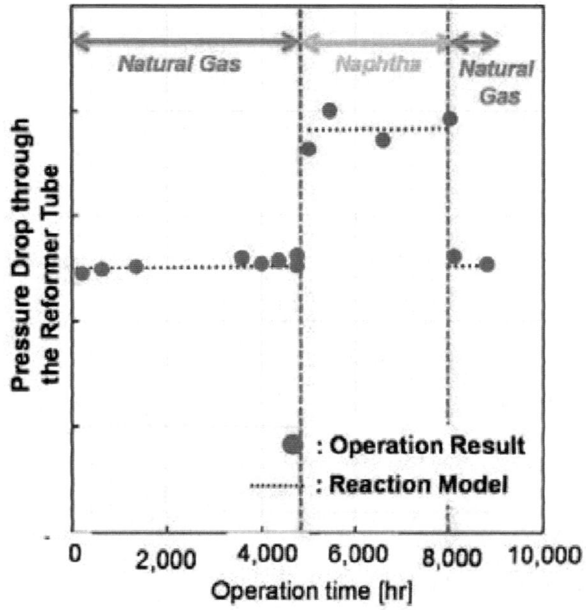

Fig. 10. Pressure difference of a reformer tube

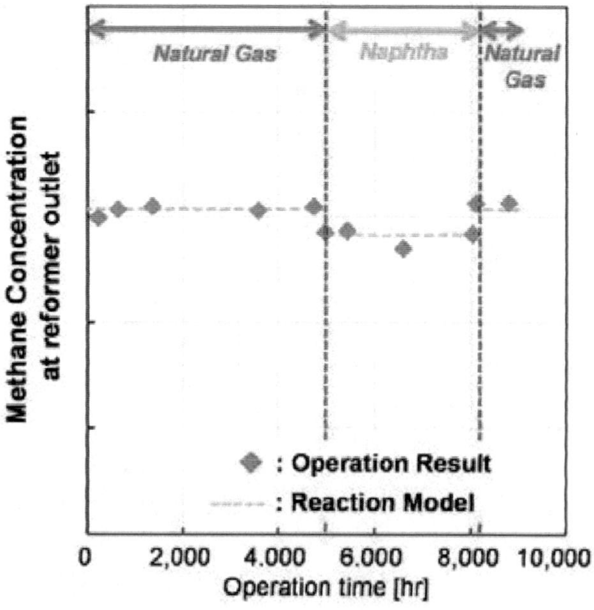

Fig. 11. Methane concentration at reformer outlet

Fig. 12 shows the comparison of the reformer duty and the amount of input natural gas between before and after replacing the CT-CO_2AR™ catalyst. By replacing the CT-CO_2AR™ catalyst, 10% of the natural gas input is reduced and this result matches the result of process study shown in section 3 (Advantages of the CT-CO_2AR™ catalyst).

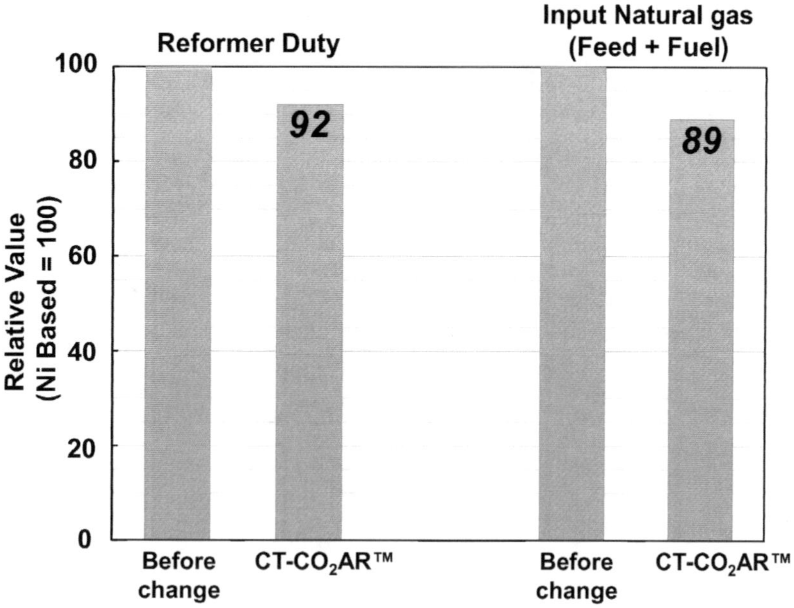

Fig. 12. Comparison result of commercial plant operation

5. Conclusion

The CT-CO_2AR™ catalyst enables a lower S/C condition operation because of its higher resistance to carbon formation.
Compared with a commercial catalyst, the CT-CO_2AR™ catalyst demonstrates high competitiveness in both manufacturing cost and CO_2 emission in the case of H_2/CO ratio = 0.5 and 1.0, especially, 0.5.
The CT-CO_2AR™ catalyst was installed in a commercial plant in 2014 and has been operating stably. 10% reduction of natural gas input was achieved in the commercial operation.

6. References

1) J.R. Rostrup-Nielsen, J. H. B. Hansen, *J. Catal.*, **144**, 38 (1993)
2) N.R.Udengaard, J.-H. Bak Hansen and D.C.Hanson, Oil & Gas J., Mar.9(1992)

Solar Thermochemical Splitting of H_2O and CO_2 using Nonvolatile Metal Oxides

(Principal Author)
Rahul Bhosale
Department of Chemical Engineering
College of Engineering
Qatar University, Doha, Qatar
rahul.bhosale@qu.edu.qa

(Other Authors)
Gorakshnath Takalkar
Department of Chemical Engineering
College of Engineering
Qatar University, Doha, Qatar

Anand Kumar
Department of Chemical Engineering
College of Engineering
Qatar University, Doha, Qatar

Fares A. AlMomani
Department of Chemical Engineering
College of Engineering
Qatar University, Doha, Qatar

Prepared for Presentation at
American Institute of Chemical Engineers
2017 Fall Meeting
Minneapolis, MN, USA
October 29 – November 3, 2017

Solar Thermochemical Splitting of H_2O and CO_2 using Nonvolatile Metal Oxides

Rahul R. Bhosale, Anand Kumar, Gorakshnath Takalkar, Fares A. AlMomani

Department of Chemical Engineering
College of Engineering
Qatar University, Doha, Qatar

Keywords: Thermochemical Reactions, Metal Oxides, Solar Fuels, H_2O/CO_2 Splitting

Abstract

In this study, a Co ferrite/ZrO_2 ceramic nanomixture was synthesized by the propylene-oxide-assisted sol–gel method. Cobalt, iron, and zirconium nitrate salts were used as the metal precursors, and ethanol and propylene oxide were used as the solvent and gelation agent, respectively. The prepared gels were aged, dried, and annealed at 1000°C for different dwell times. The annealed Co ferrite/ZrO_2 powder was analyzed by powder X-ray diffraction (PXRD), Brunauer–Emmett–Teller surface area analysis (BET), and energy-dispersive spectroscopy (EDS),. The characterization results are reported in detail.

1. Introduction

Metal-oxide-based solar thermochemical water and carbon dioxide splitting is considered a promising technology to satisfy future energy demands [1]. Zinc oxide [2-5], iron oxide [5-6], tin oxide [7-8], doped iron oxides (ferrites) [9-14], ceria-based redox materials [15-19], and perovskites [20-24] are used for splitting. Among these oxides, ferrites have been used extensively for solar thermochemical water splitting; however, their use is extremely limited for CO_2 splitting.

However, the major issue limiting the use of ferrites in solar thermochemical cycles is the increased size of the ferrite grains obtained as a result of high-temperature processing. In this study, a Co ferrite/ZrO_2 ceramic mixture was synthesized by the propylene-oxide-assisted sol–gel method [25-27], exhibiting mixing of Co ferrite and ZrO_2 with a uniform distribution. The effect of annealing time on the phase and chemical composition, crystallite size, specific surface area (SSA), pore volume, and morphology of the Co ferrite/ZrO_2 ceramic nanomixture was investigated, and the results were reported in detail.

2. Experimental

2.1 Sol–Gel Synthesis

Metal precursors of cobalt, iron, and zirconium were first mixed in a specific weight ratio and dissolved in ethanol using a sonication bath. Once a visibly clear solution was obtained, a predetermined amount of propylene oxide was slowly added, affording a gel. The gel was aged, dried, and annealed at 1000°C (at various annealing times) in air. Fig. 1 shows the flow chart for the synthesis of the Co ferrite/ZrO_2 ceramic nanomixture.

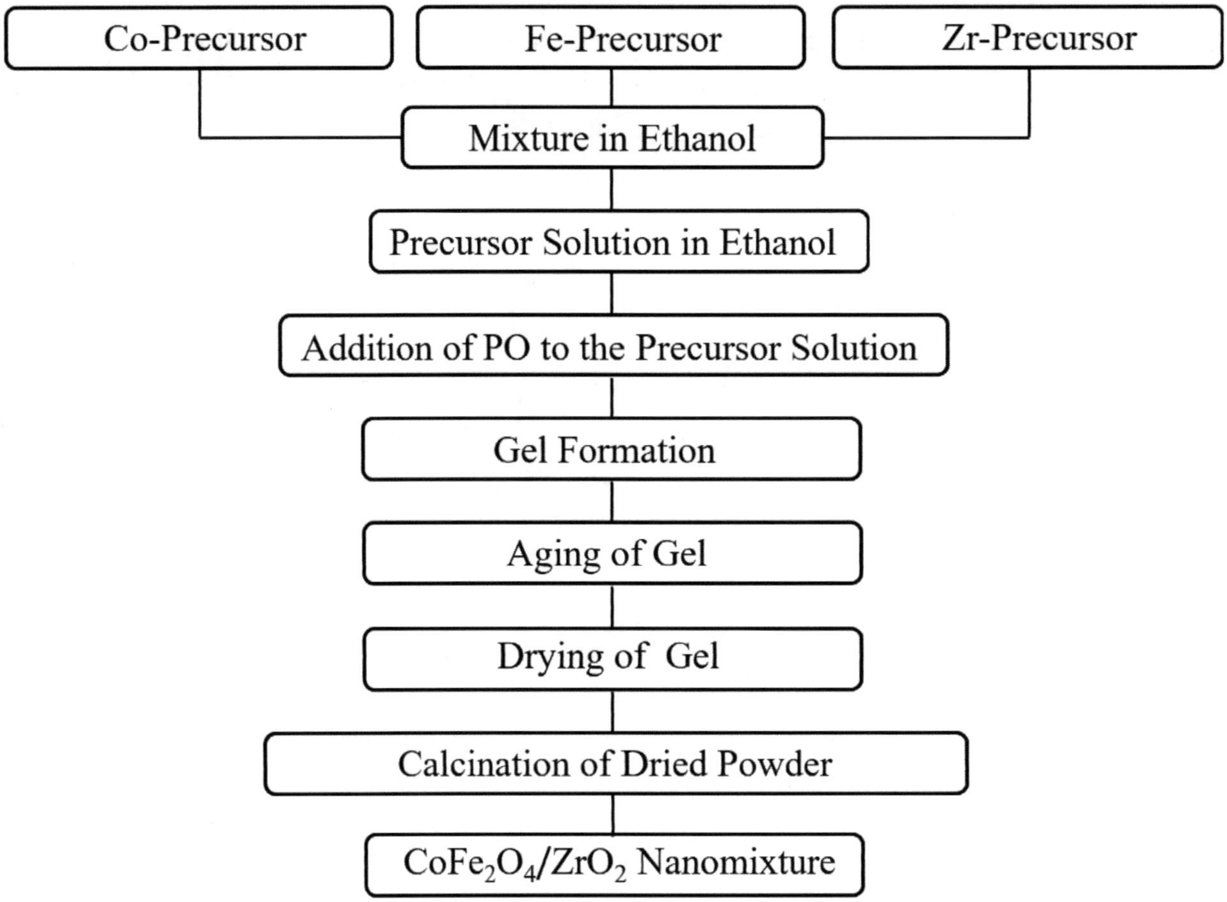

Fig. 1. Flow chart for the synthesis of the Co ferrite/ZrO_2 ceramic nanomixture by the sol–gel method.

2.2 Characterization

To investigate the effects of annealing on the physiochemical properties of the Co ferrite/ZrO_2 ceramic nanomixture, the Co ferrite/ZrO_2 powder obtained after annealing was characterized by powder X-ray diffraction (PXRD), Brunauer–Emmett–Teller (BET) surface area analysis, scanning electron microscopy (SEM) with a field-emission gun, X-ray energy-dispersive spectroscopy (EDS), and high-resolution transmission electron microscopy (HR-TEM).

3. Results & Discussion

The effect of annealing time on the physicochemical properties of the sol–gel-derived Co ferrite/ZrO_2 ceramic nanomixture was investigated. For this purpose, the Co ferrite/ZrO_2 gels (prepared using 20 mL of propylene oxide, aged for 24 h, and dried at 100°C) were annealed at 1000°C for 1–5 h. The calcined powders thus obtained were characterized by PXRD and EDS to determine the phase and chemical composition. The results shown Fig. 2 clearly indicated that the phase and chemical composition of the sol–gel-derived Co ferrite/ZrO_2 ceramic nanomixture is unaffected by the increase in the annealing time.

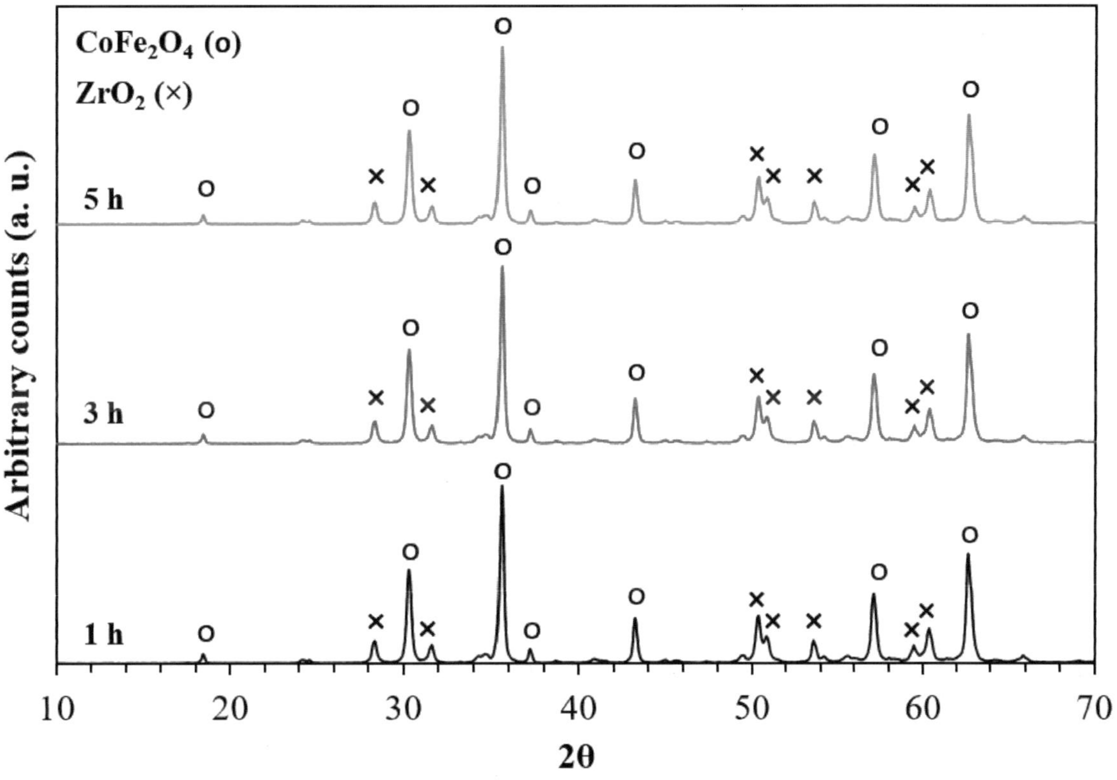

Fig. 2 – Effect of the annealing time on the phase composition of sol–gel-derived Co ferrite/ZrO_2 ceramic nanomixture.

The effect of the annealing time on the crystallite size of the sol–gel-derived Co ferrite/ZrO_2 ceramic nanomixture was also determined. Fig. 3 shows the results obtained. With the increase in the annealing time from 1 to 3 h, the crystallite size increases by 14.86 nm. With the further increase in the annealing time from 3 to 5 h, the crystallite size further increases by 22.83 nm. In summary, with the increase in the annealing time from 1 to 5 h, the size of the Co ferrite/ZrO_2 crystallites increases from 53.66 to 91.35 nm.

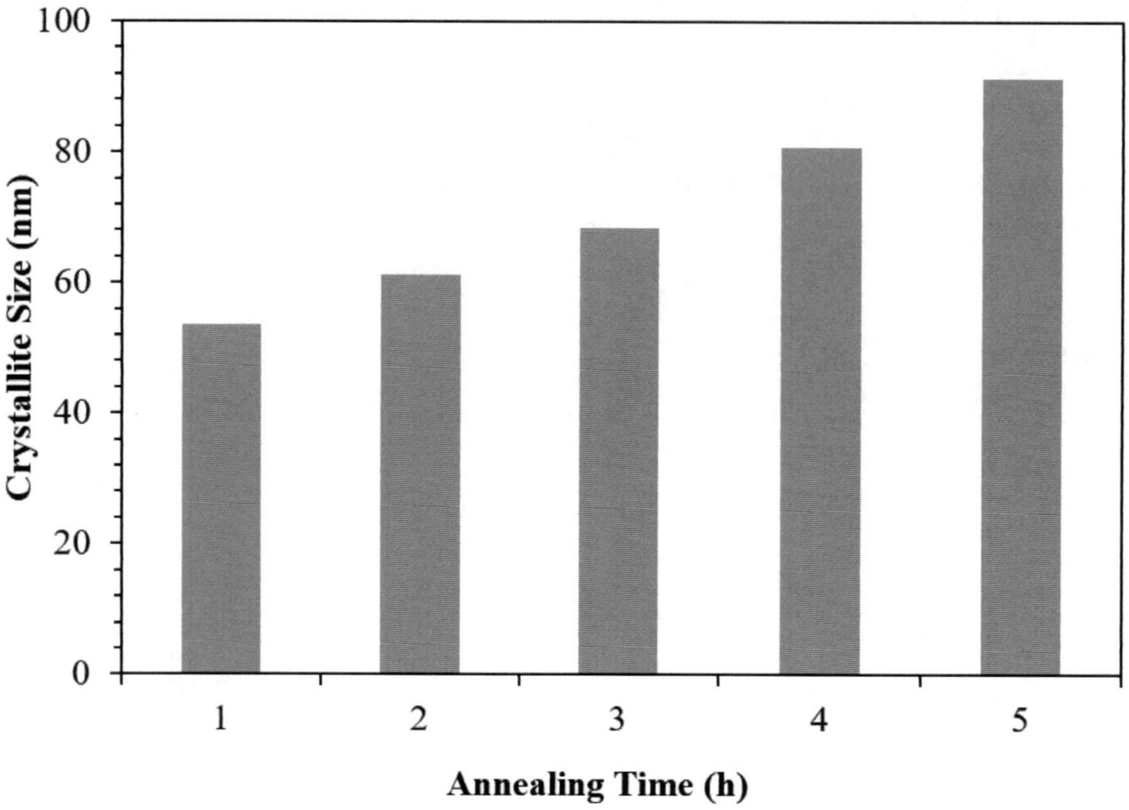

Fig. 3 – Effect of the annealing time on the crystallite size of the sol–gel-derived Co ferrite/ZrO_2 ceramic nanomixture.

In addition to the composition and crystallite size, the SSA and pore volume are also crucial parameters to be investigated. The results obtained by BET analysis indicated that with the increase in the annealing time from 1 to 3 h, the SSA decreases by 14.33%. Similarly, with the increase in the annealing time from 3 to 5 h, the SSA decreases by 13.57%. With the increase in the annealing time from 1 to 5 h, the pore volume decreases from 1.52×10^{-3} cm^3/g to 1.03×10^{-3} cm^3/g.

4. Conclusion

In this paper, a Co ferrite/ZrO_2 ceramic nanomixture, with a pure phase composition, high SSA, pore volume, nanoparticulate morphology, and a uniform distribution of Co ferrite and ZrO_2 in the mixture, was synthesized by the propylene-oxide-assisted sol–gel method. The phase and chemical composition of the Co ferrite/ZrO_2 nanomixture was not affected under all of the annealing conditions. With the increase in the annealing time from 1 to 5 h at an annealing temperature of 1000°C, the crystallite size increased by 37.69 nm. with the increase in the annealing time from 1 to 5 h, the SSA and pore volume decreased by 2.59 m^2/g and 0.484×10^{-3} cm^3/g, respectively.

5. Acknowledgments

This publication was made possible by the NPRP grant (NPRP8-370-2–154) from the Qatar National Research Fund (a member of Qatar Foundation). The statements made herein are solely the responsibility of author(s).

References

[1] J.R. Scheffe, A. Steinfeld, Oxygen exchange materials for solar thermochemical splitting of H_2O and CO_2: a review, Materials Today 17 (2014) 341-348.

[2] P.G. Loutzenhiser, A. Steinfeld, Solar syngas production from CO_2 and H_2O in a two-step thermochemical cycle via Zn/ZnO redox reactions: thermodynamic cycle analysis, International Journal of Hydrogen Energy 36 (2011) 12141-12147.

[3] A. Stamatiou, P. Loutzenhiser, A. Steinfeld, Solar syngas production from H_2O and CO_2 via two-step thermochemical cycles based on Zn/ZnO and FeO/Fe_3O_4 redox reactions: Kinetic analysis, Energy & Fuels 24 (2010) 2716-2722.

[4] R. R. Bhosale, A. Kumar, F. AlMomani, R. B. Gupta, Solar thermochemical $ZnO/ZnSO_4$ water splitting cycle for hydrogen production, International Journal of Hydrogen Energy (2017) in-press.

[5] P. Loutzenhiser, M. Gálvez, I. Hischier, A. Stamatiou, A. Frei, A. Steinfeld, CO_2 splitting via two-step solar thermochemical cycles with Zn/ZnO and FeO/Fe_3O_4 redox reactions II: Kinetic analysis, Energy & Fuels 23 (2009) 2832-2839.

[6] R. R. Bhosale, A. Kumar, van den Broeke, Leo JP, S. Gharbia, D. Dardor, M. Jilani, J. Folady, M.S. Al-Fakih, M.A. Tarsad, Solar hydrogen production via thermochemical iron oxide–iron sulfate water splitting cycle, International Journal of Hydrogen Energy 40 (2015) 1639-1650.

[7] P. Charvin, S. Abanades, F. Lemont, G. Flamant, Experimental study of $SnO_2/SnO/Sn$ thermochemical systems for solar production of hydrogen, AIChE Journal 54 (2008) 2759-2767.

[8] R. R. Bhosale, A. Kumar, P. Sutar, Thermodynamic analysis of solar driven SnO_2/SnO based thermochemical water splitting cycle, Energy Conversion & Management 135 (2017) 226-235.

[9] R. R. Bhosale, A. Kumar, F. AlMomani, U. Ghosh, P. Sutar, G. Takalkar, A. Ashok, I. Alxneit, Effectiveness of Ni incorporation in iron oxide crystal structure towards thermochemical CO_2 splitting reaction, Ceramics International 43 (2017) 5150-5155.

[10] C.C. Agrafiotis, C. Pagkoura, A. Zygogianni, G. Karagiannakis, M. Kostoglou, A.G. Konstandopoulos, Hydrogen production via solar-aided water splitting thermochemical cycles: Combustion synthesis and preliminary evaluation of spinel redox-pair materials, International Journal Hydrogen Energy 37 (2012) 8964-8980.

[11] R. R. Bhosale. A. Kumar, F. AlMomani, I. Alxneit, Propylene oxide assisted sol–gel synthesis of zinc ferrite nanoparticles for solar fuel production, Ceramics International 42 (2016) 2431-2438.

[12] M. Roeb, N. Gathmann, M. Neises, C. Sattler, R. Pitz-Paal, Thermodynamic analysis of two-step solar water splitting with mixed iron oxides, International Journal of Energy Research 33 (2009) 893-902.

[13] R.R. Bhosale, R.V. Shende, J.A. Puszynski, Thermochemical water-splitting for H_2 generation using sol-gel derived Mn-ferrite in a packed bed reactor, International Journal of Hydrogen Energy 37 (2012) 2924-2934.

[14] J.R. Scheffe, J. Li, A.W. Weimer, A spinel ferrite/hercynite water-splitting redox cycle, International Journal of Hydrogen Energy 35 (2010) 3333-3340.

[15] R.R. Bhosale, A. Kumar, F. AlMomani, I. Alxneit, Sol–gel derived CeO_2–Fe_2O_3 nanoparticles: Synthesis, characterization and solar thermochemical application, Ceramics International 42 (2016) 6728-6737.

[16] R.R. Bhosale, A. Kumar, F. AlMomani, U. Ghosh, S. Al-Muhtaseb, R. Gupta, I. Alxneit, Assessment of $Ce_xZr_yHf_zO_2$ based oxides as potential solar thermochemical CO_2 splitting materials, Ceramics International 42 (2016) 9354-9362.

[17] J.R. Scheffe, A. Steinfeld, Thermodynamic analysis of cerium-based oxides for solar thermochemical fuel production, Energy & Fuels 26 (2012) 1928-1936.

[18] W.C. Chueh, C. Falter, M. Abbott, D. Scipio, P. Furler, S.M. Haile, A. Steinfeld, High-flux solar-driven thermochemical dissociation of CO_2 and H_2O using nonstoichiometric ceria, Science 330 (2010) 1797-1801.

[19] P. Furler, J.R. Scheffe, A. Steinfeld, Syngas production by simultaneous splitting of H_2O and CO_2 via ceria redox reactions in a high-temperature solar reactor, Energy & Environmental Science 5 (2012) 6098-6103.

[20] J.R. Scheffe, R. Jacot, G.R. Patzke, A. Steinfeld, Synthesis, Characterization, and Thermochemical Redox Performance of Hf^{+4} , Zr^{+4} , and Sc^{+3} Doped Ceria for Splitting CO_2, The Journal of Physical Chemistry C 117 (2013) 24104-24114.

[21] J.R. Scheffe, D. Weibel, A. Steinfeld, Lanthanum–strontium–manganese perovskites as redox materials for solar thermochemical splitting of H_2O and CO_2, Energy & Fuels 27 (2013) 4250-4257.

[22] A.H. McDaniel, E.C. Miller, D. Arifin, A. Ambrosini, E.N. Coker, R. O'Hayre, W.C. Chueh, J. Tong, Sr-and Mn-doped $LaAlO_{3-\delta}$ for solar thermochemical H_2 and CO production, Energy & Environmental Science 6 (2013) 2424-2428.

[23] M. Gálvez, R. Jacot, J. Scheffe, T. Cooper, G. Patzke, A. Steinfeld, Physico-chemical changes in Ca, Sr and Al-doped La–Mn–O perovskites upon thermochemical splitting of CO_2 via redox cycling, Physical Chemistry Chemical Physics 17 (2015) 6629-6634.

[24] R. R. Bhosale, A. Kumar, A. Ashok, P. Sutar, G. Takalkar, M. Khraisheh, F. AlMomani, U. Ghosh, La-Based Perovskites as Oxygen-Exchange Redox Materials for Solar Syngas Production, MRS Advances (2017) in-press.

[25] R. R. Bhosale, R. Khadka, J. Puszynski, R. V. Shende, H_2 generation from two-step thermochemical water-splitting reaction using sol-gel derived $Sn_xFe_yO_z$, Journal of Renewable & Sustainable Energy 3 (2011) 063104-1 – 063104-12.

[26] R. Bhosale, R. Shende, J. Puszynski, H_2 generation from thermochemical water splitting using sol-gel derived Ni-ferrite, Journal of Energy and Power Engineering 4 (2010) 27-38.

[27] R.R. Bhosale, R.V. Shende, J.A. Puszynski, H_2 Generation From Thermochemical Water-Splitting Using Sol-Gel Synthesized Zn/Sn/Mn-doped Ni-Ferrite, International Review of Chemical Engineering 2 (2010) 852-862.

500493 Critical Issues in the Development of Commercial Natural Gas to Industrial Chemicals Bioprocesses

Abstract:

The low cost and abundant supply of natural gas make it America's greatest untapped energy resource, leading to the development of technologies for natural gas conversion to compounds of greater value. Intrexon has developed the first natural gas-to-liquids bioconversion platform powered by methantrophic fermentation, which applies synthetic biology to program bacteria and produce higher value materials. Our innovative genetic toolbox allows us to engineer the biology of the methanotroph to more efficiently convert natural gas into a wide range of valuable chemicals and fuels. These applications include isobutanol, farnesene, 1,4-butanediol, 2,3-butanediol, isoprene and isobutyraldehyde – the building blocks for industrial and consumer products like fuels, synthetic rubber, acrylics, resins and spandex.

Using a methanotrophic bioprocess to produce fuels and chemicals has several advantages. Natural gas bioconversion offers scaling opportunities beyond traditional gas-to-liquid (GTL) facilities, as well as biomass feedstock facilities. The theoretical yields of methanotroph/methane exceed that of yeast/sugar approaches by a wide margin. Biocatalytic processes that achieve high yields under milder processing conditions allow for reduced capital and operation expenditures, offering greater economic viability. To successfully bring this technology to commercial scale, consideration must be given not only to optimizing the bacterial metabolic pathway, but to key issues in the fermentor design, downstream processing, scale-up and overall plant economics.

At any scale, aerobic natural gas fermentation must overcome the challenge of gas-to-liquid transfer of low solubility gases to increase bioavailability of gas substrates to the bacteria. A low gas-to-liquid mass transfer rate reduces the bacterial productivity and process yield, which in turn increases fermentor size, capital cost, operating costs and may add additional equipment necessary to recycle gases. During commercial scale-up of gas fermentation processes, these factors can become cost prohibitive. Standard stirred tank reactors become unfeasible for use at larger scales, as the agitator power input and size become excessive. There are many fermentor designs currently available to increase the overall mass transfer coefficient of these gases. However, capital and operating expenditures of the reactor are critical metrics for evaluating commercialization feasibility. Unique fermentor configurations can greatly increase capital cost by raising the cost of fabrication and decreasing the ability to receive competitive bids. These unique fermentors also may not create a homogenous, well-mixed broth – which increases the cost of controls and instrumentation. The economic feasibility of gas transfer designs must be balanced with the biologic effects of reactor forces on the bacteria. Methods of increasing mass transfer often include increasing turbulence in the reactor, which can create detrimental effects on the bacteria through high shear forces, high pressure and vacuum areas, and dissolved gas substrate inhibition. Careful equipment selection and pilot trials are necessary to determine the various effects of gas transfer methods on mass transfer rates and cell viability.

Downstream process selection for commercialization is a balance of biologic limitations, technical feasibility, capital cost, operating cost and the risk of cutting-edge over well-established techniques. The developing bioprocess must be iterated upon the limitations of strain engineering and practicality of downstream processing techniques. Fermentation broth concentrations may be low to product inhibition on the cell, but this increases the size and energy requirement of downstream equipment. Cutting-edge technologies can increase the efficiency of low titer separation and reduce energy inputs, which can be capital intensive and unproven at commercial scales. Genetic engineering tools can increase the tolerance of the bacteria to products. However, the directed or natural evolution process requires time.

A minimum broth titer must be determined through techno-economic modeling of processes at each stage of process development and refinement. Bioprocesses also create metabolic byproducts which can add unit operations, and therefore cost, to the downstream purification process. Some byproducts may originate from the genetic pathway or from commercial feedstocks, as they are often lower in purity and have higher variability than those at the research scale. Additional downstream steps must be evaluated to determine if they are technically and economically feasible to meet final product specifications for sale, or if the byproduct genetic pathways can be manipulated to terminate production and not add cost to the overall process.

Furthermore, certain tools that aid in genetic engineering and strain development at the research and development scale must be eliminated or retooled to create commercially viable strains. During development, production pathways are often introduced to the bacteria through plasmids which carry antibiotic resistance. To avoid the material cost and downstream separation of the antibiotic at commercial scale, the genetic material must be integrated into the chromosome. This genomic integration may affect the strain's capability to produce product at previously observed yields, titers and rates. Another common development tool is that production pathways may be turned on or off using an externally applied chemical inducer. At scale, these chemical inducers can add material cost as well as affect downstream processes and product purity. Induction systems may be completely removed for commercial scale or based off an easily controllable environmental switch.

Favorable plant economics are necessary to move forward with commercialization of a technology. Scale is carefully selected based on a comparison of the production rate, fermentation productivity and yield targets to revenue, capital and operating costs. The plant location is an important selection factor, as costs are affected by utility rates in the area and/or co-location of the plant with shared infrastructure.

Traditional engineering workflows, specifically front-end loading strategy, were used to shape the commercial scale process design of Intrexon's natural gas to chemical plants. The biologic, technical and economic commercial feasibilities of the end-to-end natural gas to isobutanol process were critical to Intrexon's plant design. Laboratory scale learnings were applied to select technology that was commercially feasible at demonstration scale. This demonstration scale design was scaled down to pilot scale, where the operation will yield learnings into refinement of at-scale design, equipment selection, empirical data on heating and cooling, and controls strategies and feedback to further iterations on genetics.

495124 Direct, Single-Pass Thermocatalytic Upgrading of Biogas and Landfill Gas into Renewable Natural Gas over the Ultra-Low Loading Ru/Î³-Al_2O_3 Catalyst

Abstract:

Direct, single-pass thermocatalytic upgrading of biogas and landfill gas into renewable natural gas over the ultra-low loading Ru/g-Al_2O_3 catalyst

Yichen Zhuang and David Simakov

Department of Chemical Engineering, University of Waterloo, Waterloo, ON N2L 3G1, Canada

Converting CO_2 into synthetic renewable fuels is an attractive pathway to decrease CO_2 emissions and to reduce our dependence on fossil fuels. Thermocatalytic hydrogenation provides advantages of fast reaction rates and high conversion efficiencies (as compared to electro-catalytic and photo-catalytic routes), thus allowing for compact, high-throughput operation. One option is to convert CO_2 into synthetic methane, so-called renewable natural gas (RNG) via the Sabatier reaction ($CO_2 + 4H_2 = CH_4 + 2H_2O$). Water electrolysis using renewable or surplus electricity can be used to provide a non-fossil source of H_2 required for the reaction. There are multiple sources of concentrated CO_2-containing streams including biogas and landfill gas (up to 50 mol% CO_2). It is highly desirable to develop a technology that can convert such CO_2-rich streams directly, without costly upstream CO_2 separation.

One of the most fundamental challenges is the development of a catalyst, which is highly active, selective to CH_4 formation, stable against coking, and low-cost at the same time. The selectivity becomes a major challenge when a mixed feed (biogas or landfill gas contain up to 50 mol% CH_4 in addition to CO_2) is used instead of purified CO_2 feed. In this study, a highly selective Ru-based catalyst with ultra-low Ru loadings (less than 0.5 wt%) is reported. A series of catalysts with Ru loadings ranging from 0.05-0.5 wt% were synthesized and evaluated for their catalytic activity in selective conversion of CO_2 in the mixed CH_4/CO_2 feed. CO_2 conversions up to 80% were achieved with nearly complete selectivity to CH_4 generation (left plot in the figure below). Importantly, the catalysts showed good stability, without any notable decline in performance over 60 hours on stream at very high space velocity of GHSV = 90,000 1/h (right plot in the figure below). A variety of analytical techniques were employed in order to investigate the mechanism of the superior selectivity and activity, including electron microscopy, thermal gravimetric analysis, temperature programmed reaction, chemisorption, and X-ray photon spectroscopy. Our results point out at the strong structure-activity relationship and the importance of the active phase interaction with the support.

492217 Energy Quality Factor and Exergy Destruction Processes Analysis for a Proposed Polygeneration System

Abstract:

A new polygeneration system was established through combining coal pyrolysis with semicoke combustion, semicoke gasification and a steam Rankine cycle, and then the calculation methodology was developed to calculate the system. The energy and exergy efficiencies of the system are 65.9 % and 75.1 %, respectively; and the standard energy quality factor *Î*±Î· values of the involved 17 components and the energy quality factor *Î*± of each system stream were obtained. By comparing the differences among the standard energy quality factor *Î*±Î· values of the related 17 components, the features of the energy grade of the components at the standard conditions were elucidated. Then the effect of streamÂ ¢s composition and temperature on the actual energy quality factor was discussed. Combining the concept of energy quality factor, the processes of exergy destruction in system blocks were visually expressed via a kind of diagram; and the reasons of exergy destruction were tried to be explained from the perspective of energy quality factor.

Pacific Northwest
NATIONAL LABORATORY
Proudly Operated by Battelle Since 1965

Modularized Production of Value-Added Products and Fuels from Distributed, Waste Carbon-rich feedstocks

Robert Weber[1], Johnathan Holladay[1], Cynthia Jenks[2]

1) Pacific Northwest National Laboratory; 2) Argonne National Laboratory

Topical Conference Session: Distributed Chemical and Energy Processes for Sustainability, AIChE National Meeting, October, 2017

Overview

Valorizing waste requires distributed facilities that, in turn, require new catalysts and new reactors to achieve economic, robust operation.

- We seek to locally convert waste carbon into chemical and fuel precursors, which aggregate to about 25% of US demand.
- Throughput will come from numbering up, not scaling up, which should accelerate deployment by attenuating technical and investment risk.
- The remoteness, dispersion and heterogeneity of the resources requires development of new catalysts, reactors and intensified processes.
- For example, electrocatalytic upgrading of liquefaction bio-oil could integrate upgrading with clean-up, further "preventing a waste and turning it to profit".
- This project probes both fundamental science and the practicality of electro-upgrading of the components of bio-oil *ex* hydrothermal liquefaction.

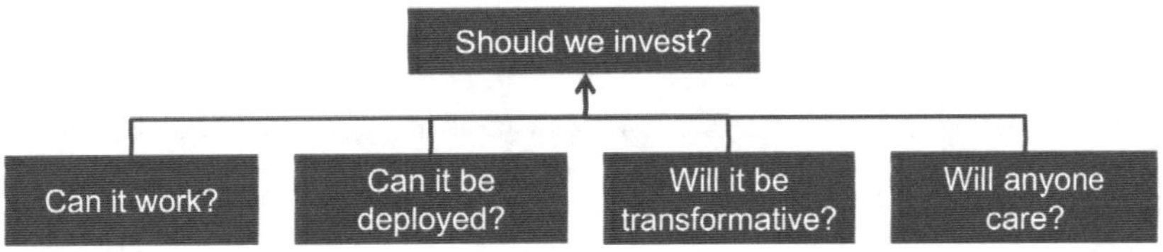

Will anyone care: *size of the opportunity*

We seek to locally convert waste carbon resources into chemical and fuel precursors, which sum to about 30% of US demand.

Type	Source	Availability (Mboe/y)	Site capacity (boe/d)	Possible Process	Tech. issues
Dry	MSW, ag & forest residue	1250	2-20	Pyrolyze, Hydrotreat	Hetero-geneity
Wet	Food, sludge manure	750	2-150	HTL, Hydrotreat	Product slate
Gas	Stranded CH_4 biogas, CO_2	50	70-1500 (CH_4) 1-10 (CO_2)	Reforming FT, Magic	Product slate
Total		**2000**			
US annual petroleum import		6900	130,000		
US annual net import of oil		1700			
Fuel demand of US cars		2800			
Energy demand of US homes		2000			

2/17

Will it be transformative: *benefits of modularity*

To attenuate technical and investment risk and to benefit from manufacturing efficiencies and linear scaling, we will number up rather than scale up.

- Replicating many (≳1000) small facilities promises more attractive investment and risk profiles than are typically found for scaling up for small scale.

Nominal risk profiles: most the risk is resolved in the first deployment for modular systems.

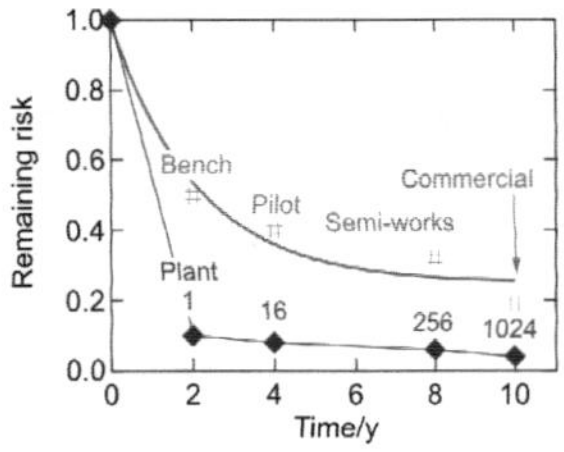

Nominal cost profiles for modular and stick-built plants each summing to 2000 bbl/day

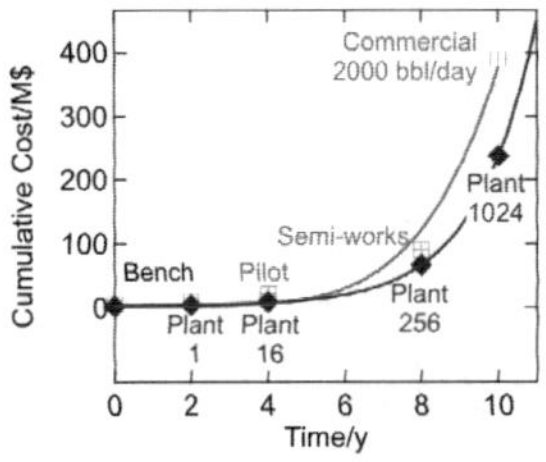

Having a large number of modular plants would benefit from economies of manufacture

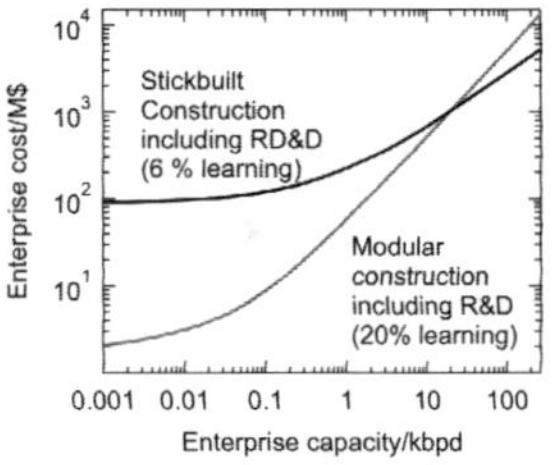

- The cross-over between scaling up & numbering up occurs at an aggregate throughput of >20 kbbl/day.
- Going from 2000 bbl/day to an aggregated 4000 bbl/day would cost ~$460 million to add 1000 more 2 bbl/day facilities; a second commercial plant (at a 6% learning rate) would cost “only” $280 million.

3/17

Can it work: *science and technology needs*

The remoteness, dispersion and heterogeneity of the resources requires development of new catalysts, reactors and intensified processes.

- The size of the facility will be dictated by the supply rate of feedstock, which will likely be neither homogeneous nor constant.
- Remoteness means we will not have a steam or H_2 grid or ways to valorize waste heat.
- The goal is to convert stranded carbon to fuel so we want to be as frugal as possible and not generate additional, even more dilute waste streams.

Challenge or technical need	Example of basic research
Separation processes and controls to facilitate conversion processes	Remote, online quantification of types and flow rates of solids and suspensions
Catalytic conversions with high efficiency, rates and selectivity under near ambient conditions	**Research on low temperature catalysis and chain reactions in liquid media, particularly in water**
Separation processes facilitating product valorization	Thermodynamic of and transport in complex mixtures
Water recovery and cleanup, again with high energy efficiency	Membrane science for separation of water from both polar and non-polar organics

4/17

Can it work: *implementing modularity*

To achieve the envisaged modularity requires changing from volume-based rates to area-based processes which downsize economically.

Modular facilities will incorporate Intensified unit operations, amenable to mass manufacture and which are cheaper than volume-based technologies at small scale

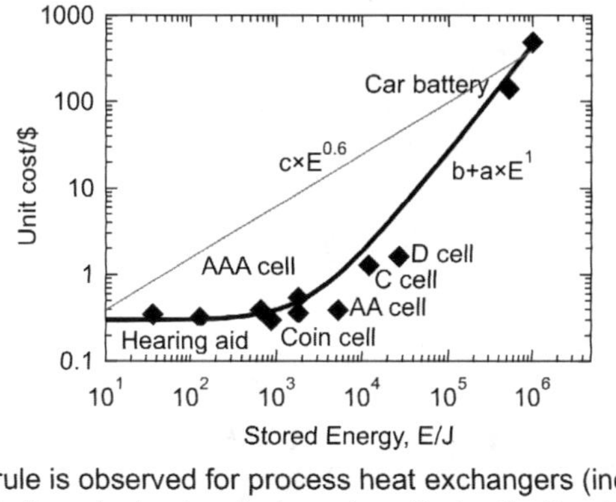

A similar scaling rule is observed for process heat exchangers (including consumer refrigerators) and other electrochemical reactors (fuel cells, flow batteries).

5/17

Can it work: *System architecture*

A complete system would include feed preprocessing, processing, upgrading and separations.

- Sites will be geographically distributed, likely without access to industrial utilities except electricity.
- The scale will be small (say, the equivalent to ~10 bbl oil/day).
- Operators may not be highly trained or closely supervised.
- The customers will be frugal.

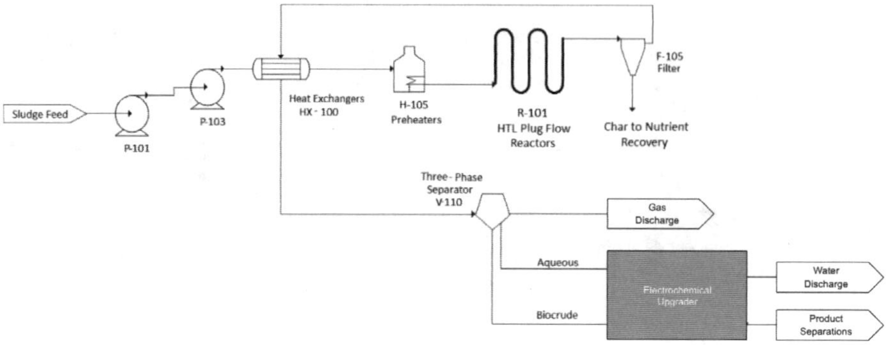

6/17

Can it work: *Scope*

We are now researching the downstream, upgrading and separations units to complement prior work at PNNL on the upstream conversion operations.

PNNL's Modular Hydrothermal Liquefaction

3 skids

- Feed prep and handling
- Hydrothermal liquefaction
- Biocrude recovery

Capability

- ~16 liters/h ≈ 2.4 bbl/day
- Solid-liquid separations at operating conditions
- Liquid-liquid separations for biocrudes that are heavier, or lighter, than water
- Product-feed heat exchange

7/17

Can it work: *electrolysis reactors*

We have constructed 3 scales of flow reactors that can provide both fundamental information and grow to practicable levels of capacity.

- Ag/AgCl reference electrode that can be positioned axially
- Carbon felt- or Carbon cloth-supported catalysts
- Automated data acquisition and control over E, q and T.

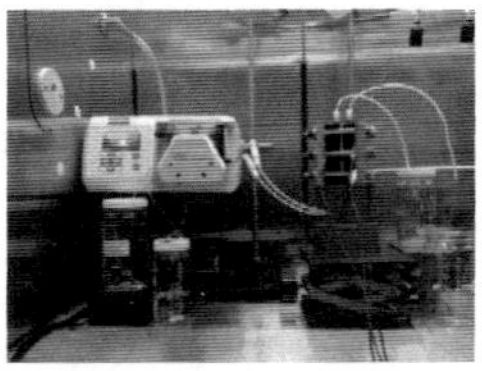
Bench: 2.4 cm × 4.2 cm

Pilot:, 5.7 cm × 10 cm

Demo: 10.7 cm × 18.7 cm

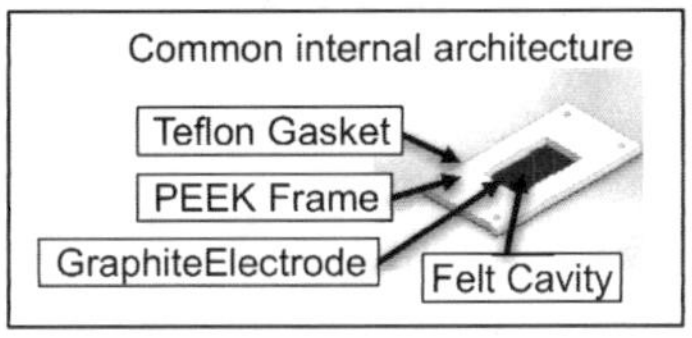

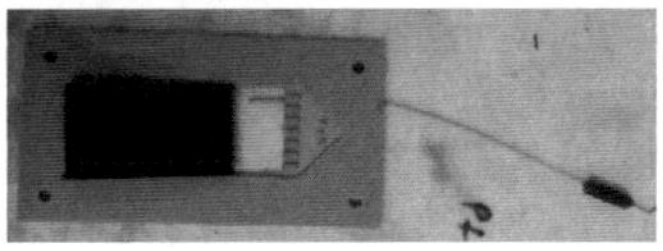
Reference electrode

8/17

Can it work: *electrolysis reactors*

We are in the process of characterizing, modeling & testing catalysts and reactors.

- The catalysts consist of redox sites and electrically conductive supports, which stabilize the redox sites and construct the solution environment.
  - E.g., metals and alloys deposited on nanoporous and fuzzy carbons; akin to PEMFC catalysts but with *high* overpotentials for making H_2
- Characterization will entail site-counting as well as site description, in both cases at least *in situ* and better *in operando*, e.g., Raman, SFG, XAS, coulometry and (with care) electron microscopy
- Modeling needs to estimate ΔG, not just ΔE, of reactions in fields $\gtrsim 10^8$ V/m
- Testing will require reactors that afford intrinsic kinetics and that are scalable.

Visualization of flow in a lab-scale electrolysis reactor using Rhodamine 6G

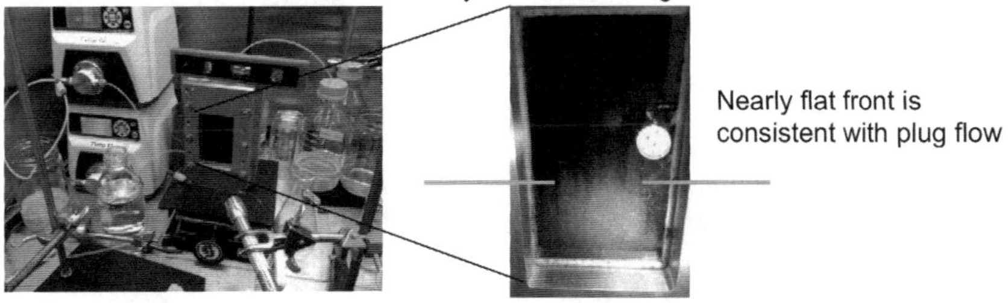

9/17

Can it work: *Kinetics*

The rates of the model reactions correlate well with Butler-Volmer kinetics.

The reactors appear to be nearly plug flow; when the reaction rate laws accord with pseudo-first order kinetics, conversion, $C_R/C_i = \exp(-k\tau)$, where the rate constant, k, derives from Butler-Volmer kinetics

$$C_R = \exp\left[-k_0\left(\left(\exp\left(\frac{-\alpha F\eta}{RT}\right)+\exp\left(\frac{(1-\alpha)F\eta}{RT}\right)\right)+C_i\exp\left(\frac{-\alpha F\eta}{RT}\right)\right)\tau\right]$$

Kinetics fit using the bench cell (V =5 mL)
$0.05 < C_0/M < 0.253$; $0.1 < E/V < 1.1$;
$150 < \tau/s < 450$; $0.01 <$ loading < 0.04

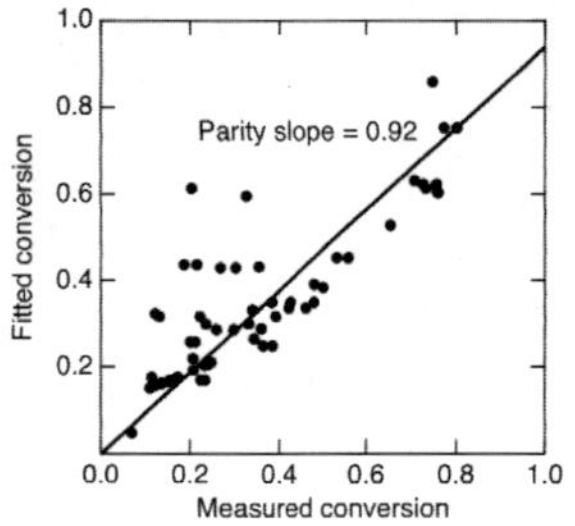

Comparison of the bench cell (5 mL) with the pilot cell (27 mL), each containing 4 wt% Cu/felt

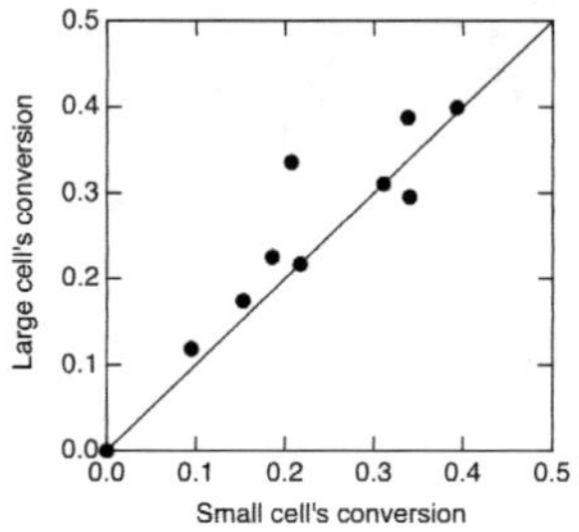

Can it work: *Cathode processes*

We have recently shown that we can electrochemically upgrade some of the functional groups present in neat bio-oil.

- Catalyst was 4 wt% Pd/C, 150 mL of whole bio-oil (ex pyrolysis of pine) recirculated for 20 h through a 6 mL cell at 1.5 V.
- The residence time τ = 20 h × 6 mL/156 mL = 0.8 h (*cf* 5 h for thermal hydrogenation).
- ^{13}C NMR and ^{1}H NMR indicate a significant decrease in carbonyl and aromatic species post electrolysis.

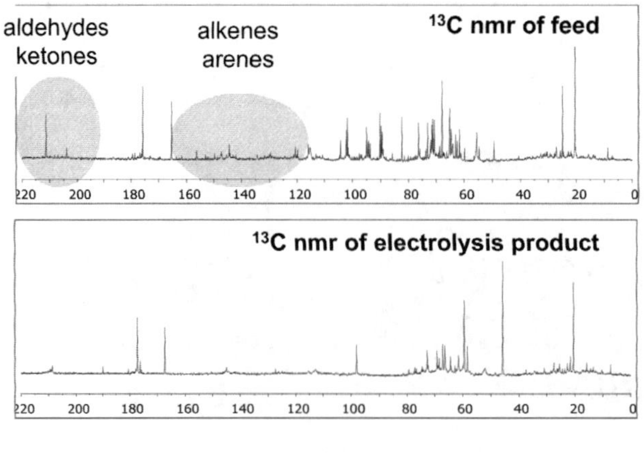

Can it work: *Anode processes*

The aqueous phase from liquefaction of wastes contains many compounds that can, in principle, be oxidized on the anode ...

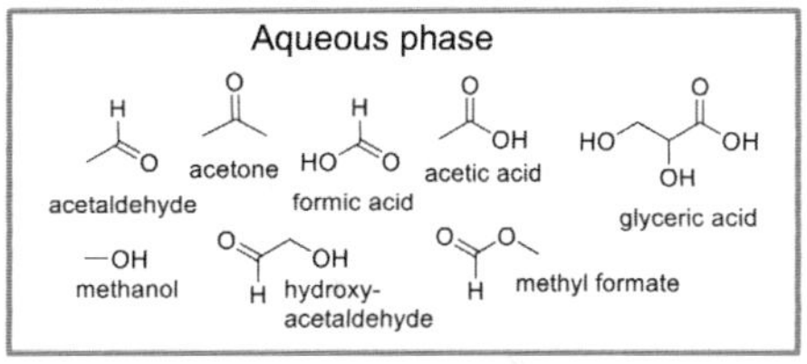

Stoichiometry of upgrading HTL bio-oil *ex* WWTP sludge (0.4 kg oil/kg sludge)

Stream	H^+ Equivalents/g_{sludge}
Oil	-0.03
Aqueous	+24

Estimated from Snowden-Swan et al., PNNL-25464 **2016**.

Can it be deployed: *process integration*

Electrocatalytic upgrading of the bio-oil could simultaneously remediate the wastewater further "preventing a waste and turning it to profit".[1]

- Electrochemistry provides two types of process intensification: high energy flux at ambient conditions (1 eV = 100 kJ/mol) and integrated operations.

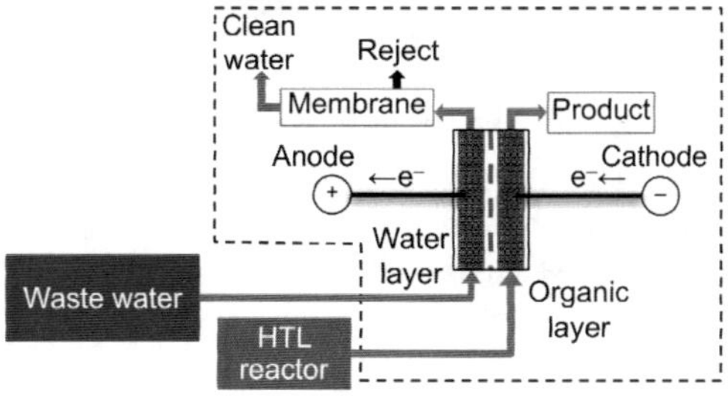

- The cathode catalyst must selectively and efficiently hydrodeoxygenate and hydrogenate the feed.
- The anode catalyst must preferentially and efficiently oxidize the dissolved polar compounds.
- The electrolysis reactor must be compact, robust and ensure good contacting between the liquid phases and their respective catalysts.

1) Arthur D. Little, *J Ind Eng Chem*, **1909**, *1*, 61-62."Every waste that is prevented or turned to profit, ... and every more effective process which is developed, makes for better living in the material sense and for more wholesome living in the higher sense."

Can it be deployed: *Techno-economics*

Experience with redox flow batteries indicate that an electrochemical upgrader would cost about 1/3rd of the target CapEx = $50,000/bpd.

- Values for the electrochemical process were based on PNNL's analysis of the hardware costs for a redox flow battery.
  - Raising the heating value of 1 T/day of pyrolysis oil (~15 MJ/kg) to that of a fuel (~45 MJ/kg) requires adding 30 GJ in 86400 s = 350 kW
  - Estimated, near term CapEx for a redox flow battery, less electrolyte, = ~300 $/kW (exclusive of any PGM catalysts)
- An electrochemical upgrader based on similar technology would cost ~105 k$/T/day = $15,000/bpd.

CapEx estimate for a Redox Flow battery

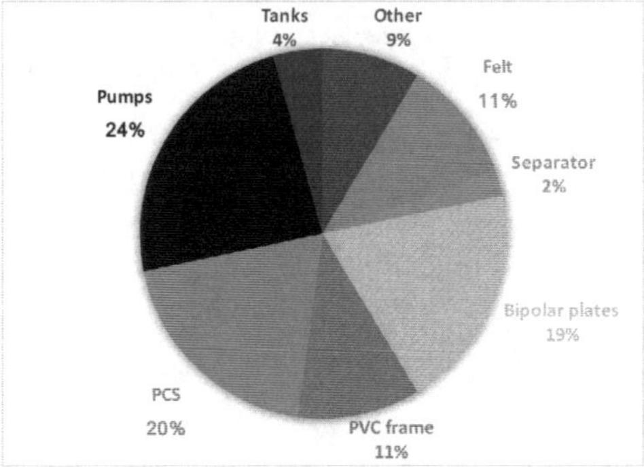

After V. Viswanathan et al., *J. Power Sources* **2014**, *247*, 1040

14/17

Summary

Downsized technology based on hydrothermal liquefaction and electrochemical upgrading appears to be a plausible way to valorize waste.

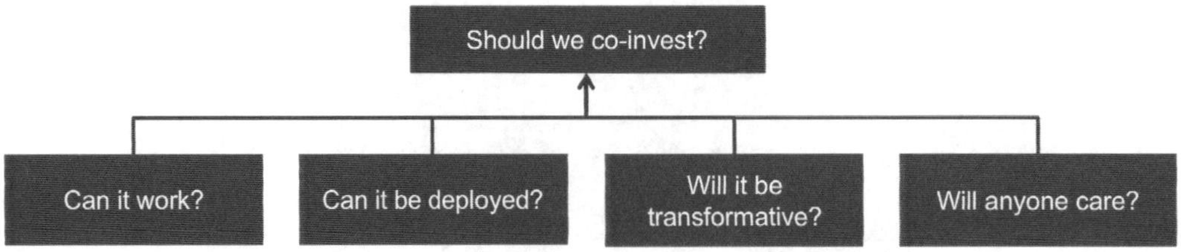

- We do need to devise new catalysts
- Reactors exist that could deploy them
- Such a new enterprise would be transformative: specifically it could convert a waste that now costs $60/ton to dispose and turn it into a profit stream.
- The new enterprise is being designed to appeal to waste-producing companies (e.g., agriculture), communities and product off-takers.

15/17

Acknowledgements

The research was supported by the Chemical Transformations Initiative at PNNL and was conducted under the Laboratory Directed Research and Development Program

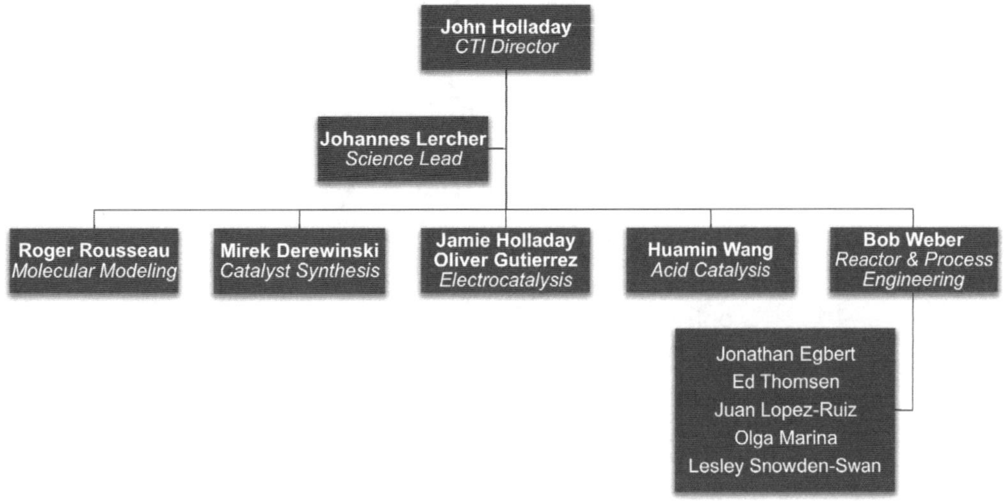

16/17

PNNL Overview

Pacific Northwest National Laboratory has been managed by Battelle for 50 years.

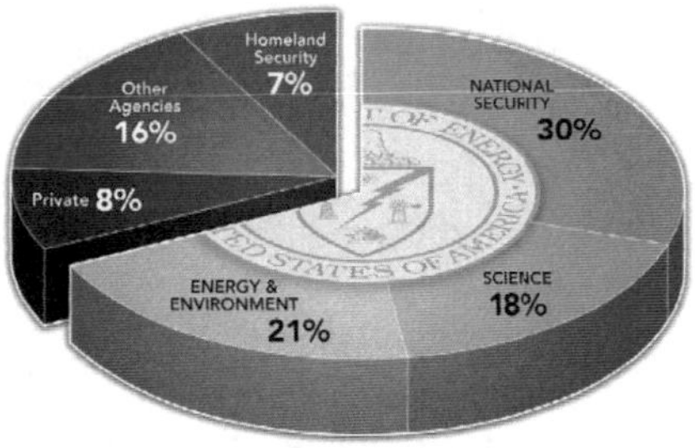

- Since 1965 operated by Battelle for US DOE
- Mission-driven collaborations with government, industry and academia
- Unique Science & Technology strengths and capabilities

- Operating budget of ~$1 B supporting a staff of 4,000
- ~900 peer-reviewed publications per year
- ~100 U.S. and foreign patents per year (2,400 since 1965)
- ~100 awards (R&D 100 and FLC)

17/17

TCP

Inclined Indirect Flaming Pyrolysis Rotary Gasifier

SERDP Project #WP-2211

Stephen Cosper, USACE-ERDC-CERL
Dr. Paul Amodeo, SUNY Cobleskill

SERDP DOD • EPA • DOE | ESTCP

Objective

- Create a design for a deployable waste-to-energy-system, with the following criteria:
  - Based on gasification
  - One to three tons/day capacity, with minimal sorting
  - Self-powered, with net-positive exportable electricity
  - Size to fit two, 20-ft containers

2

Technical Approach

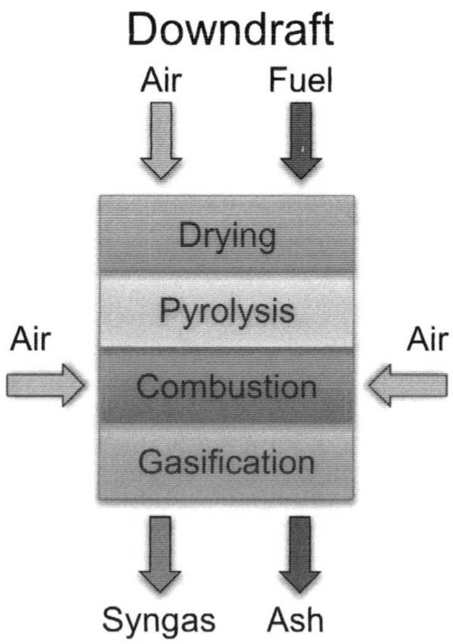

Updraft (Countercurrent)

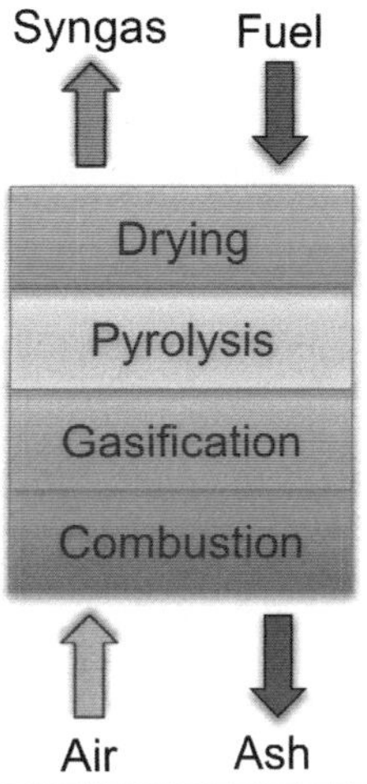

Countercurrent gasification has a number of inherent characteristics that would be beneficial for WTE systems:

- High thermal efficiency
- Less sensitive to fuel pretreatment
- Reduced particulate entrainment
- Low-reactivity ash produced from final-stage combustion

(Courtesy: UND-EERC)

3

Technical Approach – Rotary System

- System works like a modified updraft gasifier
- Waste fed from bottom, placed at hot, burning char layer
- Syngas collected through concentric tube, not directly exposed to char bed
- Rotates at ~1rpm

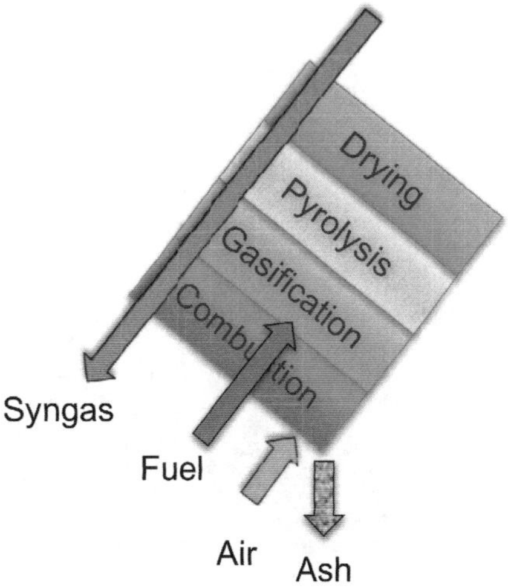

4

Initial design

- Initial hardware design and layout
- Experimental layout on two trailers to mimic portability requirements
- 1st trailer has hydraulic waste compression, gasifier, gas treatment
- 2nd has electronic control system, 60kW genset, hydraulic pump and controls, experimental electric load center

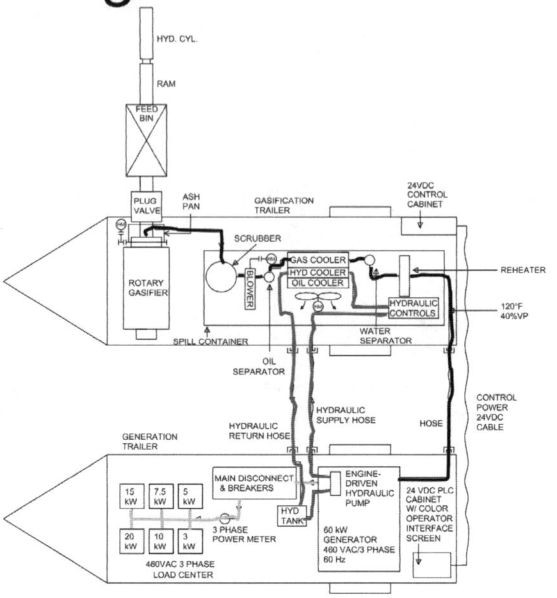

5

Waste Feed

6

Main Reactor

7

"Standard" waste composition

- Based on field surveys, 2013
- Variations included
  - 50% food
  - 50% plastics
  - 50% used oil

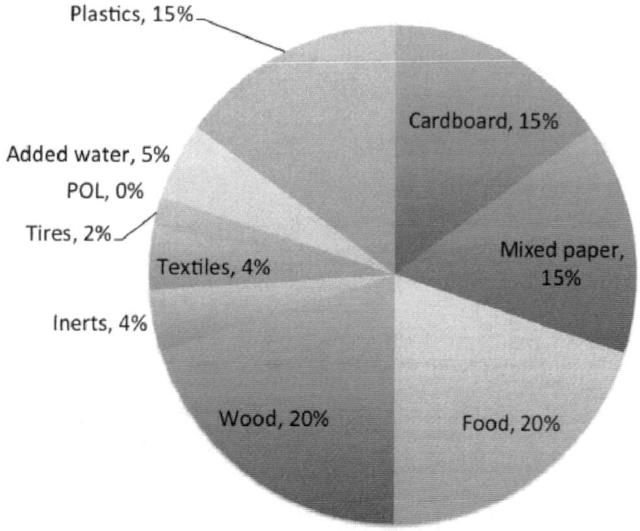

8

Temperature monitoring reactor

- Thermocouples down the length of reactor
- Steady state gas discharge temperature about 400°F
- Temperature peaks indicate time to add waste.
- High temps at end of run indicate, allowing waste layer to burn down.

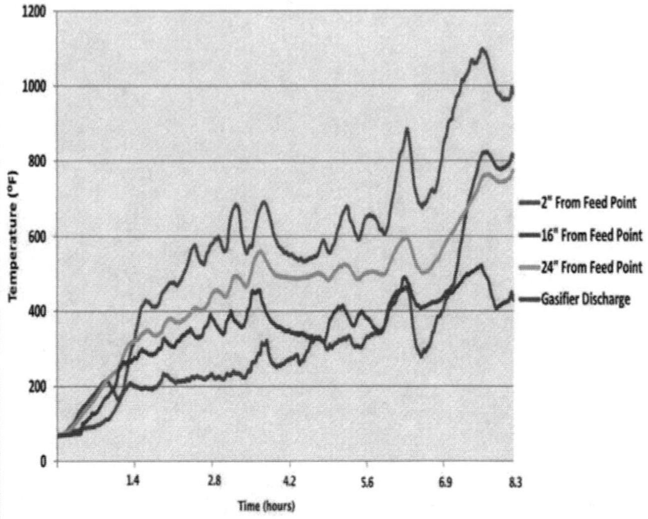

9

Quencher Tank / Pyro Oil Recycling

- Pyrolysis oils produced in reactor
- Accumulate in quencher tank
- Excess fed back into reactor
  - Eliminates a waste stream
  - Enriches syngas
  - ***Could be distilled to commercial fuels***

10

Quencher oil temperature

- Incoming syngas ~400°F
- Pyro-oil remains constant due to water emulsion
- Temp remains below boiling point of water

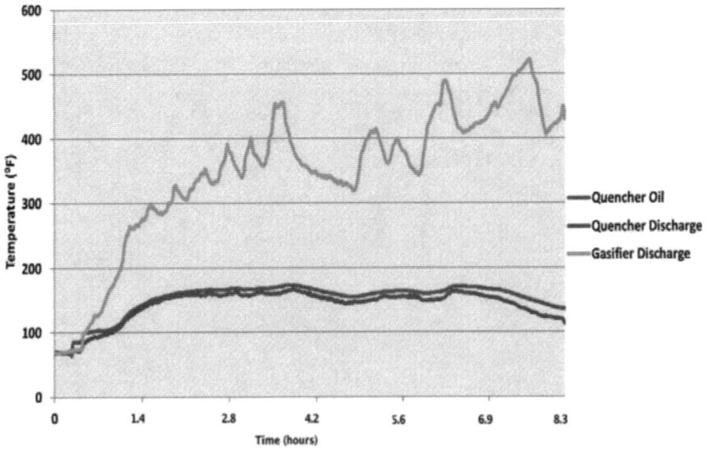

11

Comparison to Commercial Fuels

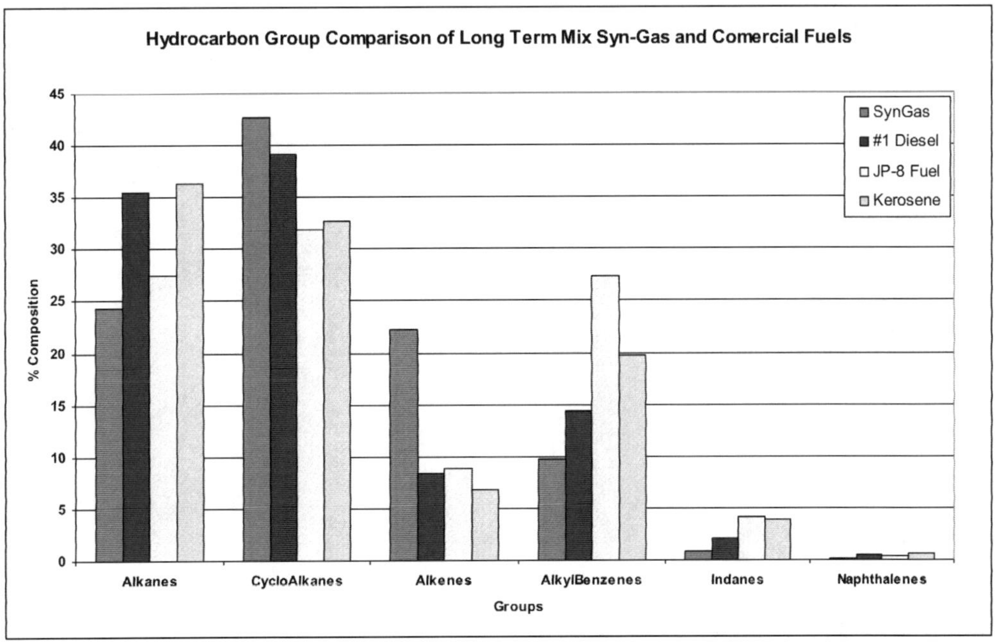

12

Exhaust gas comparison

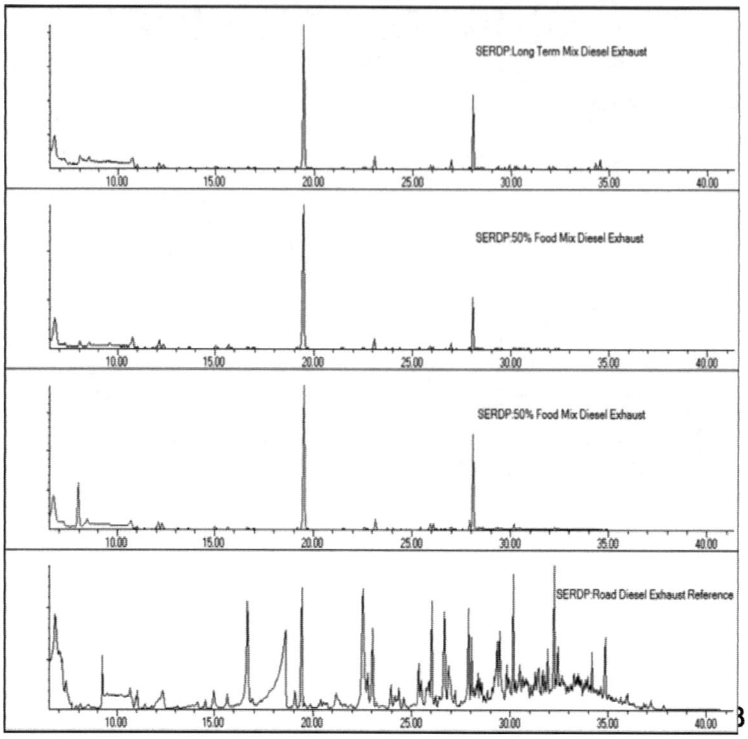

NO_X Emissions with Dual-fuel

- About 50% reduction in NO_X
- Due to:
  - less N_2 into engine
  - Cooler combustion temperatures

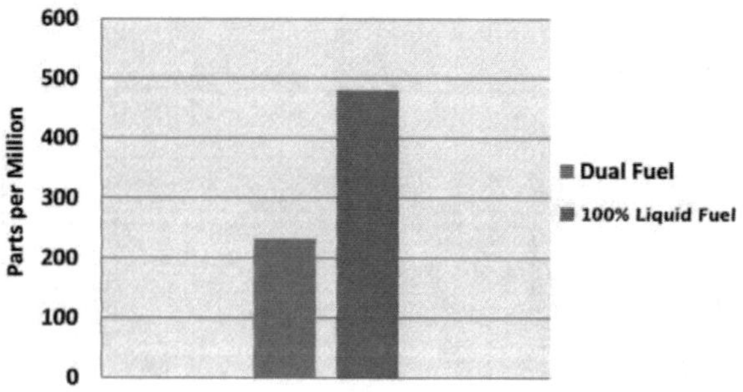

14

Status?

- Project complete
- Prototype designed, built, and tested on a variety of materials
- Final report published very soon
- Team is actively seeking funds for demonstration and/or to construct a "Mark II" automated, containerized version

15

Contact

Stephen Cosper
Army Corps of Engineers
Construction Engineering Research Lab
Champaign, IL
Stephen.D.Cosper@erdc.dren.mil
217-398-5569 M-Th
217-300-3721 F

Dr. Paul Amodeo
SUNY Cobleskill
518-255-5384
amodeopa@cobleskill.edu

https://www.serdp-estcp.org/Program-Areas/Weapons-Systems-and-Platforms/Waste-Reduction-and-Treatment-in-DoD-Operations/WP-2211/WP-2211

16

Monolithic Catalysts Coated with Hierarchical ZSM-5 for Fischer-Tropsch Synthesis

17AIChE
Annual Meeting, Minneapolis, MN

Chunxiang Zhu, David Gamliel, Julia A. Valla and George M. Bollas

Department of Chemical & Biomolecular Engineering
University of Connecticut

http://pdsol.engr.uconn.edu

πδσωλ

Motivation

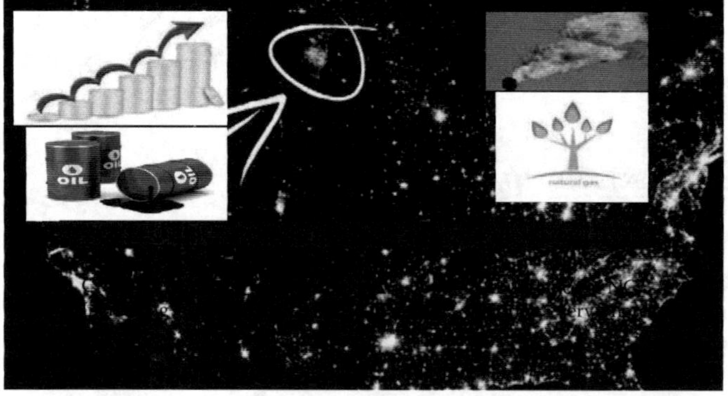

http://www.npr.org/sections/krulwich/2013/01/16/169511949/a-mysterious-patch-of-light-shows-up-in-the-north-dakota-dark
https://energyinfrapost.com/india-comfortable-crude-oil-prices-65barrel-cea/: oil barrel
https://www.123rf.com/photo_12683470_stacks-of-gold-coins-bullion-with-red-arrow-depicting-price-value-going-up-illustration.html : price goes up
https://www.google.com/search: nature gas

πδσωλ

FTS Reaction Mechanism

Reaction	Stoichoimetry	Desirable?
Alkane Formation	$nCO + (2n+1)H_2 \rightarrow C_nH_{2n+2} + nH_2O$	YES
Alkene Formation	$nCO + 2nH_2 \rightarrow C_nH_{2n} + nH_2O$	YES
Water Gas Shift	$CO + H_2O \rightarrow CO_2 + H_2$	NO
Boudouard Reaction	$2CO \rightarrow CO_2 + C$	NO

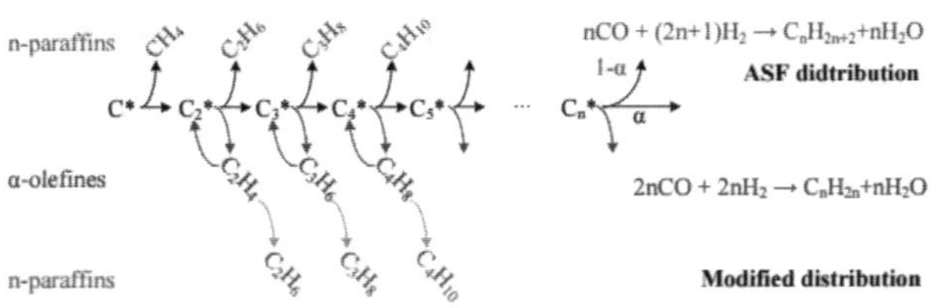

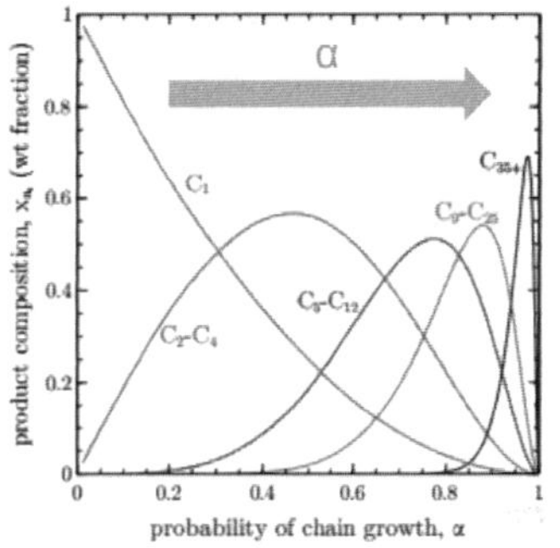

- **Low Temperature FTS (200-240 ºC)**
  - Co
  - Wax, heavy hydrocarbons
  - Low water gas shift activity

Source: F. Kapteijn, R.M. De Deugd, J.A. Moulijn, Fischer-Tropsch synthesis using monolithic catalysts, Catalysis Today. 105 (2005) 350–356.

πδσωλ

GTL Plants

Existing GTL plants[1]

Location	Operator	Start-up	Capacity (b/d)	Note
Sasolburg, South Africa	Sasol	1955	5,600	Switched to nature gas in 2004
Mossel Bay, South Africa	PetroSA	1992	36,000	
Bintulu, Malaysia	Shell	1993	14,700	
Ras Laffan, Qatar	Oryx GTL	2006	34,000	Owned by Qatar Petroleum and Sasol
Ras Laffan, Qatar	Shell	2011	140,000	Owned by Qatar Petroleum and Shell
Escravos, Nigeria	Chevron	2013	34,000	Owned by NNPC and Chevron

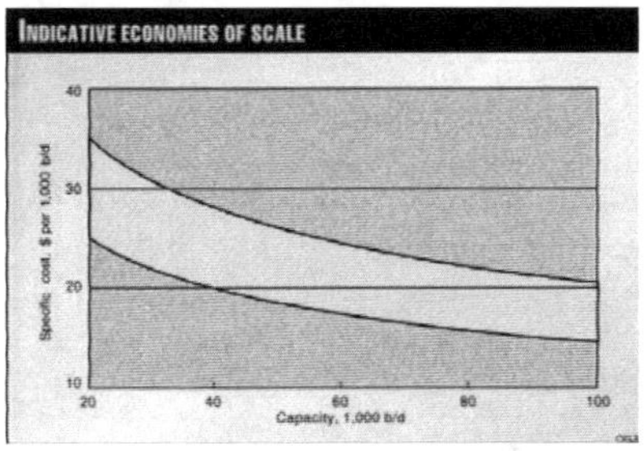

Product cost vs plant capacity.[2]

Less than 10% of the world's gas fields are large enough to sustain a 10,000+ bpd Fischer-Tropsch GTL plant[3]

1. https://www.eia.gov/todayinenergy/detail.php?id=28752
2. http://www.ogj.com/articles/print/volume-96/issue-38/in this issue/petrochemicals/gtl-technologies-focus-on-lowering-costs.html
3. Abstract, Process Economics Program Report 247 B, Small scale gas-to-liquids technology

πδσωλ

Modular GTL Plants

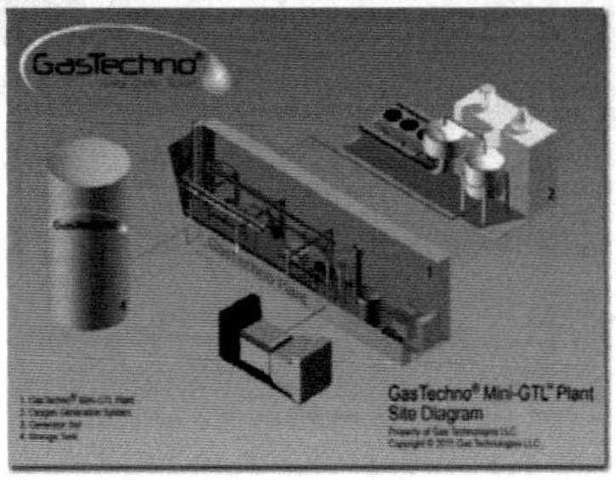

Modular GTL plant[1]

Numerous Companies Developing Mini-GTL Technology[2]

- **Reactor volume decrease**
- **High selectivity, Process equipment reduction**
- **Low capital investment, more profitable**

1. http://www.gastechno.com/gastechno-plant-scales.html
2. http://www.compactgtl.com/technology/petrobas

πδσωλ

Objectives

- **Reactor volume decrease**
  - One pot catalysts with high catalyst activity
- **Process equipment reduction**
  - High gasoline selectivity
- **This catalyst would require:**
  - High surface area
  - Excellent heat transfer
  - Improved metal dispersion
  - Presence of zeolite film to enhance secondary cracking and shape-selectivity
  - Mass transfer limitation release

Component	Function
Monolith	Rigid structure, high thermal conductivity
Cobalt-Al_2O_3	Syngas to hydrocarbons
Zeolite ZSM-5	Secondary acid cracking, shape selectivity

$\pi\delta\sigma\omega\lambda$

Properties of Mesoporous ZSM-5

Meso-ZSM-5 created using the **desilication** technique

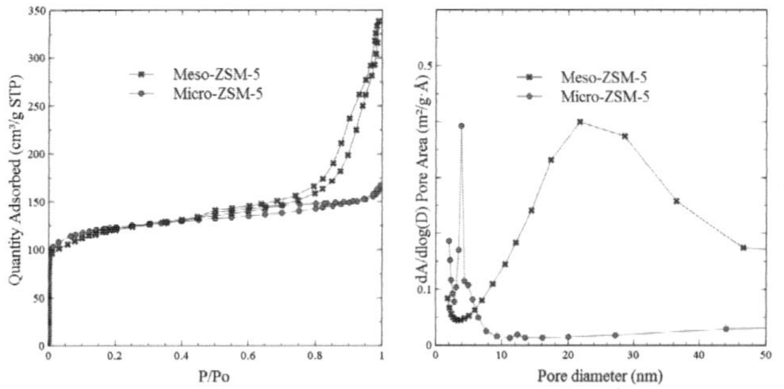

Sample	S_{BET} (m²/g)[a]	V_{micro}[b]	V_{meso}[c]	$D_{average}$[d]	ΔV_{micro}	ΔV_{meso}
Micro-ZSM-5	418.00	0.14	0.08	6.00	--	--
Meso-ZSM-5	421.00	0.10	0.36	15.00	-27.71%	353.09%

[a] Surface area obtained from Brunauer-Emmett-Teller (BET) measurements (S_{BET}). [b] t-plot micro-volume (V_{micro}). [c] BJH adsorption pore volume 1.7-500 nm (V_{meso}). [d] BJH adsorption average pore diameter($D_{average}$).

$\pi\delta\sigma\omega\lambda$

Proposed FTS Catalysis

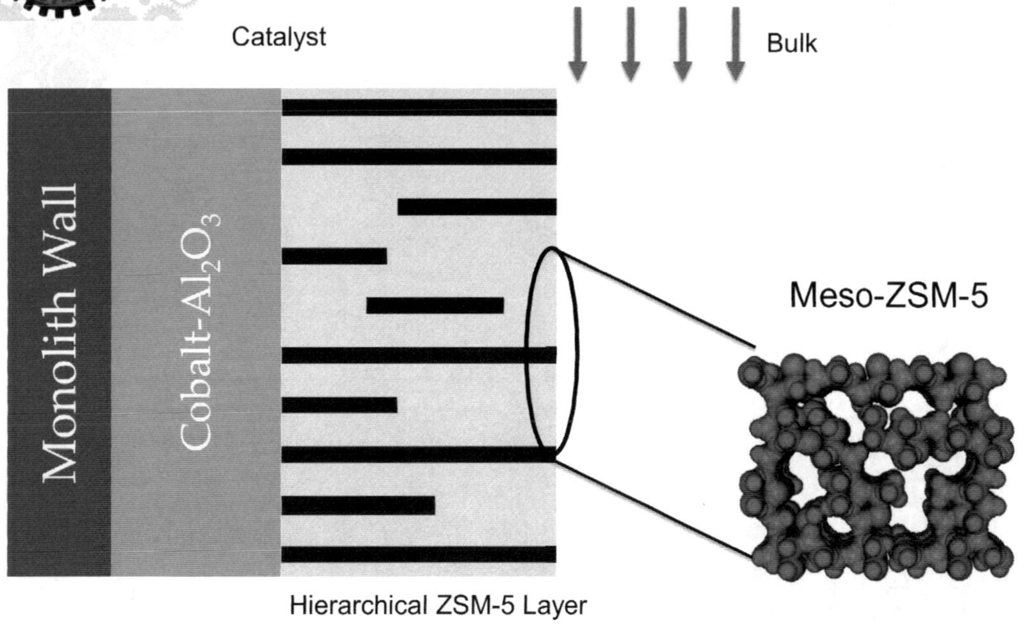

Proposed FTS Catalysis

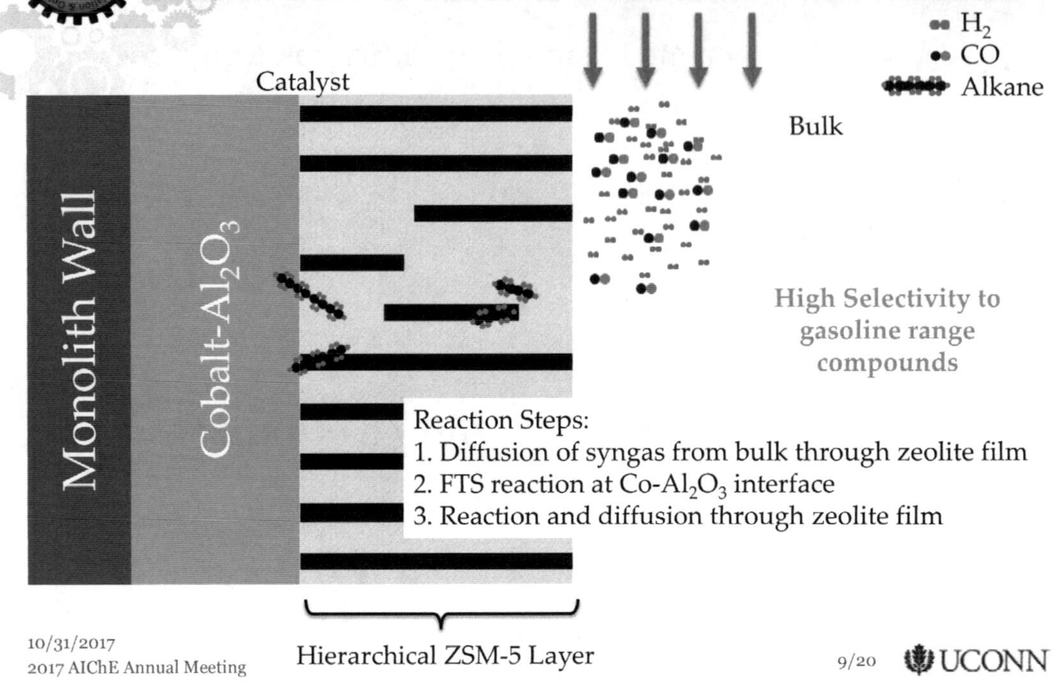

High Selectivity to gasoline range compounds

Reaction Steps:
1. Diffusion of syngas from bulk through zeolite film
2. FTS reaction at Co-Al_2O_3 interface
3. Reaction and diffusion through zeolite film

Catalysts Preparation Method

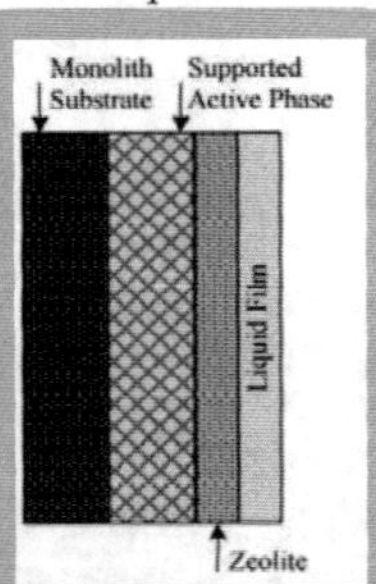

Figure 1. Schematic drawing of ZSM-5 coated monolithic catalyst

Catalysts were dried at 120 °C for 4 h and calcined at 400 °C for 12 h after each coating.

πδσωλ

Equipment Set-up

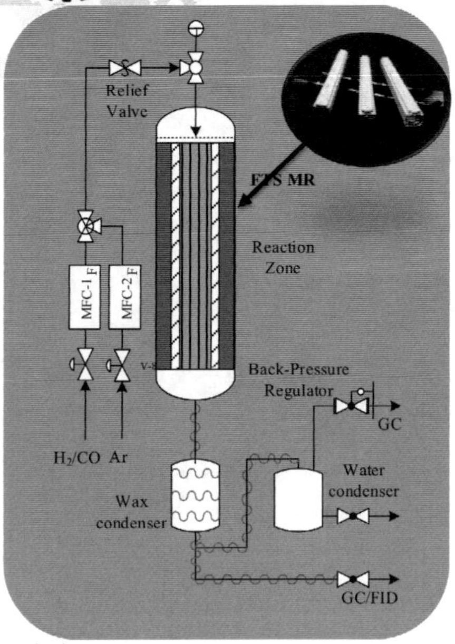

Catalysts Details

Catalyst	Catalyst weight(g)	Monolith (g)	Al_2O_3 (g)	Co_3O_4 (g)	ZSM-5 (g)
1.1 g-Micro	2.93 ± 0.03	1.42 ± 0.03	0.40 ± 0.08	0.13 ± 0.02	1.06 ± 0.04
1.1 g-Meso	3.11 ± 0.04	1.46 ± 0.03	0.50 ± 0.04	0.15 ± 0.02	1.07 ± 0.02
1.6 g-Meso	3.78 ± 0.01	1.45 ± 0.02	0.50 ± 0.01	0.16 ± 0.02	1.62 ± 0.03
1.9 g-Meso	4.01	1.43	0.53	0.16	1.89

- Premixed syngas (H_2:CO = 2:1).
- Ar was used as internal standard.
- Two-stage liquid trap. Wax trap was kept at 120 °C. Light oil and H_2O trap was at room temperature.
- Micro-GC 4900 with two columns and TCD for gases analysis.
- Agilent 6890 GC-FID with HP-5 column for wax and light oil analysis.

πδσωλ

FTS with Meso-ZSM-5 Coated Catalyst

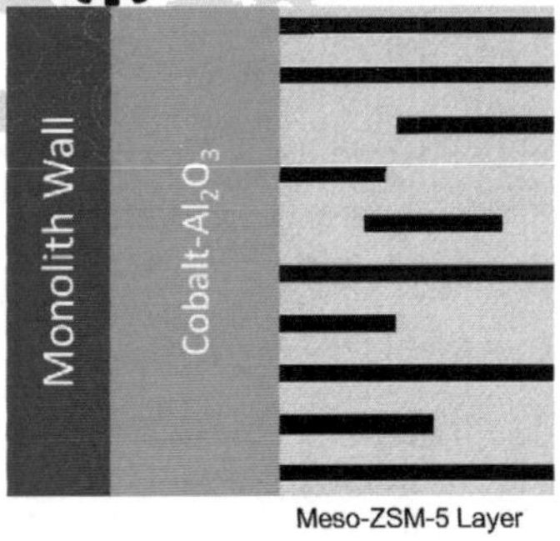

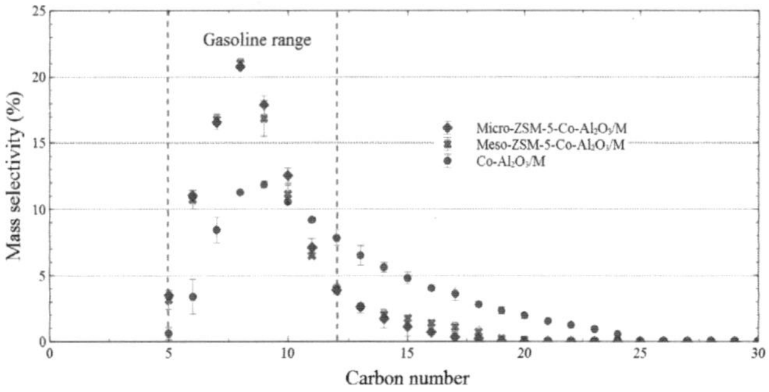

Component	CO Conv wt.%	Gas Selectivity wt.%	Liquid Selectivity wt.%	Mass Balance wt.%
Co-Al_2O_3/M	82	31.8	68.2	96
ZSM-5/Co-Al_2O_3/M	79	20.2	79.8	96
Meso-ZSM-5/Co-Al_2O_3/M	88	27.6	72.4	95

πδσωλ

FTS with Meso-ZSM-5 Coated Catalyst

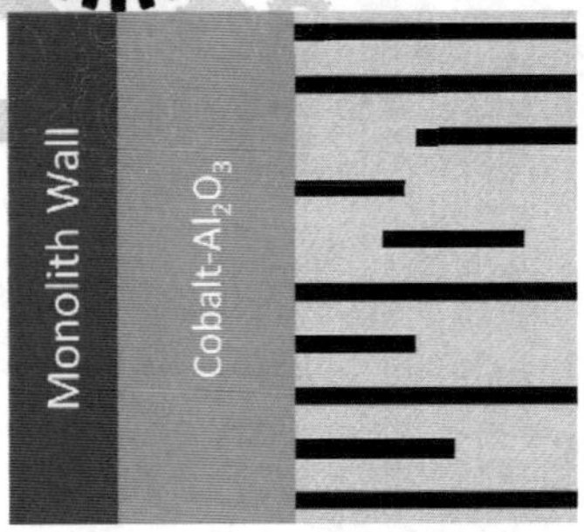

Meso-ZSM-5 Layer

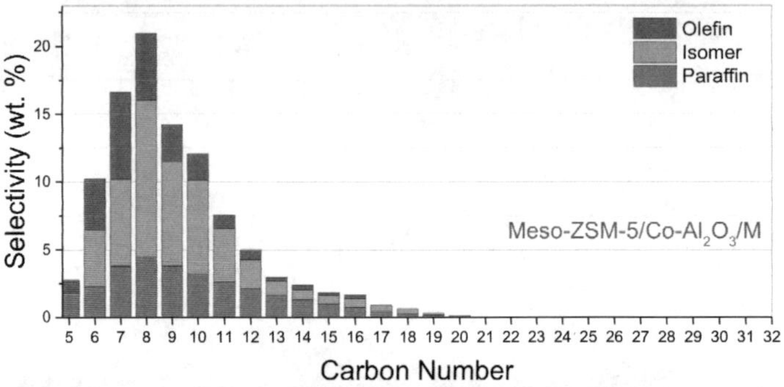

Component	Paraffin Selectivity wt.%	Isomer Selectivity wt.%	Olefin Selectivity wt.%
Co-Al_2O_3/M	69	26	5
ZSM-5/Co-Al_2O_3/M	29	50	21
Meso-ZSM-5/Co-Al_2O_3/M	29	47	23

πδσωλ

Stability Comparison

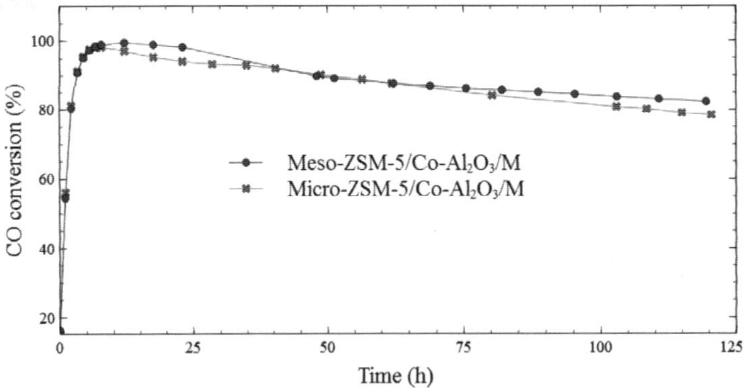

1.1 g-meso and 1.1 g-micro coated catalysts stability comparison within 120 h at 230 °C and 12 bar.

Catalysts coated with Meso-ZSM5 is more stable.

πδσωλ

Performance with Different Coating Thickness

Co-catalysts performance of 1.6 g meso-ZSM-5 at different pressures. Reaction condition: T: 230 °C; Syngas flowrate: 35 ml/min; Time on stream: 48 h

Catalyst	CO conv (%)	Carbon species selectivity (C%)		Oil phase selectivity (wt.%)			
		C_{5+}	C_5-C_{12}	C_{13+}	Paraffins	Isomers	Olefins
1.1 g-Meso	88.4 ± 0.6	77.3 ± 2.0	87.6 ± 1.3	12.4 ± 1.3	29.2 ± 1.8	44.9 ± 1.7	25.9 ± 0.5
1.6 g-Meso	87.8	80.2	93.2	6.8	27.5	50.6	21.9
1.9 g-Meso	85.1	81	92.9	7.1	26.1	50.2	23.7

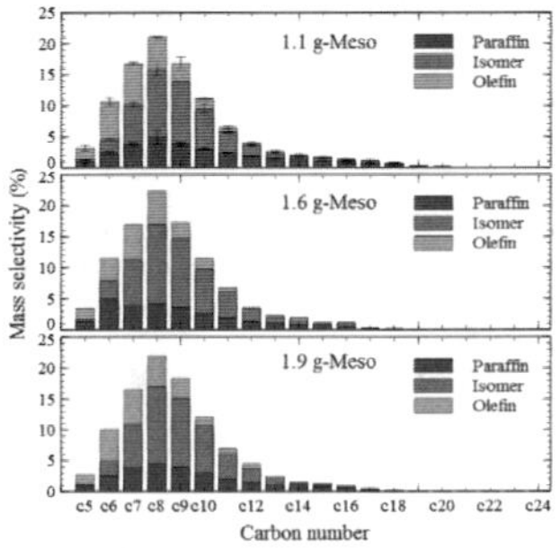

Liquid products distribution for different amount of meso-ZSM-5 coated co-catalysts.

- CO conversion didn't change too much.
- Gasoline selectivity was above 90 wt%.
- Isomers selectivity was elevated with ZSM-5 coating

πδσωλ

Performance at Different Pressures

Co-catalysts performance of 1.6 g meso-ZSM-5 at different pressures. Reaction condition: T: 230 °C; Syngas flowrate: 35 ml/min; Time on stream: 48 h

Catalyst	CO conv (%)	Carbon species selectivity (C%)		Oil phase selectivity (wt.%)			
		C_{5+}	C_5-C_{12}	C_{13+}	Paraffins	Isomers	Olefins
1.6 g -6 bar	55.3	84.6	88.9	11.1	33.1	44.3	22.6
1.6 g-12 bar	87.8	80.2	93.2	6.8	27.5	50.6	21.9
1.6 g-20 bar	92.8	78.4	92.4	7.6	28.1	50.1	21.8

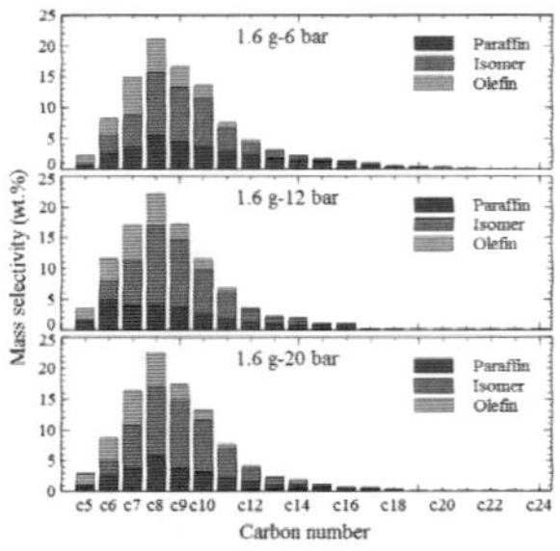

Liquid products distribution for 1.6 g meso-ZSM-5 coated co-catalysts at different pressures.

- CO conversion increased with increase of pressure.
- Gasoline selectivity reached highest at 12 bar.
- Isomers selectivity was elevated with ZSM-5 coating.

πδσωλ

Conclusions

- **Supported Zeolite/Co-Al_2O_3/monolith catalysts were created, characterized and evaluated for FTS**
- **Catalyst coated with Meso-zeolite ZSM-5 exhibits:**
  - High CO conversion
  - Excellent selectivity to gasoline range (C_5-C_{12}) compounds
  - Improved conversion to higher RON olefins and isomers
- **Recommended conditions for catalysts:**
  - Moderate loading of Meso-ZSM5: 1.6 g
  - Moderate pressure: 12 bar

10/31/2017
2017 AIChE Annual Meeting

17/20 UCONN

πδσωλ

Acknowledgements

ACS-PRF Award No. 53648-DNI5

UConn Center for Clean Energy Engineering

10/31/2017
2017 AIChE Annual Meeting

18/20 UCONN

Thank you!

Department of Chemical & Biomolecular Engineering
University of Connecticut

http://pdsol.engr.uconn.edu

10/31/2017
2017 AIChE Annual Meeting

19/20 UCONN

499690 Valorization of Natural Gas Liquids from Shale Gas

Abstract:

Recent shale gas boom has transformed the US energy landscape. Most of the major shale basins are located in remote locations and historically non-gas producing region. Therefore, many major shale basins regions are lacking the infrastructure to distribute the extracted gas into the rest of the US and particularly the Gulf Coast region. Here, we propose a new shale gas valorization process that converts natural gas liquids (NGLs) in shale gas into higher molecular weight hydrocarbon, which is more valuable to transport from these remote basins than NGL or its constituents such as ethane and propane.

In this process, natural gas liquid is transformed to higher molecular weight hydrocarbon. Compared with the existing technologies, our new approach has the following advantages: (1) This process minimizes the production of by-products. (2) This process is amenable to size reduction.

We demonstrate that, through a case study for a 96 MMSCFD shale gas feed from the Bakken region, the total capital cost for the proposed process is much cheaper than existing Gas-to-Liquid (GTL) processes such as Fischer-Tropsch and Methanol-to-Gasoline process. Also, we estimate the break-even price for our process to be lower than GTL-Fischer-Tropsch process. Therefore, this new process is an economically attractive step to upgrade shale gas in remote shale basins.

496044 Energy Sustainability Analysis of H_2 Production

Abstract:

The perspective for H_2 as an energy carrier to decarbonize energy services brings about a more and more relevant role for technologies of hydrogen production. As matter of fact, the yearly majority (>95%) of H_2 is produced from fossil fuels by steam reforming or partial oxidation of methane with only a small quantity by other routes, such as biomass gasification or electrolysis of water, hence the needs to explore new routes to produce H_2 appears evident. Stringent is the necessity to explore different technologies even at the infancy state, to produce hydrogen using different feedstocks. In this respect the energy analysis of H_2-producing technologies aimed to evaluate the net energy produced or Energy Return Of Investment (EROI) and Energy Payback Time (EPT) in order to score them towards energy sustainability is utmost of importance. To this end, the present paper is aimed to test three of the most advanced processes: *i)* steam reformer of methane, *ii)* electrolysis of water by using photovoltaic (PV) electrical energy and *iii)* dark fermentation (DF) of organic refuses (OR) by mixed microorganisms Hydrogen Producing Bacteria (HPB) mainly *Clostridium spp.* The three technologies were tested at laboratory scale for about one year, including the steam reformer owing the necessity to test the hypothesis to produce hydrogen at "the point of use" rather than trying to distribute it. In the panorama of bioH_2 production, which include heterotrophic and autotrophic microorganisms, the choice of DF is based on many reasons: *i)* reactions do not require light, so hydrogen is produced throughout the day and night, *ii)* the liquid end products of DF can be used as feedstock for Anaerobic Digestion (AD) to produce CH_4 with additional energy produced, and *iii)* because the OR are so abundant (> 70 % of wastes) and almost uniformly distributed.The data acquired by experimental campaigns were used to evaluate and scoring the energy sustainability of the three process by the application of a previous candidate procedure (Di Addario et al.,2016). The procedure is recognised into three different steps: a first screening is performed by using the Energy Sustainability Index (ESI), which takes into account the total Produced Energy under form of hydrogen referred to the direct spent energy (heat and electricity). The second step is the Analogical Model, i.e. the quantification of each chemicals, materials and energy to estimate the Useful Energy with a Life Cycle Approach (LCA), taking into account all the Indirect Energy terms. At such a step the energy embedded in the feeds, i.e. methane and water were evaluated for the steam reformer and the electrolysis process, respectively. For the DF process instead, the energy embedded in OR was assumed to be equalto zero, following the allocation procedures in LCA analysis. The third step is the evaluation of two parameters: EROI and EPT, considering all the involved energy flows. The procedure is anuseful tool to score the three processes towards sustainability: higher score correspond to high EROI & low EPT, and this permits to select the most sustainable choice among the three analyzed H_2-producing processes from an energy point of view.

Di Addario, M., Malavè, A.C.L., Sanfilippo, S., Fino, D., Ruggeri, B., 2016. Evaluation of sustainable useful index (SUI) by fuzzy approach for energy producing processes. *Chemical Engineering Research and Design*, 107, pp.153–166.

501652 Chemical Looping Based Technology for High Efficiency Production of H_2 from Ammonia (NH_3)

Abstract:

Hydrogen (H2) is widely considered to be an integral part in solving the issues surrounding the consumption of fossil fuels. As a carbon neutral fuel, H2 can be used as an energy source with minimal emissions for a wide range of uses including chemical production and transportation. One of the most advantageous properties of using H2 as a fuel source lies in its high specific energy of 39.8 kWh/kg (HHV). Despite the advantages offered by the widespread utilization of H2, market penetration is prohibited by the high transportation costs associated with its low volumetric energy density of 1.55kWh/L and thus high compression losses in the storage of H2. In an effort to overcome these limitations, the use of ammonia (NH3) as a hydrogen carrier to minimize transportation and storage costs has been proposed. This vision for improving the market penetration of H2, is contingent upon development of high efficiency H2 from NH3 processes. OSU has developed a chemical looping based NH3 to H2 (ATH) process that can operate at high thermal and H2 production efficiencies. The OSU chemical looping approach consists of two reactors to convert NH3 to H2. The first reactor, namely the reducer reactor utilizes an intrinsic O2 gradient driven by the reduction potential of NH3 and a metal oxide like Fe3O4, ZrO2, WO3 to efficiently crack NH3 to a mixture of N2, H2 and H2O. The reduced metal-oxide from the reducer reactor is re-oxidized in an oxidizer using H2O as the oxygen source. This ensures ~1.5 moles of H2 production per mole of NH3 (nearly 100% H2 production efficiency). This presentation will initially focus on the thermodynamic rationale for developing a chemical looping based NH3 to H2 process. This will be followed by the heat balance approach to achieving the maximum H2 production efficiency while maintaining a high energy efficiency values. The relevant experimental data (using Thermo-gravimetric Analyzer and a fixed bed reactor) that identifies key design conditions associated with the kinetics of the reducer and the oxidizer reactor will be presented. A series of directed sensitivity studies that identify the trends associated with the chemical looping system and the design parameters that need to be de-risked in further scale up of this technology will be discussed.

496846 Distributed Ammonia Manufacturing

Abstract:

To eliminate the dependency on fossil fuel and the carbon dioxide emissions, carbon-free ammonia can be synthesized in small wind-powered plants, and from totally renewable resources: hydrogen from electrolysis of water and nitrogen from pressure swing adsorption of air. We are making carbon-neutral ammonia in the UMN-WCROC plant, located in the agricultural region of Morris MN.[1] While this process is working, the ammonia costs twice that made conventionally, which is due to downscaling (40,000 scaled down process.) To overcome the difficulties associated with the energy demand for ammonia synthesis, and to improve the economics of the small plant, we are seeking a more efficient process to synthesize carbon-neutral ammonia.

One strategy to improve this process is to design a new class of materials for the catalysis, that allows synthesis at lower temperature and pressure. However, a century of research confirms that this is not trivial and immediately viable.[2] Essentially with the conventional catalyst, high temperature is a must in order to obtain fast kinetics, while high pressure ensures the conversion. If we can efficiently remove low concentrations of ammonia formed in the reactor, then we can keep the reaction far from equilibrium, and intrinsically we obtain rates closer to forward reaction rates. Thus the constraints of equilibrium is surpassed. We have proposed the application of solid absorber for high temperature ammonia separations.[3,4] In this process, ammonia can be synthesized at much lower pressure, with rates comparable to high-pressure conventional processes. The capital and operating costs of low pressure reaction-absorption process is improved since the work of compressing the fresh feed is significantly reduced, and the need for heat exchange also drops.

The low pressure reaction-absorption process suggests a strategy for making ammonia fertilizer locally, using wind-generated electricity. Alternatively, this ammonia made in rural area, can be used as a fuel, or a hydrogen carrier for fuel cell use in more urban areas.

References:

(1) Reese, M.; Marquart, C.; Malmali, M.; Wagner, K.; Buchanan, E.; McCormick, A.; Cussler, E. L. Performance of a Small-Scale Haber Process. *Ind. Eng. Chem. Res.* **2016**, *55* (13), 3742.

(2) Schlögl, R. Catalytic Synthesis of Ammonia—A "Never-Ending Story"? *Angew. Chemie Int. Ed.* **2003**, *42* (18), 2004.

(3) Malmali, M.; Wei, Y.; McCormick, A.; Cussler, E. L. Ammonia Synthesis at Reduced Pressure via Reactive Separation. *Ind. Eng. Chem. Res.* **2016**, *55* (33), 8922.

(4) Wagner, K.; Malmali, M.; Smith, C.; McCormick, A.; Cussler, E. L.; Zhu, M.; Seaton, N. C. A. Column Absorption for Reproducible Cyclic Separation in Small Scale Ammonia Synthesis. *AIChE J.* **2017**.

490645 Desorption of Gases from Ionic Liquids Using an Applied Electric Field

Abstract:

Ionic liquids (ILs) are defined as salts which melt at temperatures below 100 °C. ILs possess unique properties including negligible vapor pressures and molecular tunability achieved through the selection of the cation and anion with a broad range of applications in reaction engineering and separations processes. Ionic liquids have been extensively developed for selective gas separations using both physical and chemical absorption. However, alternative desorption processes to remove gas from ionic liquids remain largely uninvestigated. Desorption processes require external energy inputs to separate the gas from the ionic liquid. Conventionally this is done by applying thermal energy in the form of heat which in many cases reduces the overall efficiency and economic viability of the process. To lower the required thermal energy, we have studied the effect of applying an electric field to achieve gas desorption from ionic liquids. Preliminary results demonstrate that using electric fields to manipulate the orientation of ions within ionic liquids could be a new method for desorbing gases with industrial relevance to many gas separation processes.

511741 Green Solvent Use and Opportunities in Biomass Processing

Abstract:

As solvents comprise a large portion of processing waste and energy use for separations, the selection of alternatives with improved safety, environmental, and recyclability would significantly improve process sustainability. As new applications expand in the field of biomass refining and valorization, appropriate solvent development and usage will aid in viable implementation. This work presents an overview of green solvent applications with specific application geared towards biomass production. Particular solvent needs for biomass fractions to produce fuels and value-added products are presented. Green solvent metrics with respect to functionality, and environmental, safety, and health impacts from a process and life cycle view are also addressed and applied to conventional and neoteric solvents. Current and potential applications of various solvents for extractions and conversions are provided for consideration in a biorefinery setting with focus on research in supercritical CO_2 applications.

498338 Tempo Oxidized Cellulosic Membranes for Low Fouling Applications

Abstract:

Cellulose for membrane materials is a well developed technology. However, cellulose is still a difficult material to work with and the properties are seldom as consistent as you woudl like them to be. A variety of novel barrier membranes were cast using different concentrations of the proposed products of the tempo oxidation reaction. Film properties linked to multiple categories ranging from tensile strength to hydrophilicity were investigated. Additionally, data of physical and chemical alterations to the casting of cellulosic membranes have been examined and summarized. We found that using new methods of making tempo modified cellulose has decreased the hydrophilicity of the membrane below what has previously been reported in literature. Further, the fouling characteristics of this membranes have been improved based on the increased hydrophilicity. This paper will examine the reasoning for these improved properties and hyphotisize the possibilities of nanocellulosic membrane materials.

494908 Techno-Economic Analysis of Polymeric Membrane Systems for Post-Combustion Carbon Capture

Abstract:

The increasing consumption of fossil fuels has resulted in immense CO_2 emissions after the Industrial Revolution. Membrane processes have been suggested as an alternative to amine absorption for post-combustion carbon capture. Despite of the tremendous research efforts, CO_2-selective polymeric membranes typically exhibit a trade-off between CO_2 permeance and CO_2/N_2 selectivity. This presentation will show how to overcome the limitations as a result of this trade-off by noval membrane process designs. In the first membrane process, called the hybrid air sweep process, the flue gas flows to a vacuum membrane stage using a membrane with high CO_2/N_2 selectivity and moderate CO_2 permeance. The retentate gas then flows to an air-sweep membrane stage employing another membrane with high CO_2 permeance but low CO_2/N_2 selectivity. The CO_2-laden sweep air is fed to a power plant boiler as combustion air. A highly selective amine-containing membrane with a CO_2 permeance of 1100 GPU (1 GPU = 10^{-6} cm^3 (STP)·cm^{-2}·s^{-1}·$cmHg^{-1}$) and a CO_2/N_2 selectivity >140 is used for the vacuum stage to enrich the CO_2 to >95% purity whereas a highly permeable poly(ethylene oxide)-based membrane with a CO_2 permeance of 2000 GPU and a CO_2/N_2 selectivity of 20 is employed for the air sweep stage. This hybrid membrane configuration yields a capture cost of \$40.2/tonne CO_2 in the 2007 dollar basis for 90% CO_2 recovery, which nearly meets the target of the Department of Energy (DOE) set for 2025. In the second membrane process, i.e., retentate recycle process, the flue gas is passed to two stages employing the membrane with high CO_2/N_2 selectivity; these two stages are placed in an enriching cascade. A portion of the CO_2-depleted retentate of the first membrane stage is recycled as its own sweep gas. It is identified that the highly selective amine-containing membrane is needed for both stages and a 20% retentate recycle renders a capture cost of \$43.1/tonne CO_2. These cost calculations in the 2007 dollar basis have included both membrane module installation cost and 20% process contingency.

492718 An Outlook on Hydrogen Production from Bio-Ethanol Reforming Reactions By Membrane Reactor Technology: Opportunities and Challenges

Abstract:

Our dependence on fossil fuels as the primary energy source, has led to an energy crisis due to the depletion of fossil fuels and the subsequent warming of the globe [1]. In response to these two concerns, continuous efforts are ongoing to search for clean and renewable alternatives. In this context, hydrogen is recognized as one of the most promising energy carriers of the future. Unfortunately, hydrogen is not available in the Earth in elementary form and is currently produced primarily from steam reforming of natural gas, an energy-intensive process.

However, different methods, such as alcohol-reforming reactions, and alternative devices, such as membrane reactors, have being receiving great attention for the production and separation of high-purity hydrogen. Specifically, alcohols may be largely available from renewable sources and they could be used as â€˜bio-feedstockâ€™ for hydrogen generation via reforming reactions. In particular, bio-ethanol constitutes an important bio-source that, compared to other renewable liquid fuels such as methanol, glycerol, acetic acid, diethylether, etc., appears to be more suitable due to its very low toxicity and volatility in addition to its high hydrogen content [2]. Membrane reactor technology seems to represent a valid choice for substituting the conventional process due to its ability of combining reaction and separation in a single stage [3].

Hence, the current study reviews the most relevant scientific results on hydrogen production from bio-ethanol reforming processes performed through membrane reactor technology. The benefits and the main challenges of inorganic membrane reactors, when applied to ethanol-reforming processes, are examined and the performances in terms of hydrogen yield, hydrogen recovery, and conversion are compared to those of conventional systems.

References

[1] M. Ni, D.Y.C. Leung, M.K.H Leung, K. Sumathy, An overview of hydrogen production from biomass, *Int. J. Hydrogen En.*, 87 (2006) 461â€"472.

[2] G. Maggio, S. Freni, S. Cavallaro, Light alcohols/methane fuelled molten carbonate fuel cells: a comparative study, *J. Power Sou.*, 74 (1998) 17-23.

[3] A. Iulianelli, S. Liguori, A. Vita, C. Italiano, C. Fabiano, Y. Huang, A. Basile, The oncoming energy vector: hydrogen produced in Pd-composite membrane reactor via bioethanol reforming over Ni/CeO_2 catalyst, *Catal. Today*, 259 (2015) 368-375

495942 FO Separations with High-Performance TFC Membranes and Novel Draw Solutes

Abstract:

Forward osmosis (FO), an emerging membrane separation process utilizing the difference of osmotic pressure between the feed and draw solution as the driving force, has drawn growing attention ascribed to its unique advantages of relative low energy consumption, high water recovery, and low fouling propensity. Nevertheless, the development of a high-efficient FO system is still hindered by the lack of sustainable high-performance FO membranes and suitable draw solutes. Thin-film composite (TFC) membranes with excellent separation performance and convenient fabrication technology, have gained much attention in FO applications recently. Despite this, traditional TFC membranes formed by TMC and MPD still face problems of insufficient hydrophilicity and the permeability-rejection trade-off phenomenon. In our work, the surface modification by various hydrophilic amine monomers has been performed to improve the membrane water permeability and anti-fouling property. In-situ modification with various tertiary amines as the aqueous phase additives is also investigated, which can act as the catalyst to accelerate the interfacial polymerization. In comparison with the control TFC membrane, these modified membranes have the higher water flux, lower reverse salt flux and enhanced fouling resistance because of the smoother surface, as well as the thinner and denser PA layer with the higher crosslinking degree. In addition, the selection of proper draw solution is also crucial to guarantee a high-performance FO process with minimal reverse solute flux and easy draw solution recovery. Here, series of organic salt draw solutes with adjustable chemical structure and optimized molecular size are synthesized and applied for FO process. The results show that the introduction of carboxyl and phosphate groups to the molecular structure of organic draw solutes is a simple and effective way to enhance the osmotic pressure and FO performance of the draw solutions.

503648 The Last Quarter Century of Methods and Software for Algorithmic Process Synthesis

Abstract:

Process-Network Synthesis or PNS in short formally defined by Friedler and Fan in 1991 is the constructive procedure of building up complex processes or networks from elementary building blocks in order to achieve overall targets by the help of external and internal resources. Advanced algorithmic methodologies and computational tools developed in the last quarter of century could overcome the combinatorial explosion caused by the large number of alternative combinations of process elements having the potential to achieve the desired overall target facing industrial size problems. Moreover, aiming at finding the global optimum multiple component problems of the process-synthesis traditionally examined sequentially become solvable in integrated way, including reaction path synthesis, separation-network synthesis, heat integration, control design, and scheduling. The presentation aims at reviewing the related methods and software. Latest developments and future challenges are to be highlighted as well.

492362 Near-Term and Sustainable Carbon Dioxide Removal: Is Bio-Energy with Carbon Capture and Storage (BECCS) the Right Answer?

Abstract:

Bioenergy with CO_2 Capture and Storage (BECCS) is commonly regarded as a vital yet controversial negative emissions technology (NET) to meeting our climate goals (Fuss et al., 2014). In this contribution we present a whole-systems analysis of the BECCS value chain associated with the cultivation, harvesting, transport, pre-treatment and conversion in dedicated biomass power stations in conjunction with CCS, of a range of biomass resources â€" both dedicated energy crops (miscanthus, switchgrass, short rotation coppice willow), and agricultural residues (wheat straw). We explicitly consider the implications of sourcing the biomass from different regions, climates and land types. The water, carbon and energy footprints of each value chain were calculated, and their impact on the overall system water, carbon and power efficiencies were evaluated. An extensive literature review was performed and a statistical analysis of the available data is presented. In order to describe the dynamic greenhouse gas balance of such as system, a yearly accounting of the emissions was performed over the lifetime of a BECCS facility, and the carbon "breakeven time" and lifetime net CO_2 removal from the atmosphere were determined. The effects of direct and indirect land use change were included, and were found to be a key determinant of the viability of a BECCS project. Overall we conclude that, depending on the conditions of its deployment, BECCS could lead to both carbon positive and negative results. The total quantity of CO_2 removed from the atmosphere over the project lifetime and the carbon breakeven time were observed to be highly case specific. This has profound implications for the policy frameworks required to incentivise and regulate the widespread deployment of BECCS technology. The results of a sensitivity analysis on the model combined with the investigation of alternate supply chain scenarios elucidated key levers to improve the sustainability of BECCS: 1) measuring and limiting the impacts of direct and indirect land use change, 2) using carbon neutral power and organic fertilizer, 3) minimising biomass transport, and prioritising sea over road transport, 4) maximising the use of carbon negative fuels, and, 5) exploiting alternative biomass processing options, e.g., natural drying or torrefaction. A key conclusion is that, regardless of the biomass and region studied, the sustainability of BECCS relies heavily on intelligent management of the supply chain (Fajardy & Mac Dowell, 2017).

References

Fuss, S., Canadell, J. G., Peters, G. P., Tavoni, M., Andrew, R. M., Ciais, P., Jackson, R. B., Jones, C. D., Kraxner, F., Nakicenovic, N., Le Quere, C., Raupach, M. R., Sharifi, A., Smith, P. & Yamagata, Y. (2014). Betting on negative emissions. Nature Climate Change, 4 (10), 850â€"853. DOI:Â 10.1038/nclimate2392

Fajardy, M., & Mac Dowell, N. (2017). Can BECCS deliver sustainable and resource efficient negative emissions? Energy and Environmental Science. DOI: 10.1039/C7EE00465F

499278 A Physical Input-Output Model for the Food-Energy-Water (FEW) Nexus in Indiana

Abstract:

One of the major obstacles to understand the challenges of managing the production of food and energy while maintaining water availability and resources is the lack of physical insights into the regional economies. Physical Input-Output Table (PIOT) based models provide insights into physical interaction of various production sectors while taking advantage of structural information about the economy, thus quantifying the interdependency of food and energy production [1]. Furthermore, the environmentally extended form of PIOT allows the studying of environmental impacts from production systems by tracking the emissions to water and air by each production system. While few fully developed physical input output tables have been produced, their application has been demonstrated in measuring land footprints, analyzing the structure of the economy, and testing the resiliency of supply chains, all of which are important in FEW systems [2,3,4]. Hence, we developed an initial PIOT model for agro-based industries in Indiana to analyze the interdependency of food and bioenergy production in the state. In particular, the unidimensional flow of nitrogen was tracked as measure of production from various sectors, since N forms the basis of food production and the optimization of its use is critically important to food systems and ecosystem management. Indiana's agrobased economy is based on corn and soybean production, which together make up over 76.5% of Indiana's agricultural land [5]. Bioenergy is produced as both determining and co-product from the corn processing sectors. To build this model, we generated process models for each production unit and used the results to generate the PIOT as shown in [6]. For agricultural systems, we performed mass balances on production systems. An algorithmic approach was followed to find the statistical sources for information where available, as well as approximation methods from peer reviewed literature. Once the mass balances were completed they served as the basis for process models that fit into the PIOT building approach described in [6]. The structure of PIOT in Indiana is compared with an earlier PIOT developed for Illinois [7].
[1] Hoekstra, R., & van den Bergh, J. C. (2006). Constructing physical input–output tables for environmental modeling and accounting: Framework and illustrations. Ecological Economics, 59(3), 375-393.
[2] Hubacek, K., & Giljum, S. (2003). Applying physical input–output analysis to estimate land appropriation (ecological footprints) of international trade activities. Ecological economics, 44(1), 137-151.
[3] Altimiras-Martin, A. (2014). Analysing the structure of the economy using physical input–output tables. Economic Systems Research, 26(4), 463-485.
[4] Nakatani, J., Tahara, K., Nakajima, K., Daigo, I., Kurishima, H., Kudoh, Y., Matsubae, K., Fukushima, Y., Ihara, T., Kikuchi, Y. and Nishijima, A. (2017). A graph theory-based methodology for vulnerability assessment of supply chains using the life cycle inventory database. Omega.
[5] United States Department of Agriculture National Agricultural Statistical Service. (2017). 2016 State Agriculture Overview: Indiana. Retrieved from https://www.nass.usda.gov/Quick_Stats/Ag_Overview/stateOverview.php?state=INDIANA.
[6] Wachs, L. and Singh, S. (2016, November). Computational Approaches in Systems Modeling for Environmental Impacts of Industries Automating Physical Input-Output Tables (PIOTs) Via Aspen Process

Modeling. Poster session presented at the 2016 AIChE Annual Meeting, San Francisco, CA.
[7] Singh S, Compton J, Hawkins, T, Sobota D, Cooter E., "A Physical Input-Output Model for Illinois" In Review

501390 Interventions for Reducing Energy Impacts of Water Embodied in Domestic Food Trade: A Network Perspective

Abstract:

Integrated management of food, energy, and water resources has been advocated to mitigate unintended consequences of managing them in isolation. Food, energy, and water (FEW) systems are highly interdependent, interconnected, and exhibit numerous interactions at distinctive spatial scales. Sustainable management and availability of these resources will require quantitatively mapping their complex interdependencies and understanding their ecology. An integral component of food security and diversity involves trade of food items. Food trade also presents a pathway for transferring embodied resources and emissions associated with food production. For the United States (U.S.), irrigation adds significant value to food and feed production and represents a critical link for studying domestic FEW systems. Specifically, irrigation water withdrawals account for second largest freshwater withdrawals in the U.S. and has substantial associated energy and environmental impacts. We recently developed a network model of interstate food transfers for the U.S. The model was utilized to understand the energetic and greenhouse (GHG) impacts of irrigation embodied in interstate food transfers [1]. This analysis highlighted the high energy and GHG emissions embodied in irrigation associated with domestic food trade. We also proposed relevant network metrics to understand the structure of domestic food trade network and its implications for resilience.

In this work, we propose and evaluate a range of intervention strategies to enhance environmental sustainability and resilience of the domestic food trade network as they relate to irrigation and its embodied energy impacts. We use publicly available datasets to develop weighted and directed networks of interstate food trade and associated environmental impacts. We consider interstate food transfers of over 50 commodities including cereal grains, milled grains, livestock, and meat products. From a network perspective, the weighted food trade network consists of a few key states (Texas, California, Kansas) that control high throughput. We also observe that these influential states depend on water resources facing depletion risks (Colorado river basin, Ogallala aquifer), employ energy and water inefficient irrigation systems, and utilize electricity and diesel-based pumps. We evaluate the impact of technological, behavioral, and policy interventions using our interstate food network model to study their impact on the sustainability and resilience of the domestic FEW nexus. The investigated interventions include changes in irrigation practices, energy mixes, adjustments in diet patterns, and spatial changes in food production patterns. These strategies help in prioritizing and shedding light on interventions that aid in improving sustainability and resilience of the FEW nexus from a systems perspective as opposed to those that offer incremental advantages. The implications of these findings including their potential and limitations for enhancing the sustainability and resilience of the FEW nexus will be described in detail.

Reference Cited:

N. Vora, A. Shah, M.M. Bilec, and V. Khanna. Food-Energy-Water nexus: quantifying embodied energy and GHG emissions from irrigation through virtual water transfers in food trade, *ACS Sustainable Chemistry and Engineering,*

2017, 5(3), 2119-2128.

500119 A General Framework for Process and Utility Networks Synthesis

Abstract:

There are sub-systems in chemical plants that consist of a number of sinks with demand for a certain property and a number of alternative sources that can satisfy these properties to fulfil the sink demands. Examples of such sub-systems are heat exchanger networks [1], fuel gas networks [2], water networks [3], utility networks [4], etc. Properties that are exchanged can be energy in the form of heat as in heat exchanger networks, or in both work and heat form as in the utility networks. Although each subsystem has attracted significant attention, each synthesis problem requires a new superstructure to be postulated whenever a new problem is encountered. Similarly, separation network systems, such as membrane networks [5] and distillation sequences [6], have been a focus of attention, yet a new superstructure representation is needed for each problem. A unified superstructure representation for these different classes of synthesis problems will be highly advantageous. Recently, a systematic framework for process intensification based on building-block superstructure has been proposed based on the idea of dissecting various unit operations into fundamental building blocks [7]. The proposed method can be also used for automatic flowsheet generation.

In this work, we extend the approach to provide a general framework for the synthesis of different networks using a similar representation using building blocks. Here, each block has feed and product streams ensuring interaction with the surroundings and can be viewed as sources and sinks, respectively. Also, streams between blocks are equipped with compressors or expanders depending on the source and sink block pressures. Mass and energy flow within the superstructure are facilitated via block material and energy balances and heating and cooling requirements are decided accordingly. Separation operations, such as gas membrane separations, are represented via two neighboring blocks separated by a semi-restricted boundary. The decisions over the assignment of boundary types between blocks are enabled via use of decision variables. By using all of these features of the building block superstructure, several different process network synthesis problems can be accommodated in a single superstructure with n and m number of blocks in row and column respectively. It should be noted that for different network synthesis problem, the block superstructure size could be reduced to improve the computation performance, i.e, using $2\times m$ superstructure to represent a pooling problem without intermediate pools. The overall process network synthesis problem is formulated as a mixed-integer nonlinear optimization (MINLP) problem where the nonlinear terms are due to splitting, energy balances and/or work calculations and discrete variables are for assigning alternative operations to each block and boundary in the superstructure. In this work, we will demonstrate that the proposed method is capable of unifying the superstructure representations for different class of process network synthesis problems via a range of literature problems.

References:

[1] Yee, T.F., Grossmann, I.E. and Kravanja, Z., 1990. Simultaneous optimization models for heat integrationâ€"I. Area and energy targeting and modeling of multi-stream exchangers.Â *Computers & chemical engineering*,Â *14*(10), pp.1151-1164.

[2] Hasan, M.F., Karimi, I.A. and Avison, C.M., 2011. Preliminary synthesis of fuel gas networks to conserve energy and preserve the environment. Industrial & Engineering Chemistry Research, 50(12), pp.7414-7427.

[3] AhmetoviÄ‡, E. and Grossmann, I.E., 2011. Global superstructure optimization for the design of integrated process water networks. AIChE journal, 57(2), pp.434-457.

[4] Elia, J.A., Baliban, R.C. and Floudas, C.A., 2010. Toward novel hybrid biomass, coal, and natural gas processes for satisfying current transportation fuel demands, 2: Simultaneous heat and power integration. Industrial & Engineering Chemistry Research, 49(16), pp.7371-7388.

[5] Alnouri, S.Y. and Linke, P., 2012. A systematic approach to optimal membrane network synthesis for seawater desalination. *Journal of membrane science*, *417*, pp.96-112.

[6] Giridhar, A. and Agrawal, R., 2010. Synthesis of distillation configurations: I. Characteristics of a good search space. Computers & chemical engineering, 34(1), pp.73-83.

[7] Demirel, S.E., Li, J., Hasan, M.M.F. Systematic Process Intensification using Building Blocks. Computers & Chemical Engineering, http://dx.doi.org/10.1016/j.compchemeng.2017.01.044.

503177 Keynote - Two-step Solarthermal Water Splitting – The Path Forward

Abstract:

Using concentrated sunlight to drive water splitting for renewable hydrogen production is the ultimate goal for a truly green energy world. The seminal paper, "Hydrogen and Oxygen from Water," *Science*, 197, 1050 (1977), by E.A. Fletcher and R.L. Moen ignited a flurry of activity in the field of solar thermalwater splitting. Focused research since then has resulted in substantial progress identifying particular chemical routes, active materials and various solarthermal reactor designs. This presentation will review key findings and will focus on the path forward emphasizing active materials and processes and identifying key areas of interest for future research.

491083 Syngas Production Via Solar Chemical-Looping Reformation of Methane in a Fixed-Bed Reactor

Abstract:

The production of syngas using concentrated sunlight via solar chemical-looping reforming (SCLR) is an approach to produce sustainable transportation fuels. SCLR is a two-step thermochemical cycle that produces separate streams of syngas and high-purity H_2. In the endothermic partial oxidation step, methane reacts with oxygen released from a metal oxide to produce syngas. In the exothermic water-splitting step, steam reacts with the reduced metal oxide to produce H_2. Solar energy is stored in the products, upgrading the calorific content of the feedstock by up to 28%. Combustion of the products emits 41% less carbon per kilojoule than combustion of the productions from non-solar CLR. The cycle is thermodynamically favorable at 1273 K and 1000 suns, conditions easily achievable by commercial solar concentrated optics used currently for electricity production. Based on a thermodynamic analysis of the process, the solar-to-chemical efficiency for SCLR, defined as the net increase in the higher heating value of the gases divided by the solar thermal input to provide process heat and parasitic work for gas pumping and separation, can reach 54% with complete conversion of CH_4 to syngas and H_2O to H_2.

While the net products and chemical energy requirements match that of solar steam reforming, decoupling the partial oxidation and H_2O-splitting steps provides numerous advantages. The products of SCLR are separate streams of 2:1 H_2:CO and high-purity H_2 compared to the single stream of 3:1 H_2:CO obtained from steam reforming. The 2:1 H_2:CO syngas matches the desired composition for the synthesis of liquid hydrocarbon fuels and H_2 is valuable as a fuel for electricity production via fuel cells or as a chemical feedstock in the synthesis of various chemical commodities such as ammonia. Furthermore, conventional steam reformers must operate with excess oxidizer (e.g. $H_2O/CH_4 = 3$) to prevent catalyst deactivation by carbon deposition. With solar CLR, carbon deposition can be prevented by limiting the reduction extent of the metal oxide.

In this paper, we present measured performance for SCLR in a prototype reactor that utilizes cerium dioxide (ceria) as the oxygen carrier material. The reactor is operated in the University of Minnesota high flux solar simulator. The reactor comprises two sections: a solar receiver/reactor and a heat recuperator. Concentrated sunlight enters the cylindrical receiver/reactor cavity section through an open circular aperture and heats a sealed assembly of concentric alumina tubes, referred to as the reactive element. The annulus of the reactive element is filled with 336 g of ceria particles with a bed void fraction of 45%. The particles are 5 mm (length and diameter), 78% porous and have a specific surface area of 0.114 $m^2\ g^{-1}$. The reactive element extends beyond the solar cavity to the heat recuperator. In the heat recuperator, the annulus and inner tube of the reactive element are filled with alumina reticulated porous ceramic that is 85-90% porous and has pore densities of 5 and 10 pores per inch in the inner tube and annulus, respectively.

The cycle is accomplished by alternating the gas feed to the reactive element between CH_4 and H_2O. During the endothermic step, CH_4 flows through the inner tube, reverses direction at the closed end, and flows over the ceria particles in the annulus. Sensible heat from the product gases is transferred to the inlet flow in the heat recuperator. At the end of the partial oxidation step, the gas flow is switched automatically to H_2O. The operating conditions (temperature, reactant feed rates, reaction times) are selected based on a thermodynamic process model of the reactor. The model predicts CH_4 and H_2O conversions and H_2 and CO selectivities data from prior benchtop experiments.

Measured results include radiative input, transient temperatures of the solar cavity, reactive element, and inlet and outlet gases, upstream and differential pressures across the reactive element, and product gas composition. The axial temperature distribution along the top surface of the reactive element in the receiver/reactor and surface temperature of the reactor cavity are measured by alumina-sheathed type-B (Pt-30% Rh/Pt-6% Rh) thermocouple probes. The axial temperature distribution along the reactive element in the heat recuperator section and the gas temperatures at the inlet and outlet of the active reactive element are measured with type-K (Ni-Cr/Ni-Al) thermocouples. The high-flux solar simulator is controlled to obtain a cycle-averaged reactive element temperature of 1273 K. The radiative solar input is measured using a water-cooled black body calorimeter. The product gas composition is measured at 1 s intervals using Ramen laser gas spectroscopy. Data are evaluated to determine the solar-to-chemical efficiency of the reactor. The reactor demonstration provides an important proof of concept for SCLR and a comparison of measured efficiency to prior theoretical predictions of efficiency.

486955 Solar Gasification in a Molten Salt Reactor for Continuous Production of Syngas

Abstract:

The design and operation of a concentrated solar reactor to gasify biomass in a ternary eutectic blend of alkali metal carbonate salts are presented. Solar gasification has numerous advantages over conventional gasification. The use of concentrated solar energy as the source of process heat eliminates the need for partial combustion of the feedstock. The product synthesis gas is not diluted by excess CO_2 and N_2 nor contaminated by other combustion by-products. There is no need for an economically and energetically expensive oxygen plant that otherwise is necessary for oxygen-based partial combustion. The energetic value of the product is greater than that of the feedstockâ€"the difference is equal to the storage of solar energy in the fuel. The benefits of the molten salt solar gasifier in comparison to other solar gasifiers are the thermal mass of the salt allows continuous operation during solar transients, provides rapid heat transfer, catalyzes the gasification reaction, and reduces the amount of tars and other contaminants in the product gas stream for an operating temperature of 1200 K. In prior work, utilizing cellulose as feedstock and carbon dioxide as oxidizer, the reactor achieved a solar efficiency of 30% and converted 47% of the carbon in a continuous process. The efficiency at 1218 K was comparable to the highest achieved in other solar gasifiers operated at higher temperature, but the pneumatic feed system was prone to blockage and required a stoichiometric oxidizer ratio of eleven. The use of excess carbon dioxide increased the sensible heating requirement and exacerbated the entrainment of fine char particles in the product gas stream prior to complete conversion.

In the present study, we present a model of an improved volumetric pumping feed system and provide operational data for gasification of cellulose in the improved reactor. The feed system consists of a single fluted screw within a straight barrel, similar to that used for polymer extrusion systems. The feed enters the reactor as a densified cylinder rather than as loose powder, encouraging the formation of larger char particles which have a reduced tendency to be entrained by the product gases leaving the reactor. The design of the feed delivery system is based in part on a numerical model of the transient heat transfer and kinetics of pyrolysis and gasification of the feedstock material during transport.

The reactor was operated for CO_2 gasification of cellulose in the University of Minnesotaâ€™s high flux solar simulator. The simulator consists of seven radiation units, each composed of a 6.5 kW xenon short arc lamp close-coupled to a custom reflector in the shape of a truncated ellipsoid of revolution. The total power delivered to the aperture of the reactor was measured using a water-cooled black body calorimeter of the same aperture diameter. The feedstock is ash-free microcrystalline cellulose ($C_6H_{10}O_5$) sieved to a particle size of 0.5 mm. Temperatures of the solar cavity and within the salt melt are measured using Type-K thermocouples. The product gas composition is measured using a Raman Laser Gas Analyzer. The downstream gas lines and HEPA filter are analyzed for residual secondary products. Tars are separated from solids using a solvent wash and the particle size distribution of the solid material is measured using laser diffraction.

The reactor performance is quantified by the solar to fuel efficiency defined in rate form as the ratio of the lower heating value of the useful products and the sum of the solar input and the LHV of the carbonaceous feedstock. Carbon conversion is given by the ratio of the net gaseous carbon released from the reactor to the feedstock carbon delivered to the reactor. The measured species flow rates are compared to the equilibrium flow rates predicted using Gibbs free energy minimization. The data demonstrate the improvements in performance for the reactor with the new feed system in terms of longer operation without blockage, and higher solar efficiency.

494979 An Evaluation of Feedback and Model Predictive Control for the Rejection of Weather Transients in Renewable High-Temperature Solar Thermal Chemical Processing on a Novel Hybrid Solar-Electric Reactor

Abstract:

Concentrated solar power, which has been commercialized for electrical generation, is now of interest as a means of directly driving chemical processes. Investigations have included the use of solar heat to produce metals from oxide ores, generate renewable hydrogen for use as fuel, and to gasify biomass. These processes often operate at high temperatures where reactors and receivers are built partly or wholly from ceramics, ceramics that are vulnerable to thermal shock and fatigue. Weather transients, which can deprive a facility of solar heat within seconds, must be controlled to insure the longevity of solar-driven chemical process equipment. Supplementary electrical and/or gas heaters can be used to provide auxiliary heat when transient cloud cover would otherwise interrupt solar thermal chemical production. We evaluate different control schemes on a new solar-electric reactor prototype at the University of Colorado Boulder. These include both feedback and model predictive control paradigms, whose performance was measured by integral squared absolute error over the course of programmed weather transients on a 45kWe high-flux solar simulator. Implications for the implementation of commercial solar thermal processes will be discussed.

494571 Solar Hybrid Photo-Thermochemical Sulfur-Ammonia Water-Splitting Cycle: Photocatalyst, Thermodynamics and Plant Analysis

Abstract:

The solar driven hybrid sulfur-ammonia water splitting cycle (HySA) is a promising technology for the environmentally sustainable production of hydrogen fuel. It is a state-of-the-art process that integrates a solar-photocatalytic hydrogen production step (driven by the photonic portion of solar irradiance) with a high-temperature solar thermochemical oxygen evolution step (driven by the thermal portion) and efficient thermal energy storage as an integral part of the cycle. This work summarizes the recent findings on the two subcycles and the overall process design and optimization, following detailed experimental and numerical calculations. First, multiple photocatalysts, for the hydrogen production step, were assessed and characterized, finally, a binary band gap photocatalyst based on ZnS-CdS core shell particles doped with Pt was selected. Second, the materials and their properties for the oxygen evolution step were verified using thermogravimetric experiments, that supported a detailed thermodynamic analysis for the selection and design of the overall process. Finally, all information was assembled in one process that takes advantage of the hybrid solar photo-thermo character of the cycle and the incorporation of a portion of the process-fluids to design an integrated thermal-storage cycle. The solar to H_2 efficiency of the process was estimated up to 25% depending on reactor configurations.

495376 On the Potential for Enhancement of Solar Thermochemical Synthesis Processes By Nonequilibrium Plasma

Abstract:

The sustainable synthesis of fuels and chemicals from abundant low-value feedstock, such as carbon dioxide, methane, and water, is a key technological and global challenge. Solar thermochemical synthesis processes directly utilize the most abundant form of renewable energy, but face severe challenges due to the low reactivity of most feedstock gases to solar radiation and the intermittency of the solar resource. These challenges often lead to limited process efficiency and resiliency, which translate into systems with higher capital and operating costs. The combination of a solar thermochemical process with a non-solar co-process can be an effective option to achieve continuous operation, improve resiliency, and reduce costs. Moreover, if the concurrent operation of the solar and non-solar processes leads to greater overall efficiency, then the combined approach could have greater viability than either process alone. This talk discusses the potential of the use of nonequilibrium plasma as a co-process to solar thermochemical synthesis. The combined solar-plasma approach, denoted as Plasma-Enhanced Solar Energy (PESE), has the potential to enhance the efficiency and resiliency, and consequently improve the viability, of solar thermochemical synthesis processes.

Nonequilibrium plasmas are partially ionized gases in which the temperature of the free electrons is significantly higher than that of the heavy-species (molecules, atoms, ions). The electron temperature in atmospheric-pressure nonequilibrium plasmas are typically on the order of 1 to 2 eV, commensurate with the threshold energies in molecular bonds, whereas the temperature of the heavy-species range between room temperature to a few thousand degrees. Therefore, electrons and heavy-species are not in state of Local Thermodynamic Equilibrium (LTE), i.e., they are in a non-LTE (NLTE) state. Plasma-based chemical synthesis processes use electricity to directly modify gas chemistry and produce activated species (radicals, excited states, ions). These processes can achieve high energy efficiencies by selectively transferring energy to different channels. For example, microwave plasmas have shown over 90% energy efficiency (i.e. electrical to chemical) for the decomposition of carbon dioxide thanks to effective step-wise vibrational excitation, the most effective channel for dissociation [1]. Moreover, the reliance of plasma processes on electrical energy makes them ideal for continuous operation, but also limits their economic viability and their net sustainability benefit, even if the required electricity is obtained from renewable energy sources, such as wind and solar.

PESE chemical synthesis processes seek to combine the advantages of solar thermochemical and plasma processes. The potential advantage of PESE to improve the efficiency of solar thermochemical synthesis is due to the fact that the molecular excitation produced by free electrons in non-equilibrium plasmas can significantly increase solar photon absorption leading to the enhancement of chemical reaction kinetics. Moreover, PESE chemical synthesis processes could rely on electrical energy to compensate for fluctuations in the solar energy input (during daytime) or even substitute it (at nighttime), and therefore ensure continuous operation.

Preliminary computational analyses of solar radiation absorption through a carbon-dioxide (CO_2) mixture in a state of NLTE reveal over two orders of magnitude increase in absorption compared to the same mixture in a state of LTE, achieving 50 to 80% radiative energy absorption efficiency. The calculations were performed with the line-by-line nonequilibrium radiation code SPARTAN [2, 3] using 32 types of atomic and molecular transitions. The results show that CO_2 in LTE absorbs negligible solar irradiance (mostly UV and far IR), whereas CO_2 in NLTE displays significantly more absorption, especially in the visible range, but also emission, which diminishes the overall amount of absorbed radiation. Both findings are consistent with established understanding of radiative transport; however, the novelty is finding that the net effect is a significant increase in solar energy absorption. The results also show that CO_2 tends to behave as a perfect absorber of solar energy for very high degrees of NLTE (i.e. larger differences between electron and heavy-species temperatures). This fact, although not intuitive given that plasmas are known to be good emitters (e.g., radiation from arc lamps), can be understood by considering that a medium is prone to absorb the same type of radiation it emits. Furthermore, it is to be noted that the concept of a blackbody, a perfect absorber and emitter or radiation, corresponds to a medium in LTE. In fact, the approximation of the solar spectrum as the emission from a blackbody at $\sim$ 5780 K is a consequence that the sun's photosphere is largely regarded as being in LTE. Therefore, a NLTE medium provides the possibility to control its radiative properties by tuning its degree of thermodynamic nonequilibrium.

To evaluate the potential of PESE, a solar receiver-reactor system equipped with two types of nonequilibrium plasmas, namely a gliding arc discharge and a microwave plasma, is being built at UMass Lowell. The PESE chemical synthesis reactor system is designed to evaluate comparable solar and electrical energy inputs at the 1 kW level, spanning the whole range of combined solar-plasma operation, i.e. from 100% solar to 100% electrical energy inputs. The microwave plasma has the potential for the greatest energy efficiency, whereas the gliding arc discharge has simpler operation, more stability, and greater potential for scalability. Design challenges being addressed include the establishment of a stable plasma volume within the focal point of the concentrated solar radiation from a high-flux solar simulator [4], as well as the maximization of the plasma volume and the gas-plasma interaction region, the incorporation of a downstream catalytic monolith, among others.

Acknowledgements

The authors gratefully acknowledge support from the U.S. National Science Foundation through award CBET-1552037.

References

[1] A. Fridman, *Plasma chemistry*, Cambridge University Press, New York (2008)

[2] M. Lino da Silva, *SPARTAN - Simulation of PlasmA Radiation in ThermdynAmic Nonequilibrium*, http://esther.ist.utl.pt/spartan/

[3] M. Lino da Silva, *An adaptive line-by-line—statistical model for fast and accurate spectral simulations in low-pressure plasmas*, J. Quant. Spectrosc. and Rad. Transfer 108 (2007) 106–125

[4] S. Bhatta, D. Nagassou, J. P. Trelles, *Solar photo-thermochemical reactor design for carbon dioxide processing*, Solar Energy 142 (2017) 253–266

496114 Direct Two Step Solar Metallothermic and Electrolytic Production of Rare Earth Elements from Oxides

Abstract:

Rare Earth Elements (REEs) exhibit a range of unique electronic, optical and magnetic properties which make them attractive for many niche applications, including metal alloys, catalysts, petroleum refining, glass additives, polishing compounds, and ceramics. The demand for rare earth elements is expected to grow in the coming years, especially in uses related to critical clean energy technologies such as PV devices, batteries, high-performance magnets, and phosphors for lighting and electronics. Cerium (Ce), lanthanum (La), neodymium (Nd), and praseodymium (Pd) account for ~86% of the total global REE consumption. The current electrolytic production route through molten chlorides is quite inefficient due to the stability of Re^{+2} ion in chloride melt, whereas the oxide-fluoride route suffers from low solubility of oxides in fluorides. Additionally, these extractive processes are currently driven by fossil fuel based energy sources with very high greenhouse gas impacts and resource depletion potential.

Energy consumption and environmental concerns, coupled with the fact that the current electrolytic production method must occur at high temperatures, provides significant motivation for investigating solar driven metallothermic/ electrolytic processing routes as alternatives to conventional REE production for providing â€œcleanâ€ high temperature process heat and electricity. In this work, a new two-step REE production route consisting of direct metallothermic reduction of RE-oxides using Ca and subsequent electrolytic recovery of metallic Ca in a mixed chloride bath has been explored from the theoretical energy consumption perspective. The new process is critically evaluated using thermodynamic analysis and compared with the state-of-the-art. The two-step process shows promise in terms of significant improvement in the energy efficiency compared to the current electrolysis routes. The potential technological challenges in scale-up and impact of solar variability will be discussed as well.

497306 Economic and Energetic Analysis of Biofuel Supply Chains

Abstract:

Economic and Energetic Analysis of Biofuel Supply Chains

Rex T. L. Ng[1,2] and Christos T. Maravelias[1,2]

[1]Department of Chemical and Biological Engineering and [2]DOE Great Lakes Bioenergy Research Center, University of Wisconsin-Madison, Madison, WI

The efficient and effective transportation of lignocellulosic biomass is one of the critical challenges in biofuel production, due to its low bulk density and durability properties. Generally, in a biofuel supply chain (SC), baled biomass is either transported to a biorefinery directly, or sent to a depot for densification prior to transporting densified biomass to a biorefinery [1]. Many studies show the installation of depots in biofuel SCs offer better economic performance in large scale biorefineries [2–4]. However, from an energy point of view, depots may be infeasible if the total primary energy input (e.g., biomass harvesting and collection, depot operation, and baled and densified biomass transportation) is higher than the energy content of biofuel.

To better understand what supply chain configurations are preferable under different scenarios, we develop systematic methods to compare the cost and energy efficiency of biofuel SC under different configurations. We define a configuration as a combination of depot selection (with and without depots) and transportation modes (truck and rail). The analysis aims to provide some insights into the following questions: *When should we ship biomass directly to the biorefinery? When depot is economically and energetically feasible? Which transportation mode yields low cost and high energy efficiency? What is the trade-off between the improvement of densification technology and cost or energy efficiency?*

In this analysis, we consider corn stover which is collected from 12 Midwest states in the US. We assume that corn stover productivity is on an annual basis. A high moisture pelleting process [5] is employed at the depot, while dilute acid pretreatment, separate hydrolysis and fermentation are employed at the biochemical biorefinery [6]. Note that the analysis can be repeated using the proposed method for any feedstock, technology, and type of productivity (annual, seasonal or monthly basis). We estimate the minimum ethanol selling price (MESP) and total energy input based assuming the feedstock collection area to be circular with the biorefinery or the depot located at its center. When a combination of two or more configurations leads to better economic or energy benefits (referred to in this work as 'hybrid configuration'), the average distance from depots in the annular area to the biorefinery at the center is estimated.

The results show what reductions in cost and energy input can be achieved through a hybrid configuration (Fig. 1A). From an economic point of view, direct shipment of baled corn stover (orange region) is preferred if the distance between biorefinery and harvesting site is lower than 155 km. Depots installation and truck transportation of densified corn stover (blue region) are favorable if the distance ranges between 155 to 260 km; while rail transportation of densified corn stover (red region) will only lead to better economic benefits at larger distance (=260 km). From an energy point of view, rail transportation of densified biomass is always better than truck transportation. The depot installations with rail transportation have lower total energy input values than those with truck shipment of baled biomass if the distance is greater than 170 km. We further discuss key cost and energy efficiency drivers: corn stover productivity, biorefinery size, and densification efficiency. The MESP decreases with the increase of productivity and biorefinery size (Fig. 1B). We will show how the improvements in densification technologies can reduce transportation cost and energy consumption.

Finally, we will briefly discuss how the finding of this study can be used to formulate tractable mixed-integer programming (MIP) models for biofuel SCs operational planning, which account for non-uniform biomass availability and other practical considerations. We will show results from two case studies in southern Wisconsin.

C2

Figure 1: (A) Hybrid configuration, and (B) MESP as a function of biomass productivity and biorefinery capacity.

References:

[1] R.T.L. Ng, C.T. Maravelias, Design of biofuel supply chains with variable regional depot and biorefinery locations, Renew. Energy. 100 (2017) 90–102.

[2] D.J. Muth, M.H. Langholtz, E.C.D. Tan, J.J. Jacobson, A. Schwab, M.M. Wu, et al., Investigation of thermochemical biorefinery sizing and environmental sustainability impacts for conventional supply system and distributed pre-processing supply system designs, Biofuels, Bioprod. Biorefining. 8 (2014) 545–567.

[3] S. Kim, B.E. Dale, Comparing alternative cellulosic biomass biorefining systems: Centralized versus distributed processing systems, Biomass and Bioenergy. 74 (2015) 135–147.

[4] P. Lamers, E.C.D. Tan, E.M. Searcy, C.J. Scarlata, K.G. Cafferty, J.J. Jacobson, Strategic supply system design – a holistic evaluation of operational and production cost for a biorefinery supply chain, Biofuels, Bioprod. Biorefining. 9 (2015) 648–660.

[5] J.S. Tumuluru, K.G. Cafferty, K.L. Kenney, Techno-economic analysis of conventional, high moisture pelletization and briquetting process, in: Am. Soc. Agric. Biol. Eng. Annu. Int. Meet. 2014, ASABE 2014, 2014: pp. 4177–4189.

[6] D. Humbird, R. Davis, L. Tao, C. Kinchin, D. Hsu, A. Aden, et al., Process Design and Economics for Biochemical Conversion of Lignocellulosic Biomass to Ethanol: Dilute-Acid Pretreatment and Enzymatic Hydrolysis of Corn Stover, Natl. Renew. Energy Lab. (NREL), Golden, CO. (2011).

489815 Synthesis of Sustainable Processing Networks: Location-Dependent Biorefinery Models

Abstract:

The current and projected increase in demand of commodities, water and energy caused by the growing world population motivates innovation in process synthesis-design towards more sustainable processes. This is achieved by using alternative (renewable) raw materials, such as biomass, incorporating new process technologies and satisfying new design objectives and constraints, including sustainability.

This work describes the development, modeling and optimization of a comprehensive biorefinery superstructure with multiple product and feedstocks. Over 100 processing alternatives for converting the mentioned feedstocks into the various products are included, leading to over a hundred trillion theoretical alternative solutions.

A systematic process synthesis framework for biorefineries has been developed (Bertran et al, 2017), alongside the necessary methods and tools (such as a comprehensive database of biomass, chemicals, reaction paths, technologies, prices and many more), and implemented in a software interface named Super-O. This framework has been applied in this contribution to solve various problems derived from the biorefinery superstructure, under different scenarios. These include: (i) selection of a product or a set of products given the feedstock mix; (ii) selection of a feedstock or set of feedstocks given the product mix; (iii) selection of the processing route for a given feedstock-product combination; (iv) selection of the location given the feedstock, product and route; (v) all possible combinations of the aforementioned options. Moreover, multiple location options are investigated, by incorporating transportation options between each location, to determine the optimal distributed location-configuration for the entire value chain. Using integer cuts, a number of top-ranked solutions are generated and analyzed.

In this contribution, a strong emphasis is made on the management of the problem data through a data structure implemented as a database. Upon a specific problem definition, the relevant data is retrieved from the database and the problem is then solved via optimization.

The production of bioethanol alongside co-products (such as succinic acid, lactic acid, and lysine) from various lignocellulosic feedstocks is used as a case study to determine the optimal feedstock mix, processing route, set of locations and configuration.

References:

- M.-O. Bertran, R. Frauzem, A.-S. Sanchez-Arcilla, L. Zhang, R. Gani, Comput. Chem. Eng. (2017) DOI: 10.1016/j.compchemeng.2017.01.030.

500097 Electrochemistry As a Sustainable Alternative for Distributed Processing of Biomass

Abstract:

Electrochemical processes use electrons, a green reagent, as main driving force for the conversion of chemical energy into electrical energy of vice versa. Conventional electrochemical technologies (i.e. fuel cells and batteries) are well developed and are ideal for distributed energy generation and storage. Recent developments have extended scope to include stabilization and energy densification of biomass feedstock to produce transportation fuels. Among the main advantage of electrochemical technologies for distributed processes are that it can features rapid start up and turn off, which enable them to operate either in batch or continues operations and be directly fed by intermittent renewable energy sources, such as solar and wind. In the specific case of distributed biomass processing, electrochemical processes can be implemented the conversion of generated CO_2, upgrading of lignocellulose biomass, and biomass stabilization.

In this work the applicability of electrochemical technologies on distributed biomass processing schemes will be contextualized. Current state of the art and potential impacts will be assessed based on technology readiness, roadblocks, and viability. Case studies will be presented for the electrochemical reduction of CO_2, and the electrochemical upgrading/stabilization of biomass derived streams.

499969 Assessing Fuel and Feedstock Energy Use in the U.S. Chemical Sector: A Supply Chain Analysis

Abstract:

Assessing Fuel and Feedstock Energy Use in the U.S. Chemical Sector: A Supply Chain Analysis Scott Nicholson Alberta Carpenter Rebecca Hanes

In chemical manufacturing, fossil fuels are consumed both as an energy source and as feedstocks that are converted to higher-value chemicals. This work expands on the U.S. Department of Energy's chemical manufacturing bandwidth study, which quantified the potential for reducing direct energy consumption in the chemicals sector through the adoption of the most energy efficient manufacturing technologies available [1]. The scope of the bandwidth study was limited to fuels consumed within the plant boundary and fuels consumed for electricity generation. Fuel consumed upstream in the supply chain and fuel consumed as feedstock (defined as fuels that are converted to chemical intermediates rather than being used as an energy source) were not quantified; by one estimate, 60% of the fossil fuels consumed in the U.S. chemical manufacturing sector are consumed as feedstocks [2]. The present study quantifies, in addition to process energy, energy consumed upstream in the supply chain and energy embedded in fossil feedstocks and identifies chemical supply chains in which an understanding of the supply chain can help to identify opportunities for reducing total fossil fuel consumption. Alternative biomass-based processes, which require no fossil feedstocks but do not necessarily consume less energy, are then analyzed to determine if the alternative processes can reduce the total fossil fuel consumption in the identified supply chains. Reducing the chemical industry's reliance on fossil fuels could lead to greater long-term profitability in the industry, as crude oil prices are projected to rise steadily over the coming decades [3].

The Materials Flows through Industry (MFI) supply chain modeling tool, developed at the National Renewable Energy Laboratory, is employed to quantify supply chain energy and feedstock requirements for the chemicals analyzed previously in the bandwidth study. The MFI tool is a flexible framework for modeling material and energy usage in the supply chains of commodities such as chemicals, primary metals, and paper products [4]. MFI contains U.S.-specific data on energy and material inputs to 1,469 manufacturing processes for 707 commodities. U.S. commodities produced via multiple technologies are modeled using weighted averages of the current production technology mix based on market shares reported in the literature [5]. For this study, energy values are calculated at both the immediate process level and as a cumulative total for 10 steps up a chemical's supply chain. In MFI, a "step" consists of the material and energy inputs to the intermediate products manufactured in the previous step, which then become the products of the next step in the supply chain. Material and energy inputs beyond the 10th step have been shown to be negligible compared to the first 10 steps, so the supply chain calculations are truncated at that step.

MFI is similar to standard life cycle assessment (LCA) software packages, but the tool goes beyond these platforms in several ways. First, MFI is adaptable; processes and process weightings can be adjusted individually to model existing or future manufacturing scenarios. Modeling alternative production technologies in a particular supply chain is therefore straightforward. MFI also has the capability to disaggregate supply chain energy requirements into three types: process fuel, energy in fuels used as feedstocks, and fuel for electricity generation. For each chemical in the current analysis, the MFI model is used to calculate process and feedstock energy on a per-kilogram basis. These per-kilogram values are multiplied by annual U.S. production quantities from 2014 or later [5] to determine the energy requirements needed to produce the given chemical at the national scale. These calculations are repeated for alternative and biomass-based pathways for chemicals with feedstock-intensive supply chains, in order to compare the total fossil fuel consumption in the conventional and alternative supply chains. When comparing alternative and conventional petroleum production routes, the former may use significantly less feedstock energy, particularly for biomass-based production routes. However, alternative production routes may or may not use less process energy; in the case of advanced, non-commercial technologies, process energy is likely to increase and this increase may outweigh any savings in feedstock consumption. Both feedstock consumption and energy use must be quantified to determine which, if any, alternative technologies offer a supply-chain-level reduction in total fossil fuel consumption over the conventional technologies.

One of the chemicals in the current analysis is 1,3-butadiene, a chemical intermediate used predominantly in the production of synthetic rubber [6]. Essentially all commercial-scale production of butadiene in the U.S. derives from the byproduct streams of ethylene cracking [7]. Since ethylene is derived from fossil feedstocks, there are significant feedstock-embedded energy inputs in the conventional butadiene supply chain. Preliminary results show that supply chain feedstock consumption for 1,3-butadiene is 220 GJ/tonne, which exceeds both supply chain scale process energy consumption (21 GJ/tonne) and the onsite energy consumption reported by the bandwidth study (18 GJ/tonne) by a full order of magnitude. The alternative manufacturing processes for 1,3-butadiene are all biomass-based and begin with the fermentation of sugars [8]. These processes, because they do not involve ethylene cracking, are expected to reduce feedstock consumption in the 1,3-butadiene supply chain significantly. However, these processes have yet to be optimized for commercial scale production, and may show an increase in both onsite and supply chain energy consumption. The other chemicals analyzed in this study are caprolactam, para-xylene, and propylene glycol, all of which have biomass-based alternative processes under development [8].

This analysis quantifies the technological potential of alternative chemical manufacturing processes to reduce fossil fuel consumption. The economic implications of transitioning chemical manufacturing to next-generation energy-efficient and/or biomass-based production technologies are outside the scope of this study, as are the logistical complexities involved in shifting from a fossil-based to a biomass-based chemical sector. Still, the results of this study could be used to inform decisions around the commercialization of advanced manufacturing processes. They can be used to determine whether it is more worthwhile, from the perspective of reducing fossil fuel consumption, to upgrade an existing manufacturing process or to move to another technology entirely. Future extensions of this work may explicitly analyze the process economics of these alternative production technologies by incorporating costs of material and energy inputs. If these results show that the bio-based technologies can both reduce feedstock consumption *and* be economically competitive, then a stronger argument can be made for commercialization in the near future. In contrast, an alternative process that is significantly more expensive would still require further development. Lastly, the MFI tool accounts for energy use from resource extraction through production (cradle-to-gate); it does not currently look at use-phase or end-of-life energy requirements (i.e. the full cradle-to-grave). These latter stages are beyond the scope of this chemical sector analysis, since the downstream products manufactured from these commodity chemicals are used across many sectors —not just the chemical sector. With additional data, MFI has the ability to capture use-phase and end-of-life, and this is another potential area of future research.

References

[1] Brueske, S., Kramer, C., & Fisher, A. (2015). *Bandwidth Study on Energy Use and Potential Energy Saving Opportunities in US Chemical Manufacturing* (No. DOE/EE-1229). Energetics. Retrieved from https://energy.gov/sites/prod/files/2015/08/f26/chemical_bandwidth_report.pdf

[2] U.S. Department of Energy, Energy Information Administration. Office of Energy Consumption and Efficiency Statistics. *2010 Manufacturing Energy Consumption Survey*. https://www.eia.gov/consumption/manufacturing/data/2010/

[3] U.S. Department of Energy, Energy Information Administration. *Annual Energy Outlook 2017 with projections to 2050* (No. DOE/EIA- 0383(2017)). Retrieved from http://www.eia.gov/forecasts/ aeo/pdf/0383(2017).pdf.

[4] Hanes, R.J. & Carpenter, A. (2017) Evaluating opportunities to improve material and energy impacts in commodity supply chains. *Environ Syst Decis 37: 6*. doi: 10.1007/s10669-016-9622-5

[5] IHS (2017). *Chemical Economics Handbook (CEH)*. https://www.ihs.com/products/chemical-economics-handbooks.html.

[6] White, W. C. (2007). Butadiene production process overview. *Chemico-biological interactions, 166(1)*, 10-14. doi: 10.1016/j.cbi.2007.01.009

[7] Sriram, P., Hyde, B., & Smith, K. (2016). Butadiene. *Chemical Economics Handbook*. IHS.

[8] Biddy, M. J., Scarlata, C., & Kinchin, C. (2016) *Chemicals from Biomass: A Market Assessment of Bioproducts with Near-Term Potential*. National Renewable Energy Laboratory (NREL). Retrieved from http://www.nrel.gov/docs/fy16osti/65509.pdf

501692 Integration of Fast Pyrolysis and Electrolyzer for Deoxygenation of Biomass

Abstract:

Domestically, the US is challenged with improving the environment, energy security, and energy productivity, thus many efforts are concerted in utilizing biomass as a viable source of renewable liquid transportation fuels and contributing to a robust carbon-neutral energy production. However, achieving a high carbon, hydrogen and overall energy efficiency is an ongoing challenge for current biomass upgrading technologies. The main approach of upgrading biomass to hydrocarbon fuels is by hydroprocessing, which relies heavily on large hydrogen consumption and on large, more centralized production scales. In this work, we present a different approach of using distributed pyrolysis unit scale with low hydrogen requirement by utilizing electricity to drive the deoxygenation of molecules present in the biomass. Currently, we have demonstrated the feasibility of electrochemical deoxygenation using solid oxide electrolysis cell in small cells, multi-cell stacks, and in an integrated test with a bench-scale fast pyrolysis unit. The electrolysis process removes oxygen from the oxygenated organic molecule as well as from steam to produce hydrogen in-situ allowing for a distributed, small scale integrated upgrading unit. Specifically, we will report recent work with a fast pyrolysis unit with 1 kg/h ponderosa pine feed, 480 deg C fast pyrolysis temperature and 550-600 deg C process at the electrolyzer. These process conditions allowed us to achieve an improved quality of the bio-oil. In this presentation, we will also discuss the present challenges of electrolyzer integration with a fast pyrolysis unit as well as the characterization of products using ^{13}C NMR, ultimate analysis (CHNOS), and GC-MS.

INCREASES IN DEMAND FOR FIXED NITROGEN FROM ALTERNATIVE ENERGY AND CARBON CAPTURE SCHEMES

Dr. Luis F. Razon, De La Salle University, Manila, Philippines

Introduction

The world food supply has depended on the supply of synthetic fixed or reactive nitrogen for more than 100 years. The overuse of nitrogen fertilizer has led to numerous negative environmental impacts[1]. In addition, the treatment of nitrogen in sewage leads to increased sewage treatment costs. The manufacture of fixed nitrogen also consumes large amounts of fossil fuels and intrinsically releases large amounts of greenhouse gases. While the link between the food supply, synthetic reactive nitrogen environmental degradation is well known, the scale of synthetic fixed nitrogen requirements for alternative energy and carbon dioxide removal is not often appreciated. The present and possible future uses for reactive nitrogen are diagrammed in Figure 1.

This paper takes projected demands for the year 2050 for alternative energy and carbon dioxide removal schemes from the literature and translates these into projections for fixed nitrogen. A comparison is made with the projected demand for food agriculture. The order-of-magnitude estimates that result from these projections are important when making judgement decisions on investments in new facilities and new technologies. The following sections are arranged in the order of likelihood that the demand for fixed nitrogen for these applications will actually transpire.

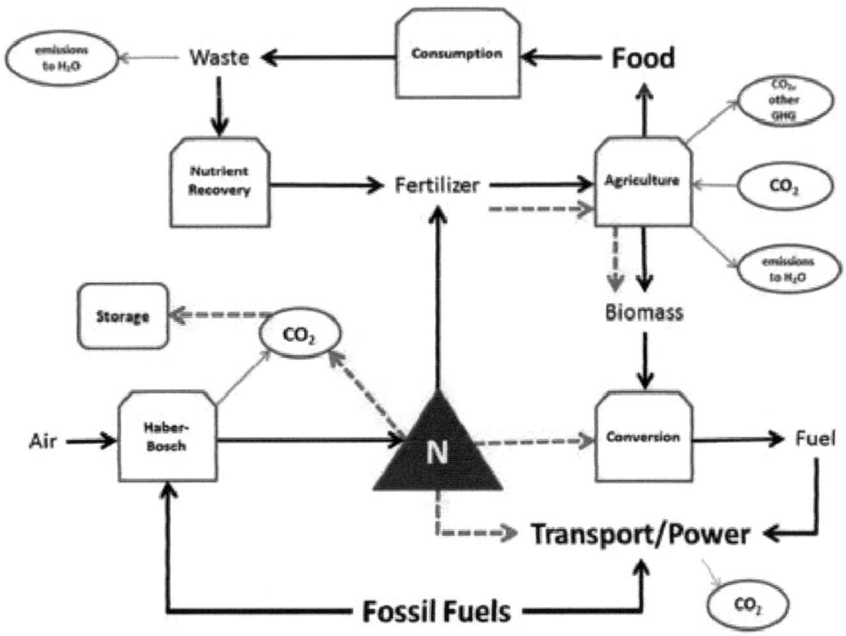

Figure 1. Uses and sources of fixed nitrogen. Solid lines are present uses. Dashed lines are possible future uses.

Demand from Food Agriculture

Fertilizer for food agriculture accounts for about 77% of the global demand for fixed nitrogen[2]. The other 23% is used for explosives, plastics and others. The consumption of N nutrient in 2014 was 109×10^9 kg N. There are many extant projections for food demand available in the literature. The projections are based on population projections and changes in total food intake and food choices based on global and regional scenarios such as those in the "Shared Socioeconomic Pathways" (SSP). These projections predict much higher demands for fixed nitrogen than would be obtained from a simple linear regression. For example, a linear regression of the reactive nitrogen demand from the FAO[3] shows that consumption would be about 187.1×10^9 kg of nutrient N by 2050. On the other hand, Bodirsky et al.[4] analyzed the Shared Socioeconomic Pathway "middle of the road" scenario (SSP2) and arrived at 232×10^9 kg nitrogen (N). Depending on the scenario the estimates of Bodirsky et al. ranged from 188×10^9 kg N under the SSP1 ("sustainability") scenario to the 290×10^9 kg N under the SSP5 ("regional rivalry") scenario. Other estimates, such as those by Galloway et al.[5], are consistent with these estimates. For the rest of the paper, I will use the figure from the SSP2 scenario, 232×10^9 kg N, as a reference figure for comparison.

Demand from Biomass Agriculture and Pretreatment

Concerns about the eventual depletion of fossil fuels and the environmental impact of their use has driven the development and introduction of biofuels such as bioethanol and biodiesel. Reactive nitrogen demand come not only from the fertilizer needed to grow the biomass feedstock but also for pretreatment of lignocellulosic feedstock.

There are not many available estimates of the reactive nitrogen needs for biofuels because of the difficulty in anticipating adoption of biofuels and the feedstocks that might be used[6]. A projection was made by Winiwarter et al.[7] using a representative concentration pathway scenario with modest population growth (RCP 2.6). Using this scenario, Winiwarter et al.[7] projected that the demand for biofuels in 2050 will be about 118 EJ. To supply the biomass for this amount of energy, they estimate that about 41.4×10^9 kg N would be needed for fertilizer. It should be pointed out that the RCP 2.6 scenario is a more optimistic scenario than the SSP2 scenario with smaller population growth because it assumes that measures will be taken such that the radiative forcing values in the year 2100 relative to pre-industrial values is only 2.6 W/m^2. For example, the food fertilizer estimate under this scenario is only 94.4×10^9 kg N as compared to the 232×10^9 kg N estimate for the SSP2 scenario. While this indicates that fundamental differences between the RCP 2.6 and SSP2 scenarios, this estimate is still useful as an order-of-magnitude estimate.

What is not commonly realized is that the reactive nitrogen needs for biofuels are not limited to fertilizer. The present "first-generation" feedstocks for biofuels like corn, sugarcane and palm oil are the same plants used for food. To sustain the expansion of biofuels, it has been proposed that the second generation of feedstocks consist of non-food plants, specifically lignocellulosic feedstocks like wood and switchgrass for making bioethanol. Lignocellulosic feedstocks require pretreatment prior to fermentation to bioethanol. There are several pretreatment methods under development. Ammonia is needed for three of these: ammonia fiber explosion (AFEX), ammonia recycled percolation (ARP) and soaking in aqueous ammonia (SAA)[8]. AFEX appears to be the closest to widespread adoption[8]. AFEX requires about 20g NH_3 per kg of lignocellulosic biomass[9]. In a scenario predicting wide adoption of biofuels to Winiwarter et al.[7] estimate that about 118 EJ of biofuels will be produced and 98% of the bioenergy in 2050 will be from lignocellulosic feedstock. On the other hand, Herve et al.[10] used the IEA Blue Map Scenario to estimate that 90% of biomass feedstock will be lignocellulosic. What is less certain is the means by which the feedstock will be pretreated in 2050. If it is assumed that in 2050, 118 EJ of energy will come from biofuels and that 50-90% of the biofuels are lignocellulosic feedstock pretreated using AFEX, which uses 20g of NH_3 per kg of biomass, then the range for N demand would range from 154×10^9 kg N to 278×10^9 kg N. Therefore, the pretreatment N requirements would be in the same order of magnitude for that of food.

Demand from Atmospheric Carbon Dioxide Removal by Amine Absorption

The rapid accumulation of CO_2 in the atmosphere has led to proposals to actively withdraw CO_2 from the atmosphere for burial in the underground chambers or in the deep ocean rather than to simply rely on limiting carbon emissions[11]. Among these proposals is CO_2 removal by chemical absorption. In this scheme, CO_2 is absorbed by a solvent which is then heated such that the CO_2 is released for further processing and ultimate disposal. The process is already used for CO_2 recovery from fermentation processes for CO_2 reuse in other products like soft drinks. The most commonly used solvent is monoethanol amine (MEA)[12]. Because it is already a mature technology, carbon dioxide removal with monoethanol amine is projected to be a significant contributor to CCS operations. In 2050, the IEA Technology Roadmap for 2013 estimates that 7×10^9t of CO_2 per year will be absorbed by solvent absorption and that "most" of it will be by amine-based absorption[13]. Ideally, all of the amine is regenerated and reused but some is inevitably lost because of evaporation and degradative reactions. About 1.5 kg MEA/t of CO_2 is required as make-up[14]. If 75% of the projected 7000 Mt is absorbed by amine absorption then about 7.9×10^9 kg MEA is required. About 2.2×10^9 kg NH_3 or 1.8 $x10^9$ kg N would be needed to manufacture this amount of MEA. This is about only a small fraction of the requirement for food, so this application for ammonia will not have a significant impact on reactive nitrogen demand.

Demand from Using Ammonia as a Diesel and Gasoline Substitute

The complete combustion of ammonia releases heat and results in a net increase in the number of moles in the mixture, so it can, in theory, be used as a fuel. The resulting emissions would be carbon-free. The idea is attractive because ammonia has superior properties when compared to hydrogen like higher energy density and lower storage pressure[15]. The US Army tested the concept as early as the 1930's[16]. There is an active NH_3 Fuel Association[17]. Uses of ammonia as a fuel range from internal combustion engines[18] to fuel cells[19] to mixed coal-ammonia power generation plants[20].

Combustion of ammonia should ideally result in N_2 and H_2O. However, incomplete combustion of results in emissions of nitrogen oxides. In addition, ammonia has a low flame speed and therefore ammonia has to be mixed with a secondary fuel if it is to be used in an internal combustion engine[21]. Therefore, only partial replacement of diesel and gasoline would be possible. Nonetheless, it can be seen in the calculations below that the projected demand from enormous.

The data of Kong et al.[18,22] indicate that, to ensure proper performance, 1 kg of NH_3 must be accompanied by 0.31 kg of diesel. On the other hand, in a gasoline engine, 1 kg of NH_3 must be accompanied by 2.5 kg of gasoline. Also, 5.64 kg NH_3 is energetically equivalent to 1 kg of diesel whereas 13.5 kg NH_3 is energetically equivalent to 1 kg of gasoline. Therefore, only 5.42% of the diesel can be replaced with NH_3 while only about 15.6% of the gasoline can be replaced with NH_3.

In 2040, the World Oil Outlook 2016[23] projects that the demand for diesel and gasoline and diesel will be 33.2 mb/d and 28.0 mb/d respectively. A simple cubic regression of the data in the World Oil Outlook 2016, provides a projection for 2050 of 34.2 mb/d and 30.0 mb/d for diesel and gasoline, respectively. This translates to 1.663×10^{12} kg/year for diesel and 1.288×10^{12} kg/year for gasoline.

The requirement for a secondary fuel puts an upper limit to the amount that can be displaced by ammonia. This works out to be a demand of about 0.51×10^{12} kg NH_3/year for the maximum diesel displacement and about 2.72×10^{12} kg NH_3/year for the maximum gasoline displacement. These amounts translate to 0.42×10^{12} kg N/year and 2.24×10^{12} kg N/year for diesel and gasoline, respectively. It is easily seen that these amounts dwarf the amounts necessary for food and the other applications.

Conclusions

The best estimate for each selection is shown in Table 1. The sources of demand for reactive nitrogen are listed in the order of likelihood that the demand may actually come true. Much depends on government policies and development of new technologies. If biofuel subsidies are abandoned, then there would be far less demand for biofuels. If climate change agreements are abrogated or are insufficiently followed, then it is likely that there will not be too much impetus for atmospheric carbon dioxide removal. Finally, the widespread use of ammonia as a transport fuel is purely speculative at the moment. There are many competing technologies including biofuels and electric vehicles. Government policies may very well drive the energy source one way or the other.

Table 1. Summary of possible reactive nitrogen demand in 2050

Source for Demand	Estimated demand in 2050 (kg N)
Food Agriculture	232×10^9
Biomass for Bioenergy Agriculture	41.4×10^9
Lignocellulosic Biomass Pretreatment	154×10^9 - 278×10^9
Carbon Dioxide Removal using Amine Absorption	1.8×10^9
Diesel Substitute	0.42×10^{12}
Gasoline Substitute	2.24×10^{12}

The scale of demand for the biomass agriculture and pretreatment are on the same order of magnitude as that for food agriculture. On the other hand, the scale of demand for the substitution for fuels is several hundred times or even a thousand times larger. This is even considering that this is only a partial substitution. Expansion of production and distribution facilities will have to be carefully planned if there is to be widespread adoption of ammonia fuel.

What is clear is that there will be a large and increasing demand for anthopogenic or synthetic nitrogen in the future, even if the demand may be only from agriculture or lignocellulosic biomass pretreatment. The manufacture of such large quantities of reactive nitrogen in the form of ammonia from the Haber-Bosch process will be accompanied by consumption of large amounts of fossil fuels and the release of large amounts of greenhouse gases. Alternatives to the Haber-Bosch process are in early stages of development[24]. More attention needs to be paid to the rapid development of these alternatives.

Acknowledgement

Dr Wilfried Winiwarter is thanked for providing more detailed information from their paper[7].

References

1. Erisman JW, Galloway JN, Seitzinger S, et al. Consequences of human modification of the global nitrogen cycle. *Philos Trans R Soc Lond B Biol Sci*. 2013. doi:10.1098/rstb.2013.0116.
2. FAO. *World Fertilizer Trends and Outlook to 2020*. Rome; 2017. https://www.google.com.ph/url?sa=t&rct=j&q=&esrc=s&source=web&cd=2&ved=0ahUKEwifpcnM88bWAhWLT7wKHaQ0AcMQFggpMAE&url=http%3A%2F%2Fwww.fao.org%2F3%2Fa-i6895e.pdf&usg=AFQjCNGYIMe1yKpnmLGmRqOcner1Es20pA.
3. FAOSTAT. http://www.fao.org/faostat/en/#data/RF. Accessed September 28, 2017.
4. Bodirsky BL, Popp A, Lotze-Campen H, et al. Reactive nitrogen requirements to feed the world in 2050 and potential to mitigate nitrogen pollution. *Nat Commun*. 2014;5(May):3858. doi:10.1038/ncomms4858.
5. Galloway JN, Leach AM, Erisman JW, Bleeker A. Nitrogen: the historical progression from ignorance to knowledge, with a view to future solutions. *Soil Res*. 2017;55(6):417-424. doi:10.1071/SR16334.
6. Searchinger TD, Beringer T, Strong A. Does the world have low-carbon bioenergy potential from the dedicated use of land? *Energy Policy*. 2017;110(December 2016):434-446. doi:10.1016/j.enpol.2017.08.016.
7. Winiwarter W, Erisman JW, Galloway JN, Klimont Z, Sutton MA. Estimating environmentally relevant fixed nitrogen demand in the 21st century. *Clim Change*. 2013;(2005):889-901. doi:10.1007/s10584-013-0834-0.
8. Tao L, Aden A, Elander RT, et al. Process and technoeconomic analysis of leading pretreatment technologies for lignocellulosic ethanol production using switchgrass. *Bioresour Technol*. 2011;102(24):11105-11114. doi:10.1016/j.biortech.2011.07.051.
9. Bals B, Wedding C, Balan V, Sendich E, Dale B. Evaluating the impact of ammonia fiber expansion (AFEX) pretreatment conditions on the cost of ethanol production. *Bioresour Technol*. 2011;102(2):1277-1283. doi:10.1016/j.biortech.2010.08.058.
10. Hervé G, Agneta F, Yves D. Biofuels and World Agricultural Markets: Outlook for 2020 and 2050. In: *Economic Effects of Biofuel Production*. Vol ; 2011:129-162. doi:10.5772/20581.
11. Meadowcroft J. Exploring negative territory Carbon dioxide removal and climate policy initiatives. 2013:137-149. doi:10.1007/s10584-012-0684-1.

12. Shakerian F, Kim KH, Szulejko JE, Park JW. A comparative review between amines and ammonia as sorptive media for post-combustion CO_2 capture. *Appl Energy*. 2015;148:10-22. doi:10.1016/j.apenergy.2015.03.026.
13. IEA. *Technology Roadmap - Carbon Capture and Storage*. Paris; 2013. doi:10.1007/SpringerReference_7300.
14. Rao AB, Rubin ES, Berkenpas MB. *An Integrated Modeling Framework for Carbon Management Technologies. Vol. 1 - Technical Documentation: Amine-Based CO_2 Capture and Storage Systems for Fossil Fuel Power Plants*. Pittsburgh, Pennsylvania; 2004. https://www.cmu.edu/epp/iecm/documentation/tech_04.pdf.
15. Zamfirescu C, Dincer I. Using ammonia as a sustainable fuel. *J Power Sources*. 2008;185(1):459-465. doi:10.1016/j.jpowsour.2008.02.097.
16. Garabedian CG., Johnson JH. *The Theory of Operation of an Ammonia Burning Internal Combustion Engine*. Warren, Michigan; 1966. http://www.dtic.mil/dtic/tr/fulltext/u2/634681.pdf.
17. NH3 Fuel Association | Practical, carbon-free fuel. https://nh3fuelassociation.org/. Accessed September 28, 2017.
18. Gross CW, Kong SC. Performance characteristics of a compression-ignition engine using direct-injection ammonia-DME mixtures. *Fuel*. 2013;103:1069-1079. doi:10.1016/j.fuel.2012.08.026.
19. Afif A, Radenahmad N, Cheok Q, Shams S, Kim JH, Azad AK. Ammonia-fed fuel cells: A comprehensive review. *Renew Sustain Energy Rev*. 2016;60:822-835. doi:10.1016/j.rser.2016.01.120.
20. Japanese utilities team on CO2-reducing tech for coal plants- Nikkei Asian Review. https://asia.nikkei.com/Business/Deals/Japanese-utilities-team-on-CO2-reducing-tech-for-coal-plants. Accessed September 28, 2017.
21. Reiter AJ, Kong S. Combustion and emissions characteristics of compression-ignition engine using dual ammonia-diesel fuel. *Fuel*. 2011;90(1):87-97. doi:10.1016/j.fuel.2010.07.055.
22. Ryu K, Zacharakis-jutz GE, Kong S. Effects of gaseous ammonia direct injection on performance characteristics of a spark-ignition engine. *Appl Energy*. 2014;116:206-215. doi:10.1016/j.apenergy.2013.11.067.
23. OPEC. *World Oil Outlook*.; 2016. doi:10.1190/1.1439163.
24. Razon LF. Is nitrogen fixation (once again) "vital to the progress of civilized humanity"? *Clean Technol Environ Policy*. 2014;17(2):301-307. doi:10.1007/s10098-014-0835-3.

FARADAY TECHNOLOGY, INC.

Pulse/Pulse-Reverse Electrodeposition of Copper Electrocatalysts for CO_2 Reduction to Ethylene

2017 AIChE Annual Meeting

Minneapolis, MN

31 Oct 2017

Brian Skinn[a], Sujat Sen[b], Rajeswaran Radhakrishnan[a], McLain Leonard[b], Stephen Snyder[a], Fikile R. Brushett[b]

[a] Faraday Technology, Inc., Englewood, OH 45315

[b] Dept. of Chemical Engineering, Massachusetts Institute of Technology, Cambridge, MA 02139

BrianSkinn@FaradayTechnology.com

17AIChE Annual Meeting, Minneapolis, MN

Outline

- Pulse/Pulse-Reverse Electrochemical Processing
- FARADAYIC® ElectroDeposition – Prior Work
- FARADAYIC® Copper Film CO_2 Electrocatalysts
  - Rationale and Electroreactor Concept
  - FARADAYIC® ElectroDeposition & Catalyst Activation
  - Electrocatalyst Performance

FARADAY TECHNOLOGY, INC.

Electrochemical Processing

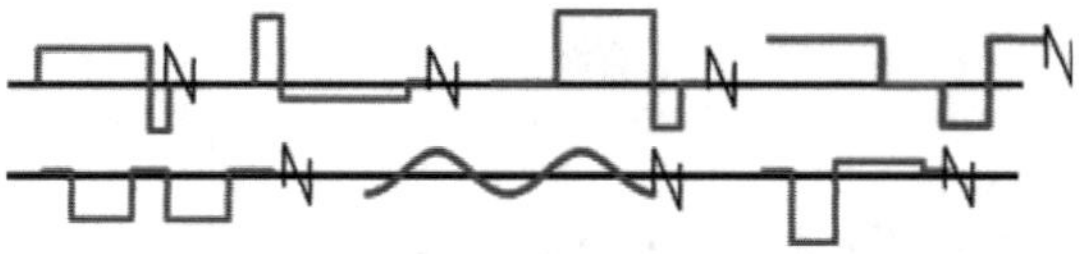

Conventional DC vs Pulse/Pulse-Reverse

Pulsed Waveforms vs DC

Know: *Linearized Concentration Profile*

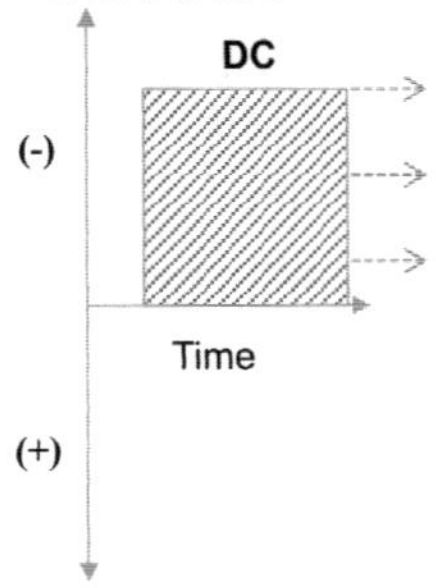

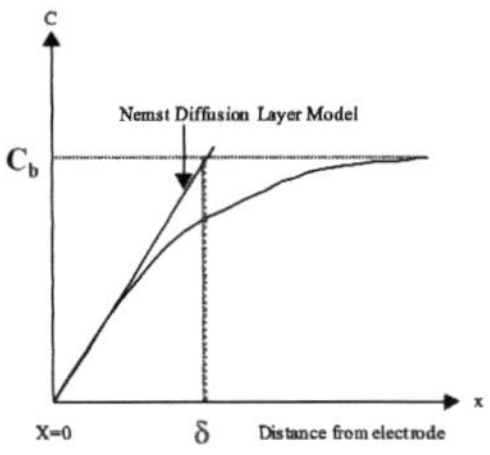

Pulsed Waveforms vs DC

Know: *Linearized Concentration Profile*

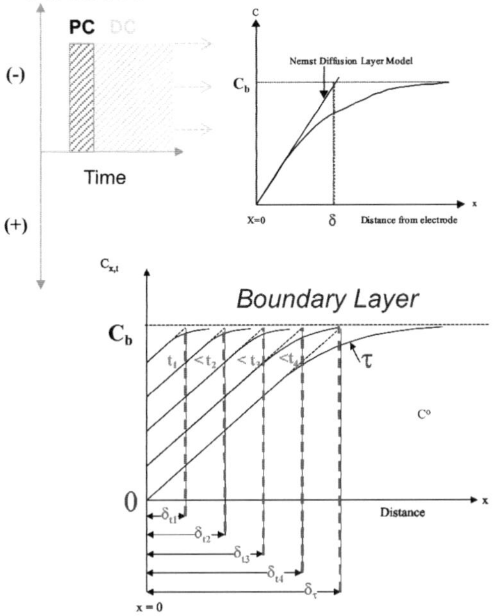

Pulsed Waveforms vs DC

Know: *Linearized Concentration Profile*

Assume:

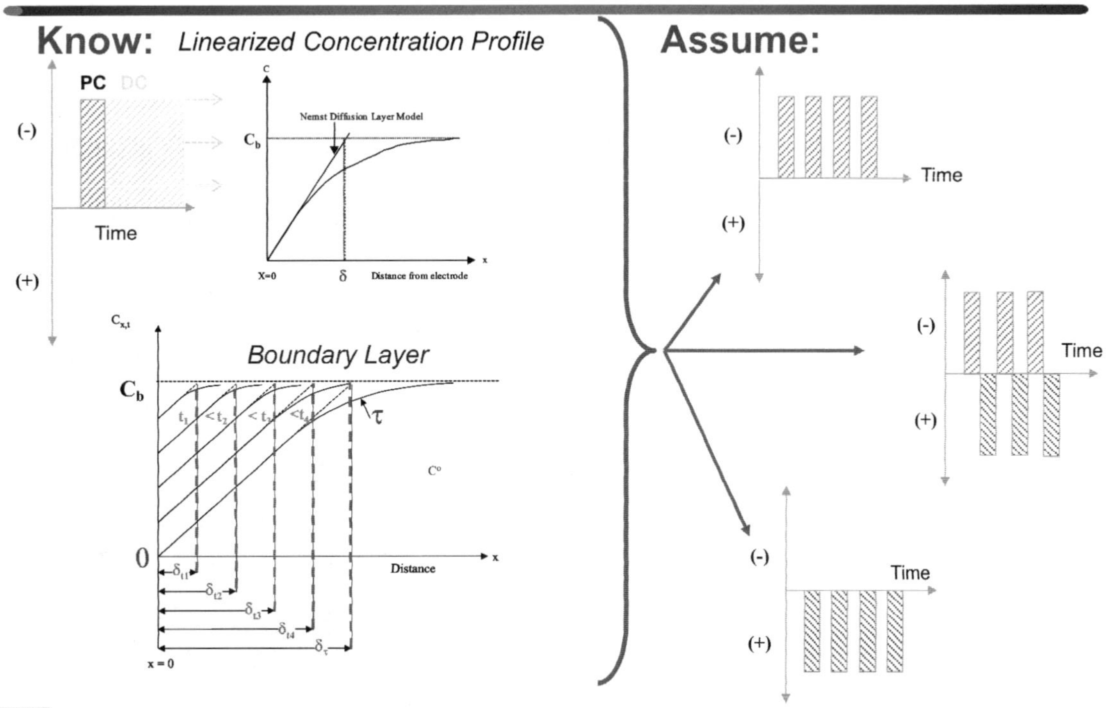

FARADAY TECHNOLOGY, INC.

FARADAYIC® ElectroDeposition

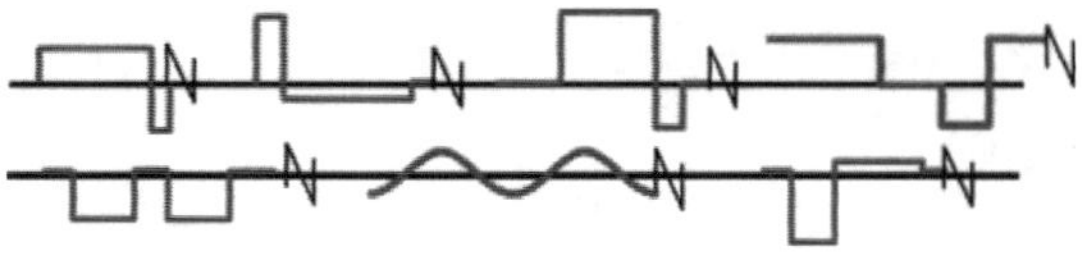

Prior Work

Prior Work – Copper Electrodeposition for PCBs

Standard DC Recipe

$CuSO_4$
H_2SO_4
"Accelerator"
"Suppressor"
"Leveler"
Other Additives

Pulsed Waveform Recipe

$CuSO_4$
H_2SO_4
Trace Cl^-
Trace PEG

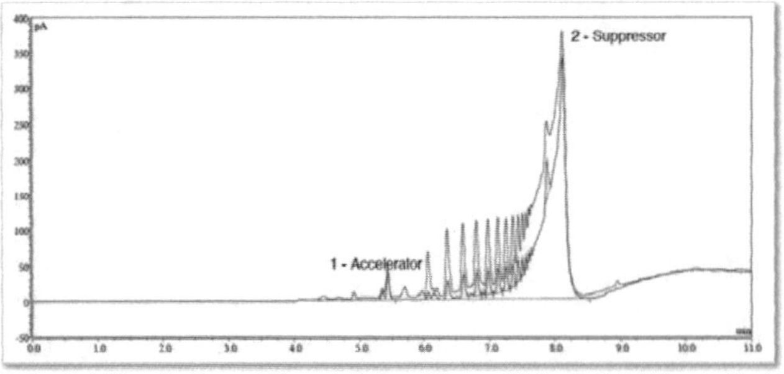

Plante, et al. "Quantization and Characterization of Copper Plating Bath Additives by Liquid Chromatography with Charged Aerosol Detection." Thermo Scientific. http://www.dionex.com/en-us/webdocs/113469-Pittcon12_1820-6_MPlante_PN70008_PlatingBaths.pdf

Z-Interconnects Cu-Plated by Pulse-Reverse FARADAYIC® ElectroDeposition

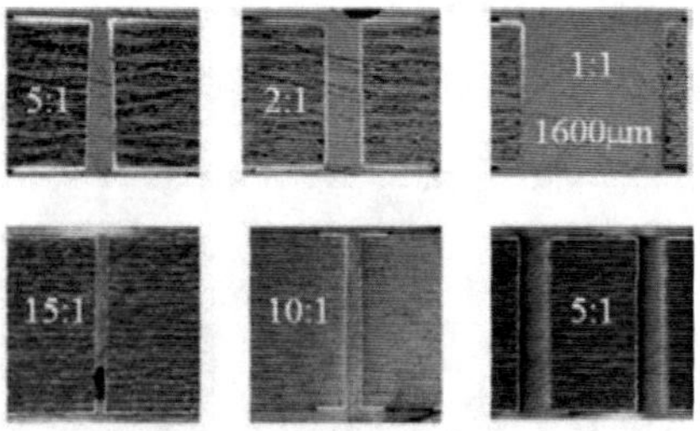

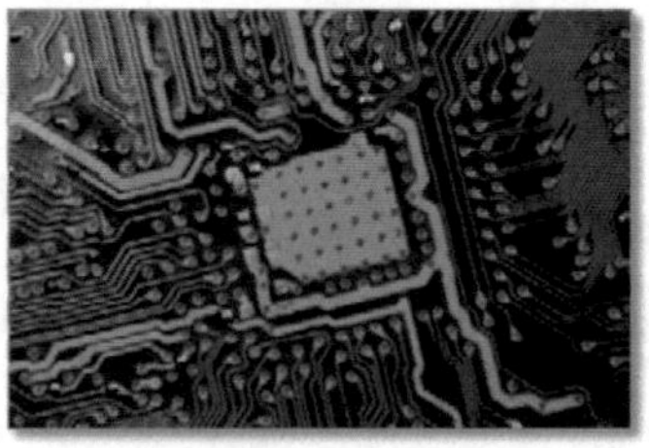

Prior Work – FARADAYIC® Lead-Free Tin

Rotating Disk Plating Tool

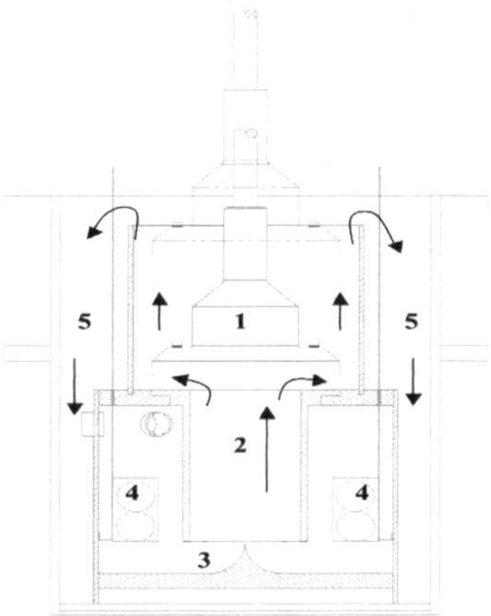

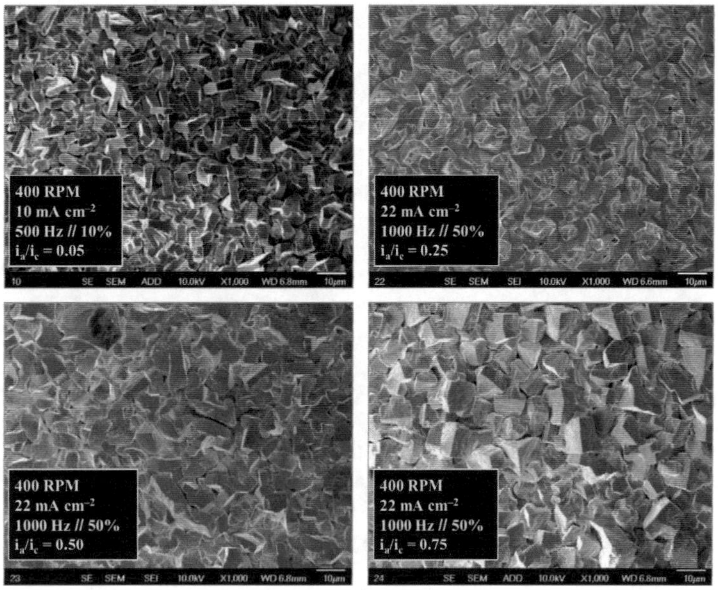

Sn Deposits on Cu Substrates — 1000× Magnification

U.S. Patent 8,603,315 Granted

FARADAY TECHNOLOGY, INC.

Prior Work – Sn GDE CO_2 Electrocatalysts

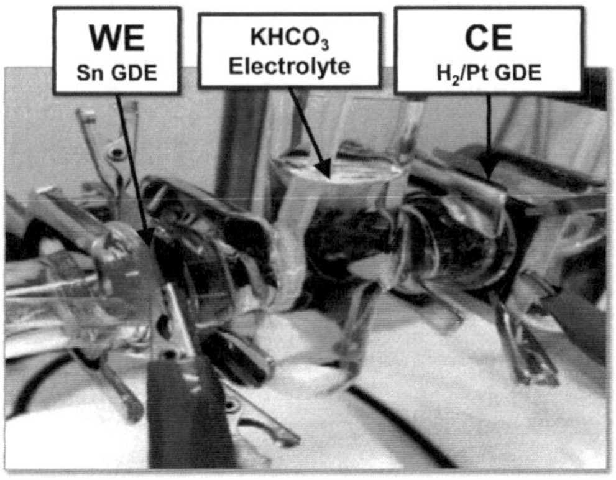

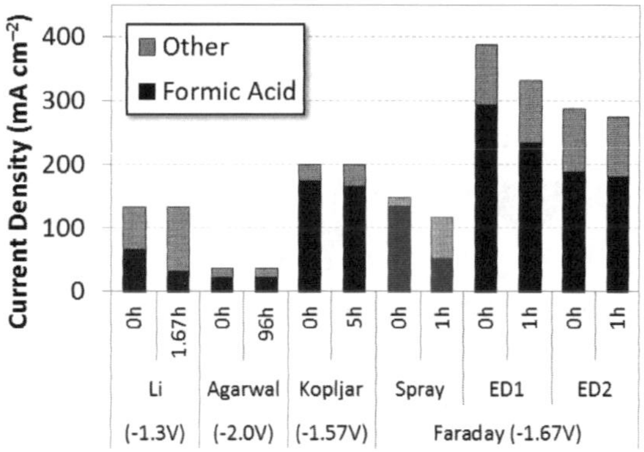

(Half-cell potentials vs SHE)

WE: $CO_2 + 2H^+ + 2e^- \rightarrow HCOOH$

CE: $H_2 \rightarrow 2H^+ + 2e^-$

Cell: $CO_2 + H_2 \rightarrow HCOOH$

Li and Oloman. *J Appl Electrochem* 35: 955, 2005.

Agarwal et al. *ChemSusChem* 4: 1301, 2011.

Kopljgar et al. *J Appl Electrochem* 44: 1107, 2014.

FARADAY TECHNOLOGY, INC.

Prior Work – FARADAYIC® ElectroCatalyzation

- Low-loading deposition of Pt fuel cell catalyst
- Beta-scale batch system (NSF)
- Pilot-scale reel-to-reel system (OH)

Batch System **Continuous Reel-to-Reel**

E.J. Taylor *et al.*, *J. Electrochem. Soc.*, 1992, 139 (5), L45; N.R.K. Vilambi *et al.*, U.S. Patent No. 5,084,144, issued 28 Jan. 1992; E.J. Taylor et al., U.S. Patent No. 6,080,504, issued 27 Jun. 2000.

TEM of Pt on Carbon

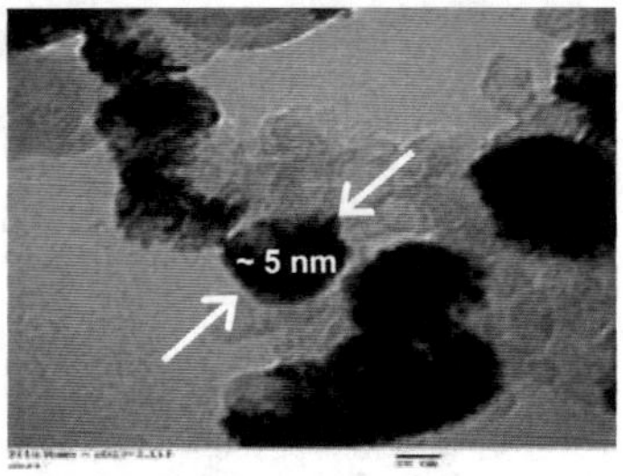

Performance Comparison

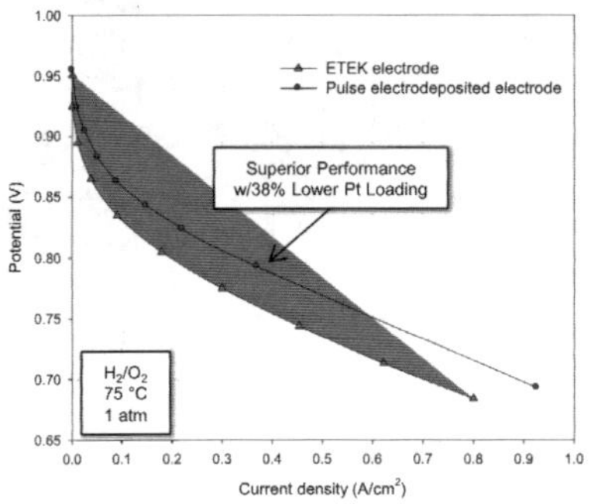

FARADAY TECHNOLOGY, INC.

FARADAY TECHNOLOGY, INC.

FARADAYIC® Copper Film CO_2 Electrocatalysts

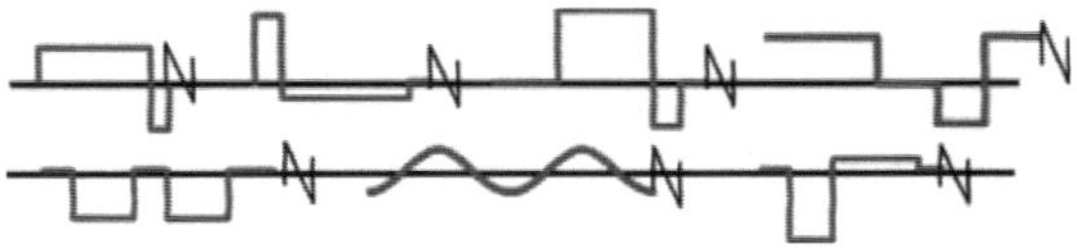

Rationale and Electroreactor Concept

CO_2 Electrocatalysis – Catalyst Selection

Product distribution over single-metal electrocatalysts as determined by Hori et al.
Electrochimica Acta 39(11-12): 1833 (1994).

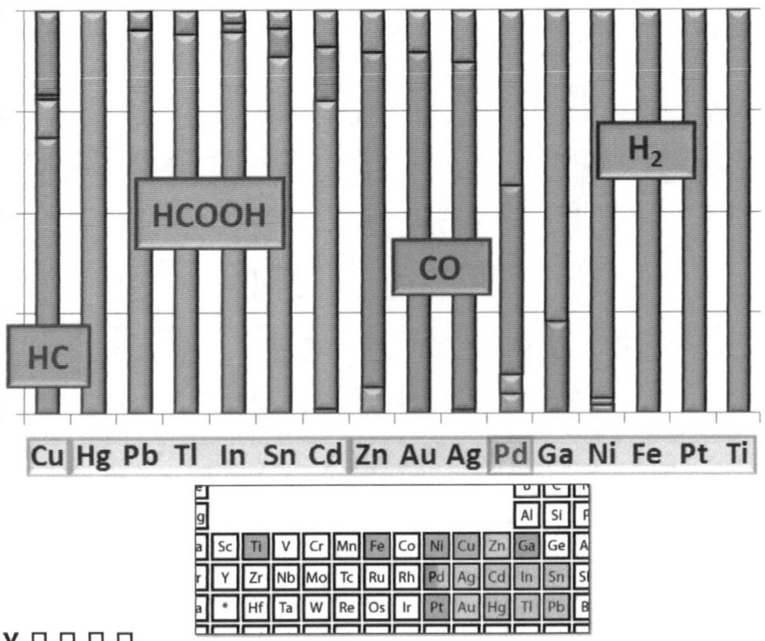

CO_2 Electrocatalysis – Form Factor Selection

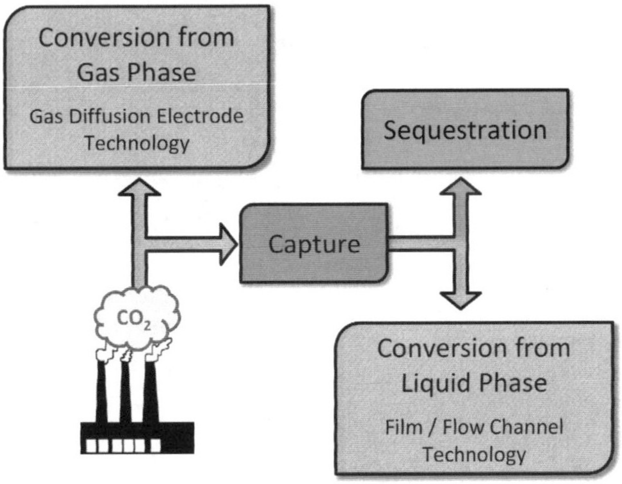

CO_2 Electrocatalysis – Form Factor Selection

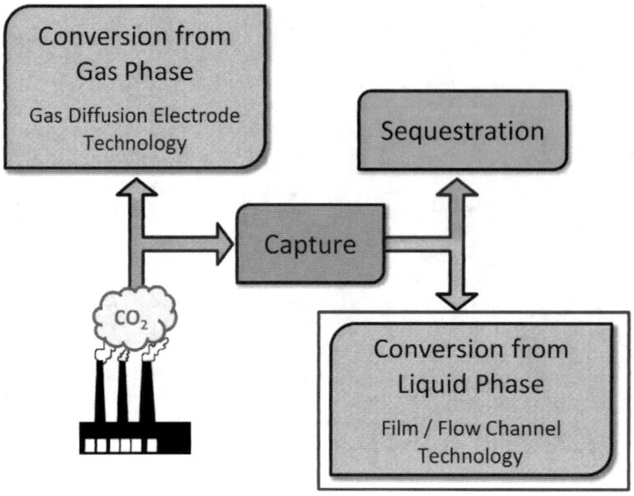

FARADAY TECHNOLOGY, INC. 17 AIChE Annual '17 - B Skinn - 31 Oct 2017

CO_2 Electrocatalysis – Flow Electroreactor Concept

- Flow channels for electrolytes
- CO_2 reduction on Cu catalyst
- O_2 evolution on commercial mixed oxide anode
- Anticipate direct use of CO_2-laden capture media
  - Aqueous amines, etc.

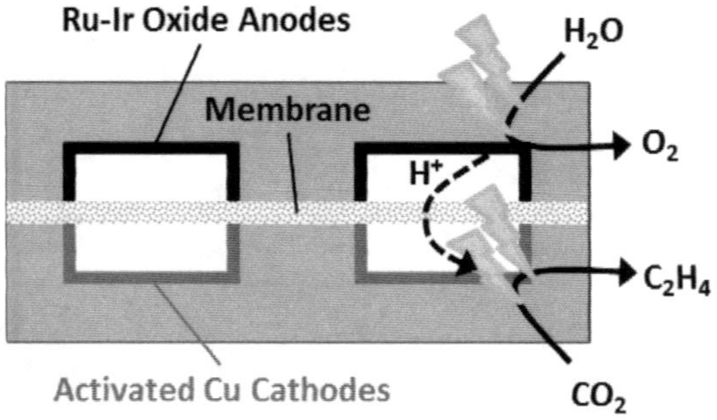

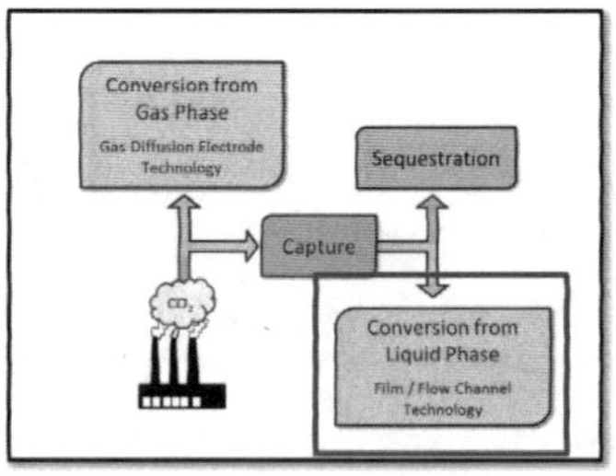

FARADAY TECHNOLOGY, INC. 18 AIChE Annual '17 - B Skinn - 31 Oct 2017

FARADAYIC® Copper Film CO_2 Electrocatalysts

FARADAYIC® ElectroDeposition and Catalyst Activation

FARADAYIC® Copper ElectroDeposition

FARADAYIC® Copper ElectroDeposition

Catalyst Activation

Deposit Oxidize Reduce

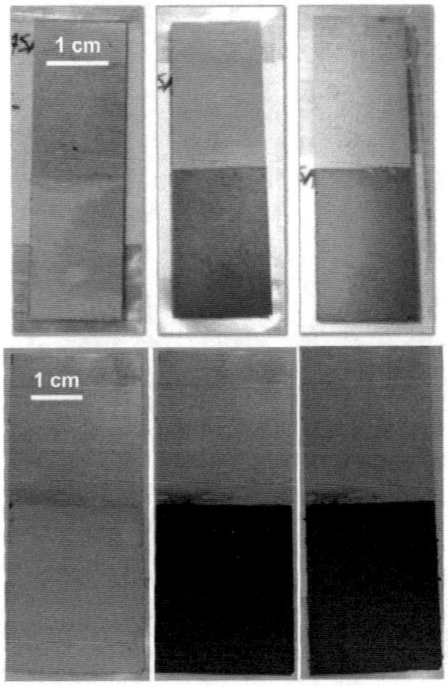

Modified from Li & Kanan, J Am Chem Soc 134: 7231 (2012)

Oxidize

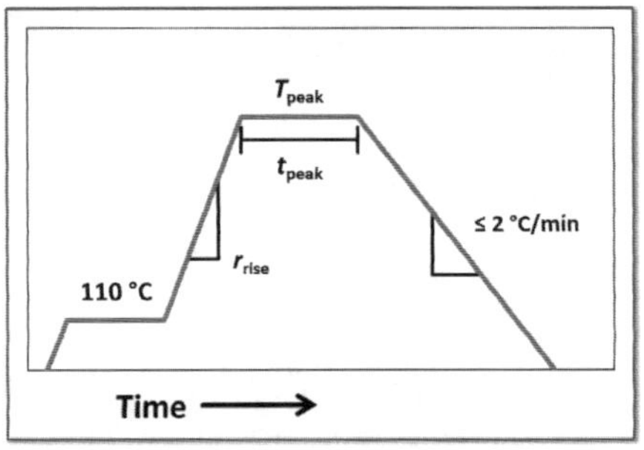

Reduce

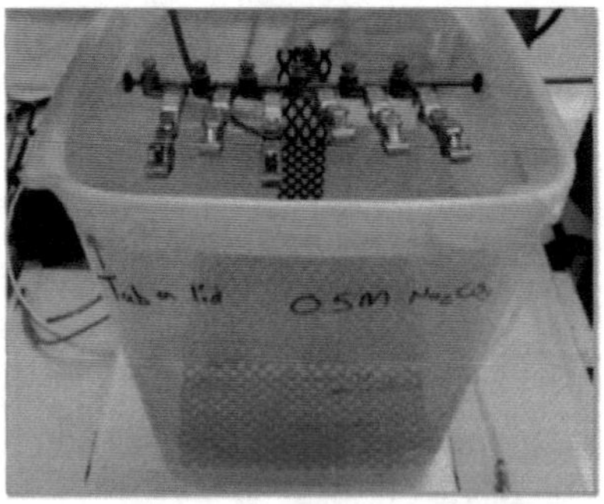

FARADAY
TECHNOLOGY, INC.

FARADAYIC® Copper Film CO_2 Electrocatalysts

Electrocatalyst Performance

Electrocatalysis Test Reactor

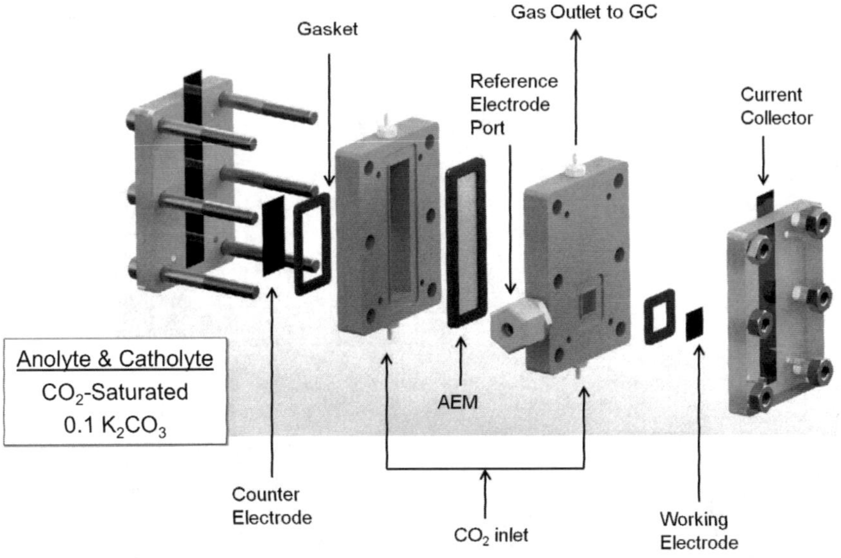

Design adapted from: Kuhl, Jaramillo, et al. Energy & Environmental Science, 5 (2012) 7050-7059.

FARADAYIC® Pulsed ElectroDeposition Waveforms

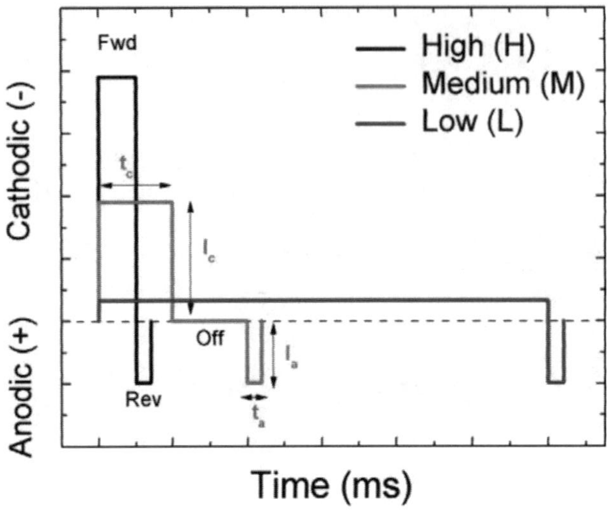

FARADAY TECHNOLOGY, INC.

Electrocatalysis – Effect of Activation Temperature

Pulsed Waveform "Low"

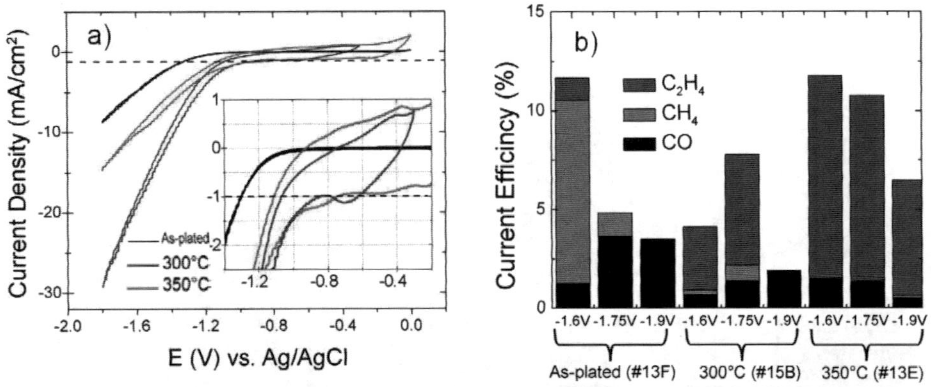

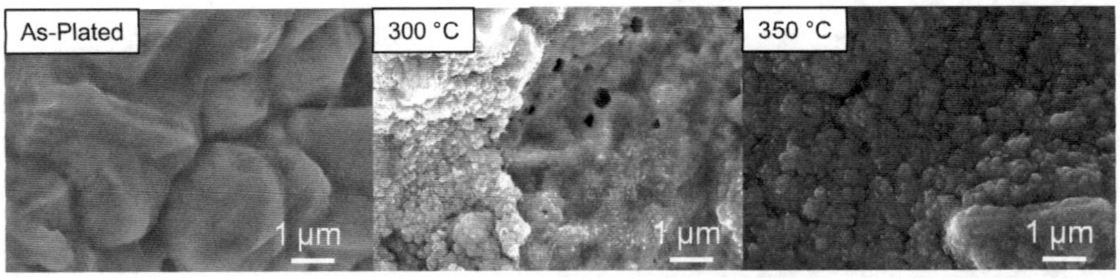

FARADAY TECHNOLOGY, INC.

Electrocatalysis – Effect of Deposition Waveform

Pulse-Deposited Samples
Activated at 350 °C

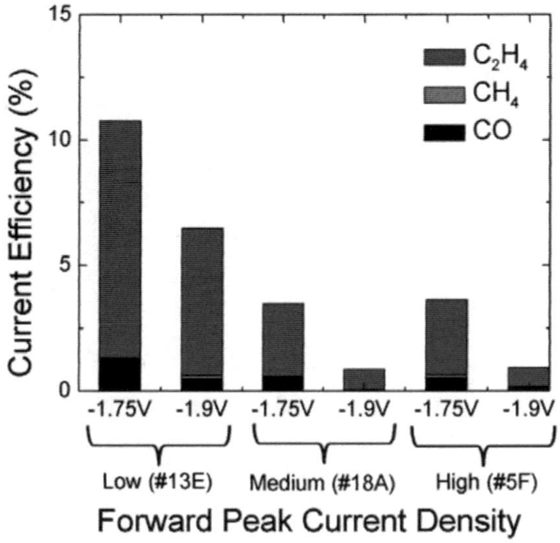

Electrocatalysis – Effect of Pulsed vs DC Waveform

DC vs Pulsed Waveform "Low"

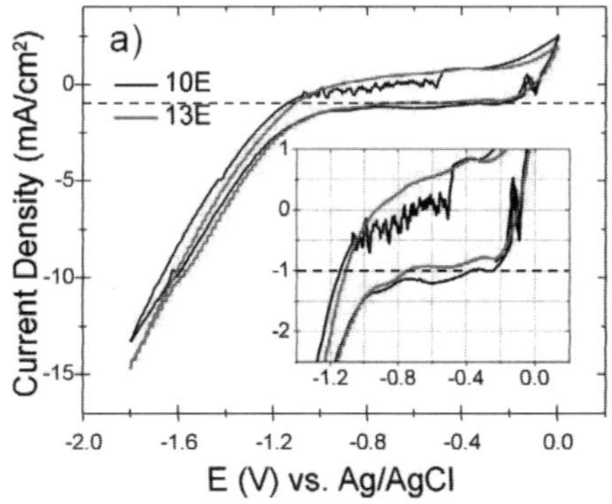

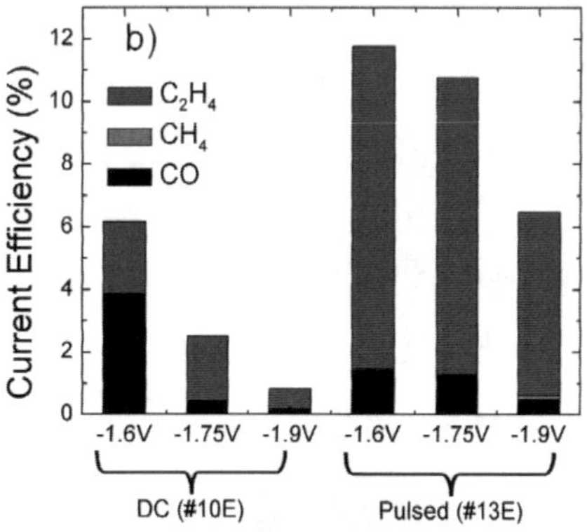

Amine Electrolyte Additives

Ionic Liquids & 1° Amines <u>Unstable</u>

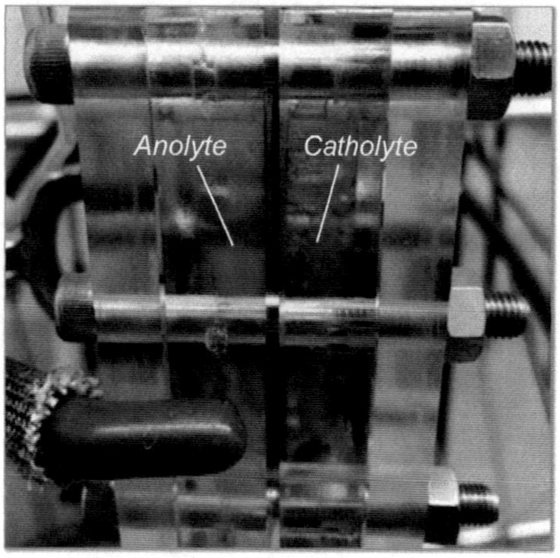

2° & 3° Amines Stable BUT Inhibit CO_2 Conversion @ RT

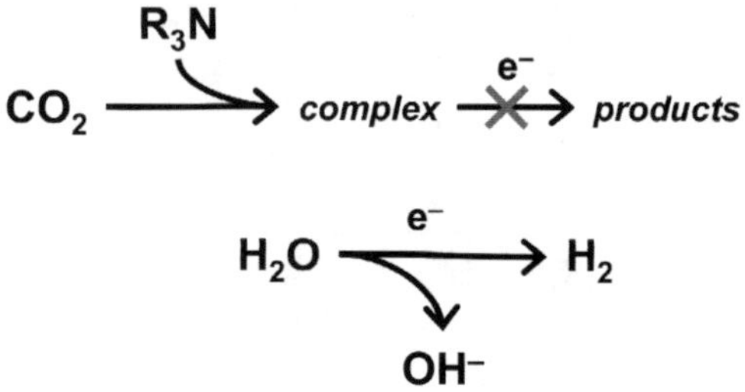

Summary

- Improved CO_2 reduction electrocatalysts via pulsed FARADAYIC® ElectroDeposition
  - Properly tuned deposition waveform superior to DC
    - Greater hydrocarbon production
    - Less cathodic onset potential
  - Activation shifts hydrocarbon selectivity from CH_4 to C_2H_4
    - Activation parameters significant
- Future development
  - Activation/deposition parameters
  - Amine additives at elevated temperature
  - Goals:
    - Increase hydrocarbon (ethylene) selectivity
    - Increase total current density

Acknowledgments

- This work was supported by the United States Department of Energy (contract #DE-SC0015812). Any opinions, findings, conclusions and recommendations expressed in this material are those of the authors and do not necessarily reflect the views of the US DOE.

- Contact Information:
  Brian T. Skinn
  Ph: +1 937 836 7749
  Email: BrianSkinn@FaradayTechnology.com

Thank You!

491354 Pilot Valve CFD Modelling

Abstract:

Pilot valves are utilized in many industrial processes. Increasing their performance efficiency as well as reducing cavitation are main objectives of the designers. We have numerically studied the motion of a 3D pilot valve to investigate the flow field, presence and location of cavitation spots as well as flow forces on the valve components. The motion of the poppet was included through the balance of spring and hydrodynamics forces. A dynamic mesh strategy was utilized to accommodate for the motion of the poppet. History of the volumetric flow rates, flow forces, and the cavitation regions are the main outputs of the simulations. The flow-rate compared favorably with the experiment where the test data was available. The outputs of this study are crucial for optimizing of the design of these valves to enhance the performance and reduce the cavitation.

496398 Multi-Objective Optimization of Solid Sorbent-Based CO_2 Capture Systems

Abstract:

This work presents a multi-objective optimization framework that uses a mixed integer nonlinear programming model (MINLP) for the optimal design and operation of carbon capture plants. A mix of first principle and surrogate-based models have been developed to characterize the process design and operation.

Coal-based power generation is one of the most important sources of energy around the globe. As such, it is important to develop cost-effective carbon capture and storage (CCS) technologies to reduce the environmental impact of power plants. Reducing emissions without affecting the overall cost of the system is difficult. Therefore, to examine the trade-off between cost and emissions, advanced energy analysis tools that consider multiple criteria are needed. The state of the art in technologies for carbon capture include solvents, membrane and solid sorbent systems. This work analyzes the solid sorbent-based adsorption system under different objectives (such as minimizing the cost of electricity and maximizing capture rate, operating vs capital cost, etc.).

In solid sorbent-based capture systems, the process is mainly driven by the interaction of the solid sorbent particles with the flue gas (mainly consisting of N_2, CO_2, and H_2O). The main process units of the solid sorbent systems are multiphase reactors, such as moving bed and fluidized beds. Given the modeling complexity and the difficulty of formulating algebraic models for such processes, previous studies either are generally limited to process-simulation[1] based frameworks or simplified models for mathematical optimization. Process simulators limit the exploration of different process configurations, while in the latest example the validation of the process behavior is limited. Our previous work proposed a mathematical optimization framework, in which both first principles-based and surrogate models[2] were developed to obtain the optimal plant design and operating conditions[3]. Carefully-tuned detailed models implemented in commercial process simulators have been exploited in the preparation and characterization of surrogate models with ALAMO[2]. Model validation and sample generation play a very important role in this mathematical framework.

The proposed mathematical model considers a set of components (CO_2, N_2, H_2O, and Carbamate), streams (utilities, gas, and solids), a set of units (adsorbers and regenerators mixers, splitters, compressors, pumps and heat exchangers), and a set of technologies (two reactor configurations for adsorption and regeneration). **Design decisions** determine the plant layout, by selecting the optimal number of parallel trains, process configuration (number of beds for adsorption and regeneration processes), and equipment design for each unit (diameter, length, area, etc.). At the **operating level**, the decisions involve the optimal flowrates, temperatures, molar fractions of gas and loading of the solids.

Making use of the reduced-order models generated from our previous work, this work uses the normal-constraint method for decision making under multiple objectives. The general idea is to obtain the Pareto frontier generating a set

of solutions by solving multiple optimization problems[4]. Finally, one of the Pareto optimal solutions will be selected based on the Utopia-tracking approach[5].

[1] Eslick J. C., Tong C., Lee A., Dowling A. W., Mebane D. S. (2015). Optimization under uncertainty with rigorous process models. 2015 AIChE Annual Meeting.

[2] Cozad A., Sahinidis N. V., Miller D. C. (2014). Automatic Learning of Algebraic Models for Optimization. AIChE Journal, 60, 2211-2227.

[3] Zamarripa M., Eslick J., Lee A., Ajayi O., Wilson Z., Sahinidis N., Miller D. (2016). Optimal Design and Operation of Hybrid CO_2 Capture Systems. 2016 AIChE Annual Meeting.

[4] Messac A., Ismail-Yahaya A., Mattson C. A. (2003). The normalized normal constraint method for generating the Pareto frontier. Structural and Multidisciplinary Optimization, 25, 86-98.

[5] Zavala V. M. (2015). Managing Conflicts among Decision-Makers in Multiobjective Design and Operations. 2015 AIChE Annual Meeting.

497637 Bioelectrosynthesis of Methane from CO_2 for Energy Storage

Abstract:

The use of electricity to drive biologically-mediated reactions (bioelectrosynthesis) is a promising technique for CO2 utilization. Unlike most industrial electrochemical means to convert CO2, bioelectrosynthesis reactions like microbial methanogensis have been shown to have high electron capture efficiency (~90%), and overall energy efficiency (~80%) (Cheng et al., 2009). Bioelectrosynthesis also has the potential to produce a highly specific product stream with limited side reactions.

As we enter an era of abundant and intermittent renewable energy, the conversion of CO2 to fuels at the utility scale is especially desirable. Low-cost electricity during peak production hours can be used to generate fuels for distribution or for storage and later use for off-peak power generation. If the CO2 comes from biological sources or the atmosphere, the resulting fuel can be sold at a premium under Low Carbon Fuel Standards like California's.

Bioelectrosynthesis is being pursued for conversion of CO2 to fuels or chemicals such as hydrogen peroxide or acetate, but the production of methane has several advantages. Methane (as natural gas) is already used as a transportation fuel and has an enormous transportation and distribution network in place. It can be inexpensively stored at large scale in underground reservoirs and can be converted back into electricity with relatively high efficiency at existing natural gas power plants.

An industrially viable scheme for bioelectrosynthesis requires a reactor with high throughput, low overpotential, and a stable biocatalyst. A recent report by Deutzmann et al. (2015) lends significant promise to bioelectrosynthetic methane production. They identified extracellular enzymes that mediate electron transfer from graphite electrodes to methanogenic microbes. The efficiency and stability (> 1 month) of these biocatalysts open a wide range of possibilities for innovative reactor design.

A challenge common among bioelectrosynthetic systems is the need to optimize the "triple junction," where electrons, protons, and substrate are all supplied to the catalyst site while products are removed away before they can further react. The catalyst, in turn, should be supported on a high-surface-area (conductive) material. The biocatalysts reported by Deutzmann et al. have so far only been demonstrated on planar electrodes. However, with extracellular enzymes to mediate electron transfer, a porous electrode can provide radically higher surface area and volumetric throughput.

In this research, we test the activity of Methanococcus maropaludis cells and extracts in a variety of cutting-edge carbon aerogel electrodes. These custom-synthesized, electrically-conductive aerogels have controlled pore network sizes and structures yielding surface areas up to 500x106 m2/m3, 100 times more than what has been used previously in bioelectrosynthesis (Zhu et al., 2015). The aerogels can be 3D-printed into innovate reactor structures that create flow channels for mass transfer while limiting distances at the triple junction.

Here, we report the potential for improved reactor throughput using the combination of enzyme-mediated electron transfer and carbon aerogel electrodes for methane production. The results have implications for reactor design with bioelectrosynthesis and for the viability of CO2 to methane conversion for energy storage.

References

Cheng, S., Xing, D., Call, D. F. & Logan, B. E. Direct Biological Conversion of Electrical Current into Methane by Electromethanogenesis. Environ. Sci. Technol. 43, 3953–3958 (2009).

Deutzmann, J. S., Sahin, M. & Spormann, A. M. Extracellular Enzymes Facilitate Electron Uptake in Biocorrosion and Bioelectrosynthesis. mBio 6, e00496–15 (2015).

Zhu, C. et al. Highly compressible 3D periodic graphene aerogel microlattices. Nat. Commun. 6, 6962 (2015).

Modeling the Sequestration and Transportation of CO_2 in Deformed Coalbed during Enhanced Coalbed Methane Recovery

Quanshu Zeng and Zhiming Wang, State Key Laboratory of Petroleum Resources and Prospecting, College of Petroleum Engineering, China University of Petroleum, Beijing 102249, China
Brian McPherson and John McLennan. Energy & Geoscience Institute at the University of Utah, Salt Lake City, UT 84108, United States

Abstract

Gas adsorption/desorption has a significant effect on coal deformation and permeability evolution during the primary recovery of coalbed methane (CBM) and enhanced coalbed methane recovery (ECBM). The objectives are to: (1) quantify the coal deformation and permeability change caused by methane (CH_4) displacement with carbon dioxide (CO_2), (2) model the sequestration and transportation of CO_2 in deformed coalbed. In this study, the coal deformation during CBM and ECBM recovery was described by substituting simplified local density (SLD) adsorption theory into a theoretical strain model. The strain model was then verified with all CH_4/CO_2 measured coal strain data published in the past 60 years. Next, the permeability evolution was further revealed by rigorously coupling the SLD theory, the strain model, and the cleat permeability model. Finally, the coupled model was further integrated into the Transport of Unsaturated Groundwater and Heat Simulator (TOUGH2) to simulate the ECBM process. The results show that the coupled model can simultaneously describe gas adsorption or desorption, coal deformation, and permeability evolution during CBM and ECBM recovery with only six parameters, including slit length, solid-solid interaction potential energy parameter, surface areas of CH_4 and CO_2, adsorption expansion modulus, and cleat parameter. The coupled model can predict the coal deformation induced by CH_4 or CO_2 adsorption at pressure up to 12 MPa, and the average relative errors between the measured and predicted results are all within 21.48%. The results also show that the adsorption capacity of CO_2 is 2-5 times as large as that of CH_4, and the volumetric strain induced by CO_2 adsorption is 3-5 times as large as that caused by CH_4 adsorption. While the stronger adsorption capacity of CO_2 on coal offers an option for CO_2-ECBM, coal swelling due to CH_4 displacement with CO_2 may narrow down or even close the cleat, significantly reducing the permeability, and thus impacting the injection efficiency. In addition, the original TOUGH2 simulator predicts similar results with several other CBM simulators. However, it's impossible that 90% of CH_4 can be displaced within 90 days. Considering the coal deformation and permeability change due to CH_4 displacement with CO_2, the modified TOUGH2 simulator shows that only 24% of CH_4 is displaced in the first 90 days, and it takes about 1800 days to displace 90% or more. Advances in the understanding of sequestration and transportation of CO_2 help to improve the ECBM technology.

Keywords

Enhanced Coalbed Methane Recovery, Coal Deformation, Gas Adsorption, Permeability Evolution

Introduction

Increasing atmospheric carbon dioxide (CO_2) concentrations are believed to be the major cause of global warming. Several ways to reduce CO_2 emissions include: improve power generation efficiency by upgrading existing plants or building new plants, rely more on renewable energy and nuclear-generated power, and finally, carbon capture and storage (CCS). Of all emission reduction mechanisms, CCS is projected to provide the largest contribution to emission reduction in the coming decade.

Carbon capture and storage is a process of capturing waste CO_2 from large point sources, transporting it to storage sites, and depositing it where it will not enter the atmosphere, normally an underground geological formation. The geological sequestration of CO_2 offers several options, including deep saline reservoirs, oil and gas fields, and unminable coal seams. Of these options, only enhanced oil recovery (EOR) and enhanced coalbed methane recovery (ECBM) provide a value-adding by-product to CO_2 injection, offsetting sequestration cost or, in some cases, making CO_2 injection profitable. This economic benefit is particularly important in the absence of regulatory controls and/or financial incentives for CO_2 sequestration.

CO_2-EOR is a relatively mature technology, however, ECBM is still in its infancy, with most of the CO_2 injection projects in coal seams not to be successful. Advances in the understanding of CO_2 sequestration and transportation may help to improve the ECBM technology. Existing coalbed methane (CBM) simulators, which are developed for primary CBM recovery, have many important features considered, such as: (1) a dual porosity system, (2) a water/gas two-phase fluid, (3) Darcy flow in the cleat system, (4) pure gas desorption in the matrix system, and (5) matrix shrinkage due to gas desorption. However, the recovery process becomes more complex with CO_2 injection. Several additional features have to be considered, such as: (1) methane (CH_4)/CO_2 mixture adsorption, (2) coal swelling due to adsorption difference between CH_4 and CO_2.

This paper is aimed at developing an internally-consistent adsorption-strain-permeability model to simultaneously quantify gas desorption, coal deformation, and permeability change during CBM and ECBM recovery. The remainder of this paper is organized as follows: modeling section presents the model and couple details of gas adsorption, coal deformation, and permeability evolution; results section presents the validation of strain model and coupled model; discussion section presents the ECBM case simulated by modified TOUGH2 simulator.

Modeling

In this section, the coupled adsorption-strain-permeability model is detailed. Gas adsorption, coal deformation, and cleat permeability characteristics are described using simplified local density theory, a theory-based strain model, and a matchstick-based permeability model, respectively.

Simplified Local Density Theory

Rangarajan[7] first proposed simplified local density (SLD) theory by combining mean-field approximation theory and density functional theory, believing that adsorption is the resultant effect of adsorbate-adsorbate and adsorbate-adsorbent molecule interactions. Several assumptions were made for SLD theory[7-10], including

(1) The chemical potential of any point above adsorbent surface is equal to the bulk phase chemical potential;

(2) The chemical potential of any point above adsorbent surface is the sum of adsorbate-adsorbate and adsorbate-adsorbent interactions;

(3) The adsorbate-adsorbent interaction of any point above adsorbent surface is independent of the temperature and the number of molecules;

(4) Coal pores are considered as perfect slits with temperature and pressure distributed uniformly;

(5) All adsorbate and adsorbent molecules are spherical, except for the adsorbate molecules touching slit walls.

Therefore, the chemical potential of any point above adsorbent surface should be the same, or

$$\mu(z) = \mu_{bulk} = \mu_{ff}(z) + \mu_{fs}(z) \tag{1}$$

where, $\mu(z)$ is the chemical potential at z position, J/mol; z is the distance between adsorbate molecule and adsorbent surface, m; μ_{bulk} is the chemical potential of bulk phase, J/mol; $\mu_{ff}(z)$ is the chemical potential caused by the adsorbate-adsorbate interaction at z position, J/mol; and $\mu_{fs}(z)$ is the chemical potential caused by the adsorbate-adsorbent interaction at z position, J/mol.

Since slit shape is assumed for coal pores while adsorbate molecules located between the parallel slit walls, adsorbate molecules have interactions with both walls (Fig. 1), or

$$\mu_{fs}(z) = N_A\left[\Psi_{fs}(z) + \Psi_{fs}(L_s - z)\right] \tag{2}$$

where, interactions between adsorbent and adsorbate molecules can be effectively represented by the Lennard-Jones potential function[11]. Specifically, the interaction between adsorbate molecules and carbon planes after the fourth layer are ignored, or

$$\Psi_{fs}(z) = 4\pi\rho_{atoms}\varepsilon_{fs}d_{fs}^2\left\{\frac{1}{5}\left(\frac{d_{fs}}{z'}\right)^{10} - \frac{1}{2}\sum_{m=1}^{4}\left[\frac{d_{fs}}{z'+(m-1)d_{ss}}\right]^4\right\} \tag{3}$$

where,

$$\varepsilon_{fs} = \sqrt{\varepsilon_{ff}\times\varepsilon_{ss}} \tag{4}$$

$$d_{fs} = \frac{d_{ff}+d_{ss}}{2} \tag{5}$$

$$z' = z + \frac{d_{cc}}{2} \tag{6}$$

and where, N_A is the Avogadro constant, 6.02×10^{23} mol^{-1}; $\Psi_{fs}(z)$ is the potential energy caused by the interaction between adsorbate molecule at z position and the left wall, J; $\Psi_{fs}(L_s - z)$ is the potential energy caused by the interaction between adsorbate molecule at z position and the right wall, J; L_s is the slit width, m; ρ_{atoms} is the carbon atom density, 3.82×10^{19} m^{-2}; ε_{fs} is the potential energy caused by the interaction between adsorbate and adsorbent molecules, J; ε_{ff} is the potential energy caused by the interaction between adsorbate and adsorbate molecules, J; ε_{ss} is the potential energy caused by the interaction between adsorbent and adsorbent molecules, J; d_{fs} is the collision diameter between adsorbate and adsorbent molecule, m; d_{ff} is the diameter of adsorbate molecule, m; d_{ss} is the carbon plane spacing, 3.35×10^{-10} m; z' is the distance between adsorbate molecule and carbon atoms of the first layer, m; and d_{cc} is the diameter of carbon atom, 1.4×10^{-10} m.

The chemistry potentials of bulk and adsorption phases can be described in fugacity forms, or

$$\mu_{bulk} = \mu_0 + RT\ln\left(\frac{f_{bulk}}{f_0}\right) \tag{7}$$

$$\mu_{ff}(z) = \mu_0 + RT\ln\left[\frac{f_{ads}(z)}{f_0}\right] \tag{8}$$

where, interactions among adsorbate molecules can be described by a fluid equation of state. The Patel-Teja-Valderrama equation of state (PTV EOS) can describe the thermodynamic properties of CH_4 and CO_2 with a reasonable accuracy[12], and thus was selected to calculate density and fugacity of the bulk phase and fugacity of the adsorption phase. The physical properties of CH_4 and CO_2 involved in this study are summarized in Table 1 and can be expressed by

$$\frac{p}{RT\rho} = \frac{1}{1-b\rho} - \frac{a\rho}{RT\left[1+(b+c)\rho-bc\rho^2\right]} \tag{9}$$

where,

$$a = (0.66121-0.761057Z_c)\frac{R^2T_c^2}{p_c} \times\left[1+\left(0.46283+3.58230\omega Z_c+8.19417\omega^2Z_c^2\right)\left(1-\sqrt{T/T_c}\right)\right]^2 \tag{10}$$

$$b = (0.02207+0.20868Z_c)\frac{RT_c}{p_c} \tag{11}$$

$$c = (0.57765-1.87080Z_c)\frac{RT_c}{p_c} \tag{12}$$

where, μ_0 is the chemical potential at reference state, J/mol; R is the gas constant, 8.314 J/(K·mol); T is the temperature, K; f_{bulk} is the fugacity of bulk phase, Pa; f_0 is the fugacity at reference state, Pa; $f_{ads}(z)$ is the fugacity of adsorption phase at *z* position, Pa; p is the pressure, Pa; ρ is the density, mol/m^3; a is the attraction term, J·m^3·mol^{-2}; b is the repulsion term, m^3/mol; c is the polarity term, m^3/mol; Z_c is the critical compression factor, dimensionless; T_c is the critical temperature, K; p_c is the critical pressure, Pa; and ω is the eccentric factor, dimensionless.

To calculate the fugacity of the bulk phase and adsorption phase, the PTV EOS can be rearranged as

$$\ln\frac{f_{bulk}}{p} = \frac{p}{RT\rho_{bulk}} - 1 - \ln\left(\frac{p}{RT\rho_{bulk}} - \frac{pb_{bulk}}{RT}\right) + \frac{a_{bulk}/(RT)}{\sqrt{b_{bulk}^2+6b_{bulk}c_{bulk}+c_{bulk}^2}} \times\ln\left[\frac{2+\rho_{bulk}\left(b_{bulk}+c_{bulk}-\sqrt{b_{bulk}^2+6b_{bulk}c_{bulk}+c_{bulk}^2}\right)}{2+\rho_{bulk}\left(b_{bulk}+c_{bulk}+\sqrt{b_{bulk}^2+6b_{bulk}c_{bulk}+c_{bulk}^2}\right)}\right] \tag{13}$$

$$\ln\frac{f_{ads}(z)}{p}=\frac{p}{RT\rho_{ads}(z)}-1-\ln\left[\frac{p}{RT\rho_{ads}(z)}-\frac{pb_{ads}}{RT}\right]+\frac{a_{ads}(z)/(RT)}{\sqrt{b_{ads}^2+6b_{ads}c_{ads}+c_{ads}^2}}$$
$$\times\ln\left[\frac{2+\rho_{ads}(z)\left(b_{ads}+c_{ads}-\sqrt{b_{ads}^2+6b_{ads}c_{ads}+c_{ads}^2}\right)}{2+\rho_{ads}(z)\left(b_{ads}+c_{ads}+\sqrt{b_{ads}^2+6b_{ads}c_{ads}+c_{ads}^2}\right)}\right] \tag{14}$$

where, the attraction term of adsorbed phase varies with the relative position of adsorbate molecules. The relation between $a_{ads}(z)$ and L_s/d_{ff} was obtained by integrating the sum of the interactions between an adsorbate molecule and all the adsorbed molecules around it[7].

For $L_s/d_{ff}\geq 3$,

$$\frac{a_{ads}(z)}{a_{bulk}}=\begin{cases}\frac{3}{8}\frac{z}{d_{ff}}+\frac{5}{16}-\frac{1}{8\left(\frac{L_s-z}{d_{ff}}-\frac{1}{2}\right)^3} & \frac{1}{2}\leq\frac{z}{d_{ff}}\leq\frac{3}{2}\\ 1-\frac{1}{8\left(\frac{z}{d_{ff}}-\frac{1}{2}\right)^3}-\frac{1}{8\left(\frac{L_s-z}{d_{ff}}-\frac{1}{2}\right)^3} & \frac{3}{2}\leq\frac{z}{d_{ff}}\leq\frac{L_s}{d_{ff}}-\frac{3}{2}\\ \frac{3(L_s-z)}{8d_{ff}}+\frac{5}{16}-\frac{1}{8\left(\frac{z}{d_{ff}}-\frac{1}{2}\right)^3} & \frac{L_s}{d_{ff}}-\frac{3}{2}\leq\frac{z}{d_{ff}}\leq\frac{L_s}{d_{ff}}-\frac{1}{2}\end{cases} \tag{15}$$

For $3>L_s/d_{ff}\geq 2$,

$$\frac{a_{ads}(z)}{a_{bulk}}=\begin{cases}\frac{3z}{8d_{ff}}+\frac{5}{16}-\frac{1}{8\left(\frac{L_s-z}{d_{ff}}-\frac{1}{2}\right)^3} & \frac{1}{2}\leq\frac{z}{d_{ff}}\leq\frac{L_s}{d_{ff}}-\frac{3}{2}\\ \frac{3L_s}{8d_{ff}}-\frac{3}{8} & \frac{L_s}{d_{ff}}-\frac{3}{2}\leq\frac{z}{d_{ff}}\leq\frac{3}{2}\\ \frac{3(L_s-z)}{8d_{ff}}+\frac{5}{16}-\frac{1}{8\left(\frac{z}{d_{ff}}-\frac{1}{2}\right)^3} & \frac{3}{2}\leq\frac{z}{d_{ff}}\leq\frac{L_s}{d_{ff}}-\frac{1}{2}\end{cases} \tag{16}$$

For $2>L_s/d_{ff}\geq 1.5$,

$$\frac{a_{ads}(z)}{a_{bulk}}=\frac{3}{8}\left(\frac{L_s}{d_{ff}}-1\right) \tag{17}$$

where, a_{bulk} is the attraction term for the bulk phase, J·m^3·mol^{-2}; b_{bulk} is the repulsion term for the bulk phase, m^3/mol; c_{bulk} is the polarity term for the bulk phase, m^3/mol; $a_{ads}(z)$ is the attraction term for the adsorption phase, J·m^3·mol^{-2}; b_{ads} is the repulsion term for the adsorption phase, m^3/mol; c_{ads} is the polarity term for the adsorption phase, m^3/mol; and ρ_{bulk} is the density of bulk phase, mol/m^3; and $\rho_{ads}(z)$ is the density of adsorption phase at *z* position, mol/m^3.

Equilibrium criteria can be eventually obtained by substituting Eqs. (2), (7), and (8) into Eq. (1), or

$$f_{ads}(z)=f_{bulk}\exp\left[-\frac{\Psi_{fs}(z)+\Psi_{fs}(L_s-z)}{k_BT}\right] \tag{18}$$

where, k_B is the Boltzmann constant, 1.38×10^{-23}J/K.

According to SLD theory, the Gibbs excess adsorption can be obtained by integrating the density difference between adsorption and bulk phases along the slit length. Since slit pore includes two walls (Fig. 1), half the surface area is considered in Eq. (19), or

$$n^{Gibbs}=\frac{A_s}{2}\int_{\frac{3d_{ff}}{8}}^{L_s-\frac{3d_{ff}}{8}}\left[\rho_{ads}(z)-\rho_{bulk}\right]dz \tag{19}$$

where, n^{Gibss} is the Gibbs excess adsorption amount, mol/kg; and A_s is the surface area of unit mass adsorbent, m^2/kg.

A fluid mixing rule[9] was implemented in the fluid equation of state to describe adsorbate-adsorbate interactions, including CH_4-CH_4, CH_4-CO_2, and CO_2-CO_2 (Fig. 2). Similar to pure component adsorption, the fugacity of CH_4 and CO_2 in both phases can be described by the PTV EOS, revised as

$$\begin{aligned}\ln\frac{f_i^{bulk}}{x_ip}&=\frac{\bar{b}_i^{bulk}\rho_{bulk}}{1-b_{bulk}\rho_{bulk}}-\ln\left(\frac{p}{RT\rho_{bulk}}-\frac{pb_{bulk}}{RT}\right)+\frac{a_{bulk}}{RT\sqrt{b_{bulk}^2+6b_{bulk}c_{bulk}+c_{bulk}^2}}\\&\times\left[\frac{\bar{a}_i^{bulk}}{a_{bulk}}+1-\frac{b_{bulk}\bar{b}_i^{bulk}+3b_{bulk}\bar{c}_i^{bulk}+3\bar{b}_i^{bulk}c_{bulk}+c_{bulk}\bar{c}_i^{bulk}}{b_{bulk}^2+6b_{bulk}c_{bulk}+c_{bulk}^2}\right]\\&\times\ln\left[\frac{2+\rho_{bulk}\left(b_{bulk}+c_{bulk}-\sqrt{b_{bulk}^2+6b_{bulk}c_{bulk}+c_{bulk}^2}\right)}{2+\rho_{bulk}\left(b_{bulk}+c_{bulk}+\sqrt{b_{bulk}^2+6b_{bulk}c_{bulk}+c_{bulk}^2}\right)}\right]\\&+\left[\frac{p}{RT\rho_{bulk}}\frac{b_{bulk}\bar{b}_i^{bulk}+3b_{bulk}\bar{c}_i^{bulk}+3\bar{b}_i^{bulk}c_{bulk}+c_{bulk}\bar{c}_i^{bulk}}{b_{bulk}^2+6b_{bulk}c_{bulk}+c_{bulk}^2}\right.\\&\left.+\frac{(-b_{bulk}+c_{bulk})\left(\bar{b}_i^{bulk}c_{bulk}-b_{bulk}\bar{c}_i^{bulk}\right)}{b_{bulk}^2+6b_{bulk}c_{bulk}+c_{bulk}^2}\right]\left(1-\frac{1}{\frac{p}{RT\rho_{bulk}}-\frac{pb_{bulk}}{RT}}\right)\end{aligned} \tag{20}$$

and

$$\ln\frac{f_i^{ads}(z)}{y_i(z)p}=\frac{\bar{b}_i^{ads}\rho_{ads}(z)}{1-b_{ads}\rho_{ads}(z)}-\ln\left[\frac{p}{RT\rho_{ads}(z)}-\frac{pb_{ads}}{RT}\right]+\frac{a_{ads}(z)}{RT\sqrt{b_{ads}^2+6b_{ads}c_{ads}+c_{ads}^2}}$$

$$\times\left[\frac{\bar{a}_i^{ads}(z)}{a_{ads}(z)}+1-\frac{b_{ads}\bar{b}_i^{ads}+3b_{ads}\bar{c}_i^{ads}+3\bar{b}_i^{ads}c_{ads}+c_{ads}\bar{c}_i^{ads}}{b_{ads}^2+6b_{ads}c_{ads}+c_{ads}^2}\right]$$

$$\times\ln\left[\frac{2+\rho_{ads}(z)\left(b_{ads}+c_{ads}-\sqrt{b_{ads}^2+6b_{ads}c_{ads}+c_{ads}^2}\right)}{2+\rho_{ads}(z)\left(b_{ads}+c_{ads}-\sqrt{b_{ads}^2+6b_{ads}c_{ads}+c_{ads}^2}\right)}\right] \tag{21}$$

$$+\left[\frac{p}{RT\rho_{ads}(z)}\frac{b_{ads}\bar{b}_i^{ads}+3b_{ads}\bar{c}_i^{ads}+3\bar{b}_i^{ads}c_{ads}+c_{ads}\bar{c}_i^{ads}}{b_{ads}^2+6b_{ads}c_{ads}+c_{ads}^2}\right.$$

$$\left.+\frac{\left(-b_{ads}+c_{ads}\right)\left(\bar{b}_i^{ads}c_{ads}-b_{ads}\bar{c}_i^{ads}\right)}{b_{ads}^2+6b_{ads}c_{ads}+c_{ads}^2}\right]\left(1-\frac{1}{\frac{p}{RT\rho_{ads}(z)}-\frac{pb_{ads}}{RT}}\right)$$

where,

$$\bar{a}_i^{bulk}=\left(\frac{\partial na_{bulk}}{\partial n_i}\right)_{\theta,V,n_j} \tag{22}$$

$$\bar{b}_i^{bulk}=\left(\frac{\partial nb_{bulk}}{\partial n_i}\right)_{\theta,V,n_j} \tag{23}$$

$$\bar{c}_i^{bulk}=\left(\frac{\partial nc_{bulk}}{\partial n_i}\right)_{\theta,V,n_j} \tag{24}$$

$$\bar{a}_i^{ads}(z)=\left[\frac{\partial na_{ads}(z)}{\partial n_i}\right]_{\theta,V,n_j} \tag{25}$$

$$\bar{b}_i^{ads}=\left(\frac{\partial nb_{ads}}{\partial n_i}\right)_{\theta,V,n_j} \tag{26}$$

$$\bar{c}_i^{ads}=\left(\frac{\partial nc_{ads}}{\partial n_i}\right)_{\theta,V,n_j} \tag{27}$$

where, attraction terms are determined by quadratic mixing rules while repulsion and polarity terms are determined by linear mixing rules, or

$$a_{bulk}=\sum_i\sum_j x_ix_ja_{ij}^{bulk} \tag{28}$$

$$a_{ads}(z)=\sum_i\sum_j y_i(z)y_j(z)a_{ij}^{ads}(z) \tag{29}$$

$$b_{bulk}=\sum_i x_i\bar{b}_i^{bulk} \tag{30}$$

$$b_{ads}=\sum_i y_i(z)\bar{b}_i^{ads} \tag{31}$$

$$c_{bulk}=\sum_i x_i\bar{c}_i^{bulk} \tag{32}$$

$$c_{ads} = \sum_i y_i(z)\bar{c}_i^{ads} \quad (33)$$

and

$$a_{ij}^{bulk} = \sqrt{\bar{a}_i^{bulk} \times \bar{a}_j^{bulk}} \quad (34)$$

$$a_{ij}^{ads}(z) = \sqrt{\bar{a}_i^{ads}(z) \times \bar{a}_j^{ads}(z)} \quad (35)$$

where, f_i^{bulk} is the fugacity of component i in the bulk phase, Pa; $f_i^{ads}(z)$ is the fugacity of component i in the adsorbed phase at z position, Pa; x_i is the mole fraction of component i in the bulk phase, dimensionless; x_j is the mole fraction of component j in the bulk phase, dimensionless; $y_i(z)$ is the mole fraction of component i in the adsorbed phase at z position, dimensionless; $y_j(z)$ is the mole fraction of component j in the adsorbed phase at z position, dimensionless; $\bar{a}_i^{bulk}$ is the attraction term of component i in the bulk phase, J·m^3·mol^{-2}; $\bar{a}_i^{ads}(z)$ is the attraction term of component i in the adsorbed phase at z position, J·m^3·mol^{-2}; $\bar{b}_i^{bulk}$ is the repulsion term of component i in the bulk phase, m^3/mol; $\bar{b}_i^{ads}$ is the repulsion term of component i in the adsorbed phase, m^3/mol; $\bar{c}_i^{bulk}$ is the polarity term of component i in the bulk phase, m^3/mol; $\bar{c}_i^{ads}$ is the polarity term of component i in the adsorption phase, m^3/mol; a_{ij}^{bulk} is the cross coefficient to calculate the attraction term in the bulk phase, J·m^3·mol^{-2}; and $a_{ij}^{ads}(z)$ is the cross coefficient to calculate the attraction term in the adsorption phase, J·m^3·mol^{-2}.

For adsorbate-adsorbent interactions, both components can be calculated by the Lennard-Jones potential function, or

$$\Psi_i^{fs}(z) = 4\pi\rho_{atoms}\varepsilon_i^{fs}\left(d_i^{fs}\right)^2\left\{\frac{1}{5}\left(\frac{d_i^{fs}}{z'}\right)^{10} - \frac{1}{2}\sum_{i=1}^{4}\left[\frac{d_i^{fs}}{z'+(i-1)d_{ss}}\right]^4\right\} \quad (36)$$

where,

$$\varepsilon_i^{fs} = \sqrt{\varepsilon_i^{ff} \times \varepsilon_{ss}} \quad (37)$$

$$d_i^{fs} = \frac{d_i^{ff} + d_{ss}}{2} \quad (38)$$

where, $\Psi_i^{fs}(z)$ is the potential energy caused by the interaction between component i at z position and the left wall (Fig. 1), J; ε_i^{fs} is the potential energy caused by interactions between component i and either wall, J; ε_i^{ff} is the potential energy caused by the interactions among the molecules of component i, J; d_i^{fs} is the collision diameter between component i and carbon atom, m; and d_i^{ff} is the molecule diameter of component i, m.

To implement the CH_4/CO_2 mixture adsorption model, each component of the mixture should satisfy the adsorption equilibrium criterion. In addition, both phases should satisfy the mole fraction constraints, or

$$f_i^{ads}(z) = f_i^{bulk}\exp\left[-\frac{\Psi_i^{fs}(z) + \Psi_i^{fs}(L_s - z)}{k_B T}\right] \quad (39)$$

$$\sum x_i = 1 \quad (40)$$

$$\sum y_i = 1 \quad (41)$$

Gibbs excess adsorption of the mixture can be calculated as

$$n^{Gibbs} = \sum n_i^{Gibbs} \tag{42}$$

where,

$$n_i^{Gibbs} = \frac{A_i^s}{2} \int_{\frac{3}{8}d_i^{ff}}^{L_s - \frac{3}{8}d_i^{ff}} \left[\rho_{ads}(z) y_i(z) - \rho_{bulk} x_i \right] dz \tag{43}$$

where, n_i^{Gibbs} is the adsorption amount of component i, mol/kg; and A_i^s is the surface area occupied by component i, m^2/kg.

Coal Strain Model

Both reservoir compaction and matrix shrinkage may occur during the primary recovery of CBM, as depicted schematically in Fig. 3. These have opposite effects on coal deformation and permeability evolution. Coal deformation due to reservoir compaction can be described by linear elastic theory, while that due to methane desorption may be predicted on the basis of Bangham and Fakhoury[13]. Both strains are assumed to be independent but superposed, or

$$\Delta\varepsilon = \Delta\varepsilon_m + \Delta\varepsilon_s \tag{44}$$

where, $\Delta\varepsilon$ is the total volumetric strain change, dimensionless; $\Delta\varepsilon_m$ is the volumetric strain change due to reservoir compaction, dimensionless; $\Delta\varepsilon_s$ is the volumetric strain change due to gas adsorption/desorption, dimensionless.

On one hand, the volumetric strain change due to reservoir compaction may be calculated on the basis of linear elastic theory and expressed as

$$\Delta\varepsilon_m = -\frac{3(1-2v)}{E}(p_n - p_0) \tag{45}$$

where, v is the Poisson's ratio, dimensionless; and E is the Young's modulus, Pa; p_0 is the initial pressure, Pa; and p_n is the current pressure, Pa.

On the other hand, Bangham and Fakhoury[13] believed that the adsorption of adsorbate will decrease the surface free energy of adsorbent, and the swell increment induced is proportional to the surface free energy decrease, or

$$\Delta\varepsilon_{sl} = \Delta\Psi \times \gamma \tag{46}$$

where,

$$\gamma = \frac{A_{ss}\rho_c}{K_{ss}} \tag{47}$$

where, $\Delta\varepsilon_{sl}$ is the linear strain change due to gas adsorption/desorption, dimensionless; $\Delta\Psi$ is the surface free energy change due to gas adsorption/desorption per unit mass adsorbent, J/kg; γ is a deformation constant, kg/J; A_{ss} is the specific surface area of adsorbent, m^2/m^2; ρ_c is the coal density, kg/m^3; and K_{ss} is the adsorbent expansion/shrinkage modulus due to gas adsorption/desorption, Pa.

According to the Gibbs adsorption equation, the surface free energy decrease may be characterized as

$$\Delta\Psi = RT \int_{p_0}^{p_n} \frac{n^{Gibbs}}{A_{ss}} \mathrm{d}(\ln p) \tag{48}$$

Combining Eqs. (46) and (48), the linear strain change due to gas adsorption/desorption may be expressed as

$$\Delta\varepsilon_{sl} = \frac{\rho_c RT}{K_{ss}} \int_{p_0}^{p_n} \frac{n^{Gibbs}}{p} \mathrm{d}p \tag{49}$$

To simplify the strain calculation, isotropy is assumed for the coal, and volumetric strain is assumed to be three times as large as linear strain[14-16], or

$$\Delta\varepsilon_s = \Delta 3\varepsilon_{sl} = \frac{3\rho_c RT}{K_{ss}} \int_{p_0}^{p_n} \frac{n^{Gibbs}}{p} \mathrm{d}p \tag{50}$$

Note that n^{Gibss} changes with pressure, and strains can be represented in differential forms and calculated according to Simpson's rule. Assuming that $f(p) = n^{Gibss}/p$, strain change due to gas adsorption/desorption may be rewritten as

$$\Delta\varepsilon_s = \frac{3\rho_c RT}{K_{ss}} \frac{\Delta p}{3} \left[f(p_0) + 4f(p_1) + 2f(p_2) + \cdots + 2f(p_{n-2}) + 4f(p_{n-1}) + f(p_n) \right] \tag{51}$$

Coal Permeability Model

Several cleat permeability models for coal reservoirs have been developed, including the Shi-Durucan (S&D) model[17-20], the Cui-Bustin (C&B) model[21, 22], the Palmer-Mansoori (P&M) model[23, 24] (and its improved version[25]). These models share three assumptions: (1) matchstick geometry, (2) uniaxial strain, and (3) constant overburden stress. Among these models, S&D model was found to have the best performance. The following section provide a brief review of S&D model.

Shi and Durucan[26] derived a permeability model accounting for two competing effects. The former effect is reservoir compaction. The effective stress will increase as the pore pressure decreases, resulting in the closure of cleats and the decrease of cleat permeability. Matrix shrinkage has an opposite effect on the cleat and permeability evolutions as pore pressure reduces and gas desorbs. Thus, an initial permeability decrease due to reservoir compaction is usually compensated by the subsequent increase owing to matrix shrinkage. Whether the permeability is higher than its initial value depends on the net effect of these two competing mechanisms.

The S&D model was derived from the constitutive equation of a linear thermo-elastic porous medium, but replaced the thermal expansion term with an analogous matrix shrinkage term, or

$$\frac{k}{k(0)} = e^{-3C_f \Delta\sigma_h^e} \tag{52}$$

where,

$$\Delta\sigma_h^e = \Delta\sigma_{h,rc}^e + \Delta\sigma_{h,ms}^e \tag{53}$$

and

$$\Delta\sigma_{h,rc}^e = -\frac{v}{1-v}\Delta p_p \tag{54}$$

$$\Delta\sigma_{h,ms}^e = \frac{E}{3(1-v)}\Delta\varepsilon_s \tag{55}$$

where, k is the permeability, mD; $k(0)$ is the initial permeability, mD; C_f is the cleat volume compressibility, Pa^{-1}; $\Delta\sigma_h^e$ is the effective horizontal stress change, Pa; $\Delta\sigma_{h,rc}^e$ is the effective horizontal stress change due to reservoir compaction, Pa; $\Delta\sigma_{h,ms}^e$ is the effective horizontal stress change due to matrix shrinkage, Pa; Δp_p is the pore pressure change, Pa.

In addition to pressure change, the adsorbate component change during ECBM will also contribute to the strain and permeability change, or

$$\frac{k_{CO2}}{k_{CH4}} = e^{-EC_f \Delta\varepsilon_s/(1-v)} \tag{56}$$

where,

$$\Delta\varepsilon_s = \varepsilon_{CO2} - \varepsilon_{CH4} \tag{57}$$

where, k_{CO2} is the permeability of carbon dioxide, mD; k_{CH4} is the permeability of methane, mD; ε_{CO2} is the volumetric strain caused by carbon dioxide adsorption, dimensionless; ε_{CH4} is the volumetric strain caused by methane adsorption, dimensionless.

Model Solution

Since adsorption in the coal strain model here is calculated by simplified local density, while matrix shrinkage in the permeability model is calculated by the strain model, these processes are rigorously coupled. The solution procedure for the coupled adsorption-strain-permeability model is depicted in Fig. 4. Input data include pressure, temperature, physical properties of CH_4 and CO_2, bulk phase mole fraction, physical and elastic properties, and adsorption, strain and cleat parameters of coal. Then, the algorithm calculates density, fugacity, and chemical potential of CH_4 and CO_2 components in both the bulk and adsorption phases, and calculates the adsorbate-adsorbent interaction and induced chemical potential. Next, loop the densities of CH_4 and CO_2 in the adsorption phase until both components satisfy a specific adsorption equilibrium criterion. Finally, adsorption, strain, and cleat parameters of coal are iteratively calculated until predicted results coincide with the adsorption isotherm, strain data, and permeability data.

Results

All CH_4 and CO_2 measured coal strain data[14, 22, 26-28] published during the past 60 years and publicly available were assembled to evaluate the performance of the proposed strain model.

Pressure, temperature, physical properties of CH_4 and CO_2, as well as the physical and elastic properties, adsorption and strain parameters of coal, must be realized prior to the coal strain calculation. The physical properties of CH_4 and CO_2 were extracted from the NIST database[29] while physical and elastic properties of coal were measured directly from coal samples, and adsorption and strain parameters were obtained by fitting the adsorption isotherm and strain data. Specifically, adsorption parameters include slit length, solid-solid interaction potential energy parameter, and surface areas of CH_4 and CO_2. The former two parameters were regressed with the shape of both isotherms while the surface area of each component was regressed with the magnitude of its isotherm. The strain parameter refers to adsorption swelling modulus, which is also the inherent nature of coal. All the input parameters required are summarized in Table 2, and the predicted adsorption and strain results are plotted in Figs. 5 and 6, respectively.

Fig. 5 suggests that CH_4 adsorption increases monotonically with pressure, while CO_2 first increases then decreases with pressure, and the inversion pressure is about 6 MPa. Fig. 5 also suggests that the adsorption capacity of CO_2 is 2-5 times as large as that of CH_4, this indicates coal has a stronger interaction with CO_2. This adsorption difference offers a possibility to deposit CO_2 in unminable coal seams, displace CH_4, thus increase CBM recovery without excessively lowering reservoir pressure.

Fig. 6 suggests that the strain model can predict coal deformation fairly accurately at pressure up to 12 MPa; the average relative errors between measured and predicted results are all within 21.48%, and the largest error occurs is associated with CO_2 adsorption on Wolf Mountain coal[22] (Fig. 6(c)). For all plotted results (Fig. 6), positive expresses expansion, while negative expresses shrinkage. Since volumetric strain is always larger than zero, this means matrix shrinkage is more obvious than reservoir compaction. The volumetric strains induced by both CH_4 and CO_2 adsorption increase monotonically with pressure. In addition, CO_2 adsorption induced volumetric strain is almost 3-5 times as large as that caused by CH_4 adsorption. This may narrow or even close the cleat, significantly reducing the permeability near the injection well, and thus impacting injectivity and limiting methane desorption.

Discussion

In this section, the coupled model was further integrated into the Transport of Unsaturated Groundwater and Heat Simulator (TOUGH2) to simulate the ECBM process.

TOUGH2 are multi-dimensional numerical models for simulating the coupled transport of water, vapor, non-condensible gas, and heat in porous and fractured media. EOS7C-ECBM, an equation of state module for the TOUGH2 program, is suitable for simulating nitrogen (N_2) or CO_2 ECBM process. CH_4/CO_2 mixture adsorption was considered, however, coal swelling due to adsorption difference between CH_4 and CO_2 is ignored. In other words, TOUGH2 fail to quantify the strain and permeability change due to gas adsorption.

A subroutine, of which gas adsorption, coal deformation, and permeability evolution was rigorous coupled, was added to the EOS7C-ECBM module of TOUGH2. The subroutine also offers an option to describe the thermodynamic properties of fluid with PTV EOS, to calculate the adsorption capacity with SLD theory, to calculate the coal deformation with the theoretical model, and to predict the permeability evolution with S&D, C&B, P&M, and improved P&M models.

An ECBM case from Law et al.[30] was selected to test the performance of TOUGH2, and to highlight the coupled adsorption-strain-permeability model. A coal seam with uniform thickness is assumed homogeneous and isotropic. No heat transfer exists between the injected gas and coal seam, the system remains isothermal. The properties of coal seam are summarized in Table 3, and coal properties are summarized in Table 4. In particular, the adsorption parameters were fitted with SLD theory, and the elastic and strain parameters were from the typical values of the San Juan basin[27].

Inverted 5-spot pattern was used in the case, with CO_2 injected to displace CH_4 in the coal reservoir. The injection well has a constant bottom hole flowing pressure of 12.5 MPa while the bottom hole flowing pressure of producing well is constant at 0.275 MPa. Since the entire seepage field is symmetrical, only a quarter was selected, and divided into $11 \times 11 \times 1$ grids, as illustrated in Fig. 7.

TOUGH2 EOC7C-ECBM was first used to simulate the ECBM process, and to compare with other simulators. After 30, 60, and 90 days of gas injection, CO_2 mole fraction distributions in the coal seam are shown in Fig. 8. It can be observed that 90% methane or more was displaced and produced within 90 days. The result is consistent with those obtained by other coalbed methane reservoir simulators, as shown in Figs. 9-12. This indicates TOUGH2 EOC7C-ECBM has a comparable performance with GEM, ECLIPSE, COMET 2, and SIMED II. However, this is not consistent with reality, and may be too optimistic.

With the EOS7C-ECBM module modified with the coupled model, and the ECBM process was re-simulated. After 30, 60, 90, 180, 360, 1800 days of gas injection, CO_2 mole fraction distributions in the coal seam are shown in Fig. 13. Taking into consider the adsorption difference between CH_4 and CO_2, and the induced strain and permeability change, only 24% of the CH_4 is displaced in the first 90 days, and it takes about 1800 days to displace 90% or more. This indicates that ECBM is even worse than primary recovery. This phenomenon was mainly due to coal swelling caused by the adsorption differences between CH_4 and CO_2. Although the matrix swell increment is limited, the cleats are easy to narrow down or even close, thus reduce the permeability near the injection well significantly and impact the injection efficiency. This help to understand the sequestration and transportation of CO_2, and may have a guidance to ECBM technology.

Conclusions

In this paper, a rigorously coupled adsorption-strain-permeability model was developed to quantify the coal deformation and permeability change caused by CH_4 displacement with CO_2, and to reveal the rules of CO_2 sequestration and transportation in deformed coalbed. The following conclusions and recommendations are suggested, based on the results of this study.

(1) With coal deformation due to matrix shrinkage calculated with gas adsorption theory, and permeability evolution calculated with theoretical strain model, these processes were rigorous coupled. The coupled model can simultaneously quantify gas adsorption or desorption, coal deformation, and permeability evolution during CBM and ECBM recovery with only six parameters.

(2) The coupled model can predict the coal deformation induced by CH_4 or CO_2 adsorption at pressure up to 12 MPa, and the average relative errors between the measured and predicted results are all within 21.48%.

(3) The adsorption capacity of CO_2 is 2-5 times as large as that of CH_4, and the volumetric strain induced by CO_2 adsorption is 3-5 times as large as that caused by CH_4 adsorption. While the stronger adsorption capacity of CO_2 on coal offers an option for CO_2-ECBM, this technology was not to be successful due to coal swelling induced by adsorption difference between CH_4 and CO_2.

(4) Considering the coal deformation and permeability change due to CH_4 displacement with CO_2, the modified TOUGH2 simulator shows that only 24% of CH_4 is displaced in first 90 days, and it takes about 1800 days to displace 90% or more.

Acknowledgments

Financial support from the National Science and Technology Major Project of China (2016ZX05044005-001), High School Subject Innovation Engineering Plan of China (111 Project: B12033), and China Scholarships Council Program (201606440099) are gratefully acknowledged.

Reference

(1) Advanced Resources International Inc. Enhanced coalbed methane recovery: Worldwide application and CO_2 sequestration potential. *International Energy Agency*, 1999.
(2) Reeves, S. An overview of CO_2-ECBM and sequestration in coal seams. *AAPG Studies in Geology* **2009**: 17-32.
(3) Godec, M.; Koperna, G.; Gale, J. CO_2-ECBM: A review of its status and global potential. *Energ. Procedia* **2014**, 63(3-4): 5858-5869.
(4) Sloss, D. L. L. Potential for enhanced coalbed methane recovery. *International Energy Agency*, 2015.
(5) Reeves, S.; Taillefert, A.; Pekot, L.; Clarkson, C. The Allison Unit CO_2 - ECBM pilot: A reservoir modeling study. *U.S. Department of Energy*, 2004.
(6) Li, X. C.; Fang, Z. M. Current status and technical challenges of CO_2 storage in coal seams and enhanced coalbed methane recovery: An overview. *Int. J. Coal Sci. Technol.* **2014**, 1(1): 93-102.
(7) Rangarajan, B.; Lira, C. T.; Subramanian, R. Simplified local density model for adsorption over large pressure ranges. *AIChE J.* **1995**, 41(4), 838-845.

(8) Fitzgerald, J. E.; Sudibandriyo, M.; Pan, Z. J.; Robinson, R. L.; Gasem, K. A. M. Modeling the adsorption of pure gases on coals with the SLD model. *Carbon* **2003**, 41(12), 2203-2216.
(9) Fitzgerald, J. E.; Robinson, R. L.; Gasem, K. A. M. Modeling high-pressure adsorption of gas mixtures on activated carbon and coal using a simplified local-density model. *Langmuir* **2006**, 22(23), 9610-9618.
(10) Zeng, Q. S.; Wang, Z. M.; McPherson, B. J.; McLennan, J. D. Theoretical approach to model gas adsorption/desorption and the induced coal deformation and permeability change. *Energ. Fuel* **2017**, 31(8), 7982-7994.
(11) Hasanzadeh, M.; Alavi, F.; Feyzi, F.; Dehghani, M. R. Simplified local density model for adsorption of pure gases on activated carbon using Sutherland and Kihara potentials. *Micropor. Mesopor. Mat.* **2010**, 136(1-3), 1-9.
(12) Valderrama, J. O. A generalized Patel-Teja equation of state for polar and non-polar fluids and their mixtures. *J. Chem. Eng. Jpn.* **1990**, 23(1), 87-91.
(13) Bangham, D. H.; Fakhoury, N. J. The translation motion of molecules in the adsorbed phase on solids. *J. Chem. Soc. Pakistan.* **1931**, 1324-1333.
(14) Levine, J. R. Model study of the influence of matrix shrinkage on absolute permeability of coal bed reservoirs. *Coalbed Methane and Coal Geology* **1996**, 109, 197-212.
(15) Megson, T. H. G. *Structural and stress analysis*; Elsevier Butterworth-Heinemann: Oxford, Britain, 2014 (ISBN No. 9780080999364).
(16) Robertson, E. P. Improvements in measuring sorption-induced strain and permeability in coal. *In Society of Petroleum Engineers - SPE Eastern Regional/AAPG Eastern Section Joint Meeting*; Pittsburgh, PA USA, October 11-15, 2008.
(17) Shi, J. Q.; Durucan, S. Drawdown induced changes in permeability of coalbeds: A new interpretation of the reservoir response to primary recovery. *Transport Porous Med.* **2004**, 56(1), 1-16.
(18) Shi, J. Q.; Durucan, S. A model for changes in coalbed permeability during primary and enhanced methane recovery. *SPE Reserv. Eval. Eng.* **2005**, 8(4), 291-299.
(19) Shi, J. Q.; Durucan, S. Exponential growth in San Juan Basin Fruitland coalbed permeability with reservoir drawdown: Model match and new insights. *SPE Reserv. Eval. Eng.* **2010**, 13(6), 914-925.
(20) Shi, J. Q.; Durucan, S. Modelling laboratory horizontal stress and coal permeability data using S&D permeability model. *Int. J. Coal Geol.* **2014**, 131(1), 172-176.
(21) Cui, X.; Bustin, R. M.; Chikatamarla, L. Adsorption-induced coal swelling and stress: Implications for methane production and acid gas sequestration into coal seams. *J. Geophys. Res.* **2007**, 112(B10202), 1-16.
(22) Bustin, R. M.; Cui, X. J.; Chikatamarla, L. Impacts of volumetric strain on CO_2 sequestration in coals and enhanced CH_4 recovery. *AAPG Bull.* **2008**, 92(1), 15-29.
(23) Palmer, I.; Mansoori, J. How permeability depends on stress and pore pressure in coalbeds: A new model. *In Proceedings - SPE Annual Technical Conference and Exhibition*; Denver, CO USA, October 6-9, 1996.
(24) Palmer, I.; Mansoori, J. How permeability depends on stress and pore pressure in coalbeds: A new model. *SPE Reserv. Eng.* **1998**, 1(6), 539-543.
(25) Palmer, I. Permeability changes in coal: Analytical modeling. *Int. J. Coal Geol.* **2009**, 77(1-2), 119-126.
(26) Ottiger, S.; Pini, R.; Storti, G.; Mazzotti, M. Competitive adsorption equilibria of CO_2 and CH_4

on a dry coal. *Adsorption* **2008**, 14(4), 539-556.

(27) Harpalani, S.; Mitra, A. Impact of CO_2 injection on flow behavior of coalbed methane reservoirs. *Transport Porous Med.* **2010**, 82(1), 141-156.

(28) Pan, Z. J.; Connell, L. D. Modelling of anisotropic coal swelling and its impact on permeability behaviour for primary and enhanced coalbed methane recovery. *Int. J. Coal Geol.* **2011**, 85(3-4), 257-267.

(29) National Institute of Standards and Technology. *Thermophysical properties of fluid systems*; http://webbook.nist.gov/chemistry/fluid/.

(30) Law, D. H. S.; Van der Meer, L. G. H.; Gunter, M. D. Numerical Simulator Comparison study for enhanced coalbed methane recovery processes, part I: Pure carbon dioxide injection. *In Proceedings - SPE Gas Technology Symposium*; Calgary, Alberta, Canada, April 30 - May 2, 2002.

Tables

Table 1 Physical Properties of CH_4 and CO_2

Fluid	T_c K	P_c MPa	Z_c -	ω_c -	d nm	ε_{ff}/k_B K
CH_4	190.56	4.599	0.288	0.011	0.3758	148.6
CO_2	304.13	7.377	0.274	0.239	0.3941	195.2

Table 2 Model Fitting Results for Coal Strain Induced by CH_4 and CO_2 Adsorption

Database	Coal	T K	L_s nm	ε_{ss}/k_B K	A_{CH4} m²/g	A_{CO2} m²/g	E MPa	v -	ρ_c m³/kg	K_{ss} MPa	δ_{CH4} %	δ_{CO2} %
Levine	Fruitland	308.15	1.50	19.5	104.2	137.3	4400	0.32	1400	1665	5.41	3.50
Bustin	Ardley	298.15	2.59	12.6	19.8	64.8	3000	0.30	1500	559	3.78	4.49
	WM	298.15	1.19	21.5	49.3	53.2	3000	0.30	1340	823	2.07	21.48
Ottiger	Sulcis	318.15	1.80	66.7	83.9	96.7	2410	0.48	1400	1812	8.31	5.06
Harpalani	San Juan	308.15	1.14	22.6	71.6	61.0	3450	0.37	1400	1484	3.37	7.61
	Illinois	308.15	2.08	18.1	51.2	97.2	2119	0.40	1400	816	7.07	9.14

Table 3 Coal Seam Properties

A_c m²	t_c m	$k_{CH4}(0)$ mD	$k_{CO2}(0)$ mD	ϕ_i -	T °C	p_{inj} MPa	p_{pro} MPa	ρ_c kg/m³	ρ_{water} kg/m³	S_{water} -	S_{CH4} -
647497	9	3.65	0.91	0.001	45	12.5	0.275	1434	990	0.592	0.408

Table 4 Coal Properties

L_s nm	ε_{ss}/k_B K	A_{CH4} m²/g	A_{CO2} m²/g	E MPa	v -	K_{ss} MPa	C_f MPa⁻¹
2.63	16.7	46.4	70.2	3450	0.37	1295	0.145

Figures

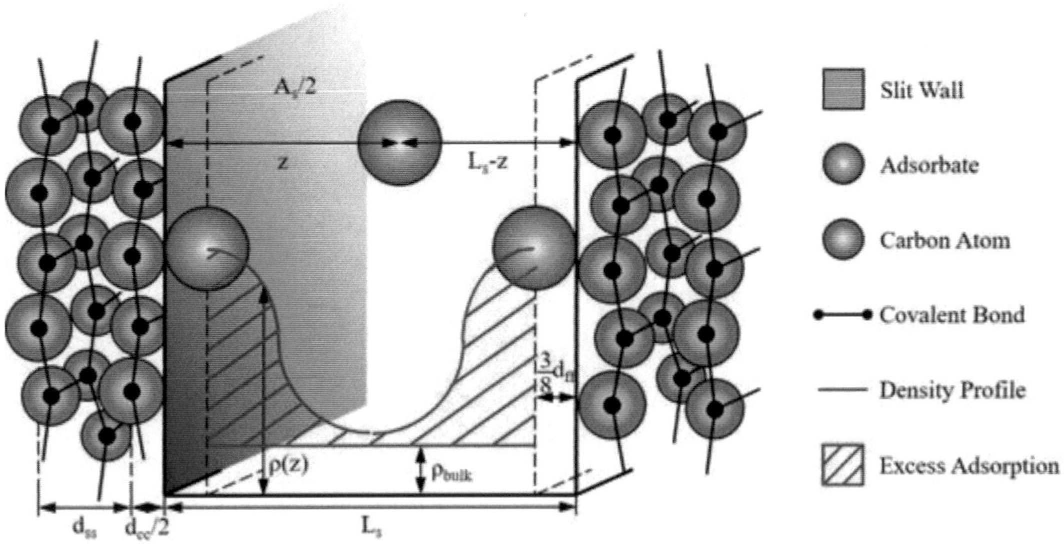

Fig. 1 Slit Pore Characterization in SLD Theory

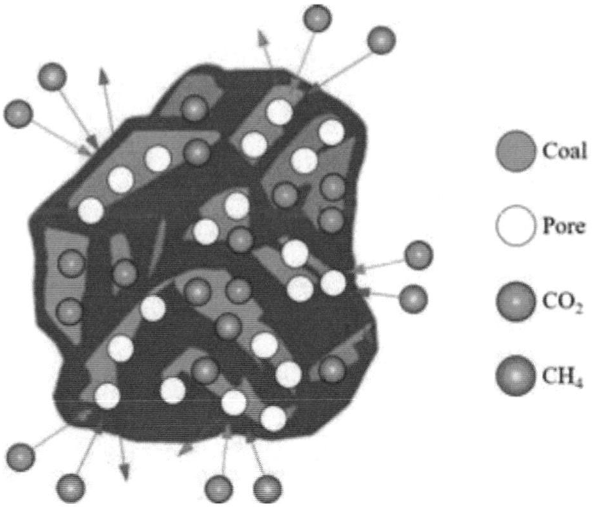

Fig. 2 Schematic Diagram of CH_4/CO_2 Mixture Adsorption on Carbonaceous Adsorbent

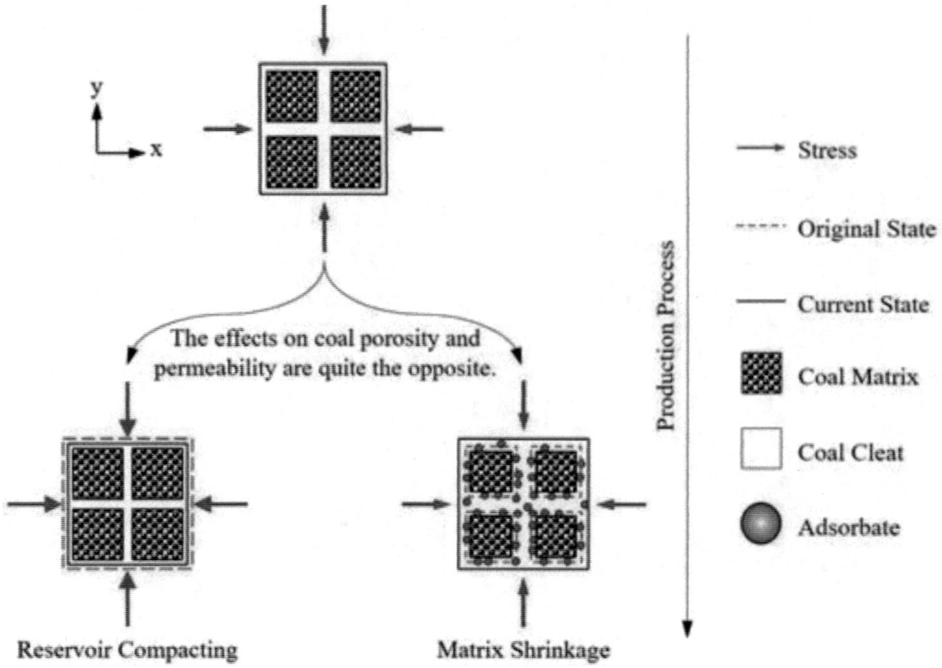

Fig. 3 Reservoir Compaction and Matrix Shrinkage Effects on Coal Permeability

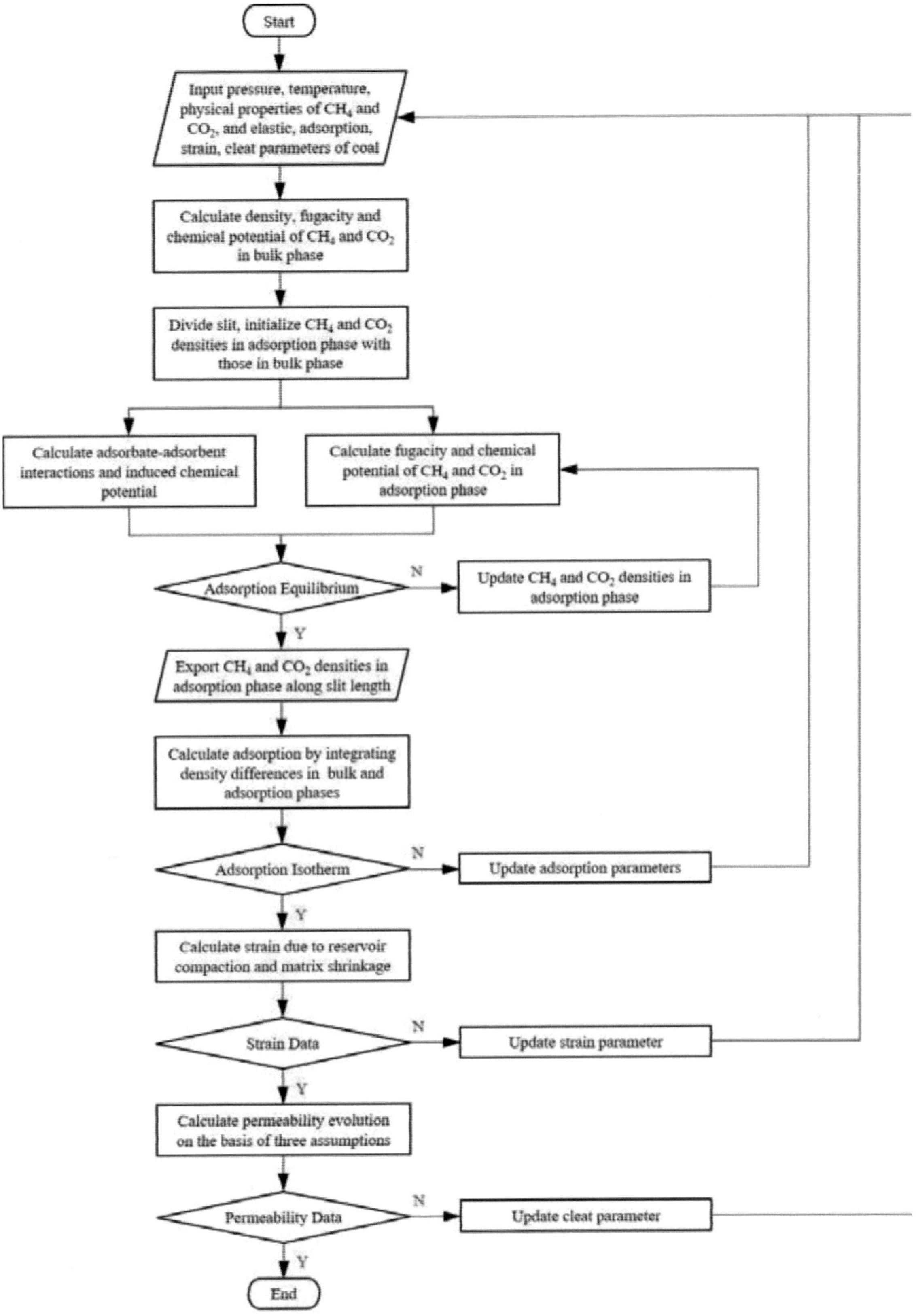

Fig. 4 Solution Procedure for Coupled Adsorption-Strain-Permeability Model

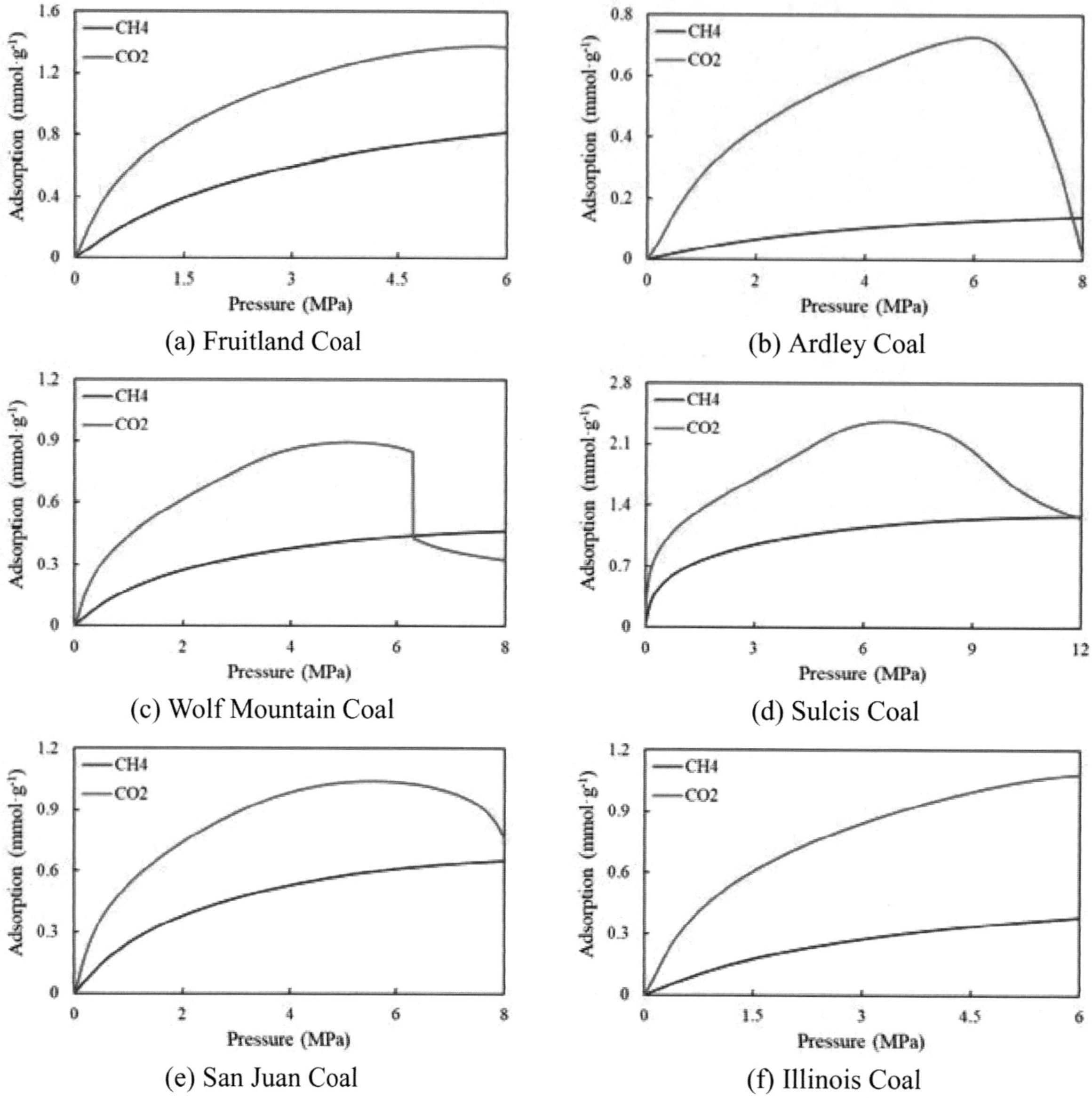

(a) Fruitland Coal (b) Ardley Coal

(c) Wolf Mountain Coal (d) Sulcis Coal

(e) San Juan Coal (f) Illinois Coal

Fig. 5 Adsorption Isotherms of CH_4 and CO_2

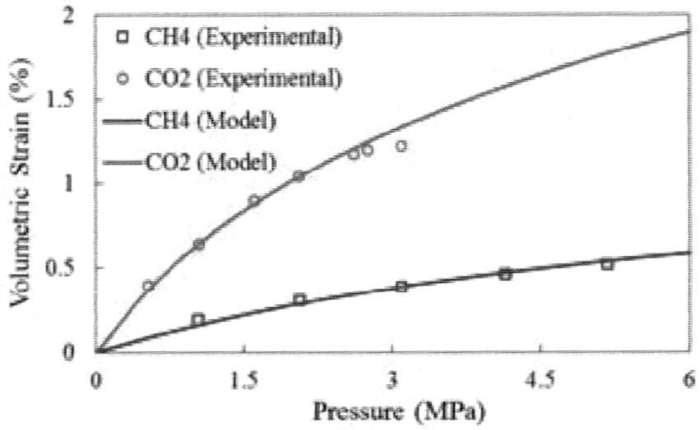

(a) Fruitland Coal

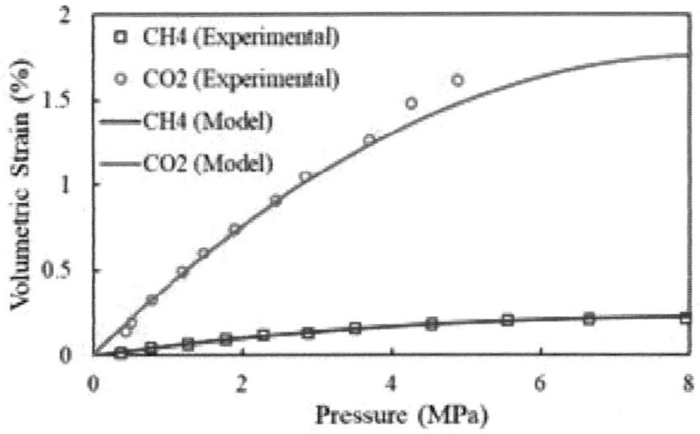

(b) Ardley Coal

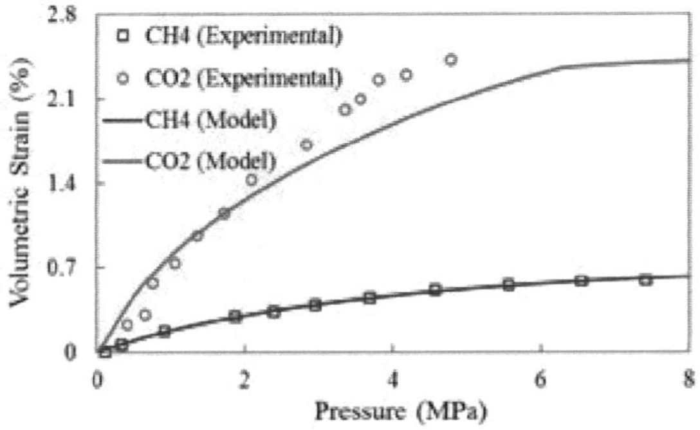

(c) Wolf Mountain Coal

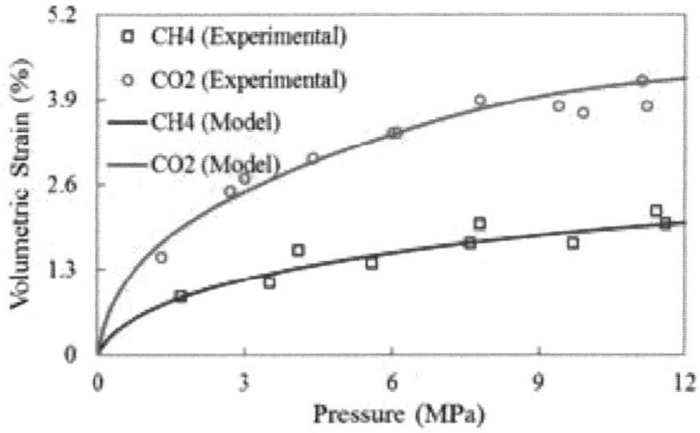

(d) Sulcis Coal

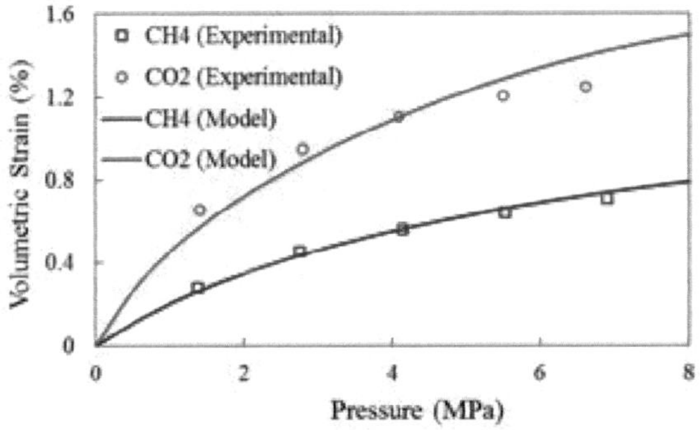

(e) San Juan Coal

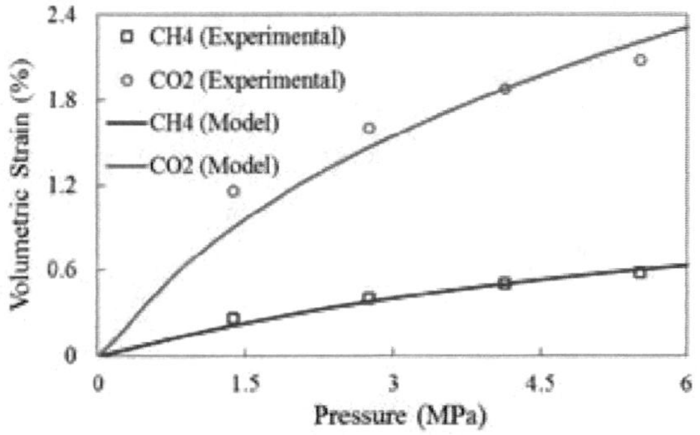

(f) Illinois Coal

Fig. 6 Volumetric Strain with Pressure

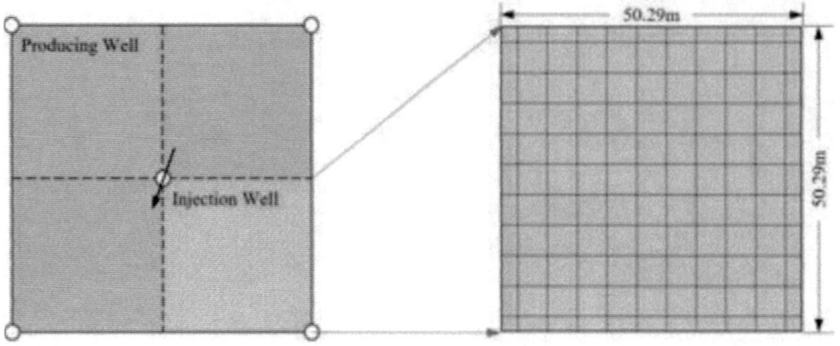

(a) Inverted 5-Spot Pattern (b) 11×11×1 Grid System

Fig. 7 Schematic Diagrams of Well Pattern and Grid System Used

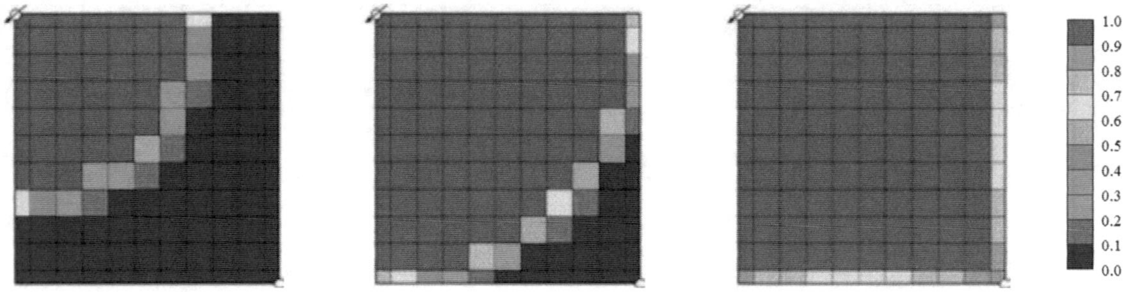

(a) 30 days (b) 60 days (c) 90 days

Fig. 8 CO_2 Mole Fraction Distributions Predicted by TOUGH2

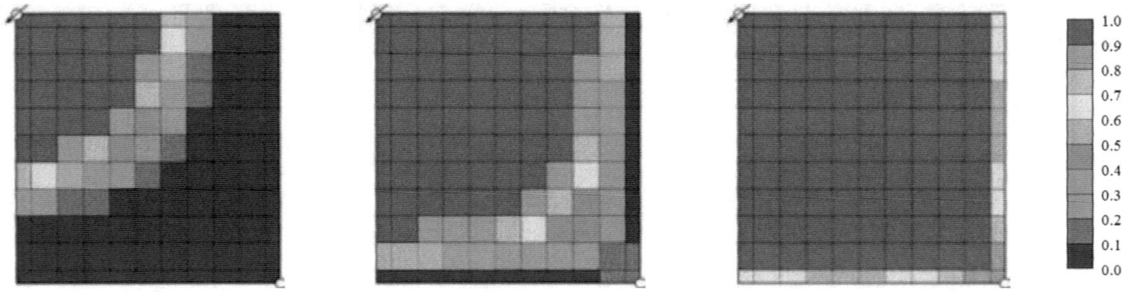

(a) 30 days (b) 60 days (c) 90 days

Fig. 9 CO_2 Mole Fraction Distributions Predicted by GEM

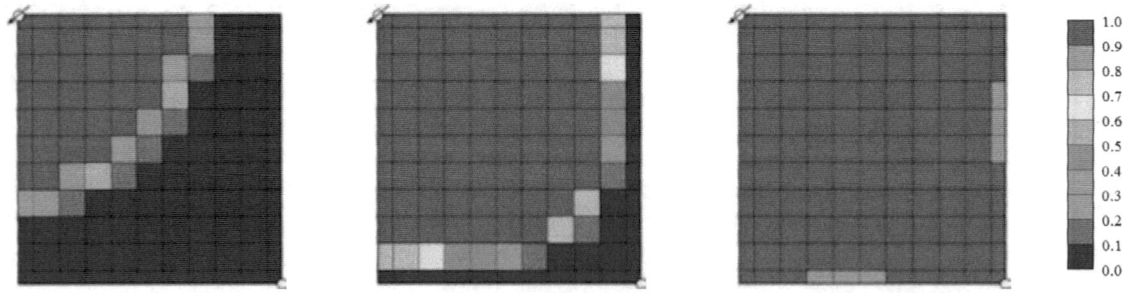

(a) 30 days (b) 60 days (c) 90 days

Fig. 10 CO_2 Mole Fraction Distributions Predicted by ECLIPSE

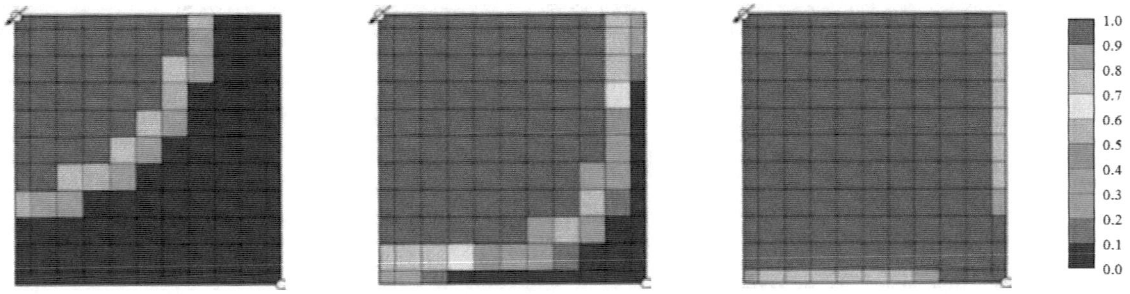

(a) 30 days (b) 60 days (c) 90 days

Fig. 11 CO_2 Mole Fraction Distributions Predicted by COMET2

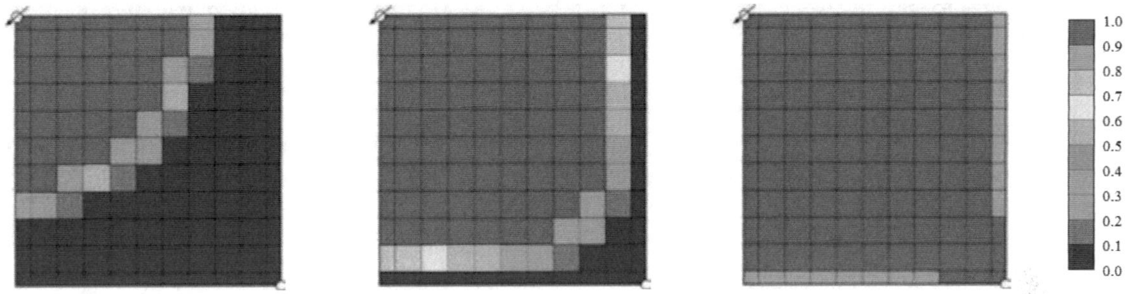

(a) 30 days (b) 60 days (c) 90 days

Fig. 12 CO_2 Mole Fraction Distributions Predicted by SIMED II

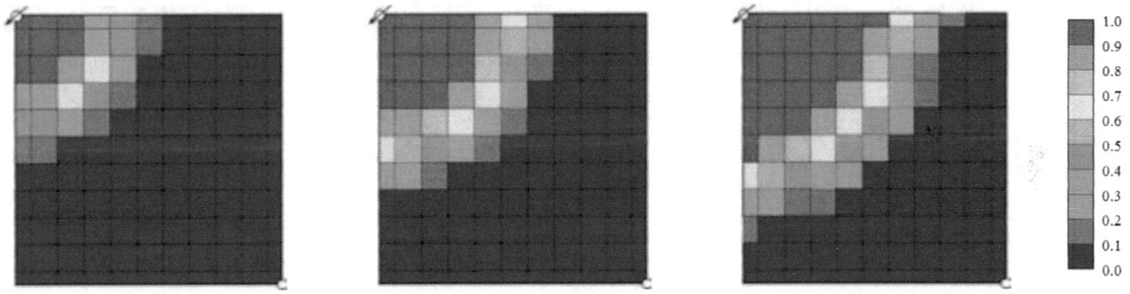

(a) 30 days (b) 60 days (c) 90 days

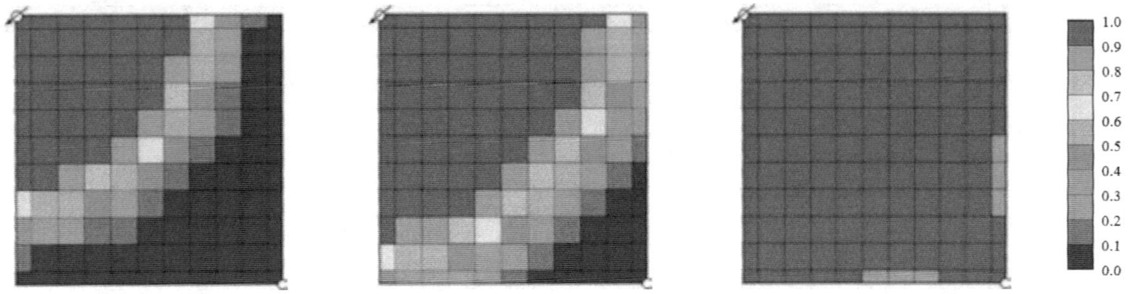

(d) 180 days (e) 360 days (f) 1800 days

Fig. 13 CO_2 Mole Fraction Distributions Predicted by Modified TOUGH2

498318 Integrated CO_2 Capture /Water-Gas Shift with Integrated HEAT Management for IGCC Applications

Abstract:

As one of the 16 transformational carbon capture technologies projects selected by Department of Energy (DOE), Southern Research is developing a combined CO_2 capture water-gas shift (WGS) process for integrated gasification combined cycle (IGCC) applications.Â Here we present a laboratory-scale combined sorbent/catalyst reactor system with advanced integrated heat management to capture 90% of CO_2 from simulated Transport Reactor Integrated Gasifier (TRIG) syngas at warm gas conditions (250 â€“ 400 Â° C). A highly efficient regenerable magnesium based sorbent was developed to remove CO_2. It offers high and stable CO_2 capacity with minimum requirement for temperature and pressure swings. Under optimized conditions, the sorbents achieved up to 6 mmol/g of CO_2 working capacity and remained stable over 250 cycles of adsorption and regeneration. The selected WGS catalyst achieved around 95% CO conversion under both continuous feed and combined cycle tests and remained stable during the test duration that lasted over 200 hrs. The approach has potential to reduce costs and increase efficiency for electricity generation from IGCC. Upon successful completion of bench-scale integrated test program, the process will be ready for pilot scale development. This presentation will examine bench-scale results on combined CO_2 capture/WGS process using an integrated heat management reactor.

504618 Grand Challenges in Sustainability for Future Energy and Chemicals

Abstract:

Providing affordable energy and chemical products for a growing world population while protecting the earthâ€™s climate, environment, and resources is a grand challenge for mankind. Abundant natural gas plus cost reduction in renewable energy together with changing societal norms will drive an unprecedented rate of change in technology, preferences, and options for transportation and mobility. Advanced manufacturing techniques together with new energy and feedstock options will drive substantial changes in the chemical industry. While concerns over greenhouse gas emissions are well documented, protection of water resources and biodiversity, management of land use, and mitigation of criteria pollutants impacting local air quality will increase in importance as the world population increases in affluence. Local economies and economic security will continue to play a strong role in consumer and stakeholder choices, which in turn may drive a diversity of preferred solutions. Development of new technology solutions must carefully consider the full breadth of sustainability impacts and drivers in selecting preferred products and pathways.

504619 Research Trends in Sustainable Chemistry and Engineering: Quantifying Progress Toward Sustainability

Abstract:

The research literature covering green chemistry and green engineering is rapidly expanding. In presenting their findings in this rapidly expanding literature, researchers often justify the relevance of their work to the sustainability of the chemical enterprise by qualitatively invoking the principles of green chemistry and green engineering or other principles of sustainability. This approach will ultimately need to be complemented by rigorous, semi-quantitative and quantitative sustainability assessments. These may include life cycle analyses, toxicity assessments, or other indicators of sustainability. More quantitative approaches to sustainability assessments can help identify the key features to be improved to make a process or product sustainable, providing valuable guidance to further process or product development. This presentation will briefly describe how one prominent research journal (*ACS Sustainable Chemistry & Engineering*) is attempting to encourage researchers to incorporate the use of quantitative sustainability metrics into their work.

504620 Hydrogen Economy Via Cost-Effective Renewable Solar Water Splitting

Abstract:

The Holy Grail of a renewable hydrogen economy is using renewable energy to split water producing hydrogen and oxygen, the two then combined via stationary or mobile fuel cells to produce electricity while regenerating water. Since more sunlight strikes the earth in 1 hr than all of the energy currently consumed on the planet in 1 yr, the use of solar energy to produce hydrogen cost-effectively from water is a Grand Challenge for sustainable energy. This presentation will provide a short overview of the options available for solar energy to drive water splitting and will then focus on the technical status for solarthermal water splitting and will discuss specific engineering challenges that need to be addressed for the process is to be commercially realized.

504621 Opportunities and Challenges Along the Path to Sustainability

Abstract:

Sustainability considerations focus attention on the impact of some proposed process, product, or activity on the environment, the economy, and society. As such, expertise across many different disciplines is required to identify and then assess adequately the different dimensions of sustainability. This brief talk will recap advances to date in assessing and progressing toward sustainability, seek to identify obstacles in the path ahead, and suggest ways they might be removed or circumvented.

Growing Scientific Leaders With Concentrated Sunlight: A Tribute to E. A. Fletcher

Bob Palumbo
Prof. of Mechanical Eng.

Luke Venstrom
Asst. Prof. of Mechanical Eng.

VALPARAISO UNIVERSITY

Growing Scientific Leaders

By Developing Deep Science and Engineering Knowledge

By Developing Communication Skills

By Developing the Passion to Serve Others

VALPARAISO UNIVERSITY

Model for Engaging Undergraduate Students

Curricular	**Senior Design:**	• Two-semester course • Design during fall semester • Fabricate and test during spring semester • Multi-disciplinary teams with faculty advisor
Curricular	**Credit:**	• Research during academic year • Students earn up to 3 credits towards mechanical engineering degree
Extra-curricular	**Summer Research:**	• 10 to 12 weeks over the summer • Includes stipend and free housing in dorm • Intentional cross-disciplinary interaction

VALPARAISO UNIVERSITY

Undergraduate Student Activity

VALPARAISO UNIVERSITY

Valpo Solar Furnace

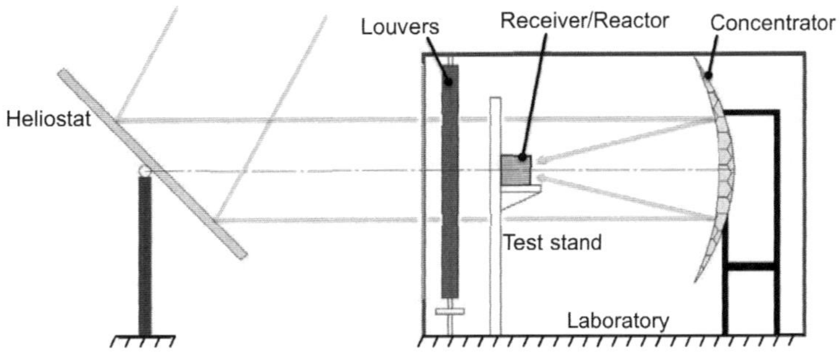

VALPARAISO UNIVERSITY

Valpo Solar Furnace

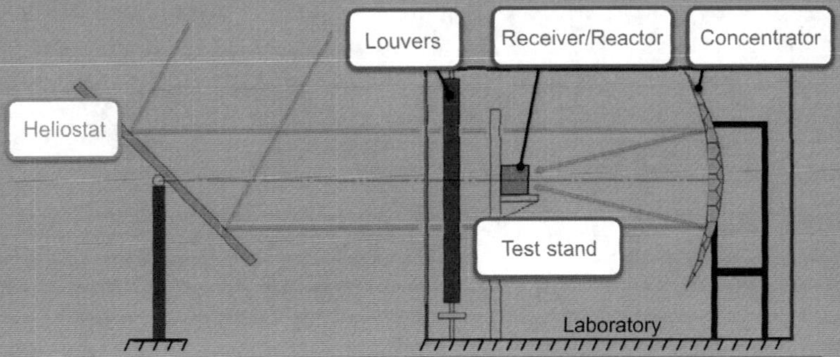

Designed and manufactured by undergraduate students

Undergraduate students contributed to design via analysis and construction of a prototype

VALPARAISO UNIVERSITY

Test Stand

Design Constraints:
- 3-axis motion accurate within .01"
- Support 250 lbf
- Integrate chilled water flow up to 10 gpm, pressurized air supply, and exhaust gas line
- Safety:
  - brake in event of power loss
  - emergency stop switches
  - laser safety scanners
- Able to hold numerous receivers:
  - Calorimeter
  - Lambertian target
  - Solar thermochemical reactors

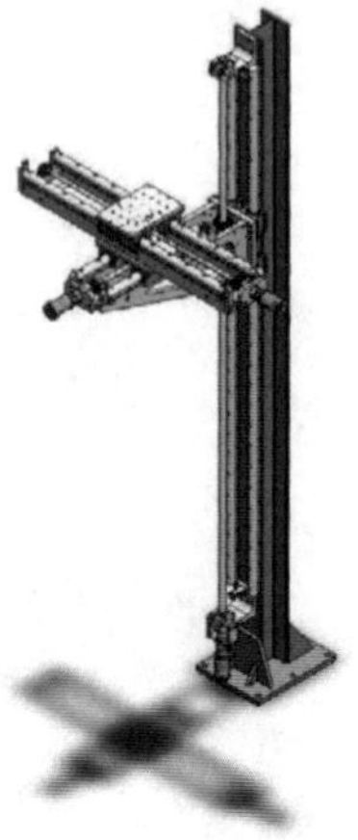

VALPARAISO UNIVERSITY

Test Stand

Analyzed by undergraduate students with FEA: Factor of safety of 11

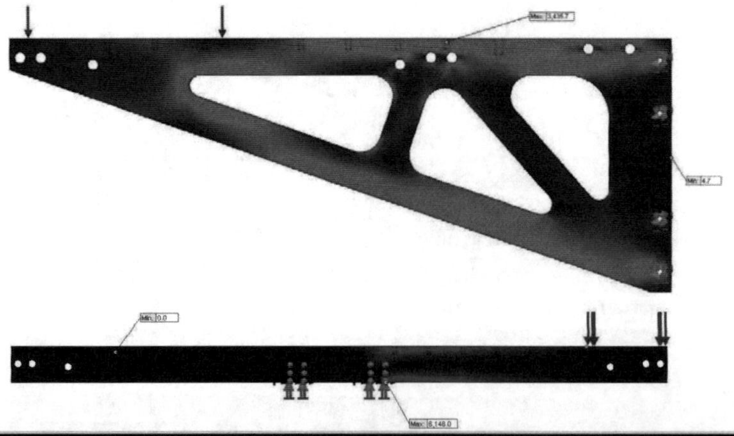

VALPARAISO UNIVERSITY

Test Stand

VALPARAISO UNIVERSITY

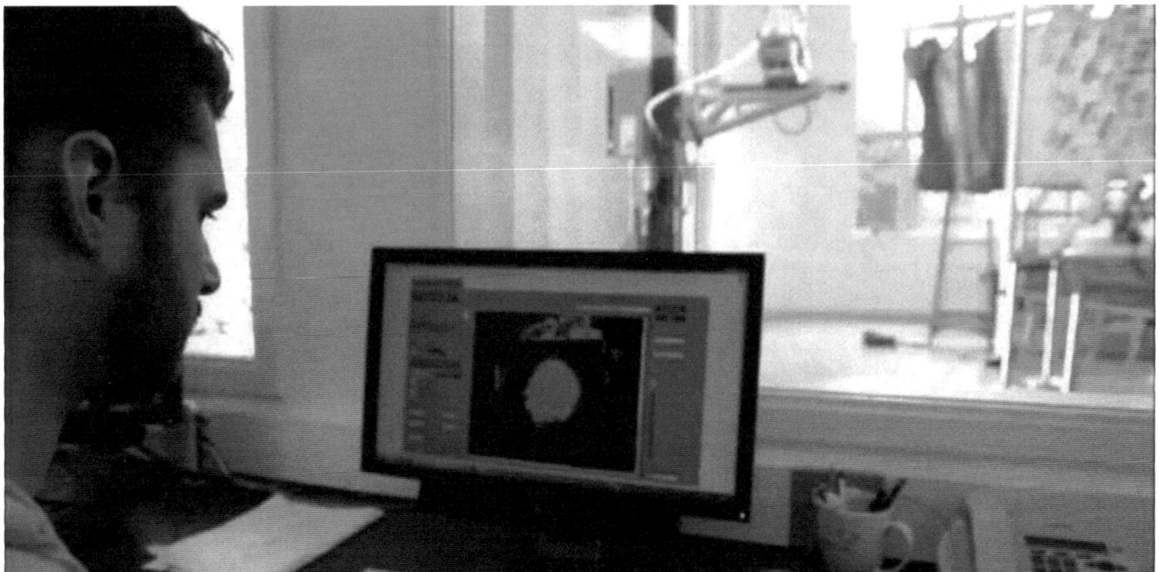

Solar Thermal Decoupled Electrolysis of Water

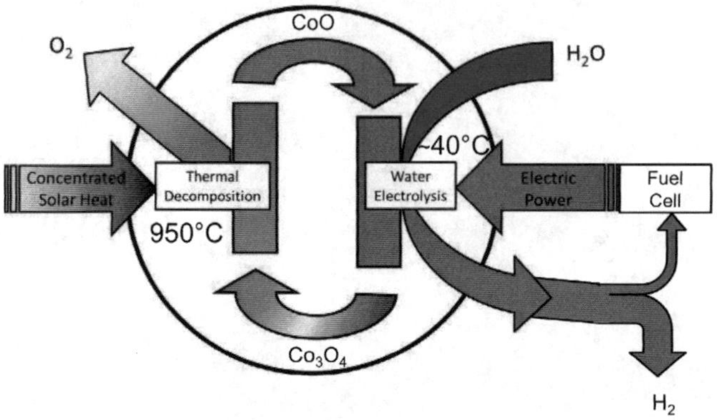

$$\eta_{cycle} = \eta_{abs} \frac{\Delta G_{H_2+0.5O_2=H_2O} - \Delta G_{Electrolysis}}{\Delta H_{Co_3O_4 \to 3CoO+0.5O_2}} = 38\%$$

Palumbo *et al.*, 2012, *Chem. Eng. Sci.*, **84**, 372–380.

VALPARAISO UNIVERSITY

Rotary Drum Reactor Concept

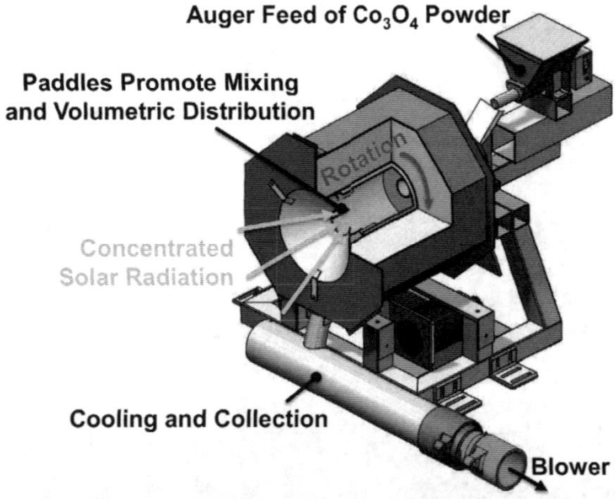

VALPARAISO UNIVERSITY

Rotary Drum Reactor Model

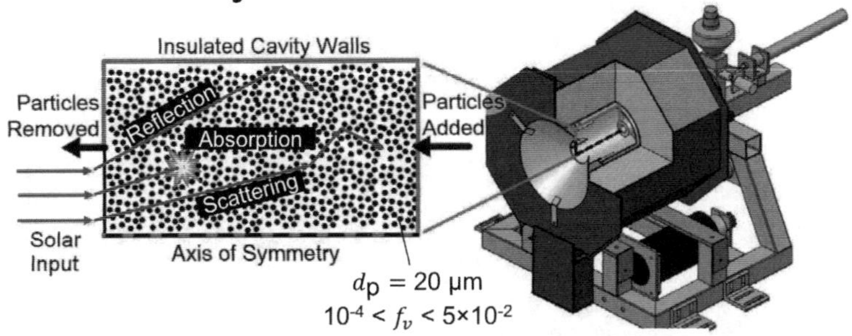

Conservation of Species (Reaction extent)

$$0 = -\nabla \cdot (\vec{G}''_{Co_3O_4}) - r'''$$

Particle motion

Rate of conversion of Co_3O_4 to CoO

Conservation of Energy (Temperature)

$$0 = -\nabla \cdot (-k_e \nabla T + \vec{G}''\hat{c}T) - q'''_{rad} - r'''\Delta H_r$$

Conduction | Particle motion | Volumetric Radiation (participating media) | Rate of energy consumption from reaction of Co_3O_4

Equations solved using the Finite Volume Method

VALPARAISO UNIVERSITY

Model Results: Impact of Feed Rate

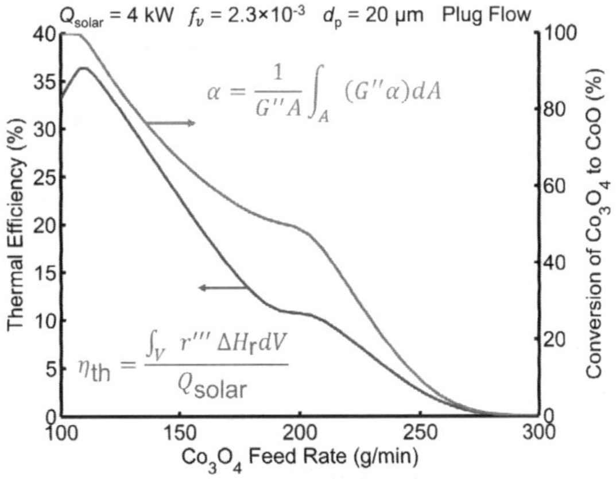

Maximum thermal efficiency achieved when the concentrated solar radiation is matched with the rate of the chemical reaction (α = 100%).

VALPARAISO UNIVERSITY

Reactor Testing

VALPARAISO UNIVERSITY

Reactor Operation

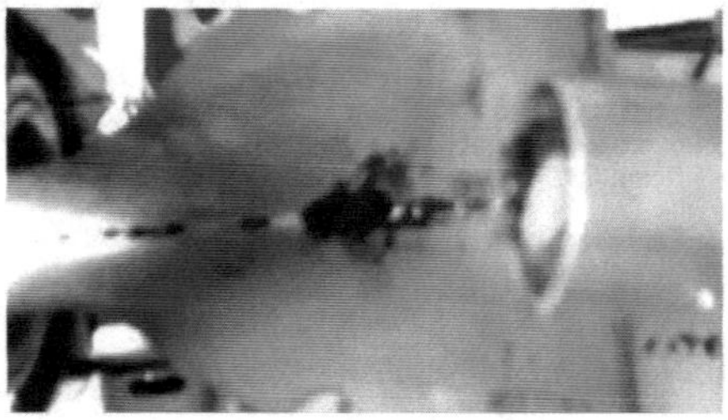

Reactor Inlet

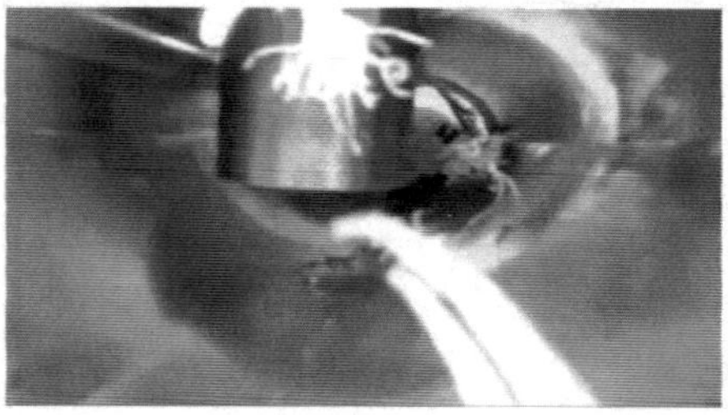

Reactor Outlet

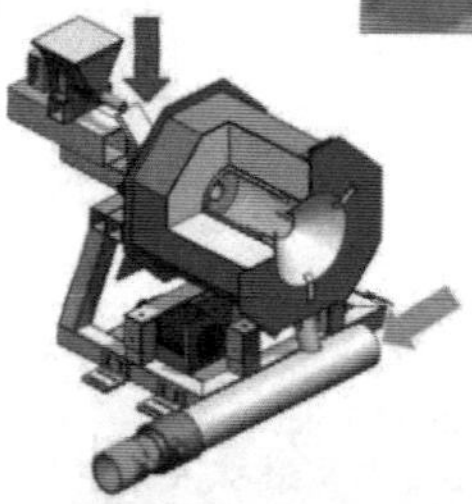

VALPARAISO UNIVERSITY

Reactor Test Results

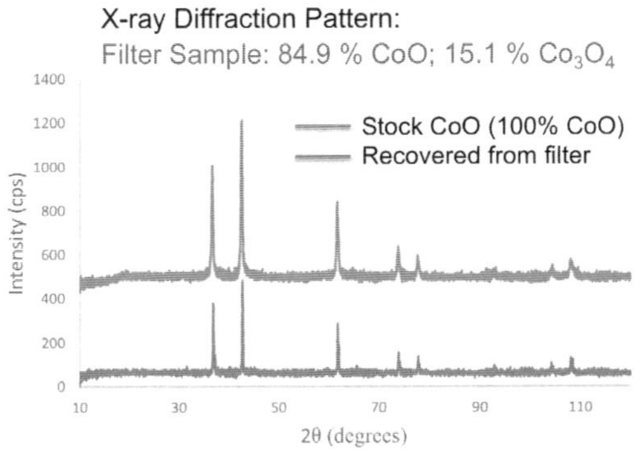

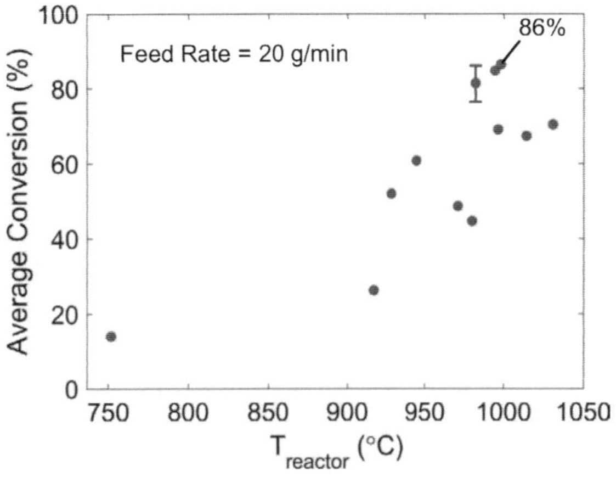

VALPARAISO UNIVERSITY

Project Roadmap ($2.3 Million Grant)

VALPARAISO UNIVERSITY

Project Roadmap ($2.3 Million Grant)

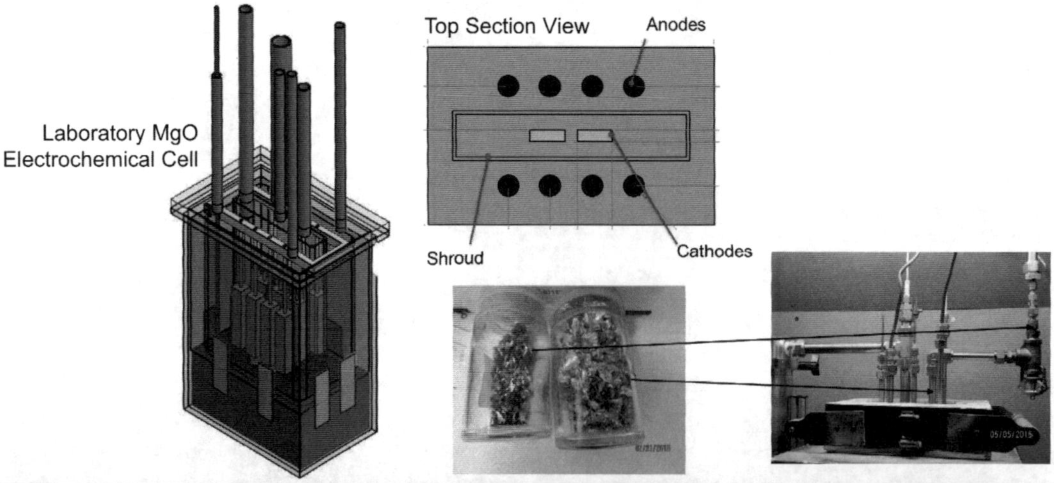

VALPARAISO UNIVERSITY

Claire Merrill

Fundamental Studies

Cathode Physics

Global Reaction: $Mg^{2+} + 2e^{-1} \leftrightarrow Mg\ (l)$

Reaction Kinetics: $j = j_o \left[1 \cdot \exp\left(\frac{(1-\alpha)F\eta}{RT}\right) - \frac{C_{Mg^{2+}}}{C^{\infty}_{Mg^{2+}}} \exp\left(-\frac{\alpha F\eta}{RT}\right)\right]$

Exchange Current Density (j_o); Transfer Coefficient (α)

Mass Transfer: $0 = \frac{\partial}{\partial t}\left(C_{Mg^{2+}}\right) - D_{Mg^{2+}} \nabla^2 C_{Mg^{2+}}$

Diffusion Coefficient ($D_{Mg^{2+}}$)

$\frac{-j}{nF} = D_{Mg^{2+}} \nabla C_{Mg^{2+}}$ — At the cathode-electrolyte interface (Zero flux at all other boundaries)

VALPARAISO UNIVERSITY

Results from 3D Finite Element Model

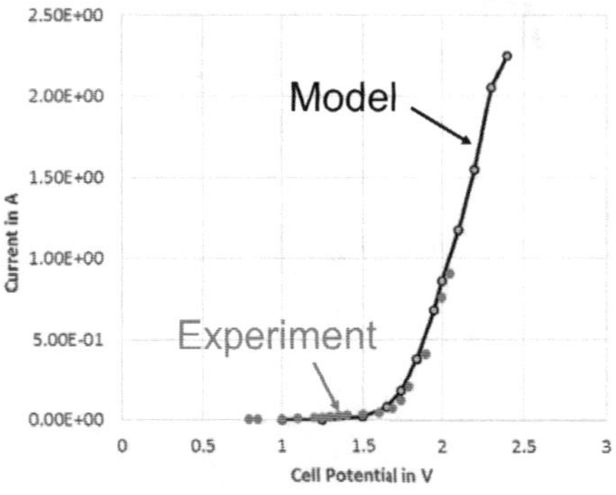

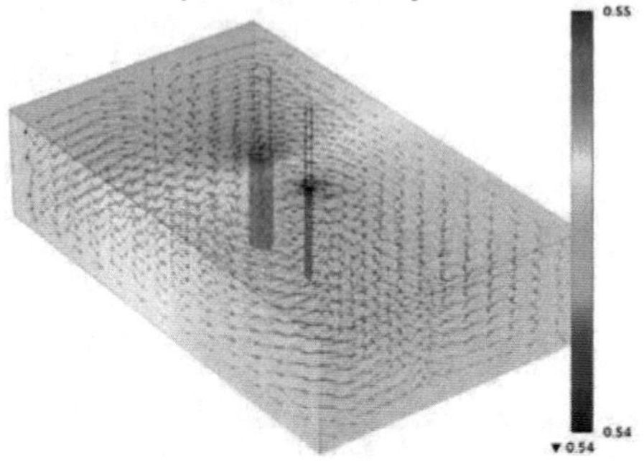

The cell is 13 cm x 7.5 cm x 2.54 cm

Anode = 4.0 cm² Cathode = 1.4 cm² Anode-Cathode Gap = 2 cm

VALPARAISO UNIVERSITY

Results from 3D Finite Element Model

Linear sweep voltammetry of Laboratory MgO Cell (20 A)

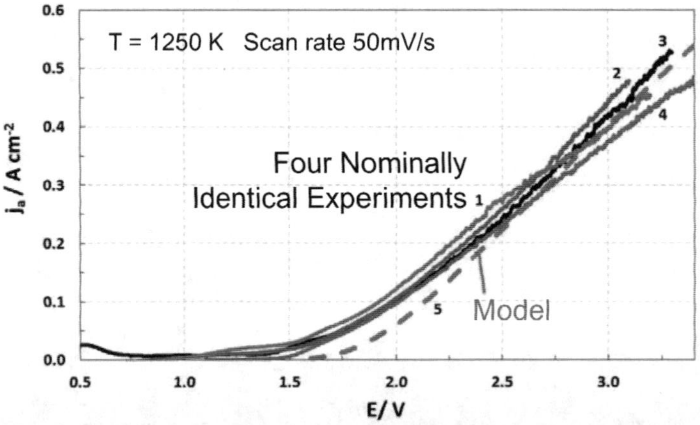

VALPARAISO UNIVERSITY

Towards Commercialization

VALPARAISO UNIVERSITY

Commercial Cell Concept

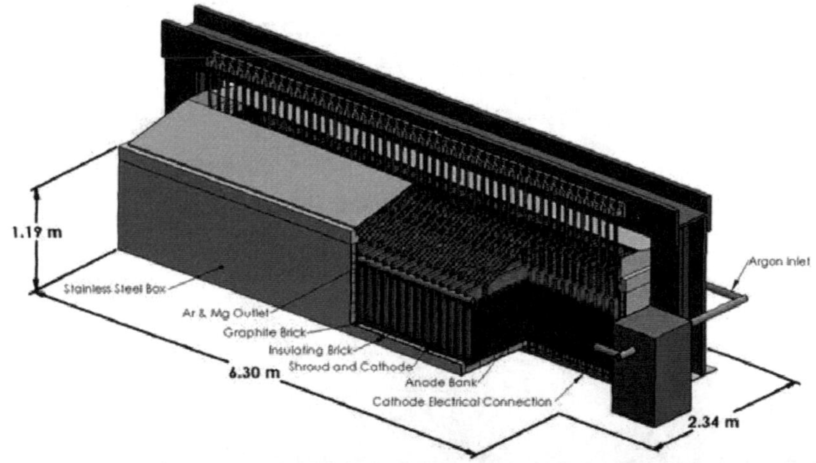

VALPARAISO UNIVERSITY

Finite Element Model of Commercial Cell Concept

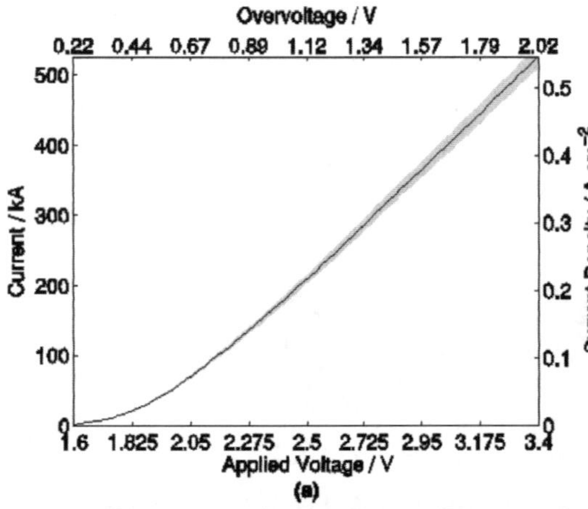

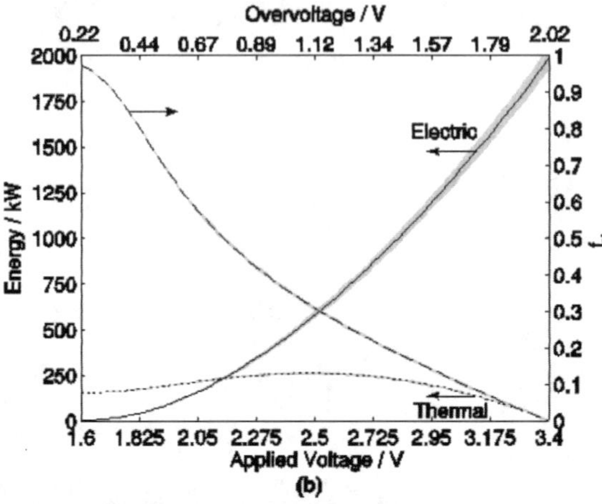

VALPARAISO UNIVERSITY

Visualizing the Solar-to-Commercial Cell Interface

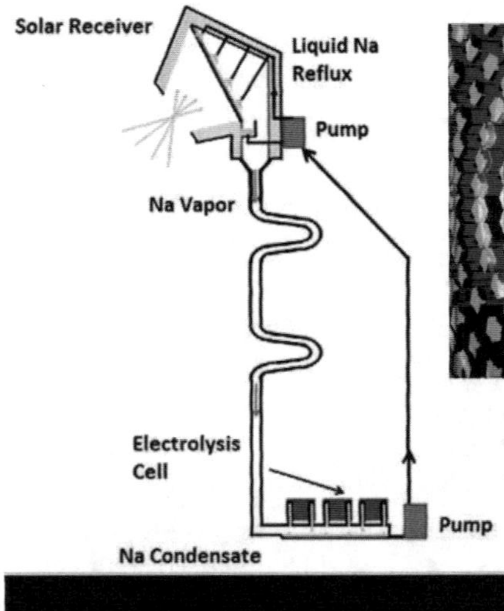

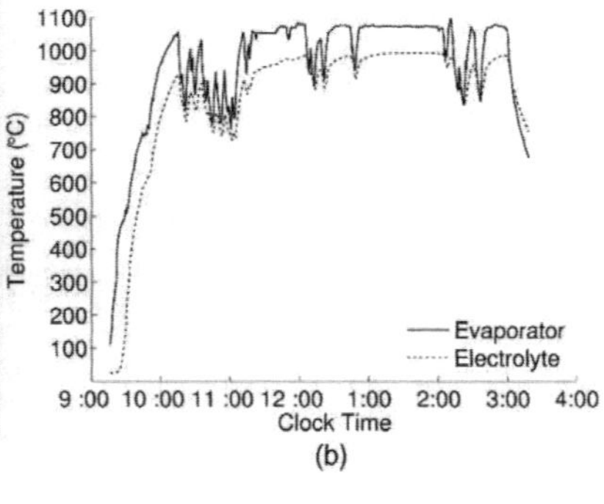

VALPARAISO UNIVERSITY

DR. NATHAN LEONARD
ELECTROCHEMICAL SCIENTIST AT GENERAL MOTORS

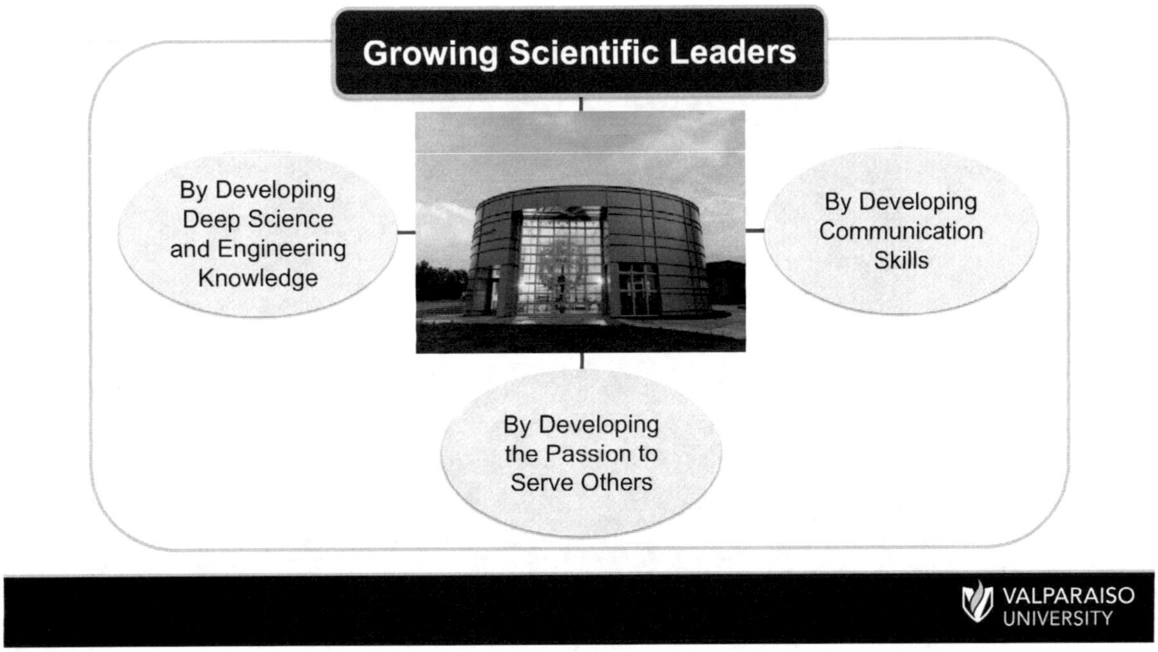
Growing Scientific Leaders
By Developing Deep Science and Engineering Knowledge
By Developing Communication Skills
By Developing the Passion to Serve Others
VALPARAISO UNIVERSITY

Undergraduate Student Co-authors

1) Leonard, N., Korenko, M., Larson, C., Blood, K., Venstrom, L.J., Nudehi, S., Duncan, S., Diver, R., Simko, F., Priscak, J. and Schoer, J., 2016, “The thermal electrolytic production of Mg from MgO: A discussion of the electrochemical reaction kinetics and requisite mass transport processes,” *Chem. Eng. Sci.*, **148**, pp.155-169.

2) Palumbo, R., Korenko, M., Larson, C., Venstrom, L. J., Duncan, S., Nudehi, S., Schoer, J., Toberman, J., Prusinski, W., Johnson, D., Robinson, B., Barkely, S., Warren, K., Diver, R., Šimko, F., Boča, M., 2015, “Thermal Electrolytic Production of Mg from MgO: Reflections on Commercial Viability”, *144th Annual Meeting & Exhibition of the Minerals, Metals & Materials Society TMS2015*, March 15-19, 2015, Orlando, FL.

3) Sheline, W., Matthews, L., Lindeke, N., Duncan, S., Palumbo, R., 2013, "An Exploratory Study of the Solar Thermal Electrolytic Production of Mg from MgO," *Energy*, **51**, pp. 163 – 170.

4) Palumbo, R., Diver, R.B., Larson, C., Coker, E.N., Miller, J.E., Guertin, J., Schoer, J., Meyer, M., Siegel, N.P., 2012, “Solar Thermal Decoupled Water Electrolysis Process I: Proof of concept,” *Chem. Eng. Sci.*, **84**, pp. 372-380.

5) Schroeder, R., Matthews, L., Leatzow, D.,Kondratko, J., Will, J., Duncan, S., Sheline, W., Lindeke, N., Neves, P., and Palumbo, R., 2011, "Solar Thermal Electrolytic Process for the Production of Zn from ZnO: The Electrolysis of ZnO from 1275-1500 K," *J. Solar Energy Eng.*, **133**(4)

6) Krenzke, P., Krueger, K., Leonard, N., Duncan, G., and Palumbo, R. D., 2010, “A Solar Thermal Electrolytic Reactor for Studying the Production of Metals from Their Oxides,” *J. Solar Energy Eng.*, **132**(3), pp. 34501-34505.

7) Venstrom, L., Krueger, K., Leonard, N., Tomlinson, B., Duncan, G., and Palumbo, R.D., 2009, “Solar Thermal Electrolytic Process for the Production of Zn from ZnO: An Ionic Conductivity Study,” *J. Solar Energy Eng.*, **131**, pp. 1-9.

8) Freid, A.P., Johnson, P.K., Musella, M., M ller, R., Steinbrenner, J.E., and Palumbo, R.D., 2005, “Solar blind pyrometer temperature measurements in high temperature solar thermal reactors: a method for correcting the system-sensor cavity reflection error,” *J. Sol. Energy Eng.*, **127**(1), pp.86-93.

VALPARAISO UNIVERSITY

Undergraduate Student Presentations

VALPARAISO UNIVERSITY

Student Outcomes: Graduate School

Graduate Schools Attended:

- Georgia Tech
- Penn State
- University of Florida
- University of Michigan
- University of Minnesota
- Harvard University
- University of Illinois at Urbana-Champaign
- Michigan State
- Stanford
- Vanderbilt
- Ohio State University
- University of Wisconsin
- Colorado State

64 percent enrolled in a graduate program (last 3 years)

4 NSF Graduate Research Fellowship winners and 4 honorable mentions (last 5 years)

VALPARAISO UNIVERSITY

A sustainability challenge: communicating science with the general public at an emotional level without compromising the science

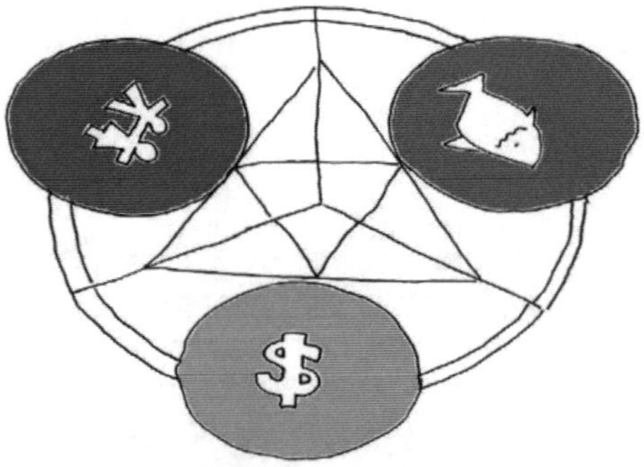

VALPARAISO UNIVERSITY

Thank you, Professor Edward Fletcher,
for helping us to Grow Future Scientific
Leaders with Concentrated Sunlight.
With Love, The VU Community
VALPARAISO
UNIVERSITY

495673 Metal Oxide Redox Materials for Solar Thermochemical Energy Storage

Abstract:

Thermochemical energy storage (TCES) systems are of great interest in concentrated solar power (CSP) applications. Storing sunlight as chemical energy during the day can enable power generation at night or during cloudy periods, effectively alleviating the inherent intermittency of solar sourced electricity. Metal oxides are among the most attractive TCES materials because they possess high-energy density and high reduction/oxidation temperatures (>800°C) suitable for driving high-efficiency thermodynamic power cycles. However, limitations involving high-temperature heat transfer, particle handling, sintering, and chemical performance of the metal oxide materials are still problematic in practical application.

Past works generally focus on evaluating pure metal oxide systems and investigate a select few mixed metal oxide systems in detail. In this work, we pre-selected over 100 high-potential mixed-metal oxide systems based on a comprehensive literature survey and then screened those materials for suitability in redox based high-temperature TCES using thermodynamic database software. The potential candidates were down-selected to the most promising contenders based on several selection criteria: high-temperature stability, large redox reaction enthalpy, and reaction temperatures of approximately 700-1100 Â°C for CSP applications. Ongoing and future work we will focus on experimental testing of the identified compositions to identify actual redox behaviours and to evaluate other thermochemical criteriaâ€”appropriate cycling physical stability, maintained chemical activity over many redox cycles, and acceptably fast reaction kinetics.

502849 SolPeD â€“ solar per demand for power generation and fuel production

Abstract:

We are challenged to develop an energy conversion method, which could gradually replace fossil fuel combustion on a worldwide scale, eventually leading to a renewabledominant global energy supply. Whereas other renewable energy resources can have important supporting roles, the only energy sources available on earth in large enough quantities to replace fossil fuels are nuclear and solar.
The talk will describe a solar energy conversion process based on several complementing novel technologies, and aimed to enable large-scale utilization of solar energy for yearround, 24/7 power generation and clean fuel production at high efficiency and affordable costs.
The major technological challenges and proposed solutions will be discussed, and ongoing effort of commercializing the technology will be reported.

498174 On the Role of Thermochemical Energy Storage in Concentrating Solar Power

Abstract:

Concentrating solar power (CSP) with thermal energy storage has the potential for grid-scale dispatchable power generation. Thermochemical energy storage (TCES), that is, the reversible conversion of solar-thermal energy to chemical energy, has high energy density and low heat loss over long periods. Previously, several reaction types have been studied as candidates, including dissociation of ammonia, hydrides, carbonates, oxides, hydroxides and methane reforming [1]. Although considerable work has been done, past research has mostly focused on experimental studies and reactor design. The limited system-level studies that are available in the literature are for specific reactions [2] and, furthermore, consider TCES in isolation, neglecting its interactions with the solar field and power generation system [3].

Accordingly, in this work, we establish a framework to systematically analyze and compare candidate reactions for TCES systems [4]. First, we develop a general process for CSP plants employing TCES systems. A general TCES system includes five tasks: reaction, storage, separation, material handling and heat recovery. We then formulate an optimization-based model for the process, which optimizes the plant design towards the minimum levelized cost of electricity (LCOE). We use this model to study ammonia and methane systems.

We identify that, for fluid phase TCES systems, the main cost driver is gas storage using pressure vessels, accounting for 77% of the TCES cost. This is because, even if reaction products are compressed and stored at 550 bar, both ammonia and methane systems require 65,000 m^3 of gas storage at a vessel price of 8900 $\$/m^3$. The main energy driver is the electricity consumption for storage compression, which consumes 18% of the generated power. Adopting underground gas storage, where products are stored at lower pressure in much cheaper underground caverns, could significantly reduce the compression power consumption (from 18% to 3%) and improve the overall energy efficiency and thus cost. The methane system combined with underground storage is identified as the most cost-efficient alternative, achieving a 13% LCOE reduction over the current two-tank molten salt CSP plants.

Finally, we discuss the impact of the reaction temperature and operating pressure. The endothermic and exothermic reaction temperature determines not only the reaction conversion, but also the receiver and turbine efficiency. We find that both reaction temperature should be in the vicinity of the reaction turning temperature (at which the equilibrium constant K=1) and separation can increase the reaction conversion, allowing lower endothermic reaction temperature. The operating pressure, including process pressure, highest and lowest storage pressure, influences both the storage volume and compression work. For pressure vessel cases, the storage tank pressure varies in a wide range (150-550 bar), reducing the storage volume at the expense of more compression power consumption. When cheap underground storage is employed, a narrow storage pressure range (135-165 bar) leads to smaller electricity consumption and larger storage volume.

The proposed framework can be used to study a wide range of TCES systems, including solid-gas phase reactions, which will be the topic of future research.

References:

[1] P. Pardo, A. Deydier, Z. Anxionnaz-Minvielle, S. Rougé, M. Cabassud and P. Cognet, *Renew. Sustain. Energy Rev.*, 2014, **32**, 591–610.

[2] A. Luzzi, K. Lovegrove, E. Filippi, H. Fricker, M. Schmitz-goeb, M. Chandapillai and S. Kaneff, *Sol. Energy*, 1999, **66**, 91–101.

[3] C. Corgnale, B. Hardy, T. Motyka, R. Zidan, J. Teprovich and B. Peters, *Renew. Sustain. Energy Rev.*, 2014, 38, 821–833.

[4] X. Peng, T. W. Root and C.T. Maravelias, *Green. Chem*, 2017, DOI: 10.1039/c7gc00023e.

500484 Combined Thermochemical and Latent Heat Energy Storage for Low Temperature Residential Applications

Abstract:

Solar thermal energy has the potential to supply clean energy, but for space heating and cooking applications, peak usage times often are misaligned with peak solar hours. Solar thermochemical energy storage systems are one promising solution. Salt-hydrate reactions are well-suited to this purpose as they have low-decomposition temperatures and high energy densities. To enhance heat transfer, improve durability, and prevent conglomeration, a solid matrix material is used, typically made of porous substances such as activated carbon, vermiculite, or expanded graphite. Here, a matrix of expanded graphite reinforced with carbonized flour was used. Certain salt-hydrates have the potential to store both thermochemical and latent heat, and one such salt was investigated through TGA/DSC analysis. A functional eutectic was found at around 130°C and freezing/melting cycles were visible as well-defined peaks. The salt-matrix complex was also investigated through both TGA/DSC and realistic application testing. The success of this new chemistry sets the stage for a new dual-release approach to energy storage.

496055 Optical Losses from Bubble Evolution in Photo-Electrochemical Reactors

Abstract:

Solar fuels, such as the production of hydrogen by photo-electrolysis of water, have the potential to revolutionise the long term future of energy production as they can overcome the issue of intermittency. The potential of this technology can only be realised by scaling up current research and lab scale demonstrations and henceforth, design of photo-electrochemical (PEC) reactors will become increasingly important.

Successful PEC reactor design will minimise resistive losses in solution, ensure the separation of the photo-electrolysis products and permit maximal light absorption by the photo-electrode(s). Methodologies for simultaneously achieving these requirements have not been exhaustively studied and in particular a comprehensive quantification of optical losses in PEC is missing from the literature.

In this work we consider the optics of PEC reactor design and its ramifications, focusing on the optical losses of bubbles on the surface of the photo-electrode. Bubble evolution is a complex process of bubble nucleation, detachment and rise and not only detrimentally reduces the effective surface exposed to electrolyte but absorbs, reflects and scatters incoming photons.

We experimentally demonstrate this effect and quantify the optical losses due to bubble evolution using a transparent conductive oxide such as fluorine doped tin oxide (FTO) glass as an anode. The optical losses are correlated against current density (and hence gas evolution rate) and the ramifications of this discussed. From this relationship, a modified maximum Solar-To-Hydrogen efficiency for different semiconductors (Shockley-Queisser limit) is derived.

Opportunities for the elimination of optical losses from bubble evolution through bubble suppression are discussed where pressurisation and forced convection approaches are investigated. A multi-physics model is presented that demonstrates the operational requirements in order to a achieve bubble suppression in a novel membrane-less flow reactor design, which has been outlined in previous work. Experimental verification of bubble suppression through hydrodynamic control is given and compared to the theoretical results and the potential scalability of this technology explored.

500252 Acrylonitrile Production from Biomass-Derived Intermediates

Abstract:

Carbon fibers are key materials in critical energy applications including light-weighting of vehicles for improved fuel efficiency and enabling larger wind turbine designs for increased power generation. Carbon fibers are produced through heat treatment of polyacrylonitrile fibers. Today, the acrylonitrile monomer is made primarily via ammoxidation of propylene with ammonia. However, price fluctuations in the feedstocks and the energy-intensive processes use to make carbon fibers have resulted in low-volume and niche adoption of carbon fibers into automotive applications, motivating the drive towards carbon fibers produced cheaply and reliably from renewable biomass sources. This talk will describe efforts in the Renewable Carbon Fiber Consortium to pursue acrylonitrile production from biomass-derived sugars via hybrid biological and chemo-catalytic processes. In particular, bench-scale efforts in two separate process schemes will be described that proceed via biological production of 3-hydroxypropionic acid or lactic acid, followed by subsequent separation and catalytic conversion to acrylonitrile.

501349 Multistream Integrated Biorefinery (MIBR) for Carbon-, Road-, and Bio-Materials

Abstract:

The success of a modern biorefinery heavily depends on the availability of diverse product streams and utilization of all fractions of input material. Traditional lignocellulosic biorefining focuses on ethanol, which has limited by-product streams that add value to the biorefinery. Essentially all current bioconversion platforms lead to a lignin-containing waste stream that needs further processing into valuable products. Although a certain amount of lignin (~30â€“40%) is needed for the thermal requirements of biofuel production, a modern cellulosic processing plant will have ~60% excess lignin. The utilization of the lignin-containing biorefinery waste as feedstock for renewable products offers a unique opportunity to achieve a multi-stream integrated biorefinery (MIBR), where the lignin-containing biorefinery waste will be utilized for various carbon-, road- and bio- materials to maximize economic return and sustainability. Specifically, we have advanced the fractionation, conversion, and processing technologies to enable different bioproduct streams. An efficient enzyme-mediator system was developed to fractionate lignin into low- and high-molecular weight fractions with more homogenous chemical characteristics. We have demonstrated that these lignin fractions are more amenable to different applications in bioconversion, asphalt binder modifier, and carbon fiber. The low molecular weight fraction is more amenable to bioconversion into PHA for bioplastics. In addition, the low molecular fraction can be used as unique asphalt binder modifiers to improve both high and low temperature performance of road pavement. The high molecular weight fraction can be used to fabricate carbon fiber with significantly improved mechanical performance. Together, MIBR will reduce ethanol production cost through the creation of high value bioproducts, produce asphalt binder modifier with unique features, develop an innovative, new pathway to quality carbon fiber; and create a means to utilize all carbon in the feedstocks.

492807 Improving Economics of Cellulosic Biofuels: An Integrated Strategy for Co-Producing 1,5-Pentanediol and Bioethanol

Abstract:

Lignocellulosic biomass is an abundant and sustainable feedstock that can be used to produce biofuels and biobased commodity chemicals. However, the low oil price and the challenges in applying biorefinery technologies at scale hamper the commercial production of biofuels. In this presentation we will discuss a strategy to integrate the production of a high-value coproduct, 1,5-pentanediol (1,5-PDO), along with a typical bioethanol production process.

1,5-PDO and two other conventional α,ω-diols (e.g. 1,6-hexandiol and 1,4-butanediol), are building blocks for polyurethane and polyester plastics. These α,ω-diols are currently derived from petroleum and, as a group, represent an annual market over $6 billion. An economically viable approach of producing 1,5-PDO from biomass-derived furfural has been developed recently (Brentzel et al., 2017; Huang et al., 2017), which offers the potential to coproduce 1,5-PDO as a high value co-product of bioethanol manufacturing process to improve cellulosic biorefinery economics.

In this work, we propose a novel strategy that integrates lignocellulosic biomass fractionation with simultaneous conversion of cellulose and hemicellulose into ethanol and 1,5-PDO, respectively. First, we fractionate biomass into its three primary constituents— hemicellulose, cellulose, and lignin—using γ-valerolactone (GVL) as a solvent (Alonso et al., 2017), which also enables excellent yields of furfural by dehydration of xylose. The purified furfural is then used to produce 1,5-PDO (Huang et al., 2017), and the cellulose portion is converted to ethanol via enzymatic hydrolysis and fermentation (Humbird et al., 2011), while the lignin is combusted to generate steam and electricity. We have developed an experimentally based process simulation model (mass and energy balances) to determine the economic potential of the integrated strategy. If we consider a plant processing to 1,000 dry metric tonne per day white birch feedstock and producing 38,520 ton per year 1,5-PDO and 19.3 MMgal per year ethanol, the integrated strategy leads to a minimum ethanol selling price of $1.52 per gallon when 1,5-PDO is sold at $3,100 per ton (Figure 1). It is shown that this strategy could lower the cost of ethanol produced from biomass by more than two dollars per gallon. Finally, we perform sensitivity analysis to identify the major economic drivers of the process and suggest future research directions.

Figure 1. Minimum ethanol selling price as a function of 1,5-PDO selling price. The "green" region (1,5-PDO selling price > $2,775 per ton) indicates that the integrated strategy has favorable economics for ethanol production compared with the biochemical conversion process used in NREL 2011 report (Humbird et al., 2011).

References

1. Alonso, D. M.; Hakim, S.; Zhou, S.; Won, W.; Hosseinaei, O.; Tao J.; Garcia-Negron, V.; Motagamwala, A. H.; Mellmer, M. A., Huang, K.; Houtman C. J.; Labbé, N.; Harper, D. P.; Maravelias, C. T.; Runge T.; Dumesic, J. A. Increasing the revenue from lignocellulosic biomass: Maximizing feedstock utilization. *Science Advances* 2017.

2. Brentzel, Z. J.; Barnett, K. J.; Huang, K.; Maravelias, C. T.; George, W.; Dumesic, J. A.; Huber, G. W. Chemicals from Biomass: Combining Ring-opening Tautomerization and Hydrogenation Reactions to Produce 1,5-Pentanediol from Furfural. *ChemSusChem* 2017, 133, 12689.

3. Huang, K.; Brentzel, Z. J.; Barnett, K. J.; George, W.; Dumesic, J. A.; Huber, G. W.; Maravelias, C. T. Conversion of Furfural to 1,5-Pentanediol: Process Synthesis and Analysis. *ACS Sustainable Chemistry & Engineering* 2017.

4. Humbird, D.; Davis R.; Tao L.; Kinchin C.; Hsu D.; Aden A. Process Design and Economics for Biochemical Conversion of Lignocellulosic Biomass to Ethanol: Dilute-Acid Pretreatment and Enzymatic Hydrolysis of Corn Stover. *NREL/TP-5100-47764*. Golden, CO: National Renewable Energy Laboratory, 2011.

498859 High Value Products from a Photobioreactor-Based Biorefinery

Abstract:

Though Algenol continues research and development activities aimed at biofuels (ethanol and biocrude), increasing emphasis is being placed on high value products and co-products that can both provide sustainable income and serve as a market entry/risk reduction strategy for eventual development of biofuels. Algenol has successfully built and operated a large pilot scale (2 acre) integrated biorefinery (IBR) based on its proprietary photobioreactors (PBRs) and genetically modified cyanobacteria to produce fuel grade ethanol via an EPA approved pathway (70% carbon footprint reduction compared to gasoline). Ethanol productivities in the range of 4000-5000 gal/acre-yr (annualized) have been achieved. However, the system is not economic under current market conditions. The IBR and R&D support infrastructures are now being applied to higher value products, including a biocrude co-product. Though several products are under active consideration, algae based soil amendment is one application that is well advanced in our product development pipline. The soil amendment contains nitrogen-fixing cyanobacteria as the active component (based on *Anabaena* and/or *Nostoc* strains). The formulated product provides enhanced yields as a biofertilizer, can be used as a soil amendment for reclamation of degraded soil, and can result in high carbon storage levels in the soil.[1,2] This talk will use the soil amendment application as an example and provide a description of the selection process for product development (e.g., techno-economics, life cycle impacts, compatibility with PBR growth systems) and the development process from the lab, through outdoor testing, and then to pilot scale deployment. In particular, the alignment of indoor and outdoor performance will be demonstrated for several strains relevant to the above application, along with a modeling system that has been developed for productivity prediction for any rational position on the earth's surface, and for various configurations for Algenol's vertically-oriented PBR arrays. Harvesting and downstream processing will be described and experimentally demonstrated for several cyanobacterial strains, including strains designed for biocrude production. A comparison of the relative productivities and economics of PBR systems versus open pond systems will be discussed.

[1]Accelergy Corporation, *"Montana Integrated Carbon to Liquids Study,"* US DOE Award Number: DE-FE0003595, December 2013.

[2]US Patent2014/0345341, November 27, 2014; *"Integrated Process for Producing Fuels and BioFertilizers from Biomass and Products Produced."*

499440 A Highly Selective Dehydration of D-Xylose and Wheat Straw C_5-Sugars into Furfural By Using Supercritical CO_2 as Catalyst: A Green and Efficient Approach

Abstract:

Crude oil is the principal feedstock for production of liquid fuels and also affords a platform for synthesis of a wide range of added-value chemicals and materials. Due to price fluctuation of crude oil-derived products coupled to increasing concerns about the impact of fossil emissions on environment, society is becoming more environmentally conscious. Thus, traditional chemical technologies for production of fuels and chemicals from biomass are being replaced with innovations. This includes the use of novel and green solvents. In this work, the selective dehydration of both D-xylose and lignocellulose-derived C_5-sugars into furfural, using supercritical CO_2 as catalyst and tetrahydrofuran (THF) as extracting solvent, is for the first time reported. The scrutinised approach involves the benefits from the use of auto-generated carbonic acid from dissolution of CO_2 in aqueous phase, assisted by THF, guiding to the improvement of furfural yield and reaction selectivity. This novel approach allowed to achieve a xylose conversion into furfural above 83 %mol, furfural yield of 70 %mol and selectivity up to 84 %mol at 180 ºC for 60 min with 50 bar of initial CO_2 pressure.[1] At aforementioned operational condition, furfural yields and selectivities up to 40 %mol and 44 %mol with respect to wheat straw hemicellulose hydrolysate were reached.[2] First, the influence of various operating conditions on pentose conversion and furfural yields were investigated. Then, various volumetric ratios of water and THF (v/v) were investigated to help define reaction conditions that favour the highest yield of furfural. In an effort to understand how addition of CO_2 improves furfural yield and its impact by the presence of THF, different pressures of CO_2were also investigated.

Acknowledgments

IF/00471/2015 grants and by BBRI - Biomass and Bioenergy Research Infrastructure (ROTEIRO/0189/2013) and the Associated Laboratory for Sustainable Chemistry–Clean Processes and Technologies–LAQV which is financed by national funds from FCT/MEC (UID/QUI/50006/2013) and co-financed by the ERDF under the PT2020 Partnership Agreement (POCI-01-0145-FEDER-007265).

References

1. A. R. C. Morais and R. Bogel-Lukasik, *Green Chem.*, 2016, **18**, 2331-2334.

2. A. R. C. Morais, M. D. J. Matuchaki, J. Andreaus and R. Bogel-Lukasik, *Green Chem.*, 2016, **18**, 2985-2994.

493990 Process Design for the Synthesis of Biodiesel, Glycerol, 3-Hydroxypropionic Acid and 1,3-Propanediol

Abstract:

The current tendency towards the use of renewable biomass feedstocks instead of the use of non-renewable resources (for example, petroleum), has created the need for development of new, innovative production processes within the chemical and biochemical industries. Therefore, alternative solutions able to reduce the consumption of non-renewable resources and consequently minimize effects on climate change, waste, while, improving product yields, should be promoted. This replacement of petroleum with biomass as the primary material for fuel and chemical production is basically the idea behind the biorefinery concept.

This project is focusing on the ways that glycerol (an intermediate or by-product in a biorefinery) could be converted into value added platform chemicals so that its surplus production is not ending up as waste. The effective utilization of glycerol will ultimately contribute to the viability of biodiesel production since it will significantly lower its already high production cost. Thus, the design and development of a sustainable process to utilize this organic raw material becomes imperative.

The case of the simultaneous production of two platform chemicals with glycerol as a starting material has been investigated, namely of 3-hydroxypropionic acid (3HP) and 1,3-propanediol (1,3PDO). This process is part of a bigger one, where, by using vegetable oil as feedstock, the aim is to produce biodiesel. However, as glycerol is obtained as a by-product, it is then subjected to biotransformation with recombinant resting cells of *Lactobacillus reuteri* in an aqueous solution. The resting cells initially grow in the fermenter at 37°C and then the biotransformation of glycerol is performed in a stirred-tank reactor. This biotransformation has 3 main products, 3-hydroxypropionic acid (3HP) and 1,3-propanediol (1,3PDO), that are of interest as well as significant amounts of 3-hydroxypropionaldehyde (3HPA) as by-product (that is not of interest and needs to be reduced).

L.reuteri is selected for biotransformation for several reasons, such as, it is a safe bacteria, has a great potential as a biocatalyst, and it cannot be used as a carbon source by glycerol, unlike other bacteria. Last but not the least, apart from 3HPA, no other by-products are formed during the bioconversion of glycerol by *L.reuteri*, thereby making the downstream process much simplified.

The goal of the project is to investigate the optimal production process towards 3HP and 1,3PDO simultaneously, in comparison to the existing methods that are oriented to the production of either only 3HP [Ramakrishnan Gopi et al.,2015] or only 1,3PDO [Nakamura and Whited, 2014]. The research is orientated towards a setup of the production process that produces equal amounts of the desired products, is economically and environmentally feasible and sustainable. In addition, the existence of significant amounts of the aldehyde by-product (3HPA) needs to be avoided. Two scenarios are investigated; one where 3HPA is isolated, purified and sold in the market as a by-product, and the other, where it is recycled back to the biotransformation reactor such that the production of 3HP and 1,3PDO are favoured.

According to the results regarding the production of 3HP and 1,3PDO, it is possible to determine the quantity of glycerol that needs to be allocated for the production 3HP and 1,3PDO and the remaining that will be allocated to the market. That is, the operation will depend on the function of the demand for glycerol, its current price, as well as the demands and prices of 3HP and 1,3PDO.

The presentation will provide the modelling and design details of the integrated process with a special focus on the increase of the atom efficiency of the processes involved, on the sustainability of the production process, the environmental impact versus the production and capital costs for different scenarios of product demand and price.

504355 Science with Purpose

Abstract:

As a company serving a multitude of industries in nearly every country on the planet, we recognize the opportunity we have: to grow our business while solving some of the worldâ€™s biggest problems. This requires collaboration with our customers, partners and stakeholders. Sustaining our business, and our planet, means protecting natural resources and empowering individuals and communities throughout the world to encourage progress. We use science to develop products that address challenges facing our customers and our global community, which include enhancing human health, promoting access to clean water, increasing energy efficiency and enabling safe work environments. We will share 3M examples and stories about who we are, how we work and what we create, and show how we use science with purpose every day.

504354 Some Thoughts on Sustainability: Food, Energy, and Water

Abstract:

The World is steadily moving into a major challenge. The challenge is how to provide for the human population within the limits of one planet, the Earth, over the long term. The effort is known as sustainability. To explore this challenge, I will focus on three closely linked issues of providing food, energy, and water. In addition, there is the issue of maintaining order and stability as necessary conditions for any kind of acceptable human existence. These issues, unfortunately, present critical but not well-bounded problems, and for that reason there are many more questions than answers. However, to explore the issues and their bounds, I will discuss what is known including some planetary limits and limits imposed by the known laws of Nature. I will lastly discuss where and how engineers can help in the effort to provide food, energy, and water for humanity. This will include some concrete examples where engineering design has helped to mitigate these challenges. The lecture will conclude with some thoughts on which approaches might stand the best chance of adoption and implementation under current and foreseeable future conditions.

Disclaimer: The views expressed in this presentation are those of the author and do not necessarily represent the views or policies of the U.S. Environmental Protection Agency.

504356 Industrial Sustainability Enhancement: A Control Perspective

Abstract:

Depletion of nonrenewable resources, global warming, air, water, and soil pollution, excessive population growth have challenged societies globally. This has demanded industries to seriously review engineering practice and find effective ways to ensure sustainable development. Over the past decades, a variety of sustainability assessment and enhancement methods and technologies have been developed for improving material and energy efficiency, environmental cleanness, human health, product life-cycle performance, social responsibility, eco-efficiency, etc. However, problem-solving solutions are mostly derived based on experience, which make them feeble in tackling complex problems involving multi-state, long-term strategic development under uncertainty. Fundamentally, industrial sustainability enhancement is a system science problem; it is featured by solutions for a series of industrial system state transitions toward sustainability. As such transitions occur in a multi-dimensional sustainability space, its dynamics and trajectory are difficult to predict, design, and control, especially when accessible data and available information are incomplete, imprecise, and uncertain. These challenges present a new opportunity for the study on system-transition-focused sustainability dynamics, assessment, and control. In this talk, we will discuss a number of fundamental aspects of sustainability control, including the controllability of system sustainability and transition. We will then introduce hierarchical control schemes for achieving sustainability goals under uncertainty. A few case studies will be presented and research directions will be explored. This talk will also discuss approaches for integration of sustainability elements and basic methods in undergraduate education, especially in the capstone process design and process control courses.

504357 Multi-Scale Life Cycle Optimization for Sustainable Engineering

Abstract:

Public interest in carbon and water footprints, ecosystem degradation and resource depletion, are just some of the environmental issues that drive the process industries to exercise greater responsibility for the environment in the design and operations of their manufacturing systems. Systems analysis and computational optimization methods from a life cycle perspective play an important role in the quest for sustainability. This talk will review some recent advances in the area of life cycle optimization for sustainability analytics. The focuses are on transformative modeling, theoretical and computational results that bridge the knowledge gaps between state-of-the-art life cycle assessment methods and emerging life cycle optimization tools. We will specifically discuss issues related to multi-scale system boundary selection, consequential perspective of the analysis, and potential presence of multiple non-cooperative decision makers in life cycle optimization for sustainable engineering problems. Applications of these general computational frameworks will be illustrated by case studies on bioenergy/biofuels, shale gas, and chemical manufacturing systems. We will conclude with a brief review of multi-scale applications of life cycle optimization methods from materials to the global scales.

498238 Selective Oxidation of Methane to Methanol in Zeolites: A Window of Opportunity

Abstract:

Despite decades of active research, selective oxidation of methane to methanol remains an unsolved 'grand challenge' of heterogeneous catalysts.[1] Although, previous experimental and computational[2-4] works have studied the mechanism in specific cation-exchanged zeolites; a universal, molecular level understanding of the methane activation and methanol extraction process is not yet available.

In this work, we use periodic DFT calculations to develop a unified framework that describes methane activation reaction and subsequently, enables us to identify strategies for improving the performance of cation-exchanged zeolites. Specifically, we explore a range of active site motifs, transition metal cations, zeolite topologies as well as binding sites for this reaction. Using the activation energy and formation energy of the active site as metrics, we obtain a 'window of opportunity' where we predict active site motifs to be both stable and sufficiently reactive to selectively oxidize methane. Encouragingly, a number of the known active sites (i.e. $[Cu\text{-}OH]^{+}$, $[Cu\text{-}O\text{-}Cu]^{2+}$, $[Cu_3O_3]^{2+}$, copper-oxide clusters and $[Fe\text{-}O]^{+2}$) lie within this region. More importantly, we also identify $[M\text{-}O]^{+}$ (M = Ni, Co, Fe), $[M\text{-}OH]^{+2}$ (M = Ni, Co, Fe), and $[bis(\text{-}O)Ni]^{2+}$as alternative target motifs for future catalyst design and synthesis. Finally, the origin of the high methanol selectivity can be explained by the (a) strong binding energy of methanol and (b) formation of surface methoxy species.

The atomistic and mechanistic insight presented in this work will guide future experimental syntheses, and will lead to the design and discovery of highly active cation-exchanged zeolite catalysts for the direct partial oxidation of methane to methanol.

References

1. Olivos-Suarez, A.I.; Szécsényi, À.; Hensen, E.J.M.; Ruiz-Martinez, J.; Pidko, E.A.; Gascon, J., *ACS Catal.* **2016**, 6 (5)

2. Zhao, Z.-J.; Kulkarni, A. R.; Vilella, L.; Nørskov, J. K.; Studt, F., ACS Catal., **2016**, 6 (6)

3. Kulkarni, A.R.; Zhao, Z.-J.; Siahrostami, S.; Norskov, J.K.; Studt, F., *ACS Catal.* **2016**, 6 (10)

4. Latimer, A.A.*; Kulkarni, A.R.*; Yoo, J.S.; Aljama, H.; Tsai, C.; Abild-Pedersen, F.; Studt, F.; Nørskov, J.K., Nat. Mater. **2016** (DOI: 10.1038/nmat4760)

497859 Computational Screening and Characterization of Single-Site Active Porous Surface for Natural Gas to Liquid Feedstock Conversion Process

Abstract:

Catalysts are needed in applications regarding the worldâ€™s fuel and chemical production. To assist these processes, catalysts are used to simplify the chemical reaction by lowering the energy required for reactions to complete. The goal of our project is to identify and characterize specific catalysts through computational research and more specifically the conversion from methane to methanol. Previous work shows that the hydrogen affinity of a structure, the energy needed to move electrons in a reaction, can be directly linked to the reactivity of a catalyst for conversions involving hydrocarbons. By recognizing the correlation between hydrogen affinity, C-H bond activation and transition state activation energy, new optimal catalysts based on a single-site active porous surface can be identified for the target conversion. Also, this work will support the idea of not only decreasing the computational costs for future simulation but also reducing the future experimental efforts by providing the necessary inputs to experiments.

501038 Selective Oxidation of Methane to Methanol or Acetic Acid on Rhodium Single-Site Catalysts at Mild Conditions

Abstract:

The selective activation of the first C-H bond of methane on a metal site at low temperatures and its functionalization to produce oxygenates in a single step using molecular oxygen has been a long-standing goal in catalysis. Recently, the renewed interest in methane utilization has led to extensive studies on several catalytic systems, including low temperature gas phase reactions [1], electrophilic C-H bond activation in acidic media [2] and aqueous- phase conversion using peroxide [3]. However, these systems either require an expensive oxidant, e.g. peroxide or sulfuric acid that poses economic and environmental concerns, or the process is non-catalytic in nature. A complete catalytic cycle of this reaction involves three elementary steps: activation of CH_4 to a M-CH_3 complex, functionalization of the methyl species with OH, halides or CO groups and desorption of the product from the catalyst surface. Fundamental understanding of each of the three steps is necessary to design and further optimize the catalyst system that can activate C-H bond of methane.

In this presentation, we will show that over single-site, isolated Rh/ZSM-5 catalysts, methane can be oxidized directly to acetic acid in the presence of molecular oxygen and carbon monoxide in aqueous solutions under mild pressures and temperatures. The reaction is found to be catalytic and heterogeneous. The Brønsted acidity of the support is crucial for the selective production of acetic acid. On the other hand, methanol formation does not require acid sites and can be realized on Na ion-exchanged ZSM-5, and even on an open oxide support. Thus, anchoring rhodium single ions on the appropriate solid support offers a catalytic system with product tunability between methanol and acetic acid. Single batch production of acetic acid can reach 13,500 μmol/g-cat at more than 4.4% conversion of methane, and a fed-batch process is designed by replenishing limiting reactants to further push the conversion.

Acknowledgment

Financial support by the U.S. Department of Energy (ARPA-e grant# SU0433) is gratefully acknowledged.

References

[1] Alayon, E. M., Nachtegaal, M., Ranocchiari, M., and van Bokhoven, J. A. Chem. Comm. 48, 404 (2012).

[2] Periana, R. A., Taube, D. J., Gamble, S., Taube, H., Satoh, T., and Fuji, H. Science 280, 560 (1998).

[3] Hammond, C., Forde, M. M., and Ab Rahim, M. H., et al. Angew. Chem. Int. Ed. 51, 5129 (2012).

486400 Synergistic Effects between CHA Zeotypes and Basic Metal Oxides in Bifunctional Strategies for Improved Lifetime in Methanol-to-Olefins Catalysis

Abstract:

Methanol to olefins catalysis by CHA zeolites and zeotypes proceeds through a chain carrier mechanism [1,2]. Impurities initiate formation of unsaturated organic moieties within the cavities of the CHA zeolite/zeotype engendering organic-inorganic hybrid chain carriers that serve as scaffolds propagating C-C formation and scission [1,2]. Active chain carriers are comprised of olefinic and monocyclic aromatic organic co-catalysts which undergo C-C scission events via cracking and dealkylation, respectively, to give light olefin products, and C-C formation events via C1 homologation with methanol close the sequence of chain carrier propagation reactions [2,3]. Formaldehyde, formed in bimolecular transfer dehydrogenation of methanol, alkylates active organic co-catalysts instigating their transformation to polycylic aromatic hydrocarbons thereby terminating chain carriers and inducing catalyst deactivation [4]. CHA zeolites and zeotypes confer dynamics of chemical and physical events that ultimately render too large the frequency of formaldehyde-mediated chain carrier termination relative to chain carrier propagation resulting in short catalyst lifetimes.

Here, we describe a bifunctional strategy to impede formaldehyde-mediated chain carrier termination that utilizes rare earth metal oxides to selectively and catalytically decompose formaldehyde, thus increasing lifetime of CHA zeolites and zeotypes for methanol-to-olefins catalysis without disrupting the inherent, high selectivity to ethylene and propylene. We demonstrate that the efficacy of rare earth metal oxides to decompose formaldehyde improves with increasing spatial proximity between the surface of the rare earth metal oxide and the chain carriers confined within the zeolite/zeotype. We evince site requirements for alkanal decomposition on rare earth metal oxide surfaces by combining kinetic, transient chemical probe, and spectroscopic techniques and develop causative and correlative relationships between dynamics of alkanal decomposition and lifetime in methanol to olefins (bifunctional) catalysis using formalisms of reaction-transport phenomena.

[1] J. F. Haw, W. Song, D. M. Marcus, and J. B. Nicholas, Accounts of Chemical Research, 36 (2003) 317-326.
[2] S. Ilias and A. Bhan, ACS Catalysis, 3 (2013) 18-31.
[3] A. Hwang, D. Prieto-Centurion, and A. Bhan, Journal of Catalysis, 337 (2016) 52-56.
[4] A. Hwang, M. Kumar, J. D. Rimer, and A. Bhan, Journal of Catalysis, 346 (2017) 154-160.

482581 Depressed Deactivation of SAPO-34 during Methanol-to-Olefins Procee By MgO

Abstract:

SAPO-34 molecular sieve is the best catalyst for methanol to olefins (MTO) reaction owing to its high ethene and propene selectivity. However, the SAPO-34 molecular sieve is deactivated rapidly due to coke deposition in its small pores, and has relatively short working-life time when used as the MTO catalyst. In order to prolong the working lifetime of the catalyst in MTO reaction, a great deal of efforts, such as optimizing the crystallite size and modifying the pores of zeolite to create hierarchical structure, have been done to enhance mass transfer and to reduce coke formation rate. What's more, modification of SAPO-34 catalyst with different cations by various methods, such as direct hydrothermal synthesis, incipient wet impregnation and ion exchange, can also improve the catalyst working lifetime. However, little attention is paid to prolong SAPO-34 lifetime through physically mixing SAPO-34 with another component.

Reported here in this paper, the rate of SAPO-34 coking was reduced by physical mixing SAPO-34 with MgO, and even the MgO, separated from the SAPO-34 by inert quartz sand layer, was also able to significantly reduce the rate of SAPO-34 coking, and therefore increase the working life time. That is to say, the catalytic performance of SAPO-34 can be significantly affected by MgO which stands "far away" from the SAPO-34. The coke content and coking deposition rate of SAPO-34 were measured, and it was shown that once the SAPO-34 was deactivated, content of coke deposited on the SAPO-34 was the same, regardless with or without the MgO as catalyst promoter. The MgO decreased the rate of SAPO-34 coking and hence increase the working life time. The CO and CO2 were formed when SAPO-34 was promoted with MgO, while no CO and CO2 were detected when pure SAPO-34 was used as the catalyst. MgO by itself shows almost no catalytic reactivity for methanol conversion. But when MgO is sandwiched between layers of SAPO-34 the catalytic activity and production distribution from methanol is significantly altered to favor carbon oxides compared to SAPO-34 by itself. MgO is only catalytically active when sandwiched between layers of SAPO-34, indicating that reaction intermediates are transported from SAPO-34 to MgO and back again. A mechanism for MgO to reduce the coking rate of SAP-34 and the formation of CO, CO2 and H2 was proposed. This research work provides not only a new route for the modification of SAPO-34, but also provided a simple, and convenient way for improving catalytic performance of SAPO-34 used for commercial MTO.

496719 Different Dependency on Copper Oxidation States between H_2 and CO_2 Production during Partial Oxidation of Methanol

Abstract:

The increasing impact of climate change and pollution necessitates the need to replace our current fossil fuel based energy supply with cleaner energy production methods, such as fuel cells. Methanol is of particular interest due to its relatively high energy density and safe handling. Methanol can be directly or inderectly oxidized to power fuel cells. The partial oxidation of methanol (POM, $CH_3OH + 0.5O_2$ â†' $CO_2 + 2H_2$) catalyzed by a copper based catalyst can provide on-board generation of high purity H_2 streams. Although copper-containing catalysts has been invesigated for decades, the reaction mechanism and active sites are still poorly understood. In particular, the chemical nature of the active phase (Cu^0, Cu^+ or Cu^{2+}) is unclear and their impacts on activity and selectivity of POM are unknown.

Results from the study of partial oxidation of methanol (POM) catalyzed by Cu/ZnO were presented here with the objective to identify correlations between POM reactivity and Cu oxidation state. A 30wt% Cu/ZnO nanoparticle catalyst was prepared by a co-precipitation synthesis. Catalyst morphology and elemental distribution on nanoscale were determined by transmission electron microscopy (TEM). The catalytic performance was measured at different O_2 /methanol molar ratios in a home-built micro-reactor. The Cu oxidation state was evaluated at different time-points via ex-situ X-ray photoelectron spectroscopy (XPS). We found that both reactivity for POM and the oxidation state of copper change with reaction time and with O_2 to methanol feed ratio. Most importantly, a strong correlation between H_2 selectivity and (metallic) Cu^0 content of the catalyst was observed. Surprisingly, the CO_2 selectivity was not significantly impacted by the oxidation state of the catalyst, but showed a strong correlation with the O_2 partial pressure. Based on the observed correlations, we propose a mechanism for POM including different bonding configurations of reaction intermediates between metallic Cu and Cu_2O surfaces. We are currently in the process of verifying key reaction steps by firstâ€"principle calculations. The knowledge we gain from this study will benefit the optimization of current Cu-based catalysts which may lead us to a promising methanol based energy economy.

491403 Continuous Hydrogenation of Furfural to Cyclopentanone Using Activated Carbon Monolith Catalysts

Abstract:

Cyclopentanone, defined as a cyclic ketone, plays a key role in synthesis of pharmaceuticals, rubber chemicals, flavors, fragrances, and aviation fuel. The current industrial process in production of cyclopentanone involves a petroleum-based feedstock. During this process, adipic acid undergoes catalytic cyclization at high temperatures to form cyclopentanone. Adipic acid is a petroleum-based chemical that is obtained from oxidization of a cyclohexanol and cyclohexanone mixture in presence of excessive amounts of nitric acid. Generating large amounts of N_2O as by-product is one of the major drawbacks of this adipic acid production process.

Furfural is a green chemical that can be hydrogenated to form cyclopentanone. Acid hydrolysis of hemicellulose-rich biomasses such as corn stover, wheat bran, or corncobs results in xylose. In a second step, furfural can be obtained through catalytic dehydration of xylose. This work focuses on continuous hydrogenation of furfural to cyclopentanone using activated carbon monolith catalysts. Activated carbon monolith catalysts (ACMC), derived from woody biomass, are impregnated with precious metals and used to form cyclopentanone from furfural in the presence of hydrogen. Multiple advantages of using ACMC, in addition to low pressure drop, high mass transfer rates and surface area to volume ratios, are first, the monolith is made from renewable carbon sources, second, high stability in aqueous/acid/base reaction medium and finally, the precious metal can be recovered after combustion of the catalyst. The kinetics of cyclopentanone formation from furfural using ACM impregnated with precious metals will be studied over temperatures of 100 to 180 °C, hydrogen pressures of atmospheric to 300 psig, liquid flow rates of 0.5 to 4 mL/min and hydrogen flow rates of 40 to 100 mL/min. The immediate goal of this work is to develop a kinetic model and rate law, perform comprehensive catalyst characterization correlating structure with function, and determine the longevity of the catalyst.

496742 Soluble Sugars from Cellulose Using Polyvinyl Chloride Derived Hydrochloric Acid As Catalyst

Abstract:

Municipal Solid Waste (MSW) contains a large portion of organic material that has the potential to be converted into valuable fuels through thermochemical conversion. In particular, cellulosic waste-derived sugars are important platform chemicals of biofuels and chemicals. In addition to organic wastes, MSW also contains significant quantities of halogenated plastics like polyvinyl chloride (PVC) that are difficult to recycle using traditional methods. When thermally converted (e.g., pyrolysis, gasification, and incineration), PVC produces corrosive hydrochloric acid (HCl) that can react with organic waste-derived products to form organohalogens and dioxins that create various environmental and health hazards. In the current work, we co-converted cellulose and PVC in tetrahydrofuran (THF) at elevated temperatures and pressures. The reactions were carried out in miniature Swagelok reactors with reaction temperature, time and PVC concentration as the experimental variables. The liquid product was analyzed and quantified with GC/MS/FID, and the solid residue was characterized by pyrolysis using a Frontier micropyrolyzer coupled to a GC/MS/FID. The results showed that HCl released during the thermal decomposition of PVC acts as an effective catalyst in the polar aprotic solvent to promote cellulose decomposition to soluble sugars. Levoglucosan was found to be the most abundant product. The highest monomeric sugar yield was 42% of cellulose at 335 ºC and 8 minute reaction time with 2.5% cellulose weight equivalent PVC. The liquid products also contained up to 46% of soluble oligomeric sugars and small amounts of hydroxymethylfurfural and furfural. Pyrolysis of the solid residue also produced a high yield of levoglucosan (~22%, residue weight basis), indicating the residue is unconverted or partly decomposed cellulose. Chlorinated organics were nearly absent in both the liquid product and the pyrolyzates of the solid residue, suggesting the chloride content in PVC is converted to the solvent-dissolved HCl. The present study aims to explore an environmentally friendly approach to convert mixed solid wastes into value added products.

495204 Effect of Temperature and Vapor Residence Time on Product Distribution of High Density Polyethylene Fast Pyrolysis

Abstract:

In search of more sustainable options for plastic waste disposal, thermochemical conversion to produce alternative fuels or chemicals appears as an outstanding alternative due to its environmental benefits such as having a lower carbon footprint than fossil fuel extraction.[1] One method of thermochemical conversion is fast pyrolysis, the thermal degradation of plastic waste in the absence of oxygen. Fast pyrolysis at high temperatures, with reaction times shorter than 10 s, has been used to produce hydrocarbon products from plastic waste that can be converted into fuel or chemical intermediates. This study analyzes the degradation products of high density polyethylene (HDPE), the most abundant plastic waste, with respect to temperature and vapor residence time. Through the use of a novel, high-temperature reactor extension to an existing micro-pyrolysis unit (CDS Analytical 5200 HP Pyroprobe), the residence time at each temperature was varied by adjusting the helium carrier gas flowrate. Residence times were varied from ~0.0 to 5.6 s (25mL/min He to 100mL/min He). The temperature of the micropyrolysis unit and the reactor extension was established at three different temperatures (650°C, 675°C, and 700°C) for each set of residence times. The pyrolysis products were analyzed in a ThermoFisher K8880181 Trace Gas Chromatograph coupled to a DSQ II mass spectrometer. The observed chemical species were alkanes, alkenes, alkadienes and aromatics and the carbon content of the pyrolysis products ranged from 2 to 32 atoms per molecule. At constant reactor temperature of 700°C, the slight variation in residence time from ~0.0 to 1.3 s greatly shifted the product distribution towards lower molecular weight compounds (LMWC, C15 and below) as opposed to higher molecular weight compounds (HMWC, C16 and above). Without extended reaction, 25% of the products were LMWC, while with extended reaction, 96% were LMWC. When the pyrolysis temperature was lowered to 675 °C the amount of LMWC decreased 6% to 90%. These results illustrated the trend that as temperature and residence time increased, there was a greater abundance of lower molecular weight compounds in the product distribution, mostly alkenes with some alkanes but almost lacking dienes. Changes in chemical species distribution were also observed. An increase in the composition of aromatic compounds was exhibited at longer residence times and higher temperatures whilst the opposite trend occurred for high molecular weight alkenes and alkadienes. These results show a promising pathway for HDPE waste as a potential source for hydrocarbon fuel fractions and chemical intermediates.

References:

1. Gracida-Alvarez, U.R.; Keenan, L.M.; Sacramento-Rivero, J.C. & Shonnard, D.R. Resource and Greenhouse Gas Assessments of the Thermochemical Conversion of Municipal Solid Waste in Mexico. *ACS Sustainable Chemistry and Engineering* **4**, 5972-5978 (2016).

497522 Effects of Acid-Pretreatment on Co-Pyrolysis of Biomass and Plastic

Abstract:

Over 521 million pounds of agricultural plastics are annually used in US farm field to protect plants and improve crop yield. Due to the non-biodegradability of the plastics, disposal of the used plastics is problematic. The agricultural waste plastics are mostly burned in farm land, buried or sent to landfill, which cause many environmental problems and health hazards. Since waste plastics and crop residues are often present together in the farm field, co-pyrolyzing the two streams of materials could be an attractive solution to address the environmental issue while promoting energy recovery from waste plastics. In the present study, the methods for improving the co-pyrolysis efficiency of biomass and plastic are explored. It is known that the inorganic impurities presenting in cellulosic biomass favor undesired ring opening and charring reactions to reduce bio-oil yield. Acid washing and acid infusion of biomass prior to pyrolysis are known as effective methods to overcome the detrimental effect of inorganic impurities to increase bio-oil yield. The effect of acid pretreatment on co-pyrolysis of biomass and plastic is studied. Corn stover biomass pretreated with sulfuric acid (both acid washed and acid infused) were co-pyrolyzed with polyethylene in a tandem micropyrolyzer with and without zeolite catalyst. During non-catalytic pyrolysis, levoglucosan yield increased from 0.5wt% with non-treated corn stover to 9.5wt% in acid treated biomass, which is expected. It was found that co-pyrolysis of acid-treated corn stover with PE can further increase the yield of levoglucosan, as well as the yields of phenolic monomers due to synergistic effect. The stronger synergistic effects were also observed when the acid treated corn stover and PE were co-pyrolyzed using HZSM-5 zeolite catalyst in comparison to co-pyrolyzing non-treated corn stover and PE using same catalyst. Due to the synergistic effects, the amount of total aromatic hydrocarbons and the selectivity of mono-ring aromatic hydrocarbons both increased significantly.

497745 Effect of Dielectric Oxides on Microwave Assisted Deconstruction of Lignin Â

Abstract:

Lignin, a natural polymer is a potential source of fuels and platform chemicals due to its aromatic structure. The conventional industrial processes which utilize plant polysaccharides, burn lignin to generate process heat. With the depletion of fossil fuels, cellulosic biorefineries would generate surplus lignin than required to fuel the biomass conversion process [1]. Conventional lignin conversion methods require temperatures greater than 200°C and longer reaction time. A non-conventional source of energy such as microwave reduces the reaction time and gives higher heating rates. The heating of reaction medium by microwaves is through the mechanism of dipolar polarization and ionic conduction. In this study, alkali lignin was solubilized in 1:1 Dimethyl Sulfoxide (DMSO)/water mixture. The DMSO/water mixture has a high polarity (0.7) and solubilizes lignin completely. Metal oxides such as lanthanum ferrate ($LaFeO_3$), iron oxide (Fe_2O_3) and zirconium oxide (ZrO_2) were synthesized by the combustion method. Experiments were carried out in a microwave reactor at different temperatures (80-100°C) and samples were withdrawn at regular intervals between 0-120min. The catalysts were suspended in the reaction media and the solution vessel was continuously stirred for the entire reaction time. The microwave treated samples were analyzed using Gel Permeation Chromatography to study the effect of catalysts on reduction in the average molecular weight of lignin. Liquid chromatography Mass Spectrometry was used for the identification of phenolic monomers produced. The formation of products such as vanillin, acetoguaiacone, homovanillic acid and coniferaldehyde were observed at temperatures as low as 80°C while using $LaFeO_3$ and Fe_2O_3 as catalysts. The effect of different catalyst properties such as structure, dielectric constant and surface area, on the yield of monomers was studied and possible reaction mechanisms were proposed.

References:

1. A.J. Ragauskas *et.al.*, Lignin valorization:improving lignin processing in the biorefinery. *Science*, 344 (2014).

494252 Spray Coating Chitin & Cellulose Nanomaterials for Enhancement of Barrier Properties

Abstract:

Chitin is an abundantly occurring biopolymer like cellulose. The annual natural chitin production is estimated to be around 10^{10} - 10^{12} tons per year (Barikani et al., 2014) and is about the same order of magnitude as cellulose, whose annual bio-production by photosynthesis is estimated to be of the order of 10^{11}-10^{12} tons per year (Klemm et al.,2004). Similar to cellulose, chitin can also occur as chitin nanofibers (ChNFs) and when it is subjected to high shear treatment, nanofibers of 10-20 nm diameter and several microns in length are obtained (Wu et al., 2014; Ifuku & Saimoto, 2012).Â Wu et al. (2014) were able to produce chitin nanofibers in suspension from purified crab Î±-chitin, and when these suspensions were cast into films the resultant films had O_2 & CO_2 gas permeabilities of 0.006 & 0.018 barrer respectively. These good barrier properties as well as chitinâ€™s natural bacterial resistance make chitin an excellent candidate for sustainable barrier packaging applications. But packaging materials are generally multilayered, and a good method for incorporating these chitin films in continuous multilayer packaging processing needs to be developed. Spray coating is one of the most versatile methods of applying liquid onto surfaces. Due to its frequent use in painting, spray coating processes are widely encountered in everyday life (McKeen, 2016). One of the main features of a spray coating process is the contactless delivery of the material to be coated onto the surface. This implies that the actual delivery process is not influenced by the surface topography and thus uneven, or delicate surfaces can be coated. The delivered load of coated material is independent of the surface and the lack of a metering gap leads to decreased sensitivity to defects and especially helps in reduction of web breaks in roll to roll process (Czerwonatis, 2008). Beneventi, et al. (2014) were able to coat slurries of micro fibrillated cellulose (MFC) onto wet highly porous papers allowing complete retention of MFC on the surface of the substrate and improved barrier properties of the final product. In the present study we demonstrate the successful spray coating of chitin nanofiber & cellulose nanocrystal (CNC) suspensions onto polylactic acid films. The spray coating of these materials was found to lead to a reduction in the oxygen barrier properties of the final composite film by as much as 75 % of its original value depending on the loading and the material used (CNC, Chitin or both). Unlike the study by Beneventi, et al. the substrate in this case is non porous and the suspensions do not wet the surface. The resulting de-wetting problems were successfully solved by heating the substrate surface. This process could open up a path for the use of more sustainable polymers with otherwise excellent properties but are lacking in barrier properties. Addition of a ChNF or CNC layer could also lead to a reduction in the amount of the more expensive polymer in the multilayer (with the sole function of providing oxygen barrier) all the while increasing the sustainability.

500752 Development of Low-Concentration Alkaline Treatment Method to Produce Thermally Stable Cellulose Nanofibrils (CNFs)

Abstract:

The thermal stability of cellulose nanofibrils (CNFs) can be improved by converting cellulose crystal structure from cellulose I to cellulose II using alkaline pretreatment method. The conventional pretreatment method requires 17.5% or higher concentrations of sodium hydroxide and large volume of water for neutralization to produce thermally stable, but interdigitated or aggregated CNFs. The main objective of this study was to develop a low concentration alkaline pretreatment method to produce thermally stable and highly dispersible CNFs. Cellulose nanofibrils with improved thermal stability were successfully prepared by treating cellulose powder with 2wt.% NaOH solution at low temperature (below 0), followed by mechanical disintegration through a French cell press with only ten passes. The thermally stable CNFs had an average diameter of 90 nm with the mean onset and maximum thermal decomposition temperature of 305 and 343 respectively, compared with those of 283 and 310 respectively, for the CNFs prepared without the NaOH treatment. The thermally stable CNFs can create new opportunities for the development of CNFs based packaging composites and high-end electronic applications.

491943 Tailored and Integrated Production of Carboxylated and Lignin Containing Cellulose Nanocrystals and Nanofibrils for Composite Applications

Abstract:

Here we demonstrate di-carboxylic acid hydrolysis for the integrated production of lignin containing cellulose nanocrystals (LCNC) and nanofibrils (LCNF) using two unbleached mixed hardwood chemical pulps of lignin contents of 3.9 and 17.2%. Acid hydrolysis experiments used maleic acid solution of 60 wt% concentration at 120°C for 120 min under ambient pressure. Yields of LCNC were low of less than 6% under this set of condition. The higher lignin content samples produced LCNC with greater height (diameter) of 25 nm but similar length of approximately 230 nm to that from the lower lignin content fibers (height of 20 nm). Interestingly, the higher lignin content sample resulted in LCNF with smaller height (diameter) of 7 nm but longer length of > 1 μm, or greater aspect ratio than the LCNF from the lower lignin fibers of height 10 nm and length < 1 μm. Lignin protected cellulose from esterification which resulted in LCNC and LCNF that was less carboxylated compared to those lignin-free CNC and CNF and therefore had lower charges. However, lignin is more hydrophobic and thermally stable than carbohydrates therefore LCNC and LCNF are favorable for composite applications.

504076 Nano Carbon Structures from Cellulosic Biomass for Use As Functional Materials Â

Abstract:

We report herein the production of hierarchical nanocarbon materials from cellulose fibers. The carbonized cellulosic fibers with branching out CNTs (CC-CNT) offer interesting surface and electrochemical properties that are of interests for chemical processing and biosensing applications. In our tests for chemical adsorption tests, CC-CNT offered highly selective adsorption of hydrocarbons from gas mixtures, especially chemicals with C=C bond and aromatic structures. More interestingly, the chemicals absorbed and accumulated can be quickly released from the surface of the materials upon application of modest electrical current. Such an electrically switchable adsorption behavior may promise applications in a broad areas including air quality control, chemical sensing and fuel storage.

502082 Methodologies for Life Cycle Inventory Generation of Chemicals and Their Implication on the Life Cycle Impact Assessment Results

Abstract:

The use of industrial chemicals is prevalent in the modern society as raw materials and intermediates making them an integral part of the life cycle of several products we use every day. Growing concerns over climate change, regulations and consumer awareness have necessitated a declaration of environmental impacts of these products. Life Cycle Assessment (LCA), the most popular tool to calculate these environmental impacts require a detailed analysis of input and output flows to the various stages of production of these chemicals. However, due to the vast universe of chemicals and hurdles such as restricted access to primary data from plants, LCA practitioners are faced with a challenging task of generating life cycle inventories for chemicals based on limited data availability in the open literature.

This study comprehensively examines eight different approaches used by LCA practitioners to incorporate chemical LCIs in their analysis based on varying data availability and the uncertainties associated with them. The implications of overestimating or underestimating the chemical LCI on the overall life cycle assessment of a product has been explored using a case study of polystyrene. The study suggests that, while approaches such as, using stoichiometry to determine the inputs and outputs in a chemical process can grossly underestimate the environmental impacts such as GWP and CED, using process simulation models without appropriately accounting for heat integration etc. can lead to overestimation of these impacts. Approaches which do not consider auxiliary operations and uses indeterminate estimations for electricity consumption could lead to misleading estimates for impact categories such as Ecotoxicity and such studies should avoid publishing these impacts.

501110 A Novel Dynamic Routing Model for the Transportation of Hazardous Materials Under Uncertainties and Potential Shipping Delays

Abstract:

Nowadays, various kinds of chemical products are daily produced and shipped out to satisfy the expanding need of modern society and life. These chemical products include not only fuel, gasoline, nature gas and propane; but also lots of intermediates, like acid, alkali, alcohol and polymers. All these chemicals are classified as hazardous materials (Hazmat) due to their dangerous properties, such as flammability, corrodibility, toxicity and even explosiveness. Any improper operations can result in extremely serious consequences to people, properties, and environment, and even leave over long-term adverse effects on health and safety. As a key stage of their life cycle, the transportation of Hazmat is very important and critical, since it has more direct and extensive impacts on human society and our environment. Therefore, great attentions have been drawn to reduce potential risks and incidents.

In previous research, different models have been developed for the routing and scheduling of Hazmat to provide safe and efficient transportation paths. However, most of works are not comprehensive, since they either employ single-criterion objective function, or ignore the concern of real-time changes and any uncertainties.[1,2] Although some studies did consider cases with uncertainties, or utilize time-dependent variables, models were solved with heuristic algorithms.

In this paper, a novel dynamic routing model is developed to route optimal shipping paths for the transportation of Hazmat, especially to dynamically route under various uncertainties. The proposed methodology contain three solving stages, identifying the risk distribution map, routing for individual shipping pairs and optimizing for the whole shipping system. Two sub-models, the individual shipping MILP model and the system shipping MILP model, are correspondingly developed to sequentially route for individual shipping pairs and for the whole shipping system, aiming at minimizing the shipping penalty, which is constructed by three criteria factors, the transportation risk, time and direction turning. When uncertainties emerge, the risk distribution map will be updated, and all affected individual shipping pairs and the whole system will be dynamically routed. In particular, if shipping time limits are violated due to uncertainties, relations between the allowable shipping time and the shipping penalty will be explored to advice stakeholders on their ultimate decisions on shipping time relaxation and shipping path selection.

The performance of the developed model has been tested by three case studies. Case I is a static routing example as base case, Case II and III are both for dynamic routing. As demonstrated, when uncertainties take place, our model can quickly response and provide an alternative shipping path with lower shipping penalty. In addition, reliable and solid information on shipping time and shipping penalty is explored in Case III for the tradeoff and path selection. Through case studies, our model displays excellent and efficient performance on the routing for the transportation of Hazmat, especially for the dynamic routing under uncertainties.

References:

[1] Erkut, E., & Alp, O. (2007). Integrated Routing and Scheduling of Hazmat Trucks with Stops En Route. Transportation Science, 41, 107-122.

[2] Fan, T., Chiang, W.-C., & Russell, R. (2015). Modeling urban hazmat transportation with road closure consideration. Transportation Research Part D: Transport and Environment, 35, 104-115.

494346 Effect of Dynamic Operation Behavior on Quantitative Risk Assessment for Gas Treatment Unit of Gas Oil Separation Plant

Abstract:

After introduction of quantitative risk assessment (QRA) to chemical process in late 1970s, the number of accident had reduced for about two decades. However, it increases more than minimum value and shows non-uniform trend, because of population growth and facilitiesâ€™ deterioration. Especially, complexity for high performance processing system have a great effect on occurrence of accidents. Thus, it is necessary to analyze risk considering complex operation condition. However, the problem of present QRA is that consequence is calculated only based on static operation conditions. Also, during past decade, dynamic risk assessment has solely focused on updating failure frequency according to failure data. Theses researches cannot reflect operation condition changes such as pressure, temperature, and flow rate.

In this study, influence of the operating condition change is focused and reflected to both frequency and consequence assessments. First of all, methodology for applying change of normal conditions (flow rate, pressure, and temperature) is presented. The minimum and maximum conditions are determined according to design basis. It is assumed that distribution of conditions follows beta distribution. This methodology is applied to existing gas treatment unit (GTU) in gas oil separation plant (GOSP). And then, QRA is conducted for each three changes. In this GTU process, distributions of pressure and temperature affect risk, and risk integral increases by 47.7% respectively. This paper contributes to recognize importance of considering dynamic operation behavior in QRA.

Unsteady Kinetics and Thermodynamics through TFE Pyrolysis Production Process based upon Inherent Safety Assessment

Yangmei Qin
School of Chemical Engineering
Sichuan University
Chengdu, China
qyangmei@stu.scu.edu.cn

Zeyi Xiao
School of Chemical Engineering
Sichuan University
Chengdu, China
mgch@scu.edu.cn

Contents

1. Idea
2. Objectives
3. Methodology
4. Results

Idea

WHY DO THIS ?

Idea

Key word: Inherent Safety

Idea

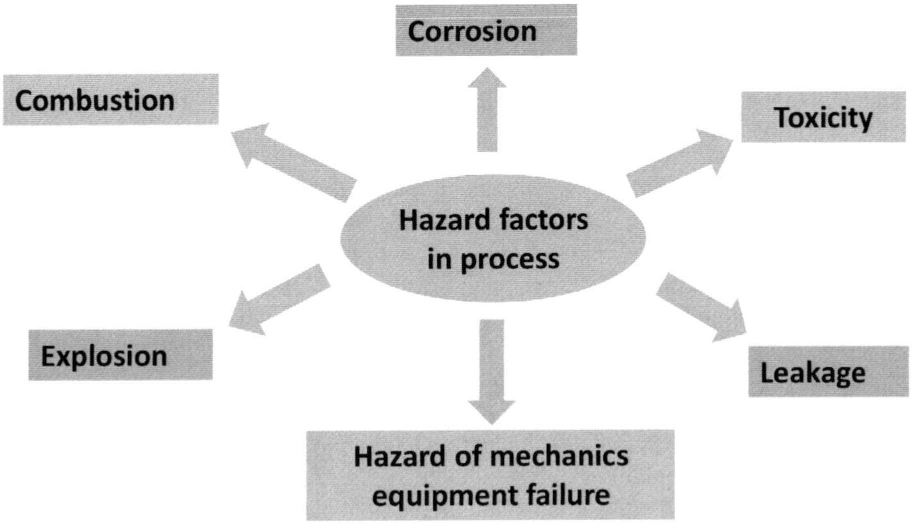

2017 AIChE

Idea

70%

Key word: Unsteady Status

Batch process

Start-stop process

Maintenance process

2017 AIChE

Idea

Inherent safety + Unsteady status

⇩

Analyze Inherent safety Of Unsteady status in Chemical process flow

2017 AIChE

Objectives

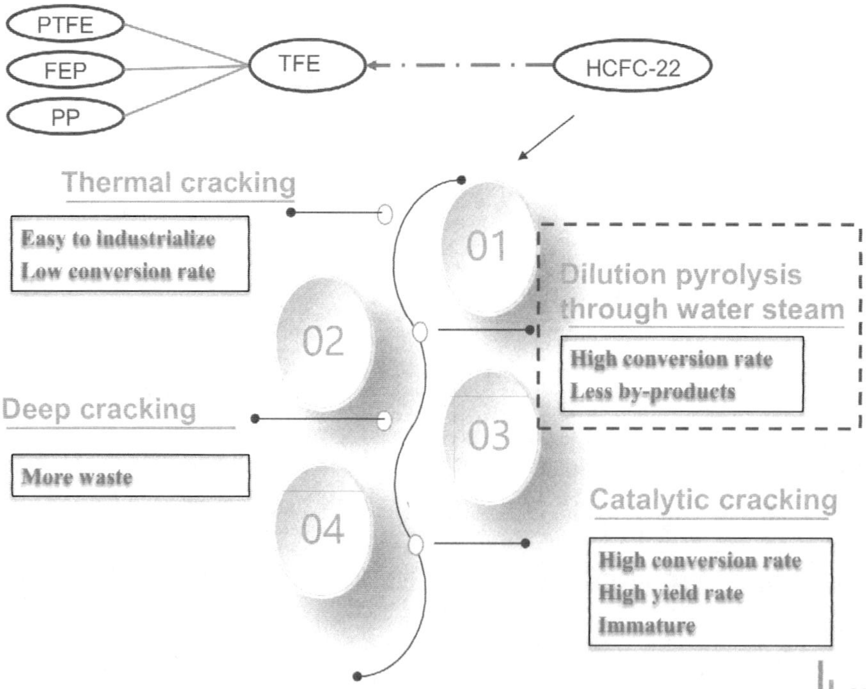

2017 AIChE

Methodology

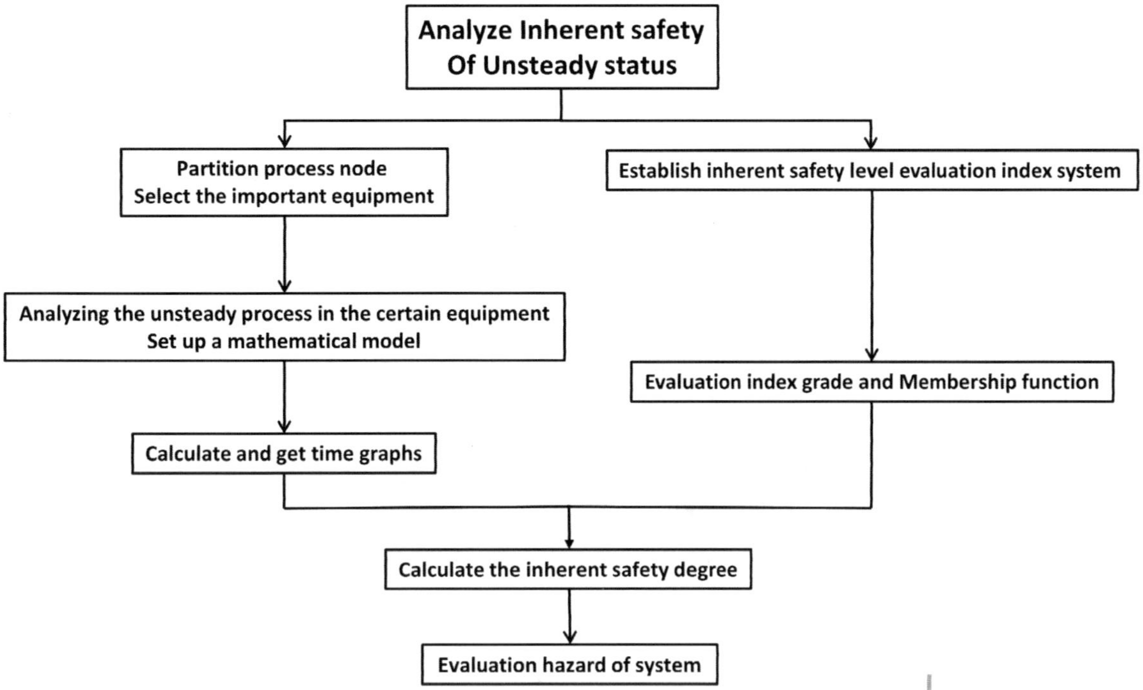

Methodology-Process flow

$$2CHClF_2 \xrightarrow{\text{diluent}} CF_2{=}CF_2 + 2HCl$$

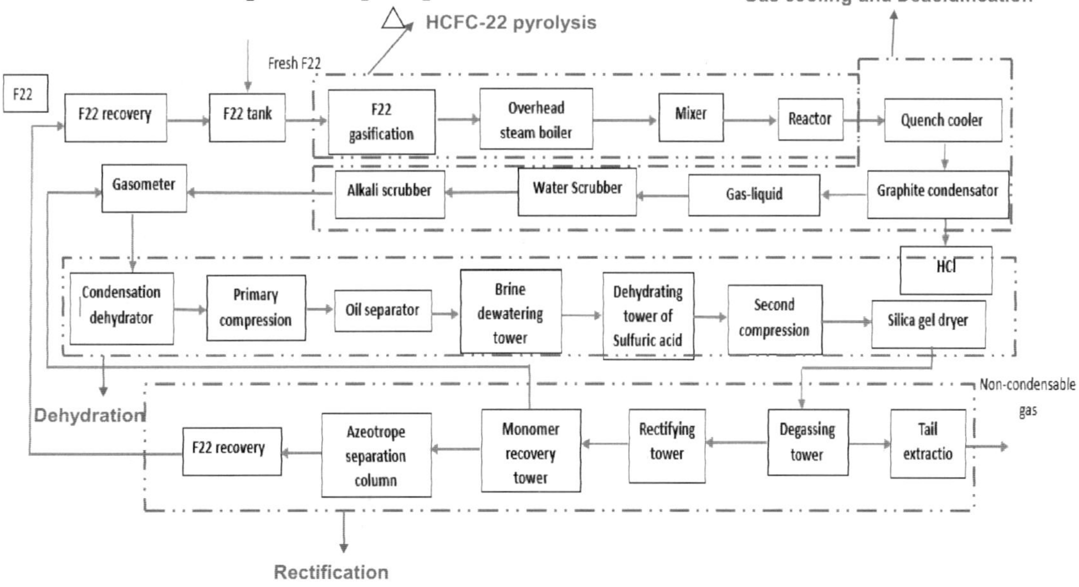

Methodology-Process node

HCFC-22 pyrolysis

NO.	Equipment	Main process indicators	Parameter	Material
1	**F22 Evaporation Tank**	Outlet temperature of F22	-10~10°C	Carbon steel
2	**Overhead Steam Boiler**	Outlet temperature of F22 Pressure of steam furnace tube Temperature of steam outlet Temperature of the flue gas	~300°C 0.3~0.5Mpa ~1200°C ~300°C	Main : Carbon steel Furnace tube: Nickel alloy
3	**Mixer**	Mixing temperature	~800°C	Stainless steel
4	**Reactor**	Pyrolysis gas outlet temperature Reaction pressure Conversion Rate of F22	600~700°C 0.02~0.05Mpa ~60%	Stainless steel

Methodology-Process node

Deacidification

NO.	Equipment	Main process indicators	Parameter	Material
1	**Quench cooler**	Outlet temperature Outlet pressure	130~170°C 0.02Mpa	Main: Carbon steel Coil: nickel alloy
2	**Steam pocket**	Pressure	0.3~0.4Mpa	Carbon steel
3	**Graphite condenser**	Outlet temperature HCl concentration of outlet	40~60°C 10~15%, pH≈2~3	Pipe: graphite Shell side: carbon steel
4	**Scrubber**	Outlet temperature Outlet pressure	20~30°C ~7	Main: Carbon steel Padding: plastic rod
5	**Alkali scrubber**	Mass fraction of NaOH solution Outlet temperature	10-15% 30~40°C	Carbon steel

Methodology-Process node

Dehydration

NO.	Equipment	Main process indicators	Parameter	Material
1	Condensation dehydrator	Outlet temperature(gas) Outlet water content Inlet temperature(liquid)	10~20°C 4748ppm ~10°C	Main: Carbon steel Padding: Polypropylene
2	Brine dewatering tower	Mass fraction of $CaCl_2$ Brine inlet temperature Outlet temperature(gas) Outlet water content	20~30% -10°C ~0°C 300~400ppm	Carbon steel
3	Compressor	Outlet temperature of 1st level compress Outlet pressure Outlet temperature of 2ed level compress Outlet pressure	20~30°C 0.2~0.4Mpa ~2°C 1.3~1.5Mpa	Carbon steel
4	Dehydrating tower of Sulfuric acid	Initial concentration of sulfuric acid The water content of the gas outlet	98% ≤100ppm	Carbon steel

Methodology-Process node

Rectification

NO.	Equipment	Main process indicators	Parameter	Material
1	**Degassing tower**	Pressure Temperature of the tower top Temperature of the kettle	~1.2Mpa ~-6°C ~17°C	Carbon steel
2	**Rectifying tower**	Pressure Temperature of the tower top Temperature of the kettle	~1.0Mpa ~-11°C ~33°C	Carbon steel
3	**Azeotrope separation column**	Pressure Temperature of the tower top Temperature of the kettle	~0.8Mpa 5~10°C 15~25°C	Carbon steel
4	**F22 recovery tower**	Pressure Temperature of the tower top Temperature of the kettle	~0.6Mpa -5~5°C 10~40°C	Carbon steel

Methodology-Modeling

F22 Evaporation Tank

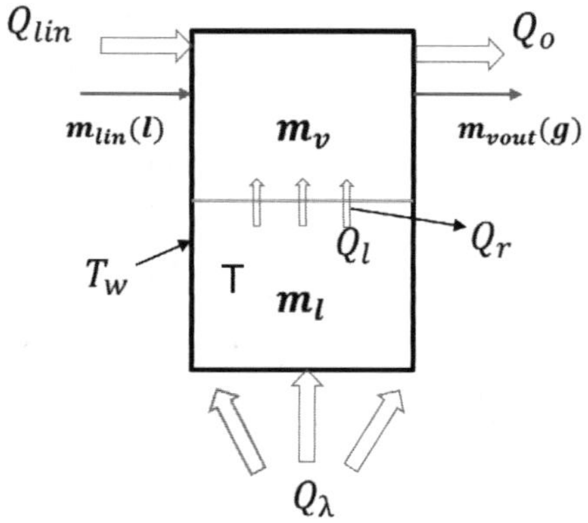

Mass balance

$m_{lin}dt =$

Energy balance

d

dQ

dQ_o

d

$$dQ_\lambda = \mathrm{KA}(T_w - T)\ dt$$

$$dQ_\lambda - dQ_{lin} = dQ_r + dQ_o + dQ_l$$

Methodology-Modeling

Condensate dehydrator

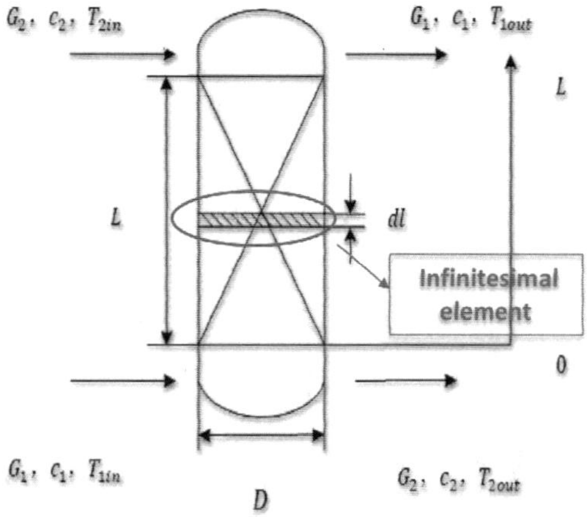

Pyrolysis gas

$$G_1c_1T_1(l,\ t) - G_1c_1\left[T_1(l,\ t) + \frac{\partial T_1(l,\ t)}{\partial l}\mathrm{d}l\right]$$

$$+ U\pi r^2 a\mathrm{d}l[T_2(l,\ t) - T_1(l,\ t)] = M_1c_1\mathrm{d}l\frac{\partial T_1(l,\ t)}{\partial t}$$

Liquor condensate

$$G_2c_2\left[T_2(l,\ t) + \frac{\partial T_2(l,\ t)}{\partial l}\mathrm{d}l\right] - G_2c_2T_2(l,\ t)$$

$$+ U\pi r^2 a\mathrm{d}l[T_1(l,\ t) - T_2(l,\ t)] = M_2c_2\mathrm{d}l\frac{\partial T_2(l,\ t)}{\partial t}$$

Methodology-Modeling

Condensate dehydrator

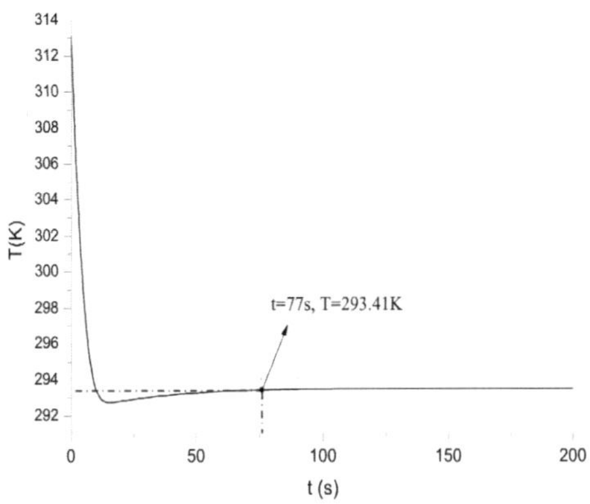

a

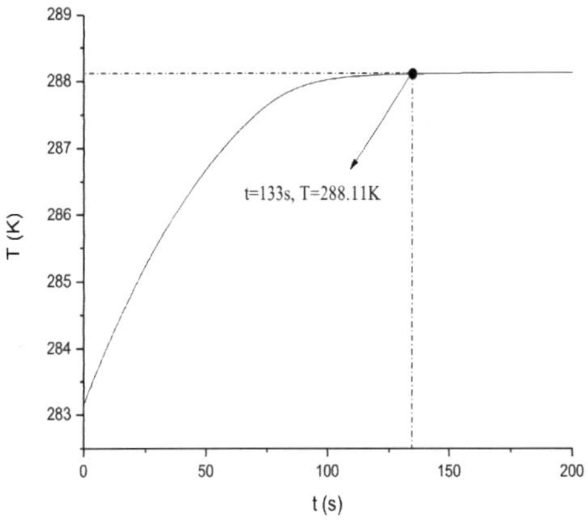

b

Methodology-Calculating

Condensate dehydrator

Condensate dehydrator	Pyrolysis gas					Liquor condensate			P/MPa
	Mass flow rate (t/h)	Inlet temperature (K)	Outlet temperature (K)	Water content of inlet (Mass fraction,%)	Water content of outlet (Mass fraction,%)	Inlet temperature (K)	Outlet temperature (K)	Mass flow rate (t/h)	
Reality	2.9595	313.15	293.15	1.5627	0.4748	283.15	288.15	2	0.1
Theoretical	-	-	293.41	-	-	-	288.11	-	-

Methodology-Inherent safety evaluation index system

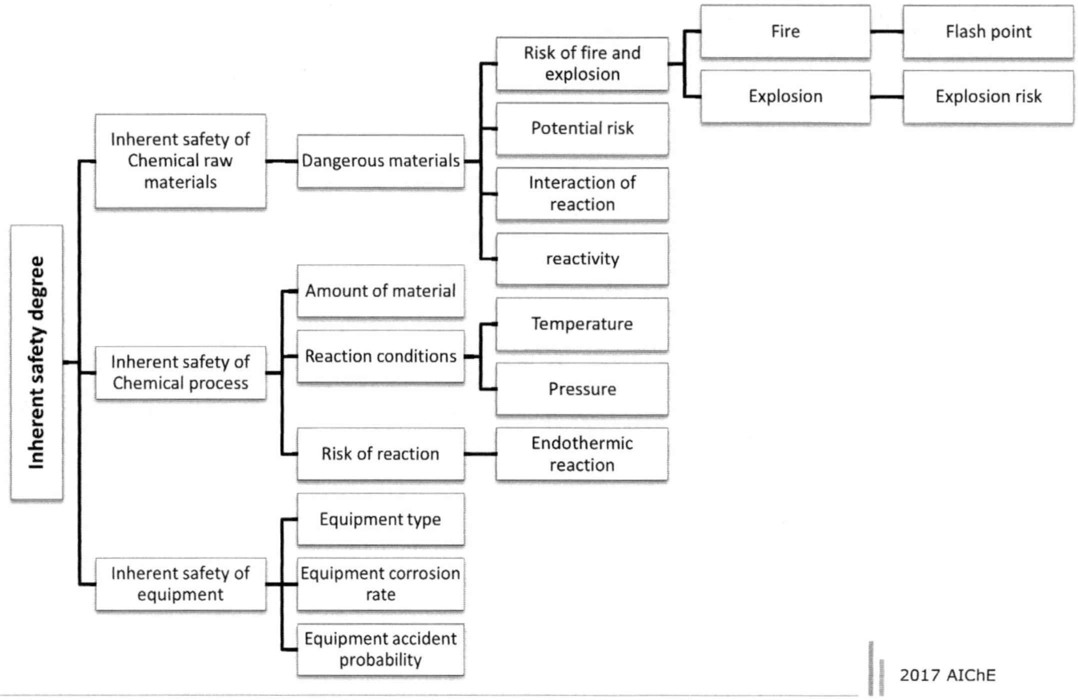

Methodology-Index parameter classification and membership function

Classification of indicator parameters

Consider the hazards caused by low temperature

Index	Inherently Safer Design（ISD）	Temperature hazard classification（T）	Classification（°C）	Shape of Membership function	Membership function
Temperature	VL	VH	[600，+∞)	Trapezoid	[450 600 +∞ +∞]
	L	H	[300，600)	Triangle	[225 450 600]
	M	M	[150，300)	Triangle	[110 225 450]
	H	L	[70，150) or (-273，0]	Triangle or trapezoid	[35 110 225] or[-273 -273 0 35]
	VH	VL	(0，70)	Triangle	[0 35 110]

Membership function curve

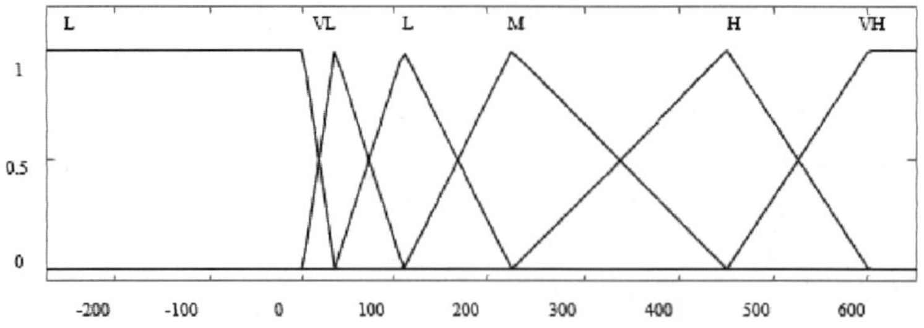

2017 AIChE

Methodology-Calculate the inherent safety degree

Formula of Variable weight

$$W_j\ (x_1,\ \cdots,\ x_m)\ =\frac{W_j^{(0)}\cdot\dfrac{\partial B}{\partial x_j}}{\sum_{j=1}^{m}|W_j^{(0)}\cdot\dfrac{\partial B}{\partial x_j}|}$$

$$B(x_1,\ \cdots,\ x_m)=\sum_{j=1}^{m}x_j^{\alpha}\ (\alpha\geq 0)$$

Inherent safety degree model

$$S=W\times I$$

S: inherent safe degree of the system

W: weight of index set

I: inherent safety degree set of each index

Results-Take the condensation dehydration node as an example

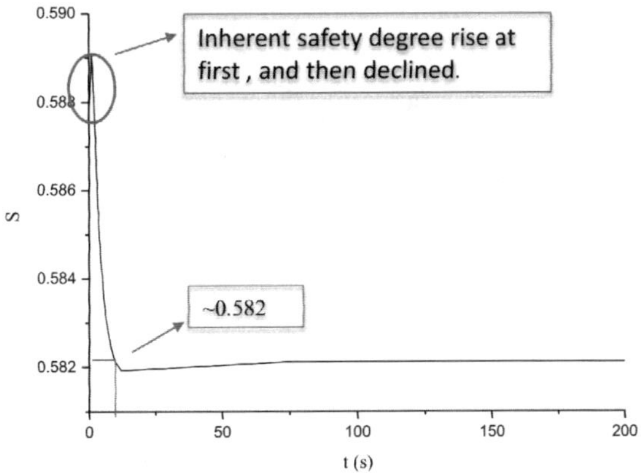

THANK YOU!

2017 AIChE

Membrane Separation and Chemical Process Group

Source-to-Customer Integration for Sustainability

Paper 527f

Proceedings 485144

Andrew W. Sloley, Principal Consultant, Advisian

Prepared for Presentation at the
American Institute of Chemical Engineers
2017 National Meeting
Minneapolis, MN, October 29 – November 3, 2017

www.**advisian**.com

Advisian
WorleyParsons Group

Today's Objectives

Objectives

1. Getting to sustainability
2. Impediments
3. Paths forward ... The future

1. Getting to Sustainability

Getting to Sustainability

- Economic unit defines trading value of differing objectives
  - Markets
  - $ £ ¥ € ₩ ₨ count
  - Revealed preference: what people do, not what they say
- 'Sustainability' consumption in fuels driven by diktat
- If we could make things attractive, mandates un-needed

Advisian / 5

Improving Existing Processes

- Even in optimistic case:
  - Oil consumption increases to 2035
- Population growth
- Increasing living standard
- More transportation
- Improving current technology important

Advisian / 6

2. Improving Existing Operations

Advisian / 7

Make Improvements Attractive

- The easiest molecule to make is the one you already have
- Molecule management
  - Get the molecules to the right place
- Increase value of products
  - One way
  - Focus here
- Minimize transformations
  - Conversions use energy
  - Add inefficiency to system

Advisian / 8

Current Position – Fuels and Downstream

- **North America/South America**: some large integrated complexes, but limited market share.
- **Asia:** large, integrated complexes more common.
- **Europe:** some integrated complexes, limited new investment.
- Market mismatch: fuels market large, downstream market smaller

Advisian / 9

3. Impediments

Advisian / 10

Integration in the Past

Cosden Refinery, Big Spring, TX:

- First aromatics extraction from crude oil in the world
- Fully integrated from crude to styrene-butadiene rubber (SBR), many other petrochemicals
- Gradually shifted to fuels-only refinery

Advisian / 11

Impediments

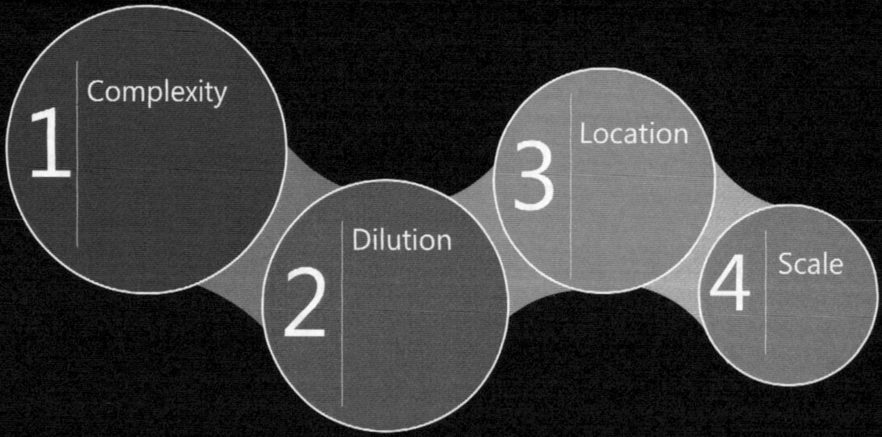

Advisian / 12

Complexity

- Feed with 000's of components
- Transportation is its own problem
- Both mineral and biological feeds have problem
- Extract molecules

Advisian / 13

Dilution

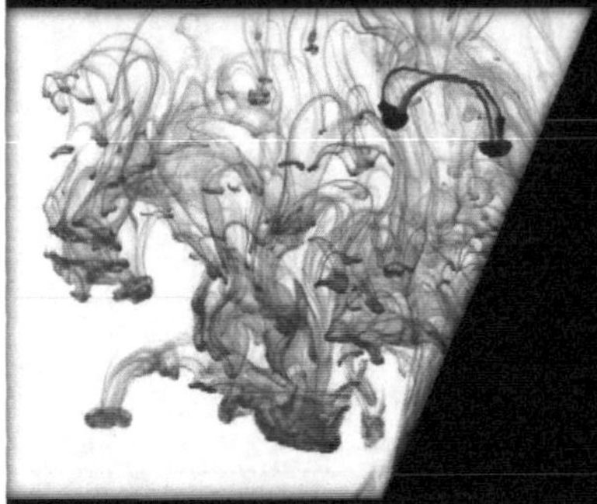

- Many components create dilution
- Low concentrations of target molecules

Advisian / 14

Location

- Resources
  - In remote locations
  - People pushing plants away
- Transportation is its own problem

Advisian / 15

Scale

As economies of scale favored larger plants...

Mismatch between scale becomes a problem

Process	Year	Economic Size mtpa	
Ethylene	1940s	10,000	Steam cracking
Ethylene	1960	150,000	Steam cracking
Ethylene	1970-1972	500,000	Steam cracking
Ethylene	2017	910,000	Steam cracking
FCC Ethylene	Today, gasoline mode	44,000	100 kbpd FCC
FCC Ethylene	Today, split mode	75,000	100 kbpd FCC
FCC Ethylene	Today, propylene mode	375,000	100 kbpd FCC

Advisian / 16

4. Path Forward ...
The Future

Advisian / 17

Success Requires – 1

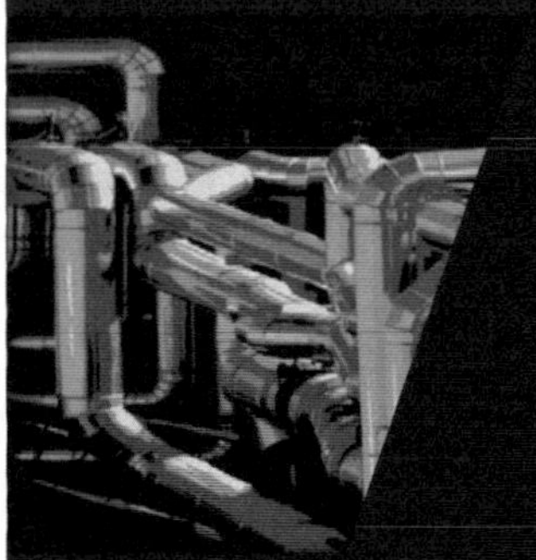

- Reduce complexity
  - Low-energy transformation
- Increase concentration
  - More efficient processes
  - Membranes?
  - Crystallization?
  - Regenerative adsorption
  - Efficient distillation

Advisian / 18

Success Requires – 2

- Deal with transportation:
  - Local supply-demand
  - Adding integration
- Deal with economies of scale:
  - Standardized packaged plants
  - Multiple plant consortiums
  - Expand to required size
- Make separate plants look integrated

Advisian / 19

Benefits and Costs

- Integrated complexes/processes capture more of the value chain
- Multiple costs:
  - More market volatility exposure
  - Long-term commitment to markets required
  - Understanding of different market types needed
  - More confusing for capital evaluation: Stock markets often have trouble with valuing integrated operations

Advisian / 20

Economic Viability

- Economics are about resource allocation
- There is always a competing resource demand
- Revealed preference – "Vox Populi, Vox Dei"

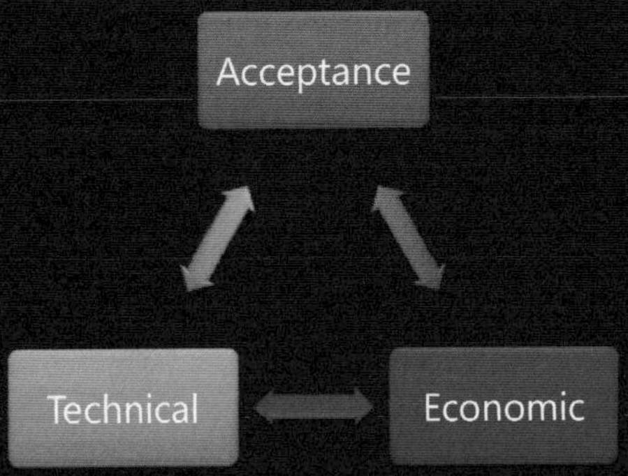

Advisian / 21

Past, Present, Future

Integration always the coming thing:

- Much talk
- Little action outside of large complexes

Predicting the future is hard:

- Uncertainty reigns
- Complex tradeoffs

Advisian / 22

What We Do – Integrate Analysis

Consulting and Engineering in

Refinery and Petrochemicals

- Molecule management
- Concept selection
- Economics and planning
- Technology selection
- Refinery-petrochemical integration
- Sustainability integration planning

Operational Excellence

- Benchmarking
- Gap analysis
- Safety and environment

Advisian
WorleyParsons Group

Infrastructure and Environmental

- Water
- Materials handling
- Transportation

Sustainability and Bioprocessing

- Analysis and integration
- Strategic planning

Management, Economics, and Business Operations

EcoNomics™

Advisian / 23

Advisian

WorleyParsons Group

502578 Process Kinetics of Mixed Bacteria Photosynthetic Hydrogen Production

Abstract:

The process kinetics of mixed bacteria photosynthetic hydrogen production is studied by experimental research and model fitting. The results showed that growth cycle of photosynthetic hydrogen production mixed bacteria is the law of the nonlinear. The fittingÂ curve of Logistic model reveal the maximum mixed bacteria dry weight is 0.4323 mg/mL, specificÂ growthÂ rate is 0.1077mg/h and the correlation coefficient is 0.9207. The modified Logistic model is used to fit the substrate consumption. The substrate consumption utilization degree of hydrogen production system which substrate concentration is 5 and 10 mg/mL is higher, but the substrate largest consumption of hydrogen production system of substrate concentration 20mg/mL of is 17.0146 mg/mLã€,When substrate concentration is 5, 10 and 20 mg/mL, the correlation coefficient of the Gompertz model fitting hydrogen production are greater than 99%, and hydrogen production rate respectively reach the maximum in 95, 101, 107 h. The substrate consumption and hydrogen production mainly concentrated in the logarithmic phase and the early stage of stable phase which is 48-120 h.

502553 5-Ethoxymethylfurfural Production from Cellulose Catalyzed By Ultra Low Mass Concentration Sulfuric Acid in One-Pot Reaction

Abstract:

5-Ethoxymethylfurfural (EMF) is a new kind of liquid biofuel,which has a high energy density and excellent combustion properties, and has been considered as one of the promising next-generation biofuels. In this paper, ultra low concentration of sulfuric acid (0.001mol L^{-1}) was used as catalyst to catalyze cellulose and ethanol to produce EMF in one-pot reaction. The effects of temperature,reaction time and substrate concentration on EMF yield were firstly studied and then the response surface methodology experiments were designed to optimize the reaction conditions. The interactions of factors and the optimum reaction conditions were obtained. The results showed that the reaction time and reaction temperature were the factors that mostly impacted the one-pot production from cellulose and ethanol catalyzed by ultra low mass concentration sulfuric acid. The interactions of two factors on the EMF yield were all significant ($P<0.005$). The maximum mean EMF yield of 14.35% was obtained under the optimum reaction conditions: reaction temperature 200 ,reaction time 60min and substrate concentration 20g/L, with the prediction error of 5.35%. This study could provide research foundation for the EMF one-pot production from much cheaper and more abundant biomass resource catalyzed by ultra low mass concentration sulfuric acid.

492488 Rapid Estimation of Life Cycle Inventories

Abstract:

Growing chemical use is a challenge to both chemical manufacturers and regulators given the emerging desire to evolve into a sustainable society. Sustainable chemicals provide benefits to society through their designated functions while minimizing their associated environmental and economic risks related to items such as resource consumption, health and safety, and profitability. As part of alternative chemical assessments, the methodological framework known as life cycle assessment (LCA) accounts for resource use and releases from a network of processes within the life cycle of a product, including extraction, production steps, use, and disposal / recycling. However, a significant challenge to using LCA is the extensive material and energy flow data needed to support assessments. This poster describes methods that can be used to determine the life cycle inventory (LCI) of necessary data to enable LCA and alternative chemical assessments.

The approach to developing LCI data is comprised of four efforts. The first effort focuses on Top-Down Data Mining, which collects data for facilities from EPA sources: Chemical Data Reporting (CDR), National Emissions Inventory (NEI), Toxics Release Inventory (TRI), Discharge Monitoring Report (DMR), etc. Steps for application of data mining and subsequent manipulation are illustrated in Cashman et al. [1]. The second effort applies Bottom-Up Simulation, where design and simulation are used to develop energy and material flows of the system. Additional techniques for estimating uncontrolled emissions from chemical processing equipment are then applied to obtain a detailed emission profile for the process [2]. This simulation method is refined by applying pollution controls to update resource use and controlled emissions. The third effort determines the LCI Chemical Lineage, which quickly determines the chemicals involved in an LCA study, defining the lineage of parents and children in the supply chain of a chemical. Finally, the fourth effort is LCI Reconciliation, which, through the combination of methods, aims to produce inventories that are qualitatively and quantitatively better than the individual methods.

[1] Cashman, S.A., Meyer, D.E., Edelen, A., Ingwersen, W., Abraham, J., Barrett, W., Gonzalez, M., Randall, P., Ruiz-Mercado, G., Smith, R.L. (2016). "Mining Available Data from the United States Environmental Protection Agency to Support Rapid Life Cycle Inventory Modeling of Chemical Manufacturing," *Environ. Sci. Technol.*, 50, 9013-9025.

[2] Smith, R.L., Ruiz-Mercado, G.J., Meyer, D.E., Gonzalez, M.A., Abraham, J.P., Barrett, W.M., Randall, P.M. (2017). "Coupling Computer-Aided Process Simulation and Estimations of Emissions and Land Use for Rapid Life Cycle Inventory Modeling," *ACS Sustainable Chem. Eng.* (Accepted).

The views expressed in this presentation are those of the authors and do not necessarily represent the views or policies of the U.S. Environmental Protection Agency.

494887 Aspects of Sustainable Production of Palm Oil in the Municipality of Teapa in Tabasco, Mexico: Evaluating the Current and Future Use of Palm Oil

Abstract:

Palm has been considered an advantageous food and bioenergy crop, due to its high fruit yields and the fruit bunchesâ €™ high oil content, approximately 40% by mass. Currently in the municipality of Teapa in the state of Tabasco, Mexico several palm plantations exist and are being expanded to accommodate the growing demand for vegetable oil markets and potential biodiesel market. This study site evaluates several different aspects of sustainability when considering palm as a feedstock as a food crop along with the future use of producing palm biodiesel. Social, economic and environmental indicators were used to evaluate the multidimensional understanding of sustainability.Â 55 palm plantations were evaluated and the data aggregated to evaluate the overall sustainability of palm cultivation. Data such as soil C, wild pollinator count, fertilizers and insecticides use, palm fruit yield, etc., were collected from a questionnaire of local plantation owners along with measurements from field ecologists. A life cycle assessment (LCA) was conducted to evaluate overall greenhouse gas emissions, water consumption and releases and air quality of the current production of edible palm oil along with the future use of palm as a renewable energy feedstock. The LCA includes direct soil C land use change, land preparation and plantation, fruit harvest and transportation and oil extraction when evaluating the current use of palm oil. When looking to future use, the LCA includes processing of the palm oil into biodiesel and final combustion within vehicles. Socioeconomic conditions were evaluated within the state of Tabasco to look at social and economic impacts of expanded palm production for palm oil and/or palm biodiesel. Indicators such as working conditions, income, and employment were evaluated within the current use of the palm oil and were extrapolated to evaluate future production of palm biodiesel. This study takes a multidimensional approach to sustainability and recommends best practices for the continued expansion of palm oil and/or palm biodiesel production.

501534 Harvesting and Extraction Technologies Contributing to Algae Biofuel Environmental Viability-Life Cycle Analysis of Naabb Developed Novel Technologies

Abstract:

There are many challenges and technical barriers to implementation of a thorough life-cycle assessment of algae-based biofuels. Biofuels made from algae must typically go through a complicated cultivation, harvesting, dewatering, oil extraction, and conversion process, involving many unit operations. Feedbacks between unit operations often depend on the particular operations being performed, so selection of a particular technology will have upstream as well as downstream consequences in the value chain. The National Alliance for Advanced Biofuels and Bioproducts (NAABB) project was established to produce and evaluate new technologies that can be implemented by the algal biofuel industry and establish the overall process sustainability. The goal of our LCA is to determine greenhouse gas (GHG) emissions and fossil energy demand associated with harvesting and extraction of algae. Four harvesting technologies and two extraction technologies within the NAABB group were evaluated. These include: Chitosan Flocculation, Electrolytic Flocculation, Membrane Harvesting, Acoustic Harvesting, Acoustic Extraction, and Wet fractionation. The scope will be limited in the sense that our focus is only on inputs of harvest and extract technologies, not including the materials and chemicals used to cultivate algae. The functional unit for harvesting technologies study is 1 kg dry algae output. Harvesting technologies will be normalized to this unit, while extraction technologies will be normalized on the basis of 1 kg algae oil output. Input/output data on the industrial operations being investigated was used to construct LCA models of the unit operations. Using the input data from multiple resources, the harvesting and extraction processes were all placed on an equivalent basis for comparison and the GHG emissions impacts were analyzed using SimaPro LCA software. In order to assess the magnitude of each technology in a similar fashion as the other unit operations, the gate-to-gate results for these unit operations were placed in the context of our previous LCA baseline modeling and the results for these unit operations are presented in g CO_{2eq} / MJ renewable diesel using the assumptions detailed in the Baseline Harmonized Report and GREET model.

495711 Comparative Techno-Economic Analysis of Algal Biofuel Production Via Hydrothermal Liquefaction: One Stage Versus Two Stages

Abstract:

Algal biofuels have been widely accepted as promising renewable energy resources and receive extensive investigations in four main research areas: algal feedstocks, cultivation methods, harvesting/dewatering technologies, and conversion processes. Conversion of algal biomass is the essential downstream step to realize high-efficient and cost-effective production of biofuels and associated co-products. Hydrothermal liquefaction (HTL) stands out in practically handling wet algal slurry and avoids toxic solvent usage as well as produces biofuels both in reasonable quantity and quality. Pacific Northwest National Laboratory (PNNL) developed a direct HTL (DHTL) and upgrading process to convert whole algal material to biofuels. This method could enable conversion of non-oil components in raw algae feedstocks such as polysaccharides and proteins to bio-crudes, thus improves overall bio-oil yield, which furthermore relieves constraints of strain-specific operation conditions. However, biofuels derived from DHTL contain high level content of nitrogen (1.5-6.0 wt.%) than crude oil (0.1 wt.%) operated at high temperature (350 â„ƒ) and pressure (20.9 MPa). SubsequentÂ denitrification process and high manufacture cost due to the sever reaction conditions would make DHTL process economicallyÂ unfeasible.

To address the drawbacks of DHTL, a novel sequential hydrothermal liquefaction (SEQHTL) for simultaneous extraction of co-products and bio-oil from algal biomass has been proposed by our group. This process highlights recovery of proteins and polysaccharides at first stage and extraction of bio-oils at second stage at much lower temperature and pressure. The pre-removal of protein substances at first stage substantially decreases nitrogen content of bio-oil obtained at second stage by 30-55% and enhanced fatty acids exaction yields additionally.

A comprehensive techno-economic analysis (TEA) of DHTL of algae associated with oil upgrading was performed by PNNL, resulting in minimum fuel selling price (MFSPï¼‰ $4.77 gal-1. However, this cost derived from DHTL cannot achieve the overall goal of 3 $/GGE established by the Bioenergy Technologies Office (BETO) by 2022. Based on sensitivity analysis, the total investment could play an important role in determining algal MFSP besides feedstock cost, thus decreases on investment cost could offer a desirable MFSP. In this case, SEQHTL process is expected to provide less investment cost due to the low operating temperature and pressure. Moreover, isolated proteins and polysaccharides can be regarded as co-products credits to further compensate production cost.

We have two phases analysis procedure. Our first goal is to compare our proposed SEQHTL process using *Chlorella* with DHTL process using another algae species, *Nannochlorpsis*. In this phase all the DHTL process data were retrieved from PNNL reports and SEQHTL data were retrived from pervious experimental work in our group. Preliminary results showed that HTL equipment cost of SEQHTL is only 60% of that in DHTL. The total capital investment is 150 million cheaper than DHTL. The sugar products indeed make some credits in operating cost. However, the final oil selling price of SEQHTL was still higher than DHTL because of the different strains applied (Oil conversion rate was 50% in PNNL, while our strain only have 30%). If we increase the oil yield up to the same value as DHTL, the selling price decreases to $4.21 gge. The second phase will be conducted to compare cost between SEQHTL and DHTL with same algal strain. Preliminay results showed that slimilar oil conversion rates were realized for these two methods, thus SEQHTL process is expected to be more economic to produce algal biofuels than DHTL in terms of capital costs, operating costs, and minimum fuel selling price.

In addition, sensitivity analysis indicated that four major factors that affecting MFSP in SEQHTL process were algal feedstock cost, bio-oil conversion rate, sugar selling price, internal rate of return. The cost of algae is the top reason that blocks the development of algal biofuels. If the cost could be lower to 70% of original price, the fuel selling price will be decreased by $1/gge. Realizing high bio-oil conversion yield is rather important, the more produced, the lower price could be sold. Additionally, the selling price of co-products plays an important role. If we could extract high value co-products as well as biofuels. The MFSSP could drop substantially.

The purpose of this study is to provide preliminary TEA result of SEQHTL process and demonstrate the potential of replacing conventional DHTL process with promising SEQHTL as a novel platform in conversion technology, especially with co-products extraction from natural biomaterials. Future R&D should be focused on enhancing bio-oil production yield and effective separation of co-products in the field of conversion technology as well as reducing the cost of algae feedstocks. Although uncertainty exists due to the lack of experimental data and scale-up experience of HTL facilities, SEQHTL process is a promising and competitive conversion technology to handle wet feedstocks and fractionate biomass into high valued co-products.

497701 Economic and Environmental Assessment for the Production of Propylene Glycol from Biodiesel Glyceroleconomic and Environmental Assessment for the Production of Propylene Glycol from Biodiesel Glycerol

Abstract:

Different countries around the world are implementing policies and legislations to promote the sustainable development of their economy (i.e. bioeconomy). Biofuels have been one of the most promoted areas in the transition towards sustainable schemes and many regulations seek their industrialization and commercialization. As a result, over the last years biodiesel production has gained track in the market, resulting in large amounts of waste glycerol being generated as low-cost and readily available feedstock.

Focusing on the enhancement of the bioeconomy and the valorization of biodiesel glycerol, we here address the economic and environmental assessment under uncertainty for different routes in the production of propylene glycol from biodiesel glycerol. More precisely, we compare three different hydrogenolysis routes based on the use of biodiesel glycerol between themselves and against the business as usual process, which uses oil-based propylene oxide as raw material. Using commercial simulators, an initial case for each alternative was first implemented and then optimized through standard heuristics, sensitivity analysis and heat integration. The economic performance was quantified using the economic potential and total annualized cost, while the environmental impact was assessed following Life Cycle Assessment principles using the methodology CML 2001. Finally, the main uncertainties affecting the calculations were evaluated via stochastic modelling.

The results revealed that there are glycerol-based alternatives that outperform the current propylene glycol process simultaneously in terms of profit and environmental impact (i.e. 90% increment in profit and 74% reduction in environmental impact at optimum process conditions).

Overall, our results aim to promote the development of sustainable biorefinery schemes to convert waste glycerol into high-value commodity chemicals, like propylene glycol, within the context of holistic bioeconomy frameworks.

489453 More Than Ethanol: A Techno-Economic Analysis of Corn Stover-Ethanol Biorefinery Integrated with Hydrothermal Liquefaction Process to Convert Lignin into Biochemicals

Abstract:

Energy being an â€œingredientâ€ for every economic activity presents challenges of providing qualitative and quantitative energy to the increasing global population. Biorenewable approach of producing energy in form of biofuels and biochemicals is the way to go. Hydrothermal liquefaction (HTL) process is an alternative to fast pyrolysis (FP) that can be used to thermal chemically convert feedstock into hydro-oil, gases, and solids. HTL compared with FP, has advantages such as low bio-oil oxygen content of 10 to 20 wt% compared to 40%, higher oil heating value of 35 MJ/kg compared to 16 -19 range MJ/kg of FP. HTL processes wet biomass whereas FP requires dry biomass thus saving on energy that would have been required for drying feedstock. Research is underway to improve biorefinery revenues by producing biochemicals and biofuels from lignin which accounts for 40% biomass energy, and 10 to 35% by weight. Besides lignin from paper and pulp industry, also ethanol plants produce lignin as a waste, thus, the objective of this work is to evaluate the techno-economics of integrating ethanol biorefineries with hydrothermal liquefaction of lignin to produce biochemicals. A 2000 metric ton per day (MTPD) corn Stover biorefinery produced 61 MMgal/yr. of ethanol and different yields of chemicals. The minimum ethanol selling price (MESP) accompanied with respective chemical were $2.70/gal (catechol), $3.17 (phenol), $3.27/gal (m-cresol), $2.80/gal (p-cresol), $3.22/gal (o-cresol), $2.91/gal (acetic acid), $2.91/gal (formic acid), $3.32/gal (furfural), and $3.36/gal (acetaldehyde). Total capital investment is M$ 479.6. The most influential factor on the minimum ethanol selling price is ethanol yield. Other critical factors are fixed capital investment, feedstock flowrate, and ethyl acetate price. Lignin derived chemicals using HTL and separation of the produced biochemicals is in the early stages of development thus more investment in terms of fixed and operating cost might be required for future commercial biorefinery plants than those obtained in this analysis.

500146 AÂ Mixed Integer Programming Model for Municipal Waste Management and Landfill Gas-to-Energy Systems

Abstract:

The interest on the optimal management of municipal solid waste (MSW) has increased considerably due to technological and industrial developments. Several technologies have been developed as an attempt to address the problem of waste disposal from a waste-to-energy perspective. Apart from conventional problems associated to solid waste management (land space usage, high costs of transportation, groundwater pollution, etc.), methane emissions from landfill sites (Landfill Gas; LFG) represent an unresolved major problem. However, despite that environmental issue, landfilling is currently the best alternative for waste management in the developing world.

Mathematical programming approaches have been proposed to address many issues concerning waste management from various perspectives. As an extension to the existing approaches, this paper presents a mixed integer multiperiod approach to optimize a MSW network, considering waste reduction processes and landfilling. An original formulation is proposed to improve the economics of the landfill operation. The model incorporates an optimal LFG-to-energy design, which involves the installation of several harnessing technologies depending upon the annual increasing or decaying of the LFG flow rate. A direct-hauling system and an optimally located off-municipality transfer station are used to optimize the transportation of the MSW to the potential destinations. Actual data from a municipality in Mexico (Celaya, Mexico) has been used to define our case-study. The resulting Mixed-Integer Linear Programming (MILP) model has been solved through the GAMS® modeling environment. Our formulation has been assessed through several scenarios, including different LFG collection efficiencies and a bioreactor landfill operation. The minimum total cost of the MSW management is achieved through the installation of an LFG-to-electricity system and a materials recycling facility.

Keywords: *Municipal solid waste management, mathematical programming, Landfill gas-to-energy.*

496211 Techno-Economic Analysis of Integrated Solid Oxide Fuel Cell - Gas Turbine - Organic Rankine Cycle Powered By Seaweed Biogas

Abstract:

Due to escalating environmental issues, alternative energy sources are required instead of fossil fuels. One of the alternatives for stationary power generation are solid oxide fuel cells (SOFC) which electrochemically convert fuel into electricity and heat. By utilizing biogas from brown seaweed (*Saccharina japonica*) as a fuel for SOFC, power generation is considered carbon neutral.

Electrical efficiency of biogas-powered SOFC is calculated to be around 45 %, while it generates 2.6 MW of heat per MW of electricity. This makes it plausible for integration into a combined heat and power (CHP) generation process. Integration of SOFC with gas turbine (GT) and organic rankine cycle (ORC) to utilize the heat, can increase the efficiency by 10-15%.

In this work, energy and economic viability of integrated SOFC, Gas Turbine (GT), and Organic Rankine Cycle (ORC) are investigated for industrial size of stationary power generation. This system is profitable with the value of corresponding minimum power selling price in the range of US average power selling price of 2016.

494516 Photocatalytic Reforming of Biomass for Hydrogen Production

Abstract:

Photocatalytic reforming of biomass for hydrogen production

Renée M. Ripken, Vincent J.H.W. de Boer, Johannes G.E. Gardeniers and Séverine Le Gac

MESA+ Institute for Nanotechnology

University of Twente, The Netherlands

Here, we describe a novel microfluidic device to determine the required bandgap for the photocatalytic reforming of biomass model substrates (ethylene glycol, glycerol, xylose and xylitol) in water. Furthermore, this device is applied to eventually elucidate the reaction mechanism of aqueous photocatalytic reforming, which is currently still a matter of debate.

Combustion of fossil fuels is one of the main causes of the climate change, due to the production of harmful CO_2. As a consequence, significant scientific efforts aim at developing cleaner and more sustainable fuel resources. Hydrogen is a promising alternative, as hydrogen has a high energy density and no harmful products are formed during its combustion. Hydrogen is currently produced from natural gas or coal by gasification, liquefaction or steam reforming [1]. However, these methods are highly energy demanding and they still require natural resources that are not renewable. In contrast, PhotoCatalytic Reforming (PhCR) only uses solar energy to convert aqueous phase biomass waste streams into H_2 and CO_2. [2]

Although PhCR appears to be a clean alternative for hydrogen production, currently used catalysts are relatively inefficient as they only use the UV-band from the entire electromagnetic spectrum. Only UV-light has sufficient energy to activate the catalyst by means of creating electron-hole pairs. To improve the efficiency of PhCR, new catalysts must be identified that are activated with a wider range of wavelengths. Therefore, we have realized a microfluidic platform to determine electrochemically the minimum required energy potential to perform the photocatalytic reforming of four biomass model substrates, ethylene glycol, glycerol, xylose and xylitol. These compounds were dissolved in water and a small amount of H_2SO_4 was added to ensure enough conductivity of the solution.

The microfluidic device includes electrodes to which a potential (up to 3 Volt) corresponding to the bandgap of a photo catalyst can be applied, mimicking an electrochemical potential. Cyclic voltammetry is used to evaluate the potential at which degradation of the substrate occurs. To study the reaction mechanism, the gases produced at the cathode and the anode sides are kept separate by an array of small channels (length: 50 Î¼m, depth and width: 2 Î¼m, spacing between the channels: 3 Î¼m) as a result of capillary pressure. The gaseous products are next extracted from the aqueous phase by means of a PDMS membrane placed on the outlets and subsequently analysed by inline GC. The

liquid phase products are collected from the device and characterized offline by HPLC.

Recorded cyclic voltammograms clearly show that a reaction is taking place for ethylene glycol and xylitol, which is also confirmed by GC and HPLC analysis. Based on the identification of the reaction products, we suspect that both proposed reaction mechanisms in the literature are taking place. [3-5] However, deactivation of the electrode surface was also observed as a result of carbon deposits, as confirmed by SEM and XPS.

In a next step, this device will be applied to help identify the optimal band gap for the photo catalyst, which would significantly improve the efficiency of PhCR.

References

1. G. W. Huber, J. A. Dumesic, *Catal. Today* **2006**, *111*, 119-132.
2. M. de Oliveira Melo, L. Almeida Silva, *Theoretical and Experimental Chemistry,* **2009**, Vol. 45 No. 6, 343-348
3. M. de Oliveira Melo, L. Almeida Silva, *Chem. Soc. Chem.* **2011**. Vol. 22, 1399-1406
4. N. Stratakia et al. *Appl. Chem. Cat. B: Environ.,* **2007,** Vol. 77, (12) 184-189
5. M. Kawai et al. *Chem. Phys. Lett.* **1984** Vol. 110 (1) 58-62

Acknowledgements

The authors wish to thank Prof. Dr. Guido Mul (University of Twente, The Netherlands) and Fang Liu (Utrecht University, The Netherlands) for all the support and help with the analysis.

Furthermore, the authors thank Remco Sanders, Stefan Schlautmann and Sip-Jan Boorsma (University of Twente, The Netherlands) for the help with the development of the experimental set-up and microreactor fabrication.

This work was supported by the Netherlands Center for Multiscale Catalytic Energy Conversion (MCEC), an NWO Gravitation programme funded by the Ministry of Education, Culture and Science of the government of the Netherlands.

499837 Experimental Analysis of Catalytic Gasification of Waste

Abstract:

(1) Title:

Experimental Analysis of Catalytic Gasification of Waste Polymers

(2) [Undergraduate] Student Researchers:

- Samuel O. Sanya
- UC Obiako
- Aliandra Barbutti
- Stephen Reeves
- Eric M. Lange

(3) Faculty Mentor:

- Dr. Jorge Gatica

(4) Abstract

Catalytic gasification, or "gasification", is one effective method which can be applied to promote low-temperature conversion of solid waste to energy. This research focuses on advancing the knowledge of a catalytic gasification process as a potential in-situ resource utilization and waste management alternative. This research has significance in a variety of engineering applications, but it is of particular relevance towards reducing landfill waste or as an in-situ resource generation system for space exploration beyond LEO.

This process evolves through a reaction mechanism consisting of two liquid-phase oxidation reactions of long-chain polymers, complemented by two gas-phase reactions. Using a laboratory grade reactor, the gasification of two model substrates, polyethylene and cellulose, was examined in the presence of ruthenium and platinum catalysts.

Gaseous products consisting of hydrogen, methane, carbon monoxide, carbon dioxide, were quantified by Gas Chromatography (GC). Calibration of the GC using chromatographic standards enabled the formulation of an equation to calculate the composition of the gaseous products for different reaction times. Fundamental reactor design equations along with stoichiometry calculations were then used for determinations of percent polymer gasified as well as selectivity of the gasification process.

Additional characterization of the solid and liquid residuals was performed using Differential Scanning Calorimetry (DSC), Thermo-Gravimetric Analysis (TGA), and Scanning Electron Microscopy (SEM). Quantification of the DSC spectra was used to correlate the thermal characterization of the residues with the unconverted (or non-gasified) substrate after the reaction was quenched. The SEM provided information on the microstructure of the residues, their atomic composition, and preliminary assessment of the possibility of catalyst recovery.

Preliminary results suggest that catalytic gasification provides potential control for the liquid-phase oxidations. Under appropriate conditions, oxidation was found to follow the incomplete oxidation pathway which resulted in significant yields of hydrogen and methane, which are potentially high value products as they can be used for energy recovery.

Comparisons between the gasification rates for the two substrates are drawn. Similarly, comparative studies of different catalysts are also presented. The results are used in conjunction with a kinetic model for the gas-phase to formulate a methodological model for the gasification of solid waste. The results will be presented along with a preliminary extension for a continuous gasification process.

500200 Investigation of Closed Loop Bioponic Irrigation Systems for Urban Agriculture

Abstract:

Bioponics is a novel take on traditional hydroponics farming in which petrochemicals are replaced by naturally derived nutrients in the form of "teas". The irrigation system is virtually the same as in hydroponics farming minus the inclusion of petrochemical fertilizers. What is used instead is a biologically active, nutrient "tea" that is brewed by aeration of fresh compost and water and is then pumped through the irrigation system to crops. The main factor in the effectiveness of these fertilizer teas is the large population of active bacteria. These bacteria are the same species of bacteria found in healthy topsoil and there is thought to be many links between these bacteria and markedly improved crop yields within a bioponics system (Vessey, 2003).

Traditionally, a hydroponics system are a single pass system. Meaning that large amounts of water are passed through the system into the plant containers and then passed out of the system to not be used again. This causes a large amount of waste water to be produced which is difficult to process especially in urban settings where such systems may be used. This also poses numerous environmental problems since this waste stream contains unused fertilizers that pose a threat to vulnerable ecosystems if not properly treated. The model-based study of lab and pilot-scale systems could prove beneficial in the search for more sustainable options for low-impact urban farming or even large scale, soilless agriculture.

495345 A Mechanistic Model for the Product Distribution of Fast Pyrolysis of High Density Polyethylene Waste

Abstract:

Fast pyrolysis emerges as an interesting alternative for the recycling of plastic waste on an industrial scale, despite incomplete information about the fundamental principles governing this process. Among the most important research areas is the development of an accurate kinetic model to predict the formation of different hydrocarbon classes when plastic is subjected to high temperatures and different residence times in an inert atmosphere. Several researchers have developed kinetic models that attempt to estimate the pyrolysis products distribution at different temperatures and residence times. One of the first models was proposed by Bockhorn et al.[1], it predicted the thermal degradation of polyethylene and polypropylene by using mass spectrometry measurements of pyrolysis products under isothermal conditions. Costa et al.[2] established a simplified model by grouping the products of polyethylene pyrolysis into four classes. This study idealized a series of chemical reactions that gave a kinetic model with only nine elemental first order reactions. Other models attempted to predict the product distribution of over 100 intermediate and hydrocarbon products using over 10,000 reactions.[3,4] Nevertheless, these models require complex methods for solving the differential equations that are not easily accessible. Mastral et al.[5] developed the sole kinetic model for fast pyrolysis of polyethylene under short residence times (0.52-2.07 s) and high temperatures (500-1000 °C). This model predicted the production of 8 groups of linear hydrocarbon and aromatic products. Random chain scission has been proposed as the mechanism responsible for the degradation of polyolefins into shorter hydrocarbon molecules through a set of free radical-based chain reactions. In this mechanism, polymers are broken down through a series of initiation, propagation, and termination steps. The complexity of these sequential reactions suggests that new insights may be gained by isolating the primary reactions from the secondary reactions. Therefore, this project attempts to develop a kinetic model to estimate the product distribution of primary and secondary products of the fast pyrolysis of waste HDPE. Primary reactions are carried out in a micro-pyrolysis CDS Analytical 5200 HP unit while secondary reactions occur in reactor extension attached to the micro-pyrolyzer. Products of waste HDPE degradation will be tested at high temperatures (650 â€“ 700 Â°C) and short residence times (~0.0 to 5.3 s) where no char is produced and be analyzed using a gas chromatography-mass spectrometry equipment. The developed model will be based on the random chain scission mechanism described previously and will describe the product distribution of four classes of linear hydrocarbon compounds and aromatics. Possible reactions that describe the formation of upcoming hydrocarbon species will be discussed.

References

1. Bockhorn H, Hornung A, Hornung U, Schawaller D. 1999. Kinetic study on the thermal degradation of polypropylene and polyethylene. Journal of Analytical and Applied Pyrolysis 48(2):93-109.
2. Costa PA, Pinto FJ, Ramos AM, Gulyurtlu IK, Cabrita IA, Bernardo MS. 2007. Kinetic Evaluation of the Pyrolysis of Polyethylene Waste. Energy & Fuels 21(5):2489-2498.
3. Németh A, Blazsó M, Baranyai P, Vidóczy T. 2008. Thermal degradation of polyethylene modeled on tetracontane. Journal of Analytical and Applied Pyrolysis 81(2):237-242.
4. Levine SE, Broadbelt LJ. 2009. Detailed mechanistic modeling of high-density polyethylene pyrolysis: Low molecular weight product evolution. Polymer Degradation and Stability 94(5):810-822.
5. Mastral JF, Berrueco C, Ceamanos J. 2007. Theoretical prediction of product distribution of the pyrolysis of high density polyethylene. Journal of Analytical and Applied Pyrolysis 80(2):427-438.

495712 A Model-Based Approach for Sustainability Assessment of Biomethane from Anaerobic Digestion of Food Waste and Manure Mixtures

Abstract:

Abstract:

Anaerobic digestion (AD) is a process involving a series of biological reactions (Hydrolysis, acetogenesis, acidogenesis and methanogenesis), which converts the biodegradable organic materials into biogas comprised mostly of CH_4 and CO_2. As an earlier study on lifecycle assessment of Bio-CH_4 has shown (Ankathi et al. 2017) [1], disposal of the food waste in landfills and conventional animal manure management has significant methane emissions contributing 20% to the U.S. national total. When these wastes are managed by AD, avoiding the landfilling and conventional manure management emissions, it is estimated that life cycle greenhouse gas emissions are -3.5 kg CO_2 eq./kg Bio-CH_4. The major advantage of AD on converting the food waste and manure into Bio-CH_4 is the displacement of fossil natural gas. On full-scale application of U.S. food waste / manure, AD saves 0.74 % of current annual natural gas consumption through large-scale Bio-CH_4 production.

A model-based approach involving AD modelling and Aspen process simulation for sustainability assessment is proposed based on an existing model by Batstone et al 2002 (ADM1) [2]. Aspen simulation includes different unit operations for Bio-CH_4 production; mixing, heating, anaerobic digestion, membrane separations, and gas compression. This kind of approach helps to study the effect of heat integration in the process and to predict the variations in the output concentrations of AD gases based on input compositions of food waste and manure. Through this approach, the economic costs and profits and environmental impacts will be analyzed and optimized.

Â References

1. Ankathi Sharath K., James S. Potter, David R. Shonnard, â€œCarbon footprint and energy analysis of Bio-CH_4 from a mixture of food waste and dairy manure in Denver, Colorado.â€ The manuscript is currently under review in *Environmental Progress and Sustainable Energy*. The manuscript number is EP-17-129.
2. Batstone, D. J., et al. (2002). "The IWA anaerobic digestion model no 1 (ADM1)." Water Science and Technology **45**(10): 65-73.

500196 Efficient Saccharification of Softwoods By an Integrated Thermochemical and Biological Process

Abstract:

Softwoods have great potentials as renewable feedstocks for biorefineries because of their high cellulose content. However, softwoods are more recalcitrant compared to herbaceous plants and hardwoods. Hence, the use of softwoods in biorefineries has been viewed as limited. To overcome their recalcitrance, we applied a cellulose solvent-based fractionation process on southern yellow pine and douglas fir at a mild condition: 50 ºC and 1 atm for 90 min. Pretreated softwood samples were hydrolyzed fast by cellulase enzyme (15 FPUs) and > 85% glucose yield was achieved by 24 h. Cellulose accessibility to cellulase of pretreated samples was characterized by the fusion protein and was found to increase by 39 folds, suggesting that enhanced surface accessibility via cellulose solvent-based fractionation was responsible for the high glucose yield. After fractionation, the degree of crystallinity of pretreated samples decreased and fibrous structure appeared to be disrupted as observed by solid-state 13C Cross-Polarization Magic Angle Spinning (CP/MAS) nuclear magnetic resonance (NMR) and scanning electron microscopy. No changes in lignin functionalities after fractionation were observed by 2D Heteronuclear Single-Quantum Correlation (HSQC) NMR, giving rise to the potential use of lignin as a co-product in biorefineries.

495604 Single-Objective Versus Multi-Objective Optimization of Integrated Fermentation and *in situ* Product Recovery Based on Time-Dependent Fermentation Models

Abstract:

Process simulation provides a platform that can be used to systematically analyze the process at hand, evaluate the feasibility of different process alternatives, and provide the most informative scenarios that can be experimentally investigated to aid in process design and optimization. The batch fermentation is characterized by low final product concentration, yield, and reactor productivities as a result of product toxicity to the microorganisms. Approaches to reduce product toxicity include integrated batch fermentation and *in situ* product recovery techniques, such as gas stripping, adsorption, membrane separation, etc. A broad range of experimental conditions in terms of gas recycle flow rate relative to fermentation volume, initial substrate concentration, gas stripping start time relative to fermentation start times, etc. have been investigated experimentally for fermentation and *in situ* gas stripping processes. Process optimization will reveal the optimal conditions of the process to guide the design of high performing integrated fermentation and *in situ* gas stripping processes across the broad range of experimental conditions investigated.

Typically, there exist a trade-off among the performance parameters of integrated fermentation and *in situ* gas stripping processes. For example, the total product produced increases with increasing gas flow rate whereas the concentration of products in the stripped stream (condensate) decreases with increasing gas flow rate in an integrated fermentation and *in situ* gas stripping process. The trade-off in these performance parameters stem from the fact that the parameters are conflicting; conflicting parameters mean as one parameter is enhanced, another parameter is simultaneously made worse. This phenomenon typically leads to conflicting objectives in optimization of the process. Single-objective optimization translate a multiple objective system into one objective system by either using relative weights to combine different objective functions into one objective function or choosing one objective function as the main objective function while transforming other objective functions into additional constraints. In contrast, multi-objective optimization is able to optimize multiple objectives simultaneously even in the presence of conflicting objectives, revealing multiple optimal solutions relative to the single-objective optimization that finds only one optimal solution.

To this end, this study will use and compare (a) single-objective optimization and (b) multi-objective optimization approaches for fermentation and *in situ* gas stripping processes under different conditions. An unsteady state batch fermentation and *in situ* gas stripping (simulated with a batch reactor in Aspen Plus linked to a Fortran user kinetics subroutine) will be connected to the multi-objective genetic algorithm in MATLAB for the multi-objective optimization. This study will highlight the advantages of multi-objective optimization to reveal and understand the trade-off that exist among conflicting objectives over traditional single-objective optimizati

496186 Enzyme Cocktail Design of β-Agarase Enzymes for Complete Hydrolysis of Agarose in Ionic Liquid-Pretreated *Gelidium amansii*

Abstract:

Red macroalgae *Gelidium amansii* is a prospective renewable resource for producing D-galactose and 3,6-anhydro-L-galactose due to its ubiquity and high carbohydrate content. Its main component agarose, is a linear polysaccharide with alternate repeating units of D-galactose and 3,6-anhydro-L-galactose linked via β-1,4- and α-1,3-glycosidic bonds. Enzymatic saccharification using β-agarase enzyme system with subsequent acetic acid or thermal pre-hydrolysis is currently the most popular method to produce the monomeric sugars from agarose. However, accumulation of 5-hydroxymethylfurfural (HMF) as sugar dehydration product, and generation of agarotriose as a byproduct that is recalcitrant to complete enzymatic hydrolysis, have been typically encountered.

Ionic liquid (IL) pretreatment with subsequent enzyme cocktail hydrolysis using β-agarases is a facile strategy to effectively deconstruct and hydrolyze the agar into its monomeric sugars, and to eliminate the generation of unwanted products. Moreover, the use of an IL system as an effective and eco-friendly solvent could alleviate enzyme activity losses in unfamiliar solvent phases. Four ILs namely [EMIM]Cl, [BMIM]Cl, [EMIM]OAc and [BMIM]OAc were employed to dissolve *G. amansii* from which [BMIM]OAc was found most suitable. The enzyme cocktail of β-agarase I (endo-acting), β-agarase II (exo-acting), and neoagarobiose hydrolase (NABH) were subsequently used for D-galactose and 3,6-anhydro-L-galactose production from the IL-dissolved *G. amansii*. Performance optimization of IL-enzyme cocktail system is currently under investigation. This work was supported by the Korea Research Fellowship Program through the NRF funded by the Ministry of Science, ICT and Future Planning (2015H1D3A1062172 and 2016R1C1B1013252) and the Ministry of Education (2009-0093816).

492357 Development of Bottom-up Life Cycle Inventory Methods for Chemical Reaction Systems

Abstract:

Bottom-up life cycle inventory (LCI) methods involve the use of process models to obtain the material and energy flows at various stages within an overall product life cycle. Process models of chemical reactions utilize the reaction rate, physical and chemical characteristics of the reaction mixture, and desired reaction extent to determine the reactor vessel configuration. Material flows of interest to life cycle inventory for the chemical reaction system include the reactant and product flows, byproducts, catalysts, and other reaction aids required. Energy balances for the reactor provide an estimate of the energy requirements. Energy flows include addition or removal of heat from the reaction system to maintain proper reaction conditions. The overall life cycle inventory must include an enumeration of the energy requirements and ancillary material flows, as well as the reactor infrastructure. Releases from the reactor, including fugitive emissions must be considered as part of the life cycle inventory. This presentation will focus on the development of life cycle inventory data from models of chemical reaction systems. The approach will include shortcut methods for sizing equipment and fugitive emission estimation.

UNIVERSITY OF LOUISVILLE.
CONN CENTER FOR RENEWABLE ENERGY RESEARCH

National Institute of Technology Tiruchirappalli
राष्ट्रीय प्रौद्योगिकी संस्थान तिरुचिरापल्ली

SIMULTANEOUS PRODUCTION OF BIOGAS AND HYDROGEN USING MICROBIAL ELECTROLYSIS CELL INTEGRATED WITH ANAEROBIC DIGESTER

Samsudeen N[a], Josh Spurgeon[b] and Jagannadh Satyavolu[b, *]
[a]Department of Chemical Engineering, National Institute of Technology, Tiruchirappalli, India
[b]Conn Center for Renewable Energy Research, University of Louisville, Kentucky, USA
Corresponding e-mail: Jagannadh.Satyavolu@Louisville.edu

Background

Microbial Electrolysis Cell (MEC) utilizing anaerobic digestion (AD) technology is a green technology for sustainable hydrogen production from wastewater. Simultaneous production of biogas as well as biohydrogen by integrating MEC in to an AD unit can increase the COD degradation rate in waste water as well as improve the overall economics of fuel production. The main aim of the present objective is to enhance the hydrogen and biogas production by integrating MEC with AD. The integrated system performance in terms of hydrogen and biogas generation and waste water treatment efficiency is investigated in this study

Reaction Mechanism

Biogas reaction $C_2H_4O_2 \rightarrow 2CO_2 + CH_4$

Anode : $C_2H_4O_2 + 2H_2O \rightarrow 2CO_2 + 8e^- + 8H^+$

Cathode: $8H^+ + 8e^- \rightarrow H_2$

Materials and Methods

Mixed inoculum was collected from MSD plant, Louisville, KY. The integrated system was designed with an anode chamber capacity of 450 mL and a cathode chamber capacity of 25 ml – both separated by a Nafion 117 membrane. Carbon cloth and Pt mesh used as anode and cathode electrode. Synthetic wastewater used as substrate in MEC. Potentiostat is used to apply potential to the MEC to produce hydrogen. Biogas and Hydrogen produced in MEC were analyzed using Gas chromatography.

Results

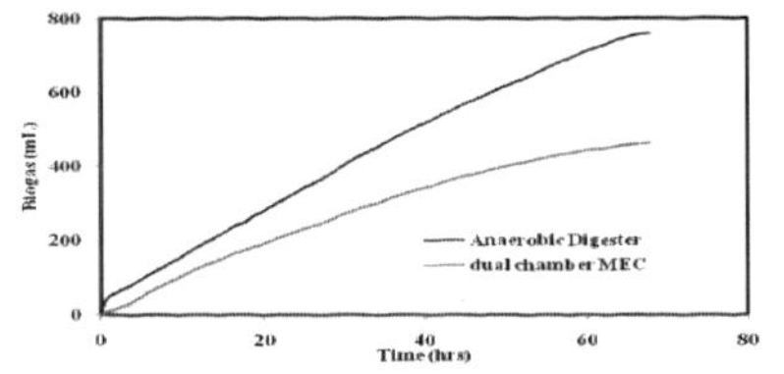

Fig. 1. Biogas production comparison in AD and MEC

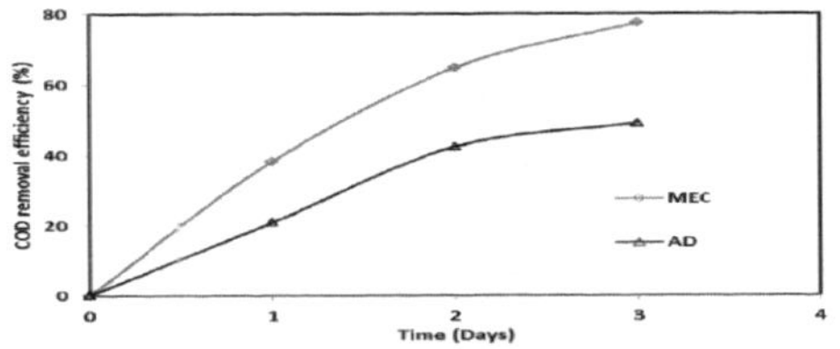

Fig.2. COD removal efficiency in AD and MEC

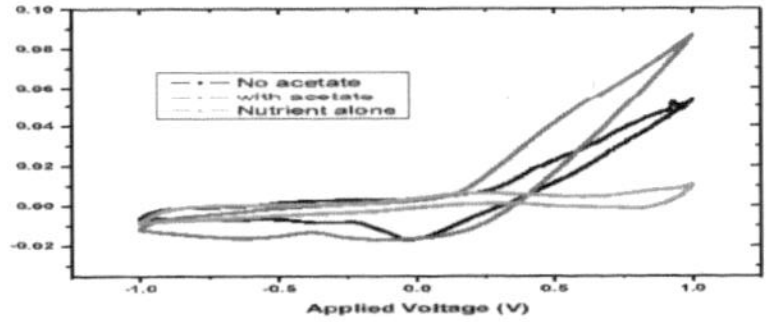

Fig. 3. CV analysis showed an effect of acetate in MEC

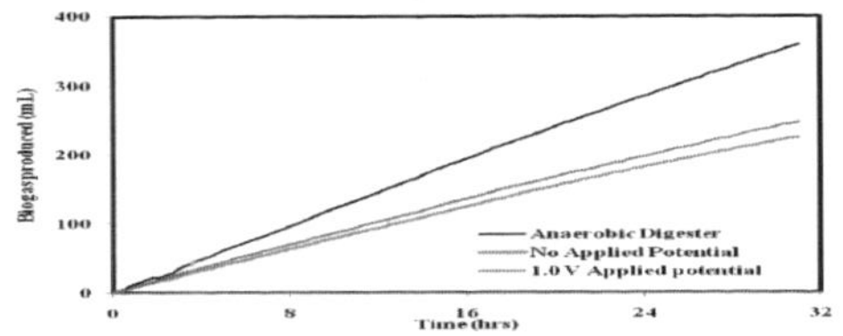

Fig. 4. Effect of voltage apply for biogas production

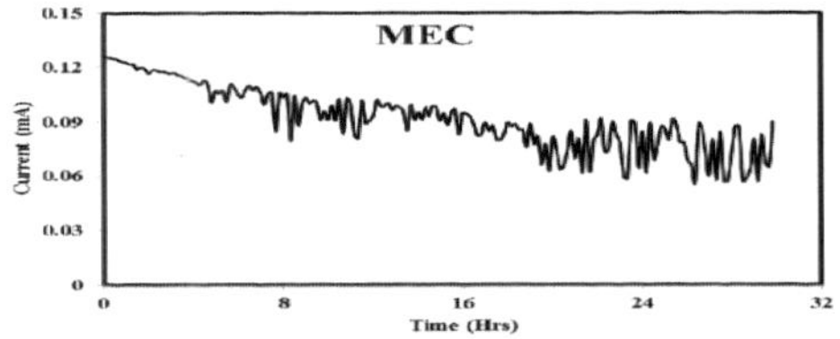

Fig. 5. Current variation during hydrogen production

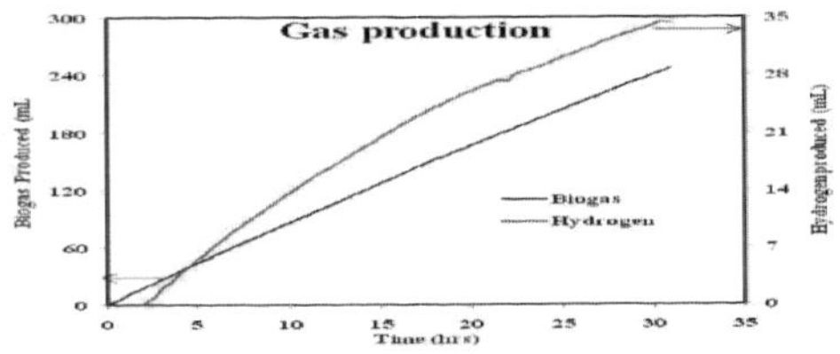

Fig. 6. Biogas and hydrogen production in MEC

Conclusions

- Biogas and hydrogen production by integration MEC into AD was studied.
- A maximum biogas and hydrogen of 232 mL and 32 mL were produced in 32 hours.
- Biogas and Hydrogen gas was confirmed by GC analysis.
- Cyclic Voltammetry (CV) showed a significant effect of presence and absence of acetate in MEC.

Acknowledgement

The authors thank Dr. Mahendra Sunkara of University of Louisville and Mr. Robert Bates of MSD treatment plant, Louisville, KY for their support during this work. Dr. Samsudeen also acknowledges the support from the Department of Biotechnology, Indo-US Science and Technology Forum (*IUSSTF*) for financial support for the B-ACER Award.

References

1. **Oh & Logan (2005).** Hydrogen and electricity production from a food processing wastewater using fermentation and microbial fuel cell technologies. *Water Research*, 39(19), 4673–4682.
2. **Kadier et al., (2016).** Recent advances and emerging challenges in microbial electrolysis cells (MECs) for microbial production of hydrogen and value-added chemicals. *Renewable and Sustainable Energy Reviews*, *61*, 501–525.

Rowan University

Novel Bio-Based Polyesters & Polycarbonates from Xylochemicals

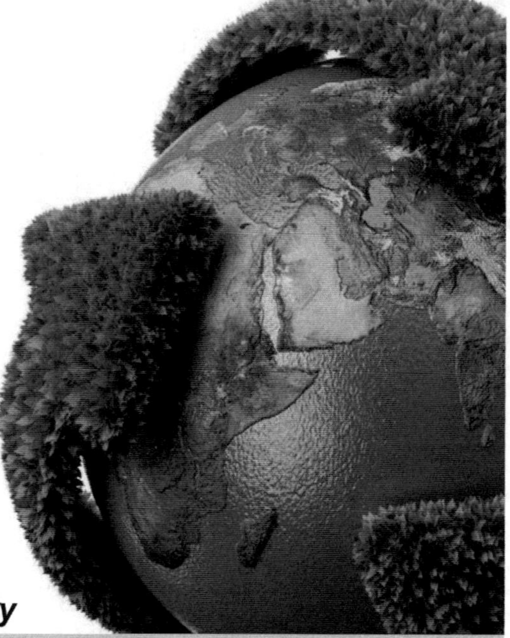

2017 AIChE Annual Meeting
Minneapolis, MN October 29 – November 3 2017
"Bioplastics, Biocomposites and Value-Added Uses of Biofuel Coproducts for Sustainable Manufacturing"

Joseph F. Stanzione, III, Ph.D.
Department of Chemical Engineering
Sustainable Materials Research Laboratory

 1

Plastics: The Facts

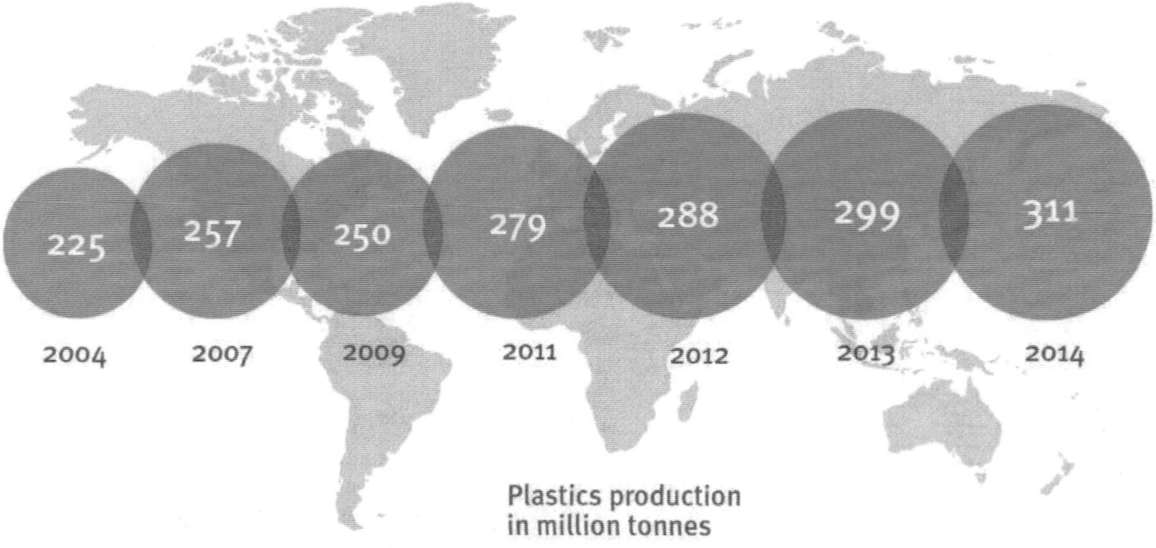

- **Plastic production is growing globally[1]**
- **Currently around 330 Mtonnes**

[1] PlasticsEurope, *Plastics - the Facts* (2015)

 2

Plastics: The Facts

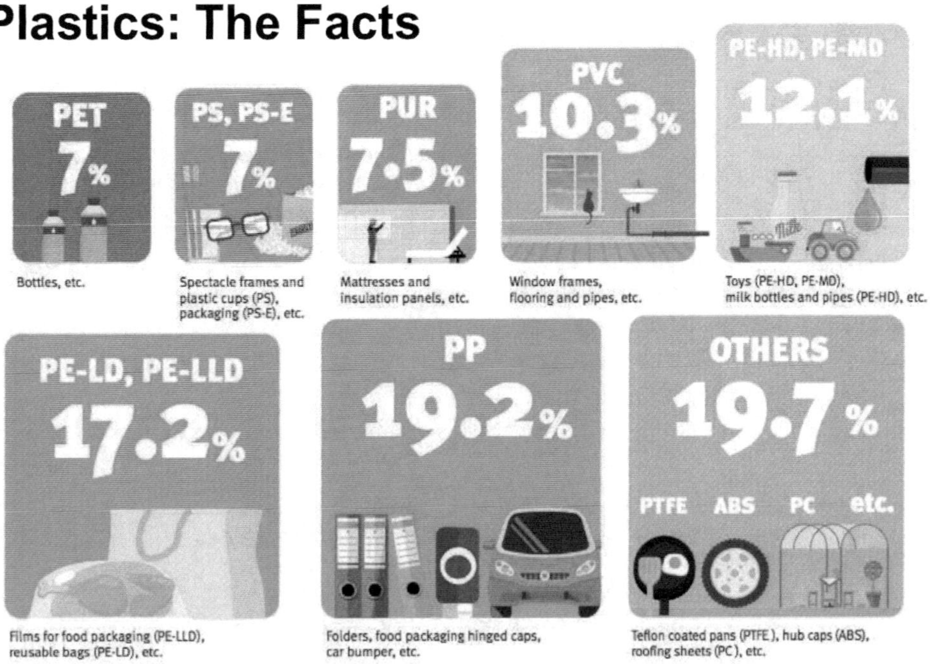

Less than 1.5% are renewably sourced[2]

[2] European Bioplastics, *Bioplastics facts and figures* (2017)

Plastics: The Future

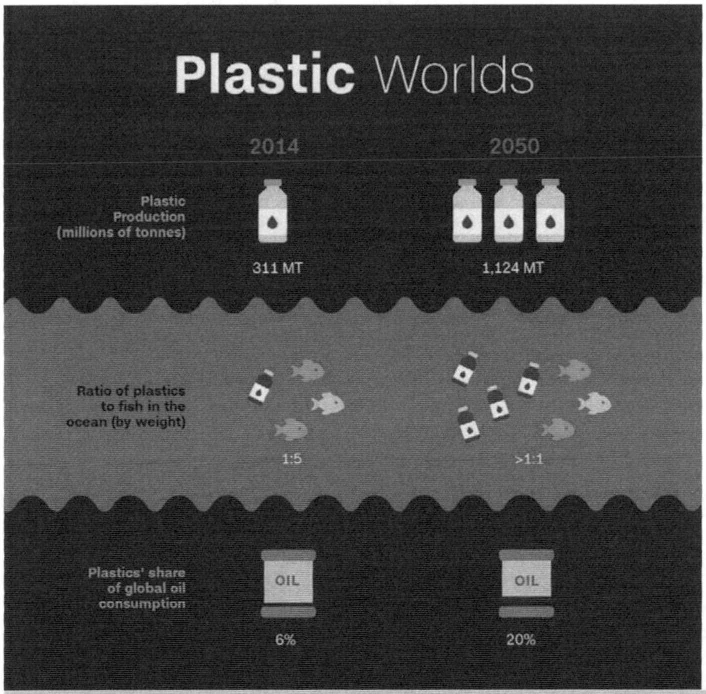

By 2050:

- **Expected to triple**
- **More plastic than fish in the ocean**
- **20% of global annual oil consumption**

[3] World Economic Forum (2016)

Trying to Replace PET

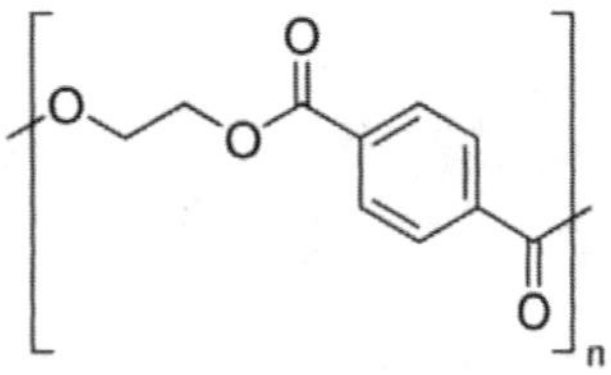

- **Very good mechanical properties**
- **T_g = (67-81) °C, $T_m \approx$ 267 °C**
- **Petroleum-derived (green routes in progress)**
- **Not easily degraded**

High Tg Bio-Based Aliphatic Polyesters from Bicyclic D-Mannitol
Cristina Lavilla, Abdelilah Alla, Antxon Martínez de Ilarduya, and Sebastián Muñoz-Guerra*
Departament d'Enginyeria Química, Universitat Politècnica de Catalunya, ETSEIB, Diagonal 647, 08028 Barcelona, Spain

- Sugar derived
- T_gs range from -40 to 70 °C

Renewable (semi)aromatic polyesters from symmetrical vanillin-based dimers†
Audrey Llevot,[a,b] Etienne Grau,[a,b] Stéphane Carlotti,[a,b] Stéphane Grelier[a,b] and Henri Cramail*[a,b]

- Lignin derived
- T_gs up to 140 °C
- Insoluble in any solvent

Bio-based alternatives in the synthesis of aliphatic–aromatic polyesters dedicated to biodegradable film applications
Nicolas Jacquel [c], René Saint-Loup [c], Jean-Pierre Pascault [a, b], Alain Rousseau [a, b], Françoise Fenouillot [a, b, *]
[a] Université de Lyon, F-69003, France
[b] INSA-Lyon, Ingénierie des Matériaux Polymères, IMP, UMR5223, F-69621 Villeurbanne, France
[c] Roquette Frères, Division Recherche, F-62080 Lestrem, France

- Sugar derived (isosorbide)
- Negative T_gs, high T_ms

etc…!

Additional Concerns: Toxicity

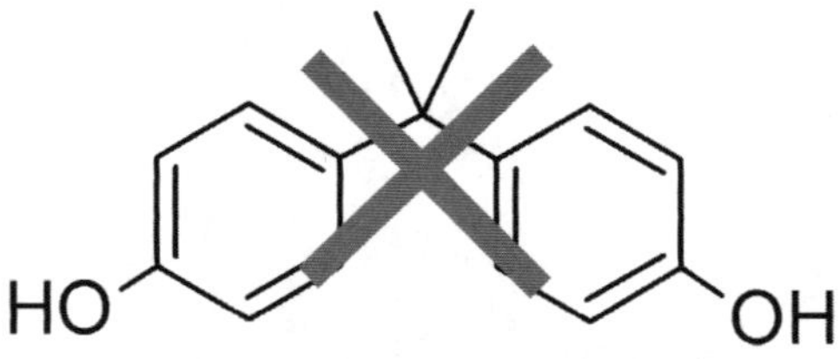

Bisphenol A (BPA)

Bisguaiacol (BG): A Lignin-Derived Bisphenol

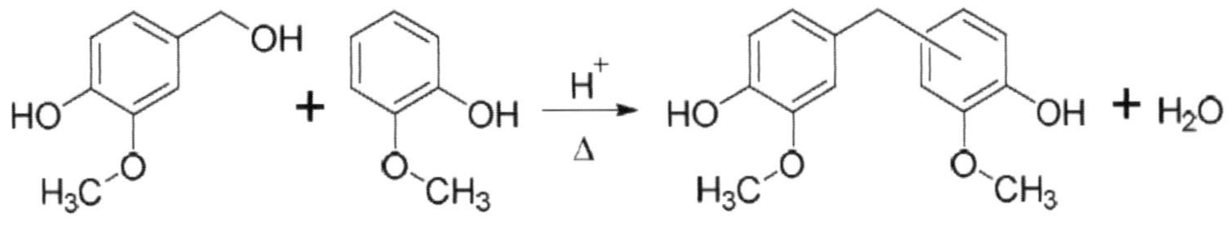

- **Similar structure to BPA[4]**
- **Fully renewable**
- **Reduced toxicity[5]**

- **Isomeric distribution**
  - **82%** ***p,p'***
  - **15%** ***m,p'***
  - **3%** ***o,p'***

[4] Hernandez *et al.*, *ACS Sustain Chem Eng* (2016)
[5] Hong *et al.*, *Int. J. Environ. Res. Publ. Health(* 2016)

 7

Q: Can we use BG to synthesize high-performance renewable thermoplastics?

 8

BG-based Polycarbonates

Joseph Mauck,
M.S. CHE

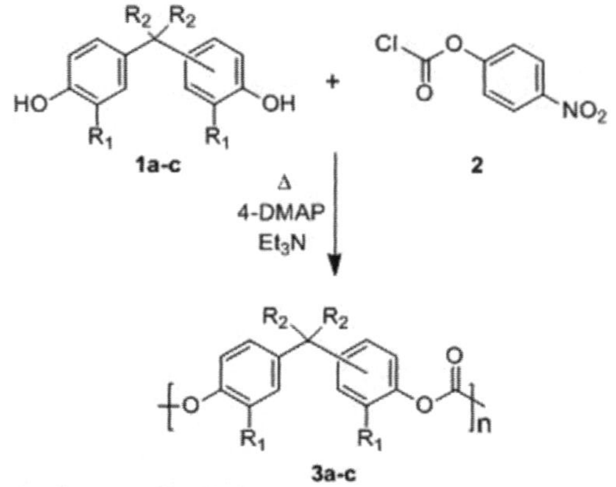

1a,3a: R_1=H, R_2=CH_3
1b,3b: R_1=H, R_2=H
1c,3c: R_1=OCH_3, R_2=H

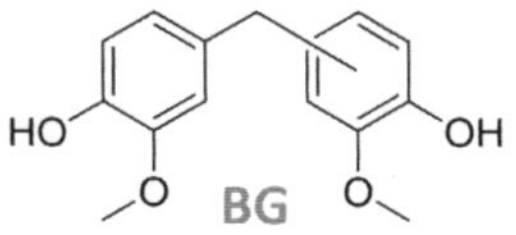

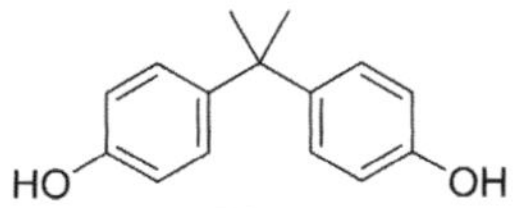

BPA

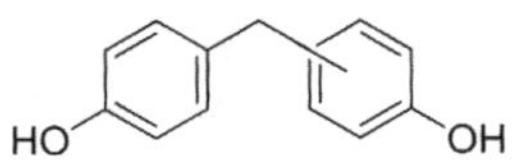

BPF

[6] J. Mauck *et al.*, *J Biobased Materials & Bioenergy* (2017)

BG-based Polycarbonates: Thermal Properties

Sample	*IDT* (°C)	$T_{50\%}$ (°C)	T_{max} (°C)	Char Content (%)	T_g (°C)
BG-PC	375	455	450	24.2	100
BPA-PC	440	543	545	23.2	150
BPF-PC	369	546	501	40.5	119

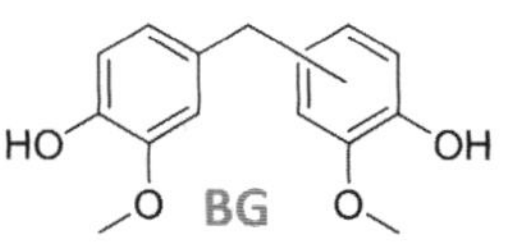

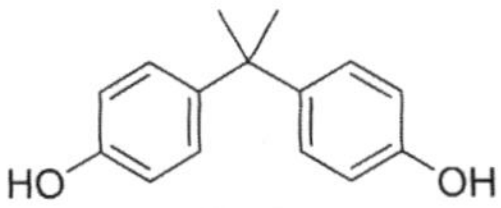

BPA

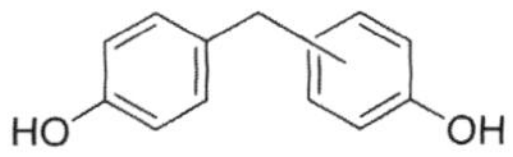

BPF

[6] J. Mauck *et al.*, *J Biobased Materials & Bioenergy* (2017)

Q: Can we use BG to synthesize polyesters?

Preparation of BG-based Polyesters

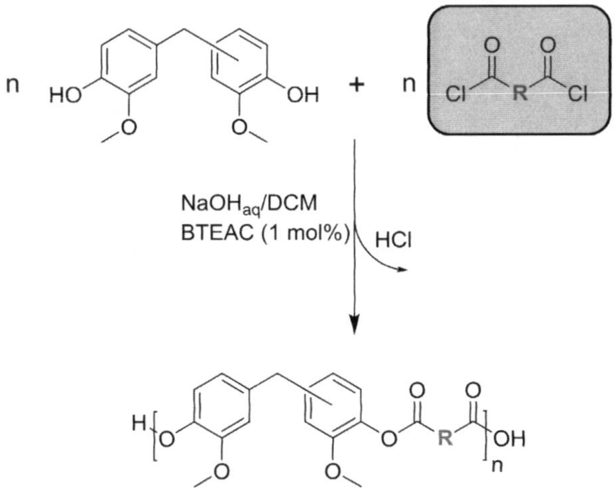

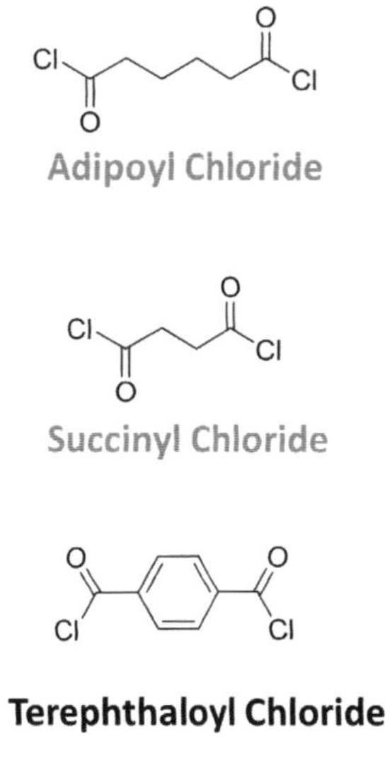

[7] S. Curia *et al.*, *In preparation* (2017)

BG-based polyesters

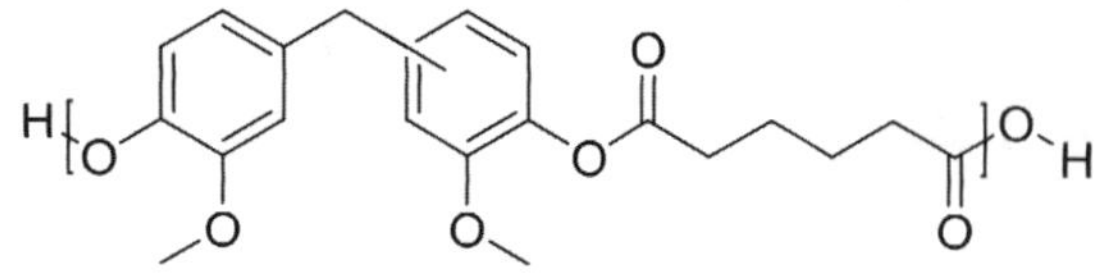

Poly(BG adip)
M_n=6.6 kg mol^{-1} (Đ=1.82)

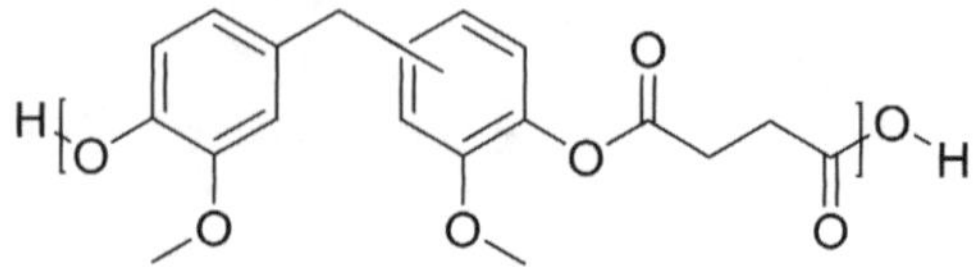

Poly(BG succ)
M_n=7.2 kg mol^{-1} (Đ=2.16)

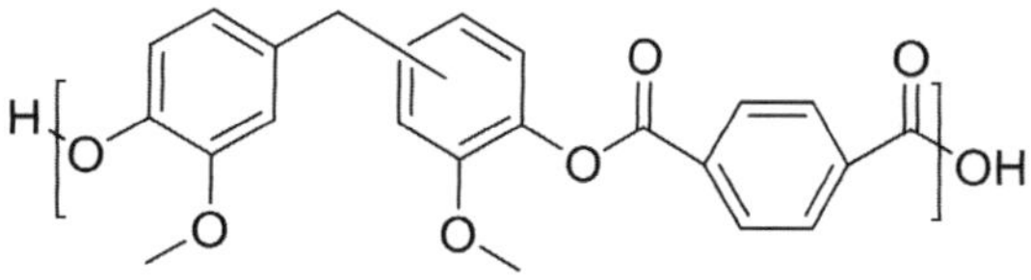

Poly(BG tere)
M_n=26.9 kg mol^{-1} (Đ=2.29)

[7] S. Curia *et al.*, *In preparation* (2017)

IR Analysis

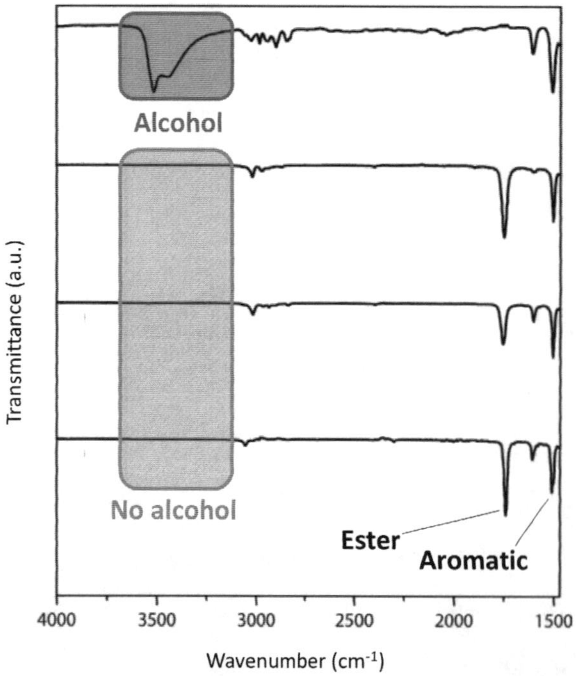

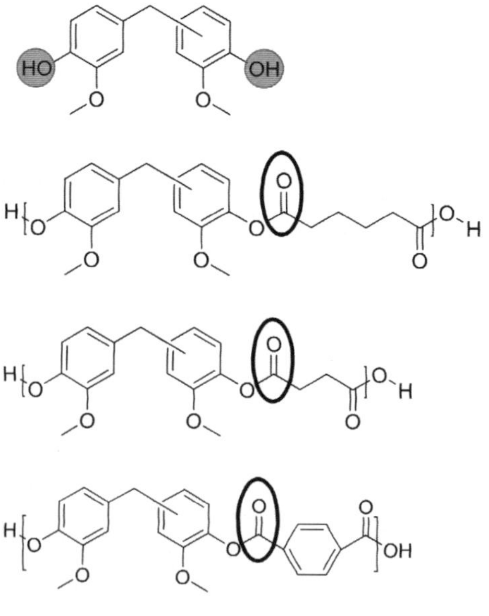

¹H-NMR Analysis (in $CDCl_3$)

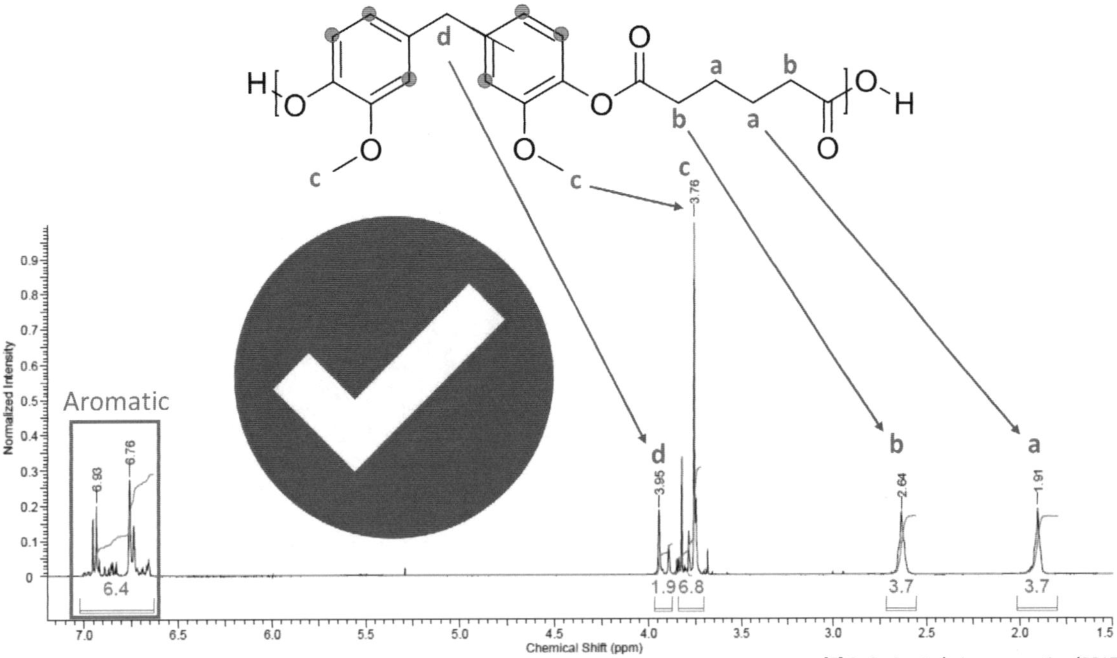

[7] S. Curia *et al.*, *In preparation* (2017)

BG-based Polyesters: DSC Analysis

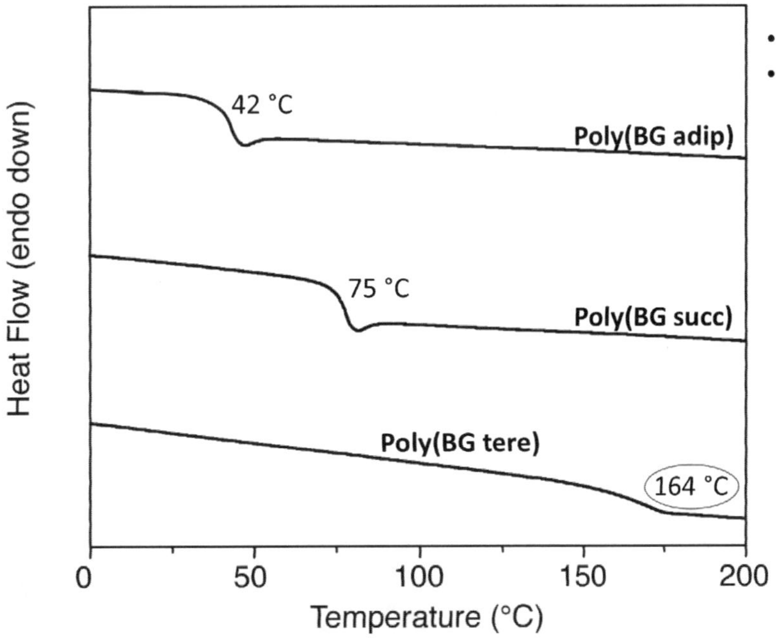

- **10 °C min^{-1}**
- **N_2 50 mL min^{-1}**

[7] S. Curia *et al.*, *In preparation* (2017)

BG-based Polyesters: Thermal Stability (N_2)

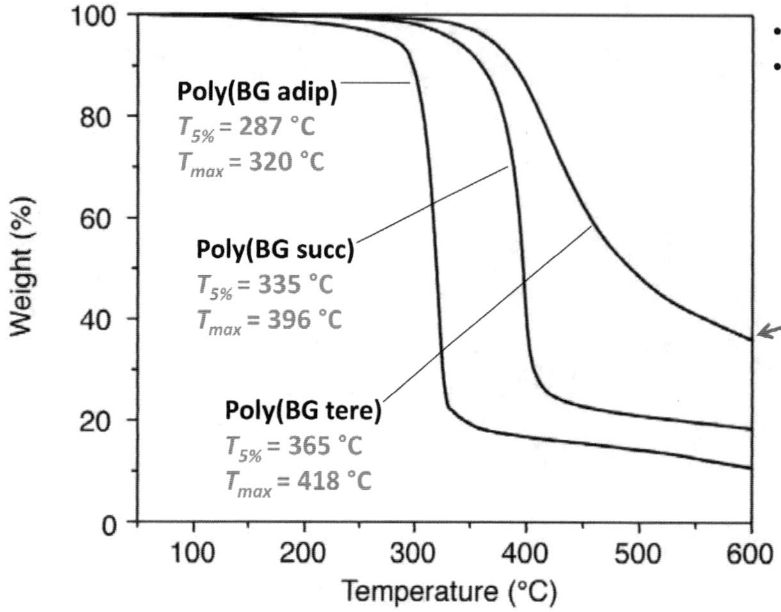

- 10 °C min^{-1}
- N_2 25 mL min^{-1}

36% char
@ 600 °C

[7] S. Curia *et al.*, *In preparation* (2017)

BG-based Polyesters: Thermal Stability (Air)

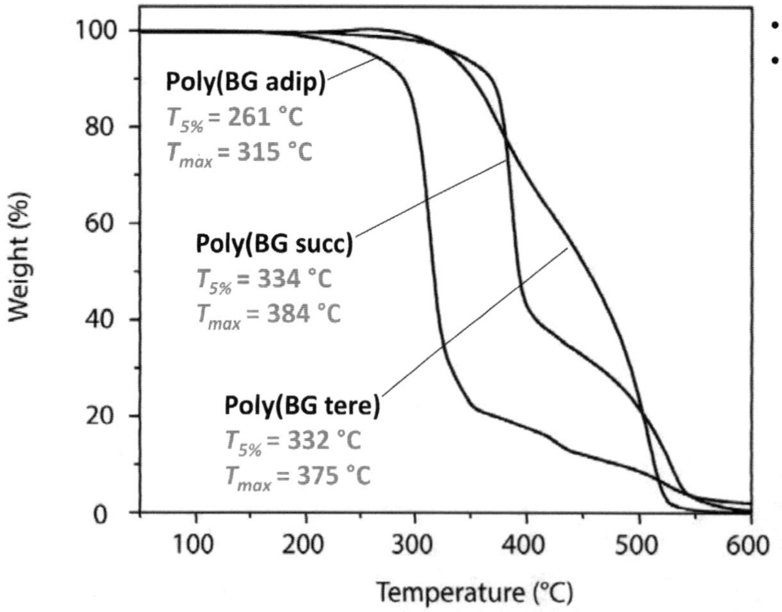

- 10 °C min^{-1}
- Air 25 mL min^{-1}

T_{max} still above 300 °C

[7] S. Curia *et al.*, *In preparation* (2017)

BG+FDCA: a Fully Renewable Polyarylate

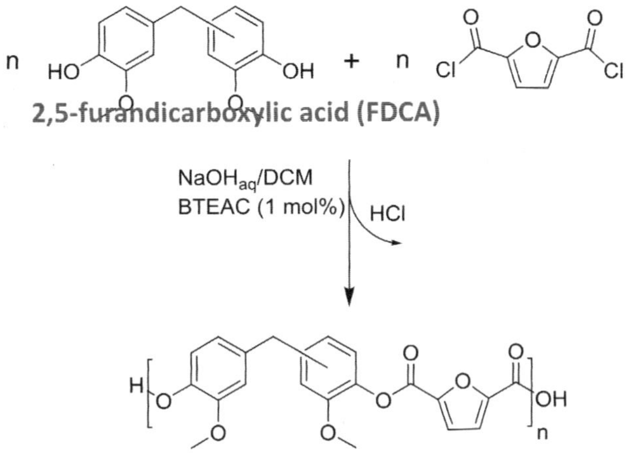

2,5-furandicarboxylic acid (FDCA)

Poly(BG FDCA)
M_n=6.3 kg mol^{-1} (Đ=1.57)

[7] S. Curia *et al.*, *In preparation* (2017)

Poly(BG-FDCA): DSC

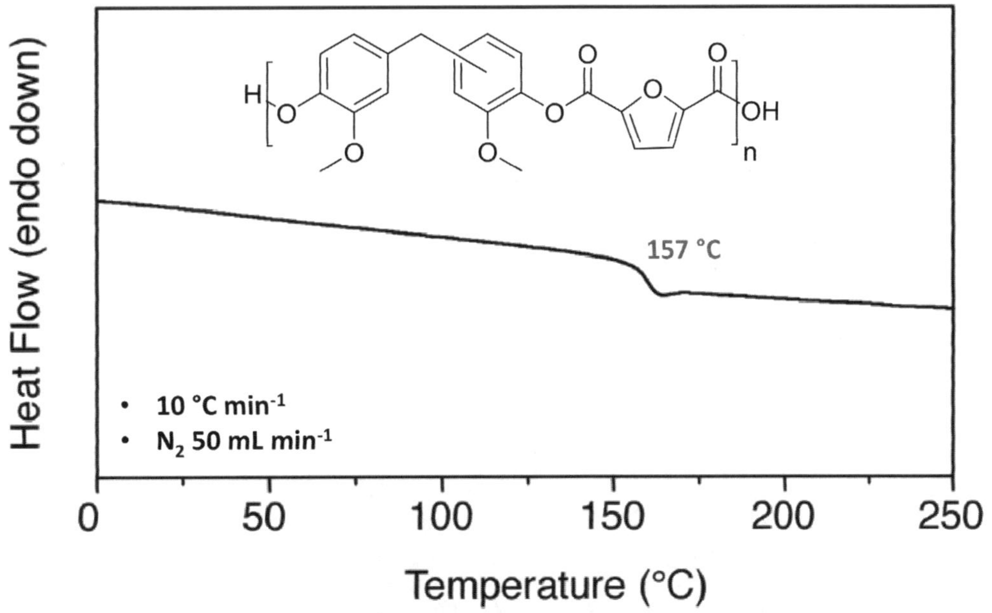

[7] S. Curia *et al.*, *In preparation* (2017)

Poly(BG-FDCA): TGA N_2 *vs* air

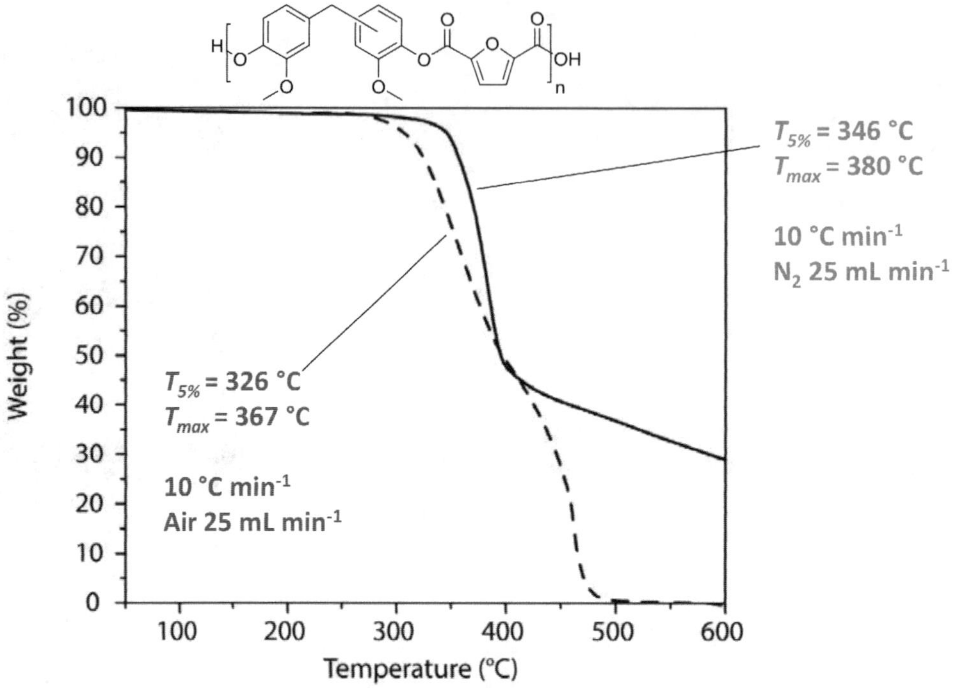

[7] S. Curia *et al.*, *In preparation* (2017)

Conclusions

- **BG** can be used to synthesise **useful polycarbonates & polyesters**
- **BG-based polyesters** show **high thermal resistance** and **tuneable T_g**
- **Poly(BG FDCA)** has a remarkably **high T_g (157 °C)**

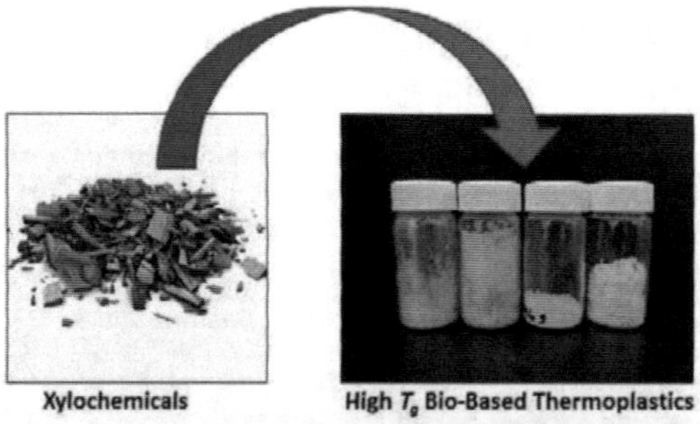

Future Investigations

- Hydrolytic degradation & biodegradation tests
- End-cap with functional molecules
- Chain extension, melt polymerization (achieve higher MW)
- Gas permeability/diffusivity
- Mechanical tests

Xylochemicals High T_g Bio-Based Thermoplastics

Acknowledgements

- Sustainable Materials Research Laboratory
- Joseph R. Mauck
- Silvio Curia, Ph.D.
- John J. La Scala, Ph.D. (ARL)
- Rowan University
  - ❖ Department of Chemical Engineering
  - ❖ Department of Chemistry & Biochemistry
- U.S. ARL for Financial Support via Cooperative Agreement W911NF-16-2-0225

Silvio Curia, Ph.D.

Rowan University

ARL U.S. ARMY RDECOM

South Jersey Technology Park at Rowan University

Thank you!
Any questions?

Crater Lake National Park

Rocky Mountain National Park

503981 Light Weight Biocomposite from Toughened Polyolefin and Biocarbon

Abstract:

In converting biomass to bioenergy through thermochemical processes, a considerable amount of solid residue generates in each conversion cycle (up to 35 wt.%) which mainly consists of renewable carbon (biocarbon). However, the biocarbon is considered as a byproduct of the pyrolysis units. Unless a high-volume application exists for the biocarbon, the significant amount of this byproduct can challenge the long-term sustainability of pyrolysis units. In this project, biocarbon is intended to be used as filler in plastic industry targeting toughened polypropylene (PP) based automotive parts. In order to achieve a suitable balance of mechanical properties, biocomposites were compatibilized using a maleated polypropylene. Furthermore, the PP phase was nucleated using a Î² nucleating agent to increase the impact toughness without significant loss in the stiffness. Tensile properties and notched Izod impact strength were measured as response variable. Atomic force microscopy (AFM) was utilized to analyze the PP-biocarbon interface in presence and absence of the compatibilizer. The effect of the nucleating agent was investigated by differential scanning calorimetry (DSC) and correlated to the improvement of biocomposites impact toughness. The results suggest that the morphological changes caused by the compatibilizer and toughening effect of Î² crystals can generate high impact toughness in composites with 20 wt.% loaded biocarbon. Therefore, the biocarbon can be utilized in high loading levels similar to current petroleum-based composites with comparable mechanical properties. At the end, substituting conventional mineral filled composites with biocarbon-based biocomposites can provide the required mechanical properties with added benefits of lower density (â‰ˆ15%), 30% higher bio-based content and cheaper price. These benefits would translate into improved fuel economy of cars and reduction in greenhouse gas production. Considering the high-volume usage of plastic parts especially in automotive industry, this approach would make the biomass to bioenergy conversion more sustainable.

Keywords: Biocarbon, Toughened biocomposites, Polypropylene.

Acknowledgement: We gratefully acknowledge the financial support of the Ontario Ministry of Agriculture, Food and Rural Affairs (OMAFRA)/University of Guelph - Bioeconomy for Industrial Uses Research Program Theme Project # 030055; Ontario Research Fund, Research Excellence Program; Round-7 (ORF-RE07) from the Ontario Ministry of Research, Innovation and Science (MRIS) (Project #052644 and 052665), and the Natural Sciences and Engineering Research Council (NSERC), Canada Discovery Grant Project # 401111 to carry out this research work

497443 Secondary Fermentation of Corn Ethanol Co-Products for Improved Amino Acid Qualities

Abstract:

Secondary Fermentation of Corn Ethanol Co-Products for improved Amino Acid Qualities

In 2016 5.28 billion bushels of corn were used to produce about 14.79 billion gallons of ethanol in the United States. As a result, about 36 million tons of Dried Distillers Grains with Solubles (DDGS) were manufactured and fed to livestock (U.S. Dept. of Energy, 2017). DDGS are a common feed supplement in cattle rations as it is inexpensive and has positive feeding characteristics. One of the drawbacks of DDGS is that it is lacking in in key amino acids such as tryptophan, arginine, and lysine. Historically these amino acids have been supplemented by external addition of feed grade amino acids to rations. The research carried out attempts to fortify DDGS with higher amounts of tryptophan through secondary fermentation of Whole Stillage (WS) by bacteria. When cultured on WS the bacteria secretes extracellular tryptophan, which serves to close the amino acid gap in corn ethanol co-products. The bacteria used is Generally Regarded as Safe (GRAS) and has been used to produce feed ingredients historically. Because of the research, profitability of corn ethanol operations will be increased due to the improved feeding value of the co-products produced.

References

United State Department of Energy, â€œEthanol Fuel Basics,â€ Energy Efficiency & Renewable Energy http://www.afdc.energy.gov/fuels/ethanol_fuel_basics.html 03/16/2017

500583 Extraction High-Value Chemicals from Ethanol Co-Products: A Feasibility Assay on Phytate Extraction with Life Cycle and Techno Economic Assessment

Abstract:

Background

The well-established dry-grind process today accounts for also over 80% of the operating ethanol facilities in the United States, producing not only ethanol, but also distiller's grains. Distiller's grains, commonly commercialized as dried distiller's grains with solubles (DDGS) accounts for most of the heavy fraction from the ethanol distillation and rectification operations, and is processed through a complex downstream processing after the fermentation is unloaded. Despite being a nutritious animal feed, inclusion of DDGS as feed ration is still limited due to compositional properties, such as the high presence of non-starch polysaccharides, which are non-digestible to monogastric animals, the fatty acid profile, and the overdose of phosphorus, which is largely non-available to some animals. Phosphorus behavior in ethanol downstream processing is of particular interest, not only because it can lead to environmental issues, but also because it can potentially be an additional economic source to ethanol producers. Over 40% of the phosphorus present in ethanol co-products is present as phytate, the main form of organic phosphorus in cereals and grains, which is not completely hydrolyzed during the whole ethanol process. Phytate has a high market value, and today is a platform chemical utilized in a wide array of industries, and it is mostly exploited because of its high chelating potential (He et al. 2017). Phytate is also the raw material to produce *myo*-inositol, a cyclic polyol with a variety of uses by pharmaceutical and nutraceutical companies.

Phytate is found to be dissolved on the ethanol co-products(He et al. 2017). During the dry-grind downstream processing, thin stillage, i.e., the liquid-rich fraction derived from the centrifugation of whole stillage, accounts for the highest concentration of phytate among all the co-products. Recent research (He et al. 2017) has proposed an extraction method using ion exchange separation, which can efficiently separate phytate from thin stillage, with little to no effect on the residual nutritional profile of the final DDGS. Even though the concentration of phytate being relatively low on ethanol co-products, at around 1 g L^{-1} in thin stillage, due to the enormous volumetric production by dry-grind plants, the production leveraging costs in extracting phytate might become an industrial alternative.

The high volumes of DDGS produced by ethanol plants, alongside with its limited dietary inclusion, pose a challenge for higher revenues to ethanol plants. The second-most important co-product from dry-grind ethanol facilities, distiller's corn oil, represent today a mature product suitable for biodiesel production and other uses of low-grade vegetable oil(Reis et al. 2017). Removal of distiller's corn oil, on the other hand, has been criticized by animal nutritionists claiming a loss in energy value from DDGS. Despite still representing about 20% of the total revenue stream from dry-grind ethanol plants, DDGS are still ranking in low positions when compared to higher end animal feeds, such as soybean meal. The extraction of high-value chemicals through mature sound technologies may represent, thus, an alternative for higher economic security of ethanol plants. This current research describes a techno-economic analysis and a life-cycle assessment of a phytate extraction process on a large-scale dry-grind corn-to-ethanol plant in the United States, and its posterior *in-loco* conversion to *myo*-inositol.

Methods

The scope of this work analyzed the utilization of corn-to-ethanol thin stillage produced through the dry-grind process as raw material to extract phytate using an ion exchange process, followed by hydrolysis to generate *myo*-inositol. The process was modelled according to He et al. (2017) , and by additional empirical analysis, when required. Thin stillage was obtained from a dry-grind ethanol plant in the state of Iowa, USA. The process of converting thin stillage to a phytate brine solution is divided into the major steps: (1) filtration, (2) flow through ion exchange column, (3) desorption of phytate using brine solution, (4) reconditioning of column using sodium hydroxide, and (5) column washing using water. The conversion of phytate brine solution to calcium phytate is done by (6) addition of calcium chloride and sodium hydroxide, followed by product filtration, and production of *myo*-inositol is achieved by (7) hydrolysis of phytate brine solution, (8) precipitation of phosphate with calcium chloride and sodium hydroxide, and (9) evaporation and drying of *myo*-inositol.

The life cycle assessment of calcium phytate extracted from corn-to-ethanol thin stillage was estimated using GREET model. Calcium phytate extraction, and posterior conversion to *myo*-inositol were implemented on two stationary models on GREET – Dry mill ethanol production w/ corn oil extraction, and Dry mill ethanol production w/o corn oil extraction. Techno-economic assessment for ethanol plants was considered using the BuGal model, with added market and technical data for the modified phytate extraction and conversion units. Since the basis of the BuGal model for techno-economic analysis has been modified to analyze both conditions (with and without distiller's corn oil extraction), the two stationary models on GREET to describe corn-to-ethanol process were utilized.

Results and discussion

LCA results indicate that the proposed phytate extraction does not impact significantly most of the environmental impact factors of ethanol. As a gallon of ethanol was considered as functional unit, the results indicate that for a system with phytate extraction and oil extraction, CO_2 emissions will increase from 3.21 kg to 3.28 kg per gallon of ethanol, whilst for a system without distiller's corn oil extraction, 3.23 kg of CO_2 will be increased to 3.30 kg of CO_2 per gallon of ethanol produced. NOx, SOx, N2O, and other gases also increased with phytate extraction by a factor of about 1%. Analyses were carried out for techno economic assessment in 4 different cases, (1) with distiller's corn oil extraction and phytate extraction, (2) with distiller's corn oil extraction and without phytate extraction, (3) without oil extraction and without phytate extraction extraction, (4) without oil extraction and with phytate extraction, and were consistently all across different cases.

Unlike major recent upgrades to the dry-grind ethanol industry, such as distiller's corn oil removal, extraction of phytate and production of *myo*-inositol are related to products found in dilute concentrations. The low concentration of phytate in ethanol co-products requires low energy input on an ethanol basis, while still providing a high-value co-product. BuGal model calculates the operating expenses per gallon of denatured ethanol sold as USD 0.2673 for the control conditions, i.e., without phytate extraction, and USD 0.2686 for a system considering 20% of thin stillage being redirected to phytate extraction, due to increase in energy consumption for additional processing. Regarding chemical costs, phytate extraction and conversion to inositol likely increased the value from USD 0.1201 to USD 0.1226 per gallon of denatured ethanol, mainly due to the requirement of NaCl for phytate desorption from the ion exchange process, additional process water, and calcium chloride addition. These costs are balanced by the calcium phytate, *myo*-inositol, and calcium phosphate as fertilizer (derived from the phytate hydrolysis) sales, at prices estimated at USD 12, USD 10, and USD 0.20 per kg, respectively. BuGal estimates an average large-scale dry-grind operation, with factor of equity and debt being 0.7 and 0.3, respectively, with an interest rate charged on debt as 0.07, and the rate of return required by investors on the plant's equity as 0.12. Even though these numbers are variable, the economic considerations on such assumptions are beyond the scope of this work, and remain constant through all assumptions. Following similar numbers as Kwiatkowski et al. (2006), 2.795 gallons of ethanol are estimated to be produced from each bushel of corn, and 17.5 lb of

DDGS, and CO_2 are produced on a bushel basis. Denatured ethanol sales are estimated to be at USD 1.70 per gallon, an estimate within current market prices , and that phytate extraction would not affect the price of DDGS, thus, being marketed at a price of USD 161 per ton, also within current market estimates.

Conclusion

The removal of phytate from thin stillage may represent a novel process in the dry-grind ethanol industry. The removal of phytate via ion exchange does not increase significantly (around 1% of increase) the emissions involved on the production of one gallon of ethanol, and that it can be a source of profit if operated within a certain range of thin stillage processed within a plant. The results also presented herein describe the potential of converting the extracted phytate to myo-inositol. Myo-inositol has a more flexible market when compared to phytate, and could also represent another source of revenue for ethanol plants.

References

Kwiatkowski JR, McAloon AJ, Taylor F, Johnston DB (2006) Modeling the process and costs of fuel ethanol production by the corn dry-grind process Industrial crops and products 23:288-296

He Q, Rodrigues Reis, CE, Wang F, Hu B (2017) Phytate extraction from coproducts of the dry-grind corn ethanol process. RSC Advances. doi:10.1039/C6RA27409A

Reis CER, Rajendran A, Hu B (2017) New technologies in value addition to the thin stillage from corn-to-ethanol process Reviews in Environmental Science and Bio/Technology 16:175-206 doi:10.1007/s11157-017-9421-6

500197 Towards the Design of Sulfur-Tolerant CO_2-Reforming Catalysts

Abstract:

Mixed metal oxide catalysts, combining late transition metals (TMs) doped into earth oxides (REOs), offer hydrocarbon conversion (e.g., tar reforming) activities and selectivities that meet or exceed supported metal systems,[1] while also offering continued catalytic performance in the presence of sulfur poisons and sintering agents such as water.[2] These materials also have potential as sulfur tolerant CO_2-reforming or tri-reforming catalysis. The objectives of the proposed work are to acquire a "proof of principle" data set towards the development of sulfur- and water-tolerant (especially the former) TM-doped rare-earth oxysulfide catalysts for high temperature reforming, and 2) develop a mechanistic rationale for how surface sulfidation alters reforming and coke formation kinetics. Integrated experimental and computational efforts will provide a mechanistic rationale for designing sulfur- and water-tolerant catalysts. Density functional theory (DFT) methods examine C-H and C-C bond activation and coke formation over partially sulfided metal oxide surfaces, identifying materials promising for retaining, or possibly enhancing, conversion in the presence of sulfur. Synthesis, testing, and characterization evaluate the computationally predicted catalyst performance and provide feedback and further hypothesis development to aid the computational efforts.

To date, we have found that Ce-Zr supports at the 1:1 or higher elemental ratio are superior to the Ce-La supports at similar elemental ratios. The latter are relatively inactive. We have also found that coking rates in the absence of water are strongly dependent upon stoichiometric equivalence between CO_2 and CH_4. Structurally, both the Ce-Zr and Ce-La families of supports show no post-mortem evidence of separate TM (Ni, Mn, Fe) or TM oxide phases, at least for TM:REO ratios kept below thresholds determined in past work on tar reforming. Characterizations of the used catalysts have included XANES, XAFS, Raman, XRD and chemisorption techniques.

Extended coking rate measurements by DSC-TGA show that for the TM-doped REO catalysts a coking plateau is attained, while for simpler Co- and Ni-supported Al_2O_3 catalysts the coking rates are nonzero and almost constant. This suggests that regardless of their short-term performance, the latter family of catalysts supported on Al_2O_3 will ultimately deactivate more rapidly.

1. R. Li, A. Roy, J. Bridges and K.M. Dooley, "Tar Reforming in Model Gasifier Effluents: Transition Metal/Rare Earth Oxide Catalysts," Ind. Eng. Chem. Res. 53, 7999–8011 (2014)
2. M. Krcha, K. Dooley, and M. Janik, "Alkane Reforming on Partially Sulfided CeO_2 (1 1 1) Surfaces," J. Catal., 330, 167-176 (2015)

500717 Development of Catalytic Process for CO_2 Utilization

Abstract:

The utilization of CO2 as a raw material in the synthesis of chemicals and liquid energy carriers offers a direct way to mitigate the increasing CO2 build-up in the atmosphere. However, high-energy reagents and harsh reaction conditions are typically required for transforming CO2 into other chemicals. Renewable hydrogen is a high-energy material that can be stored by hydrogenating CO2 to valuable secondary energy carriers, such as methane, methanol, and formic acid. Indeed, formic acid/formate has been widely used in agriculture and industry and is the potential hydrogen storage. Herein, we developed a process of hydrogenating liquid amine captured CO2, the intermediates in the Bioenergy with Carbon Capture and Storage (BECCS) process, over supported palladium nano-catalysts, to produce formates. We further developed the rechargeable â€œhydrogen batteryâ€ based on the redox equilibrium between amine captured CO2 / formate amine adducts. The stored hydrogen can be catalytically released under mild conditions as the feed to fuel cells for regenerating electricity. The strategies of the process development has been focused on the efficiency (rate of CO2 hydrogenation and dehydrogenation reactions) and energy use (the amount of energy required to utilize CO2 in existing products). This development also includes gaining insight into the relationships among the various structures of amines, solvent effect, distribution of intermediates, catalytic performance, and reaction mechanism.

487340 Catalytic Methanation of Carbon Dioxide with Ni/MgO Catalyst in Dual-Step Reductive Calcination

Abstract:

The heterogeneously catalyzed gas-phase conversion of carbon dioxide with hydrogen to produce methane has gained major interest in recent years [1]. One of the reasons for this increased interest is the quest for hydrogen storage systems. Renewably produced hydrogen is stored via methanation of the greenhouse gas carbon dioxide. Methane can partially substitute natural gas in an established gas grid. Our research focuses on process integration in this context and addresses potential carbon dioxide sources. Calcination is a widely used preparatory step in mineral carbonate beneficiation. During calcination carbon dioxide is released in high concentration into the off-gas. If calcination is performed in hydrogen atmosphere (= reductive calcination), a highly concentrated (CO_2/H_2) process gas stream suitable for direct gas phase methanation can be prepared. Our research [2,3] revealed that reductive calcination is a low temperature process compared to calcination in oxidizing atmosphere, and carbon dioxide is partially converted into methane and/or carbon monoxide in a single-step process. The yield of methane and carbon monoxide can be partially controlled by reaction temperature and pressure. Selective methanation, however, is not achieved in the calcining step. Therefore, methanation of the process gas from reductive calcination by catalytic methanation at ambient pressure with a Ni/MgO catalyst was investigated [4]. Although methanation of CO_2 according to the Sabatier reaction has been investigated since decades, little information about process performance at ambient pressure is available. At moderate temperature (534 K) and ambient pressure CO_2-conversion close to equilibrium conversion beyond 90% is possible. Adsorption of reactants and products does not affect kinteics. At moderate temperature conversion does draw advantages from interaction of carbon dioxide with MgO. Latter effect was confirmed by direct methanation of $MgCO_3$ with hydrogen without catalytic support.

1. Bailera, M., Lisbona, P., Romeo, L.M., and Espatolero, S. (2017) Power to Gas projects review: Lab, pilot and demo plants for storing renewable energy and CO2. Renew. Sustain. Energy Rev., 69 (January 2016), 292–312.

2. Baldauf-Sommerbauer, G., Lux, S., Aniser, W., and Siebenhofer, M. (2017) Synthesis of carbon monoxide from hydrogen and magnesite/dolomite. Chemie Ing. Tech., 89 (1–2), 172–179.

3. Baldauf-Sommerbauer, G., Lux, S., Aniser, W., and Siebenhofer, M. (2016) Reductive calcination of mineral magnesite: hydrogenation of carbon dioxide without catalysts. Chem. Eng. Technol., 39 (11), 2035–2041.

4. Baldauf-Sommerbauer, G., Lux, S., and Siebenhofer, M. (2016) Methanation of carbon dioxide with a nickel/magnesium oxide catalyst. in: AIChE Annual Meeting 2016, San Francisco, USA.

501005 Glycerol Transfer Hydrogenation of CO_2 Using Ir and Ru Carbene Organometallics Immobilized on Active Hydrotalcites in a Packed Bed Flow Reactor

Abstract:

Utilization of CO_2 has become a focal point for reducing carbon emissions. A long-standing method involves direct hydrogenation and transfer hydrogenation (TH) using isopropanol as a hydrogen donor to convert CO_2 to value-added commodity chemicals such as formic acid. Less explored is the use of glycerol, a low value byproduct of biodiesel synthesis made from trans-esterification of triglycerides, as a substitute for isopropanol due to thermodynamic and kinetic challenges. Using CO_2 as a substrate and glycerol as a hydrogen donor for TH yields formate and lactate under basic condition following a Cannizzaro reaction. Herein we explored the use of ruthenium and iridium N-heterocyclic carbenes (NHC) homogeneous catalysts immobilized on a tunable hydrotalcite (HT) supports for CO_2 TH, converting CO_2 to potassium formate and glycerol to potassium lactate in a packed bed flow reactor. HT was synthesized using 3:1 ratio of Al:Mg in flow, and composition was varied by incorporating Cu or Fe in place of Al up to a total of 15 wt. %. Comparison of the heterogenous catalysts to their respective homogenous catalysts indicated that the HT is non-innocent during the reaction. The use of a flow reactor allowed for continuous data collection of kinetic parameters as well as catalytic efficiencies within a single reaction setup and demonstrated that immobilized catalysts are stable for over 300 hours on stream with no sign of deactivation in the presence of 0.25 - 2.0 M potassium hydroxide (KOH) at 150-225ËšC. Precise control over the gas-to-liquid ratio within the reactor provided insight into the efficiency of catalysts at utilizing CO_2, giving scale-up details on required gas stream purity and operating pressure requirements. Reactions were initially carried out in 45 bars of CO_2 in a batch reactor and then scaled to a flow reactor using less than 35 bar CO_2. Upon depletion of accessible carbonate in the solution, the formate yield stabilizes while lactate/lactic acid yield continues to form under acid conditions at a reduced rate. Due to similar turn-over-frequencies (TOF) in 0.25-2.0 M KOH exceeding 4.4 K and 1.5 K per hour for lactate and formate, respectively.

489777 Combining CO_2 Reduction with Propane Oxidative Dehydrogenation over Bimetallic Catalysts

Abstract:

Combining CO_2 reduction with propane oxidative dehydrogenation over bimetallic catalysts

Elaine Gomez and Jingguang G. Chen*

Department of Chemical Engineering, Columbia University, New York City, NY 10027, USA

*jgchen@columbia.edu

Abstract: Propylene is one of the most diverse petrochemical building blocks used for the production of many chemicals (e.g. polypropylene, propylene oxide, and acrylonitrile). The inherent variability and insufficiencies in the co-production of propylene from steam crackers has raised concerns regarding the global propylene production gap and has directed industry to more on-purpose propylene technologies[1]. The oxidative dehydrogenation of propane by CO_2 (CO_2-ODHP) can potentially fill this gap while consuming a greenhouse gas. Utilizing propane and CO_2 presents several advantages over the conventional processes of direct or O_2 assisted propane dehydrogenation, including lower operating temperatures, less coke formation, and higher equilibrium conversions.[2] The reactions of CO_2 with propane may occur through two distinct pathways, oxidative dehydrogenation ($CO_2 + C_3H_8 \rightarrow C_3H_6 + CO + H_2O$) and dry reforming ($3CO_2 + C_3H_8 \rightarrow 6CO + 4H_2$). The two reactions should occur simultaneously at temperatures around 823 K and above with considerable conversions, allowing the formation of both dehydrogenation products (propylene) and reforming products (synthesis gas). Thus, it is of great interest to identify catalytic systems that can selectively break either the C-H bond to produce propylene or the C-C bonds to produce synthesis gas ($CO + H_2$).

Non-precious Fe_3Ni and precious Ni_3Pt supported on CeO_2 have been identified as promising catalysts for CO_2-ODHP and dry reforming, respectively, in flow reactor studies conducted at 823 K. For oxidative dehydrogenation, nonprecious metal based Fe_3Ni had higher stability and propylene production rates. Fe_3Ni showed distinct C_3H_6 yield compared to its parent metals and the Ni_3Pt catalyst showed both greater activity and stability than the corresponding monometallics. In-situ X-ray absorption spectroscopy measurements revealed the oxidation states of metals under reaction conditions and density functional theory was utilized to identify the most favorable reaction pathways over the two types of catalysts. Additional flow reactor studies have been performed on the Fe_3Ni and Ni_3Pt bimetallic systems with other oxide supports (TiO_2, ZrO_2, SiO_2, γ-Al_2O_3, and V_2O_5) to examine the effects of strong support metal interactions, surface defects, inert surfaces, irreducibility, as well as alkane activation on both the CO_2-ODHP and DRP pathways. Overall, two types of bimetallic catalysts have been identified for the $CO_2 + C_3H_8$ system; one is selective toward reforming (C-C bond cleavage) and the other favors dehydrogenation (C-H bond scission).

References

(1) Plotkin, J. S. *ACS News Industrial Chemistry & Engineering*. 2016, pp 1–4.

(2) Wang, S.; Zhu, Z. H. *Energy and Fuels* **2004**, *18* (4), 1126–1139.

494828 CO_2 Splitting Using Miec Membranes

Abstract:

In this presentation, we discuss fuel production using membrane-supported CO_2 splitting. A mixed ionic-electronic conductive (MIEC) $La_{0.9}Ca_{0.1}FeO_{3-\delta}$ (LCF-91) membrane is utilized. At intermediate temperatures (850 – 990°C), CO_2 undergoes thermochemical reduction on the feed-side producing CO, while oxygen incorporates into the perovskite as lattice oxygen. Oxygen diffuses to the sweep side, driven by the chemical potential gradient across the membrane generated in this case by fuel (either H_2 or CO in this study) oxidation on that side.

LCF-91 membranes have been examined in our laboratory before for air separation [1-3] and water splitting [4, 5]. In this work, we report a detailed study on CO_2 reduction on the same membrane. We examine the dependence of CO_2 splitting rate on various operating conditions such as CO_2 concentration on the feed-side, fuel (H_2 or CO) concentration on the sweep side and temperatures. CO_2 splitting rate is proportional to the oxygen flux across, and the measured maximum flux is 0.19 μmol/cm^2•s on a 1.3 mm membrane at 990°C, when 9.5% H_2 is added to the sweep side (balance with argon gas). Long term study shows almost a constant flux, i.e., stable performances for 106 hours. Following the experiment, morphology changes, carbonate formation and calcium/iron enrichment are observed on the feed-side surface, while no obvious changes are found on the sweep-side surface as revealed by SEM, EDS and XRD. A resistance-network model is developed to describe the oxygen flux, in which CO_2 direct-incorporation mechanism is used on the feed side, and the Mars-van Krevelen (MvK) mechanism for the oxidation reaction on the sweep side (for both H_2 and CO) [6]. Using the flux measurements, surface kinetics parameters are derived, which shows that H_2 has larger oxidation rate constant than CO. The model also shows that in CO_2 splitting, the rate limiting step changes from the CO_2 splitting reaction on the feed-side to the fuel (H_2 or CO) oxidation reaction on the sweep-side at higher temperatures. This is different from the H_2O splitting cases, where fuel oxidation is always the limiting step within the same temperature range tested. This is supported by the change in the slope of the Arrhenius plot of the oxygen flux. Moreover, the transition temperature for H_2 is higher than that of CO (both used as fuels).

Porous layers made from the same materials are added on both side of the membrane to enhance the oxygen flux. Results shows that the flux is raised by about 0.02 μmol/cm^2•s when CO is used as a fuel. And the same transition temperature is observed in the Arrhenius plot of oxygen flux with or without porous layers.

Reference

[1] Hunt, A., Dimitrakopoulos, G., and Ghoniem, A. F., 2015, "Surface oxygen vacancy and oxygen permeation flux limits of perovskite ion transport membranes," J. Membr. Sci., 489, pp. 248-257.

[2] Dimitrakopoulos, G., and Ghoniem, A. F., 2016, "A two-step surface exchange mechanism and detailed defect transport to model oxygen permeation through the $La_{0.9}Ca_{0.1}FeO_{3-\delta}$ mixed-conductor," J. Membr. Sci., 510, pp. 209-219.

[3] Dimitrakopoulos, G., and Ghoniem, A. F., 2017, "Developing a multistep surface reaction mechanism to model the impact of H_2 and CO on the performance and defect chemistry of mixed-conductors," J. Membr. Sci., 529, pp. 114-132.

[4] Wu, X. Y., Chang, L., Uddi, M., Kirchen, P., and Ghoniem, A. F., 2015, "Toward enhanced hydrogen generation from water using oxygen permeating LCF membranes," PCCP, 17(15), pp. 10093-10107.

[5] Wu, X. Y., Ghoniem, A. F., and Uddi, M., 2016, "Enhancing co-production of H_2 and syngas via water splitting and POM on surface-modified oxygen permeable membranes," AIChE J., 62(12), pp. 4427-4435.

[6] McFarland, E. W., and Metiu, H., 2013, "Catalysis by Doped Oxides," Chem. Rev., 113(6), pp. 4391-4427.

494005 Molten Metal Catalysts for Methane Pyrolysis

Abstract:

Molten metals were investigated as catalysts for the thermochemical pyrolysis of methane. Conversion of methane to hydrogen and solid carbon is demonstrated in bubble column reactors where the solid carbon can be readily separated and removed enabling continuous processes. Twenty-one molten metal systems consisting of elemental metals and binary alloys are compared for methane pyrolysis activity. Nickel-bismuth was found to be the most active catalyst, and 95% methane conversion was observed at 1060 °C in a 1.1 meter bubble column. Initial characterization of the carbon product indicated predominately sp^2 graphitic carbon as the solid product. Although the hydrogen produced from partial oxidation of methane by pyrolysis retains only approximately half the combustion energy potential as complete methane combustion, pyrolysis produces CO_2-free hydrogen, and catalytic molten metals may provide a means for performing reactive separation of the hydrogen from the potentially valuable carbon co-product.

500486 Increasing Activity, Reducing Coking, and Promoting Unexpected Reaction Pathways during Methane Steam Reforming By Applying Uniform Electric Fields in a Scalable Reactor

Abstract:

It has been shown both computationally and in small-scale studies using STM tips for single-molecule analysis that very high electric fields can change both the kinetics and thermodynamics of chemical reactions. This could be an exciting and unexplored new area for reaction engineering, but generating uniform electric fields in a larger scale continuous flow reactor has proven to be a significant challenge. We have designed a system capable of generating uniform electric fields over Ni catalysts for methane steam reforming, which has been compared to density functional theory computations by our collaborating group in the past. This talk focuses on the experimental observations, with special focus on the properties of the catalyst itself after methane steam reforming in the presence of electric fields. We report both expected findings and a new mystery â€" and our attempts to unravel said mystery. On the expected side, our DFT calculations correctly predicted that, in a positive electric field, reaction rates would increase consistent with the lowering of the energy barrier of the first methane C-H bond scission; H_2O would bind more readily to the surface; H_2O would have a more difficult time decomposing; and that coke would have a more difficult time forming. Unexpectedly, the negative field apparently promotes alternate reaction pathways which may lead to C-C coupling reactions to form larger unsaturated hydrocarbons without coking. Both post-reaction catalysts were scrutinized using visual observations including macroscopic photographs and SEM imaging, x-ray photoelectron spectra, temperature-programmed oxidation, and x-ray diffraction patterns.

499357 Design of CO_2 Conversion to Dimethyl Carbonate By the Process-to-Planet Multiscale Modeling Framework

Abstract:

Total U.S. greenhouse gas (GHG) emissions have increased by 3.5 percent from 1990 to 2015 and CO_2 emissions from the fossil fuel combustion account for approximately 77 percent of the GHG emissions [1]. To close a chemical carbon cycle, technological strategies, such as carbon capture and conversion, have been developed [2]. CO_2 conversion technologies not only utilize CO_2 captured from industrial emissions, but also produce valuable products [3]. But, those solutions might cause unintended harm to the environment because processes for the carbon capture and conversion cause additional CO_2 emissions. The Process-to-Planet (P2P) multiscale modeling framework has been developed to design engineering process models by accounting for their upstream life cycle up to the entire economy scale [4,5]. Solving design problems with the P2P modeling framework prevents the environmental impacts from shifting outside the systems boundary.

In this work, the P2P modeling framework is applied in designing a dimethyl carbonate (DMC) synthesis process. DMC is an environmental-friendlier substitute reagent for toxic substances such as phosgene and dimethyl sulfate [6]. Since conventional DMC synthesis routes use or produce toxic substances, alternative routes that are environmentally friendly have been developed. Kongpanna et al. screened various DMC synthesis routes and identified potentially sustainable routes that utilize CO_2 [3]. In this study, the DMC synthesis route from urea and methanol is selected for the application of the P2P modeling framework. Urea is prepared from ammonia and CO_2 that is captured from industrial activities. Life cycle inventory (LCI) data are collected from the U.S. LCI database [7,8] to develop the process-based life cycle (value chain scale) model. The economy scale model is developed using the U.S. industrial input-output data [9] and U.S. GHG emissions data [1]. Based on the flow diagram from the reference article [3], the engineering process (equipment scale) model is developed and its design variables are identified. To connect multiscale models, cutoff flows across different scales are identified. The P2P model for the DMC synthesis process is constructed using the General Algebraic Modeling System (GAMS) to optimize design variables. Optimal designs are identified in consideration of a trade-off between maximizing net present value and minimizing environmental interventions.

The Life Cycle Assessment (LCA) approach has been developed to account for multiple spatial scales. The hybrid LCA approach combines value chain scale and economy scale models [10]. The P2P modeling framework integrates the hybrid LCA approach with the engineering process model. In this work, the difference between each approach in total environmental interventions and optimal values of design variables is discussed. Also, the insights gained by applying the P2P modeling framework to designing the CO_2 conversion process are discussed.

Although CO_2 conversion processes utilize CO_2 to produce desired products, in most cases, their net CO_2 emissions are still positive. Also, the demand for products from CO_2 conversion process is limited. Therefore, technological solutions only account for a small portion of total GHG emissions that need to be reduced. To close the carbon cycle toward sustainability, other strategies, such as ecological and agricultural solutions, need to be combined with technological strategies. In such integrated systems, since food, energy, and water (FEW) flows interact across multiple scales, the FEW flows need to be quantified and considered as optimization objectives to account for the nexus of FEW systems. For example, an integrated model with the P2P modeling framework and an agro-ecological model, such as the Soil and Water Assessment Tool (SWAT) model, could be developed to assess multiple flows at the nexus of FEW systems toward closing the carbon cycle. The integrated model prevents the shifting of environmental impacts not only across spatial scales, but also across multiple flows.

References

[1] USEPA. Inventory of U.S. greenhouse gas emissions and sinks: 1990-2015. U.S. Environmental Protection Agency, Washington, D.C., 2017.

[2] George A. Olah, G. K. Surya Prakash, and Alain Goeppert. Anthropogenic chemical carbon cycle for a sustainable future. Journal of the American Chemical Society. 133 (33): 12881-12898, 2011.

[3] Pichayapan Kongpanna, Varong Pavarajarn, Rafiqul Gani, Suttichai Assabumrungrat. Techno-economic evaluation of different CO_2-based processes for dimethyl carbonate production. Chemical Engineering Research and Design. 93: 496â€“510, 2015.

[4] Rebecca J. Hanes and Bhavik R. Bakshi. Process to planet: A multiscale modeling framework toward sustainable engineering. AIChE Journal. 61 (10): 3332-3352, 2015.

[5] Rebecca J. Hanes and Bhavik R. Bakshi. Sustainable process design by the process to planet framework. AIChE Journal 61 (10): 3320-3331, 2015.

[6] Pietro Tundo and Maurizio Selva. The chemistry of dimethyl carbonate. Accounts of Chemical Research. 35 (9): 706â€“716, 2002.

[7] U.S. Life Cycle Inventory Database. National Renewable Energy Laboratory, 2012.

[8] Rolf Frischknecht et al. The ecoinvent database: Overview and methodological framework. International Journal of Life Cycle Assessment, 10(1):3â€“9, 2005.

[9] U.S. Bureau of Economic Analysis. Interactive data: Industry data input-output. http://www.bea.gov/iTable/index_industry_io.cfm, accessed November 1, 2016.

[10] Clark W. Bullard. Net energy analysis: Handbook for combining process and input-output analysis. Resources and Energy. 1 (3): 267-313, 1978.

491762 Optimal Design of Biomass Supply Chain for Deploying Bioenergy with CCS (BECCS) in the UK

Abstract:

The UK drafted the Paris Agreement to tackle climate change under the UN negotiations in December 2015, which aims to limit warming to well below 2 Â°C. The UKâ€™s 2050 target is to reduce emissions at least 80% from those of 1990, and options for zero CO2 emissions include greenhouse gas removals from afforestation, bioenergy with carbon capture and storage (BECCS) and wood in construction. Among them, BECCS is reported to produce 50TWh/yr of power generation with -47 Mt CO2e in 2050 by the Committee on Climate Change in the UK. BECCS deployment is achievable by 2030 since it could be cost competitive by 2030s. Both domestic and international feedstocks need to be considered. Second generation (2G) biomass, such as miscanthus, short rotation coppice (SRC) and short rotation forestry (SRF) on arable land or grassland sites could provide both greenhouse gas (GHG) emission savings and wider biodiversity. Analysis of the UK Energy Technology Institute suggests a planting rate of 30, 000 hectares per annum for biomass production while converting up to 1,000 hectares of 2G bioenergy crops per annum. Meanwhile, recovering energy from waste wood (e.g. construction and demolition) has been addressed in Waste Strategy for England 2007 to generate electricity and save CO2e emissions simultaneously. In this work, we address the problem of the optimal design of biomass supply chain for deploying BECCS in the UK with a mixed integer linear programming (MILP) model. Waste wood, municipal solid waste (MSW), miscanthus and SRF poplar are available to be selected as raw material within the UK. Wood pellets can be imported from international markets to the current available ports under their loading capacities. Possible geographical locations of future power plants, pellet plants and biomass farms are also provided. The total annual cost of the whole supply chain is minimised by choosing the locations of new BECCS power plants and pellet plants subject to the CO2 reduction targets and supply chain constraints associated with the availability of raw materials and mass balances. The key decision variables include raw material supply, pellet production and electricity generation rates, flows of materials between cells and imported pellets, and fuel burn rates. The results indicate that MSW and waste wood are consumed as the basic raw material supplies for the BECCS in the UK from 2025 to 2055 resulting from their low prices. Poplar is preferred over miscanthus as the virgin biomass supply and more BECCS power plants and pellet plants are selected as the CO2 reduction targets increase. Based on this work, no wood pellets need to be imported to the UK to fulfil the carbon emission reduction target for the BECCS during this 30 years.

492828 Water Management within Energy Systems

Abstract:

Energy sources, including fossil fuels and renewable energy sources, have supported and will continue to support human prosperity across the globe. The International Energy Agency (IEA) systematically assesses and presents three different scenarios, i.e. New Policies, Current Policies, and 450 scenarios, to forecast global energy demand by 2040. The 450 scenario refers to an energy pathway that curbs greenhouse gas emissions so that their atmospheric concentration do not exceeds 450 parts per million of CO_2 in order to keep global warming below 2°C. Indeed, according to the projections of the IEA, in the Current Policy, New Policy, and 450 scenarios, world primary energy demand is expected to growth by 45%, 32%, and 12% from 2013 to 2040, respectively. Likewise, global electricity demand is projected to increase over the same time period by 86%, 71%, and 49%, for the same three scenarios, respectively [1]. However, the development of energy sources relies on water, which is subject to physical limitations and government regulations that can and do constraint its availability and accessibility.

This paper presents a broad and comprehensive summary of the relevance and effectiveness of decision-support tools to help in the water management associated with the development of energy sources as well as with the design, planning, and operation of energy systems. We provide a discussion of water intensity of energy sources and systems in addition to the application of decision-support tools for the corresponding water management. Specifically, we first present a global context of the water-energy nexus in term of national/regional energy demand and spatial distribution of water risk and stress level. Then, we review data on the water intensity of different primary energy sources and power generation technologies. Subsequently, decision elements associated with the water management in energy systems are outlined. A shale gas supply chain design and planning model [2–4] is used to investigate the impact of water availability on the optimal water management strategy and the economics associated with the exploitation of shale gas resources. Finally, needed advances and continuing challenges in water management related to energy systems are identified.

References

[1] International Energy Agency. World Energy Outlook 2015. 2015. doi:10.1787/weo-2015-en.

[2] Calderón AJ, Guerra OJ, Papageorgiou LG, Siirola JJ, Reklaitis G V. Preliminary Evaluation of Shale Gas Reservoirs: Appraisal of Different Well-Pad Designs via Performance Metrics. Industrial & Engineering Chemistry Research 2015;54:10334–49. doi:10.1021/acs.iecr.5b01590.

[3] Guerra OJ, Calderón AJ, Papageorgiou LG, Siirola JJ, Reklaitis G V. An optimization framework for the integration of water management and shale gas supply chain design. Computers & Chemical Engineering 2016;92:230–55. doi:10.1016/j.compchemeng.2016.03.025.

[4] Guerra OJ, Calderón AJ, Papageorgiou LG, Reklaitis G V. Strategic Design and Tactical Planning for Energy Supply Chain Systems. Advances in Energy Systems Engineering, vol. 50, Springer International

Publishing; 2017, p. 47–74. doi:10.1007/978-3-319-42803-1_3.

496723 Considering Ecosystem Services in US Bioenergy Supply Chains

Abstract:

Ecosystems provide essential services – termed ecosystem services (ES) – to humanity that make our lives possible, such as nutrient cycling within a forest or flood management from mangroves.[1] However, despite these critical services that ecosystems provide, many ecosystems around the world are either degraded or destroyed in the name of economic growth.[1] Usually, the inherent value of the degraded or destroyed ecosystems is not considered. It is possible that the economic development that takes the place of existing ecosystems does not provide as much value as the original ecosystem, resulting in a net loss of value. However, if the original ecosystem's value is not considered, then this net loss to society would instead be mistakenly interpreted as a net economic gain. Industries that require land, mainly agriculture, can quickly and efficiently destroy ecosystems. Perhaps the most well-known example in the US is the Dust Bowl, caused by rapid expansion of poor farming practices in the Great Plains, resulting in degraded soil that was easily picked up by wind. Biofuels rely heavily on agriculture to produce the industry's feedstocks. Thus, biofuels could pose a significant threat to ecosystems around the globe if feedstock sourcing is not responsibly planned.

In this work, we consider ES values for the first time into a bioenergy supply chain design and optimization model. There is a clear need to consider ecosystem service values into the design of sustainable bioenergy systems[2] and energy systems in general.[3] However, few biofuels value chain design studies consider ecosystems and the value of the ES they provide, though some works consider ES at the process design level.[4] Some important initial studies address qualitative impacts of bioenergy supply chains on ES. For example, Guo et al. (2016) employ estimated ES impact factors in a large-scale, spatial-temporal bioenergy value chain model.[5] The vast number and different relative impacts of ES is part of the challenge in considering ES values in the design of biofuels value chains. To address this challenge, we turn to the field of ecological economics which characterizes, classifies, and places a monetary value on ES.[6] The Economics of Ecosystems and Biodiversity (TEEB) study contains a database of hundreds of ES values from hundreds of different studies of varied ecosystems in many regions around the globe, the largest database of its kind.[7] There is enough data (over 1,310 data points) in the TEEB database to make reasonable, spatially-explicit estimates on the monetary value of ecosystems at a regional or global scale.[8] We propose to add these ES values into a US bioenergy supply chain design and optimization model. By considering both natural and nominal economics, we can arrive at a more sustainable – both economically and environmentally – biofuel value chain.

We formulate a mixed integer nonlinear program (MINLP) that considers ES values explicitly within an economic objective function. The model's goal is to identify the most economically favorable biomass sourcing and biofuel/bioenergy production strategy that meets renewable energy targets for the US, considering traditional supply chain capital and operating costs as well as loss of ES values due to feedstock production. We identify 2,033 US counties that are fit for agriculture as potential candidates for either biomass sourcing, biofuel production, or both via geo-spatial USDA data.[9] The 50 counties with the most population are selected as possible destinations of the finished fuel, with no county able to accept more fuel than it can use. Several different biomass feedstocks and conversion technologies, including biochemical and thermochemical techniques, are considered in the model. To manage model complexity, we introduce a branch-and-refine algorithm with piecewise linear approximations to model the nonconvex capital cost terms, shortening computation times.[10] We introduce overall supply chain cost (to be minimized) and Green GDP[11] (to be maximized) as economic objectives, demonstrating how the supply chain design shifts when considering ES values economically.

References

1. Assessment ME. Ecosystems and Human Well-being: Biodiversity 26. *Synthesis. World Resources Institute, Washington, DC.* 2005.

2. Garcia DJ, You F. The water-energy-food nexus and process systems engineering: A new focus. *Computers & Chemical Engineering.* 2016;91:49-67.

3. Gong J, You F. Sustainable design and synthesis of energy systems. *Current Opinion in Chemical Engineering.* 2015;10:77-86.

4. Gopalakrishnan V, Bakshi BR, Ziv G. Assessing the capacity of local ecosystems to meet industrial demand for ecosystem services. *AIChE Journal.* 2016;62(9):3319-3333.

5. Guo M, Richter GM, Holland RA, Eigenbrod F, Taylor G, Shah N. Implementing land-use and ecosystem service effects into an integrated bioenergy value chain optimisation framework. *Computers & Chemical Engineering.* 2016;91:392-406.

6. De Groot R, Brander L, van der Ploeg S, et al. Global estimates of the value of ecosystems and their services in monetary units. *Ecosystem services.* 2012;1(1):50-61.

7. Van der Ploeg S, De Groot R. The TEEB Valuation Database–a searchable database of 1310 estimates of monetary values of ecosystem services. *Foundation for Sustainable Development, Wageningen, The Netherlands.* 2010.

8. Costanza R, de Groot R, Sutton P, et al. Changes in the global value of ecosystem services. *Global Environmental Change.* 5// 2014;26:152-158.

9. Boryan CG, Yang Z, Willis P, Di L. Developing crop specific area frame stratifications based on geospatial crop frequency and cultivation data layers. 2016.

10. Garcia DJ, You F. Multiobjective optimization of product and process networks: General modeling framework, efficient global optimization algorithm, and case studies on bioconversion. *AIChE Journal.* 2015;61(2):530-554.

11. Boyd J. Nonmarket benefits of nature: What should be counted in green GDP? *Ecological economics.* 2007;61(4):716-723.

498028 Robust Multi-Period and Multi-Objective Strategic Planning of Hydrogen Networks

Abstract:

As renewable energies gain more traction due to improving efficiencies and decreasing costs, there will be a gradual shift away from dependence on fossil fuels toward the deployment of carbon-neutral energy technologies. One such energy vector is hydrogen, which has enormous potential as a renewable fuel for vehicle mobility and electricity generation. However, its integration into the energy landscape is currently strained by the dearth of enabling infrastructure. Several options exist for the production, storage, distribution and retailing of hydrogen to end users, but it is unresolved how to develop this infrastructure given different short- and long-term technological and economic actualizations.

What is the most energy efficient, environmentally benign, and cost effective pathways to deliver hydrogen to the consumer considering these uncertainties? To answer this question, several groups have investigated multi-objective supply chain optimization for hydrogen infrastructural development [1-3]. A few have considered a multi-period approach [4-6], and even fewer have considered uncertainty in the supply chain [7, 8].

Here, we propose a multi-period and multi-objective mixed-integer programming approach to the optimal planning of hydrogen infrastructures. By incorporating recent advances in robust optimization [9] in the strategic planning model, we address uncertainty by designing robust hydrogen networks that are protected against different technological and economic realizations. In the United States, California is taking the lead with more than 20 hydrogen retailing stations spread across the state [9]. To demonstrate our method, we present a scenario analysis using California as a case study.

1. De-León Almaraz, S., C. Azzaro-Pantel, L. Montastruc, L. Pibouleau, and O.B. Senties, Assessment of mono and multi-objective optimization to design a hydrogen supply chain. International Journal of Hydrogen Energy, 2013. 38(33): p. 14121-14145.

2. Hugo, A., P. Rutter, S. Pistikopoulos, A. Amorelli, and G. Zoia, Hydrogen infrastructure strategic planning using multi-objective optimization. International Journal of Hydrogen Energy, 2005. 30(15): p. 1523-1534.

3. Almansoori, A. and N. Shah, Design and Operation of a Future Hydrogen Supply Chain. Chemical Engineering Research and Design, 2006. 84(6): p. 423-438.

4. Liu, P., A. Whitaker, E.N. Pistikopoulos, and Z. Li, A mixed-integer programming approach to strategic planning of chemical centres: A case study in the UK. Computers & Chemical Engineering, 2011. 35(8): p. 1359-1373.

5. Almansoori, A. and N. Shah, Design and operation of a future hydrogen supply chain: Multi-period model. International Journal of Hydrogen Energy, 2009. 34(19): p. 7883-7897.

6. Li, Z., D. Gao, L. Chang, P. Liu, and E.N. Pistikopoulos, Hydrogen infrastructure design and optimization: A case study of China. International Journal of Hydrogen Energy, 2008. 33(20): p. 5275-5286.

7. Guillén, G., F.D. Mele, M.J. Bagajewicz, A. Espuña, and L. Puigjaner, Multiobjective supply chain design under uncertainty. Chemical Engineering Science, 2005. 60(6): p. 1535-1553.

8. Almansoori, A. and N. Shah, Design and operation of a stochastic hydrogen supply chain network under demand uncertainty. International Journal of Hydrogen Energy, 2012. 37(5): p. 3965-3977.

9. Guzman, Y.A., L.R. Matthews, and C.A. Floudas. New a priori and a posteriori probabilistic bounds for robust counterpart optimization: I. Unknown probability distributions. Computers & Chemical Engineering, 2016. 84: p. 568-598.

10. EPA, C., Hydrogen Fuel Cell Electric Vehicle Deployment and Hydrogen Fuel Station Network Development, in AB 8, A.E. Report, Editor. 2016: California.

502272 A Stochastic Techno-Economic Model for Quantifying the Economic Costs of Cellulosic Bioenergy Pathways in the Northeast U.S

Abstract:

New York State has established a Clean Energy Standard that requires 50% of its electricity to be derived from renewable resources by 2030. The closing of substantial nuclear power capacity in the near future is expected to create additional demand for renewable baseload power in the interim. Upstate New York is characterized by large tracts of unproductive former pasture and cropland on which shrub willow has been demonstrated as a potential cellulosic bioenergy feedstock that has the additional potential to rebuild soil carbon via net carbon sequestration. Shrub willow is used as a feedstock for baseload heat and power at small commercial-scale projects throughout New York State. Researchers at SUNY College of Environmental Science & Forestry have developed a stochastic techno-economic analysis model of shrub willow plantations in the U.S. Northeast to quantify the economic feasibility of the biomass as a bioenergy feedstock. This research, which is being funded by the USDAâ€™s Biomass Research and Development Initiative, is part of a broader project that will determine the economic and environmental feasibility of multiple forms of cellulosic bioenergy in the region in pursuit of state energy goals (Massachusetts has also implemented ambitious renewable electricity targets). This proposed oral presentation within the Biomass Availability and Logistics technology area will provide an overview of the project's energy and environmental motivations, the design and building of the stochastic techno-economic model as well as results, and the modelâ€™s potential applications in regions outside of the U.S. Northeast.

495994 Techno-Economic Comparison of an Industrial Scale Pyrolysis of Seaweed for Liquid Fuel Production: H_2 Production vs. H_2 Purchase Scenarios

Abstract:

In this study, process design, simulation and technoeconomical assessment is performed using Aspen plus software and experimental data for indirectly heated pyrolysis of *Saccharina japonica* seaweed. The scale of the process was set to 350,000 tons per year of dry feed. For the upgrading and stabilization of the liquid pyrolysis product, hydrogen is required. Experimentally determined molar ratio of hydrogen to feed is 5.5 which at this scale translates to 220 kmol/hr. The base case process design includes on site hydrogen production through natural gas reforming and pressure swing adsorption, which presents a significant capital cost investment of estimated 55% compared to other parts of the process. This study compares the base case with hydrogen purchase and provides economical insight in terms of expected lower capital cost investment and lower Minimum Product Selling Price.

498430 Techno-Economic and Life Cycle Analysis for the Production of Renewable Acrylonitrile from Non-Food Biomass

Abstract:

The industrial sector anticipates an 11-18% annual increase in the market for carbon fiber, specifically driven by the motivation to reduce weight for vehicles. Currently, the highest performance carbon fibers are produced by heat treatment or controlled pyrolysis of polyacrylonitrile (PAN). The precursor acrylonitrile (ACN) monomer is currently produced via ammoxidation of propylene, a high volume commodity chemical subject to the volatility of the petrochemical market. Propylene accounts for approximately 70% of the total ACN production cost. The goal of this work is to develop a cost-effective, low environmental impact process for production of ACN from biomass-derived sugars, such that the ACN meets all chemical specifications of and is a direct replacement for conventional propylene-based ACN.

Southern Research (SR) has developed the biomass to ACN (B2ACN) process under a cooperative agreement with the Department of Energy during Phase I. The process is currently being designed for Phase II pilot studies at a scale of kg/hr. Briefly, B2ACN is a multi-step catalytic process for the conversion of sugars from non-food biomass to ACN at mild conditions. The process utilizes known pretreatment methods for recovery of sugars from any type of biomass. In the first reaction step, sugars are converted to glycols/glycerol using a novel multi-functional catalyst. After desired separation, glycerol is then converted to a gas phase intermediate acrolein in the second reaction step, which is followed by ammoxidation of acrolein to acrylonitrile in the final reaction step. Critical performance attributes of the acrylonitrile product are compared to a propylene-derived standard for production readiness in the PAN and carbon fiber processes. The process has been optimized for heat recovery and separations. Phase I experimental results and initial estimates show a significant reduction in greenhouse gas (GHG) emissions of up to ~37% and cost reduction of up to ~15-22% compared to propylene-derived ACN. This paper will discuss the techno-economic and lifecycle impact on the production of ACN with respect to key variables, including feed materials, impurity/by-product separations, catalysts, utilities, water utilization, and green field capital expenditures tailored to downstream applications. The motivation of this study is to provide a guide to end consumers and brand owners to realize the potentials and pitfalls of the process for further improvements and adoption.

499367 Relative Sustainability of Natural Gas Assisted High-Octane Gasoline Blendstock Production from Biomass

Abstract:

Biomass-derived hydrocarbon fuel technologies are being developed and pursued for better economy, environment, and society benefits underpinning the sustainability of transportation energy. Increasing availability and affordability of natural gas (NG) in the US can play an important role in assisting renewable fuel technology development, primarily in terms of economic feasibility. When a biorefinery is co-processing NG with biomass, the current low cost of NG coupled with the higher NG carbon conversion efficiency potentially allow for cost competitiveness of the fuel while achieving a minimum GHG emission reduction of 50% or higher compared to petroleum fuel. This study evaluates the relative sustainability of the production of high-octane gasoline blendstock via indirect liquefaction (IDL) of biomass (and with NG co-feed) through methanol/dimethyl ether intermediates. The sustainability metrics considered in this study include minimum fuel selling price (MFSP), carbon conversion efficiency, life cycle GHG emissions, life cycle water consumption, fossil energy return on investment (EROI), GHG emission avoidance cost, and job creation. Co-processing NG can evidently improve the MFSP. Evaluation of the relative sustainability can shed light on the biomass-NG synergistic impacts and sustainability trade-offs associated with the IDLâ€™s high-octane gasoline blendstock production.

495797 *In situ* Catalytic Fast Pyrolysis of *Nannochloropsis sp.* using Co-Mo Catalysts for Value Added Chemicals

Abstract:

Growing global energy need demands an urge for exploitation of renewable and environmentally friendly energy resources. In order to reduce our dependency on fossil fuels, carbon neutral sources such as biomass are seen as potential source of energy. Besides lignocellulosic biomass, fast growing plants such as microalgae are looked upon as a prospective source of energy. Microalgae do not require arable land for its cultivation, and so they do not compete with food crops. Thermochemical conversion of microalgal biomass is a promising strategy to convert them into valuable chemicals and fuel molecules. Fast pyrolysis of microalgae is recognized as a promising technique to produce bio-oil. However, high oxygen and nitrogen content of bio-oil limit its direct application in engines or transportation fuels. The various oxygenates like aldehyde, ketone, carboxylic acid and nitrogen containing nitriles, pyrroles, indoles result in low heating value, poor storage stability and low thermal stability. The problem of complex composition of bio-oil can be addressed by deploying catalysts for enhanced selectivity with an aim of promoting deoxygenation, denitrogenation and dehydration resulting in a stable product.

In this study *Nannochloropsis oculata* (*N. oculata*), a high lipid and protein containing microalgae, was fast pyrolyzed in the absence and presence of catalysts. The effect of temperature on catalytic and non-catalytic fast pyrolysis and the effect of catalyst to algae ratio on product distribution was studied. The algae was characterized by proximate, elemental, heating value and thermal stability analyses using various analytical techniques such as thermogravimetric analyzer (TGA), CHNS analyzer and bomb calorimeter. Fast pyrolysis experiments were performed in an analytical micropyrolyzer and the products were analyzed in an online gas chromatograph-mass spectrometer (Py-GC/MS) at different temperatures from 400-600°C. Bifunctional Co-Mo supported on γ-Al_2O_3 was prepared and characterized before using it as a catalyst for Py-GC/MS experiments. Thermal stability, pore size distribution and catalyst composition were studied using various characterization techniques like TGA, porosimetery and ICP-OES (Infrared coupled plasma – optical emission spectroscopy), respectively. This is the first study to investigate the effect of Co-Mo catalysts on *N. oculata* under the fast pyrolysis conditions.

Aromatic and aliphatic hydrocarbons were the major products (c.a. 50%) produced during the non-catalytic fast pyrolysis at 500°C. The major products were benzene, styrene and alkenes (C_{14} – C_{19}), nitriles (C_4 – C_6). The yields of oxygenated and nitrogen containing compounds were low in the pyrolysates. Catalyst: algae ratios of 1:9, 1:3, 1:1 and 2:1 were tested for catalytic fast pyrolysis. The use of catalyst resulted in the formation of 1-isocyanobutane and dimethylketene (combined yield of 25-40%) as the major compounds. Long chain nitriles, long chain alkanes and alkenes, polyaromatic and monoaromatic hydrocarbons were the major product groups. Pyrolysates obtained from catalytic fast pyrolysis had lower amount of oxygen containing organic compounds (< 3%) as compared with non-catalytic fast pyrolysis (c.a. 21%). This was evident from the increase in the approximate HHV (33-40 MJ kg^{-1}) of organics in the pyrolysates from 18 MJ kg^{-1} (raw dried *N. oculata*). Highest deoxygenation was observed with CoMo

catalyst to algae ratio 1:3. Mo and support (γ-Al_2O_3) as catalysts resulted in long chain nitriles. The plausible reaction mechanism for the conversion of microalgae to important chemicals was proposed in this study. More interesting results on (i) effect of metal loading, (ii) effect of temperature and (iii) effect of catalyst: algae ratio will be discussed during the presentation.

501045 Wood Adhesive As a Coproduct of Cellulosic Alcohol: Crosslinking and Bond Performance

Abstract:

The conversion of woody biomass to fermentable sugars through enzymatic saccharification (hydrolysis) often leaves behind solid residues that are rich in lignin. Our laboratory had demonstrated that these residues can be converted into a thermosetting wood adhesive without pre-purifying for lignin. This presentation reports the evolution of crosslinking during thermal curing of the adhesive. Aspen wood flour was enzymatically hydrolyzed, milled, and blended with a biobased crosslinker before application onto wood sticks to form lap shear specimens. The lap shear assemblies were hot-pressed for different durations of time and then mechanically tested to infer bond strength development. Variously cured adhesives were also examined for characteristic functional groups of the crosslink using infrared spectroscopy and conductometric titration. Results showed that the amount of functional groups was closely correlated to the lap shear bond strength of the cured and partially cured adhesive. This finding suggests that the bond performance of the adhesive can be potentially optimized by manipulating crosslinking conditions during thermal curing. This outcome signifies an improved understanding of saccharification residues for adhesive coproducts aiming at increasing the economic viability of the bioenergy system.

497553 Design, Fabrication, Testing, and Operation of a Continuous Reactor for Hydrothermal Carbonization

Abstract:

Hydrothermal carbonization (HTC, thermal hydrolysis, or wet torrefaction) is a treatment process which converts moist feedstocks into homogenized, carbon rich, and energy dense solid fuel, called hydrochar. One of the main advantages of HTC compared to other thermochemical treatment processes is the use of moisture as reaction medium and reactant, thereby precluding the need for drying prior to treatment. Thermodynamic properties of water change greatly in the subcritical region from 180-280 °C, and as a result, subcritical water behaves as a non-polar solvent and mild acid and base catalyst simultaneously. When subjected to HTC biomass undergoes rapid hydrolysis and other reactions, releasing oxygen-containing volatiles and producing a highly hydrophobic hydrochar.

Although HTC offers a relatively simple and straightforward solution to process diverse biomass feedstocks, the requirements of high pressure and high temperature make the process complex and potentially costly to design and operate. The lab-scale batch process has already been demonstrated in various laboratories around the world, but batch process is not cost-effective for industrial-scale deployment. A continuous process offers a relatively small footprint, energy recovery and efficiency, and notably, economies of scale.

In this study, a bench-scale continuous HTC reactor system has been designed, constructed, commissioned, and operated with cellulose and with whole biomass feedstocks. The maximum temperature and pressure of the reactor are 250 °C and 40 bar. The throughput of the reactor system was maintained at 0.3 L/min with solids loading of ~5%, while the reaction time was maintained at 5 min. The basic reactor configuration is a section of straight pipe, and numerous engineering challenges were encountered during design and commissioning. The primary challenges are related to instrumentation durability, temperature control, and flow control.

Results are presented for hydrothermal conversion of microcrystalline cellulose. Both solid and liquid products were tested for their physico-chemical properties and compared with the corresponding products from batch process generated in a 2L batch Parr reactor. It was found that temperature and pressure were stable during operation and products are relatively similar to that of batch process. Similar results were found for HTC of cow manure, when comparing solid yield, char fuel value and total organic content in the aqueous phase, when corrected for water volume.

495138 Fate of Nitrogen during Hydrothermal Treatment of Septage

Abstract:

Septic tanks are a popular method in rural areas to treat human waste. Waste enters the septic tank and settles so that solids collect in the bottom of the tank. Here organic matter is aerobically digested. Digestion only occurs partially, and is variable with each design of tank and the conditions of the tank (such as temperature and extent of cleaning). Septic tanks offer many advantages, such as the containment of the digestate and the containment of human waste. However, the resulting solid and liquid products must be pumped and disposed of, or further treated. One concern with the use of septic tank is the long term effects of untreated pollutants that are only present in the microgram/liter range, which have been shown to have serious impacts on aquatic ecology of surrounding water systems [1],[2]. Failure of septic tanks can also result in contamination of surrounding water sources with the various macronutrients present in the digestate and effluent of the tank.

Hydrothermal treatment has been investigated as a possible method for the treatment of municipal solid waste and sewage sludge [3]. Septage, or the combination of both septic sludge and effluent, can also be treated by hydrothermal means. Hydrothermal treat is an attractive option for additional septage treatment due to its high water content; there is no need for any drying or separation for solid treatment. More so, the high nitrogen content of septage can be recovered and reused as fertilizer. In this study septage was collected from local health department and hydrothermally treated. A full nitrogen balance was performed between soluble nitrogen species, as well as solid nitrogen content determined by CHNS. Experimental trials for this treatment used a batch reactor with a reaction time of 30 minutes. It was found that hydrothermal treatment at temperatures between 180-260 °C results in increased nitrogen in the process liquid compared to the untreated septage. In addition, treatment results in increased ammonia concentration in the liquid phase and a pH between 6 and 8. This generation of soluble nitrogen species is a result of the solid phase decomposing as it is heated.

References

[1]. Schwarzenbach, René P., et al. "The challenge of micropollutants in aquatic systems." *Science* 313.5790 (2006): 1072-1077.

[2]. Withers, Paul JA, et al. "Do septic tank systems pose a hidden threat to water quality?." *Frontiers in Ecology and the Environment* 12.2 (2014): 123-130.

[3]. Zhao, Peitao, et al. "Clean solid biofuel production from high moisture content waste biomass employing hydrothermal treatment." *Applied Energy* 131 (2014): 345-367.

491306 Hydrothermal Treatment of Paper Mill Sludge

Abstract:

Each year the pulp and paper industry generates large amounts of effluent treatment sludge. The sludge is made up of three different fractions namely, primary, secondary, and tertiary sludge. Depending on the layout of the sludge handling in the individual mill, these three sludges can either be extracted separately or mixed into a stream called mixed sludge. Usually secondary and tertiary sludge are mixed together. The primary sludge, often rich in fiber material and easy to dewater, can easily be used as fuel source. However, water is difficult to remove from mixed sludge.[1-2] Moreover, a low fuel heating value of mixed sludge results against economical combustion. Therefore, a new technology that converts mixed sludge to viable solid fuel may be required for mixed sludge valorization.

Hydrothermal carbonization (HTC) has received considerable attention during the last several years as a prominent alternative for upgrading biomass feedstock ranging from wood and algae to agricultural and forest residues and industrial or municipal wastes. One of the main advantages of HTC is the use of aqueous solutions under high temperature (180-260 °C) and autogenously generated pressure (5-50 bar).[3] As a result, moist and wet feedstocks are favorable for HTC.[4]

This goal of this study was to determine the effects of HTC conditions on paper mill sludge dewaterbility and fuel value. A 100 ml Parr batch reactor was be used for HTC. Reaction temperature was varied from 180-260 °C, while reaction time was varied from 30 min-1h. The solid hydrochars were analyzed for energy content, ultimate analysis, FTIR, and NMR. Meanwhile, the process liquid is characterized with GCMS, HPLC, UV-VIS, NMR, and pH. The energy content of the hydrochar was increased with reaction temperature and longer reaction time as a result of the decrease in oxygen content. The resulted hydrochar is more hydrophobic than the feedstock. Therefore, dewaterbility is enhanced with the hydrothermal treatment of paper mill sludge.

References

1. Makela, M.; Benavente, V.; Fullana, A., Hydrothermal carbonization of industrial mixed sludge from a pulp and paper mill. *Bioresource Technol* **2016,** *200*, 444-450.

2. Makela, M.; Yoshikawa, K., Simulating hydrothermal treatment of sludge within a pulp and paper mill. *Appl Energ* **2016,** *173*, 177-183.

3. Reza, M. T.; Andert, J.; Wirth, B.; Busch, D.; Pielert, J.; Lynam, J. G.; Mumme, J., Hydrothermal Carbonization of Biomass for Energy and Crop Production. *Applied Bioenergy* **2014,** *1* (1).

4. Reza, M. T.; Coronella, C.; Holtman, K. M.; Franqui-Villanueva, D.; Poulson, S. R., Hydrothermal Carbonization of Autoclaved Municipal Solid Waste Pulp and Anaerobically Treated Pulp Digestate. *Acs Sustain Chem Eng* **2016,** *4* (7), 3649-3658.

486662 Structural Analysis of Humins Formed in the BrÃ¸nsted-Catalyzed Dehydration of Fructose

Abstract:

Humins are carbonaceous and polymeric by-products formed during acid-catalyzed hydrothermal processing of sugars to 5-hydroxymethylfurfural (HMF). These are formed via cross-polymerization of HMF with sugars in acidic solutions, and cause significant carbon loss and unfavorable process economics of biorefineries. The deposition of these solid particles on the reactors and processing equipment also increases the maintenance costs. Currently humins are combusted to supply heat in biorefineries. Minimization of humins formation or their valorization is essential for improving the process economics of the biorefineries. In this work, we developed methods to understand microstructure of humins through solubility experiments. First, we have conducted dissolution experiments in various solvents and correlated the solubility data to molecular properties of solvents to develop suitable descriptors. By doing this, we have identified solvents for high solubility of humins at room temperature and provided insights into the mechanism of humins dissolution. Next, Liquid Chromatography- Mass Spectrometry (LC-MS) was used to determine and compare the masses of components in the humins solubilized in different solvents. Finally, a six stage dissolution experiment was done to investigate the structural inhomogeneity of humins. Both Fourier Transform Infrared Spectroscopy (FT-IR) and LC-MS were used to characterize the solubilized and insoluble humins obtained at each stage. Based on our results, an inhomogeneous structure where macromolecular and molecular components are connected through weak forces, was proposed.

Catalysis Center for Energy Innovation

UNIVERSITY OF DELAWARE

Super Oleophilic and Water Repellent Graphene Oxide Sponge for Catalytic Coupling of Furans and Oil Absorption

Saikat Dutta

Advisers: Basu Saha, Dion G. Vlachos

Catalysis Center for Energy Innovation

University of Delaware

Newark, Delaware

Session: Hydrothermal Carbonization

November 2, 2017

AIChE Annual Meeting 2017, Minneapolis, MN

U.S. DEPARTMENT OF ENERGY | Office of Science

CCEI is an Energy Frontier Research Center funded by the U.S. Department of Energy, Office of Basic Energy Sciences

Catalysis Center for Energy Innovation

UNIVERSITY OF DELAWARE

Why Graphene Sponge

- Graphene is generally hydrophobic, can be tuned to hydrophilic

Hydrophobic properties

Graphene@SiC
CA 92°

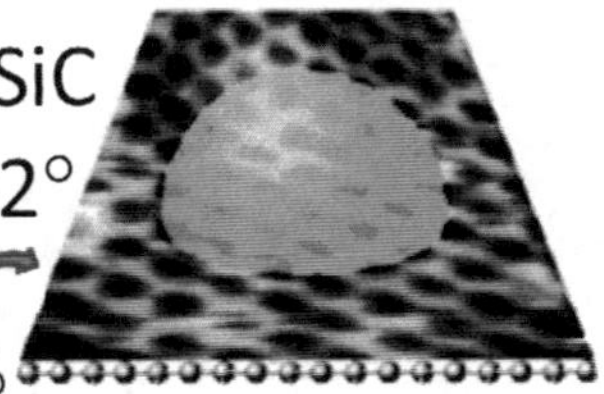

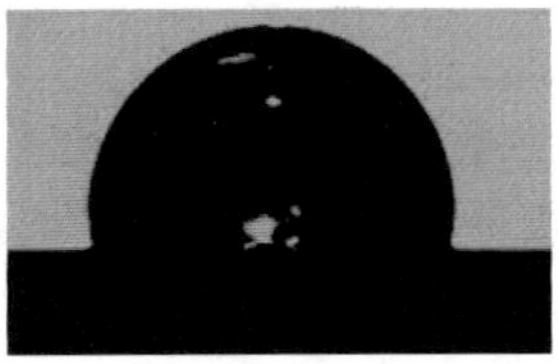

SiC CA 69°

O_2 plasma

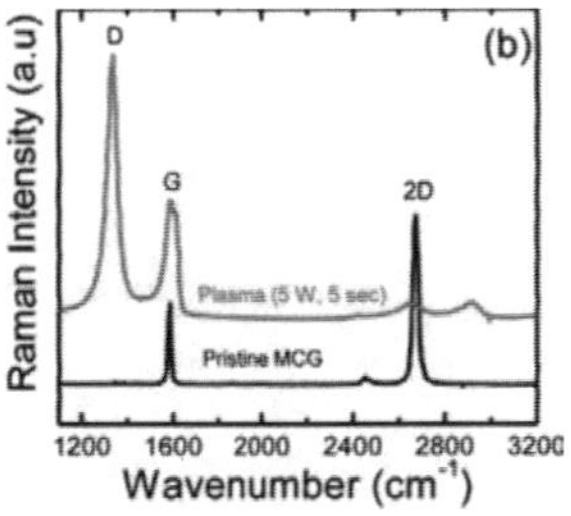

- Thickness Independent Hydrophobicity

Energy Environ. Sci. **2012**, *5*, 7908

Langmuir **2010**, *26*, 3798

- More defects, more hydrophilic
- GO sponges are not explored extensively

2

Sponge Materials: Why Is It Important

Marine Oil Spill

Threats to the planet

- High frequency
- Long-lasting adverse environmental impacts
- Significant economic impact

Current Methods

- Skimming on calm water
- Chemical dispersants spread

Sponge Materials

carbon-based aerogels

Adv. Mater. **2013**, *25*, 2554

porous boron-nitride

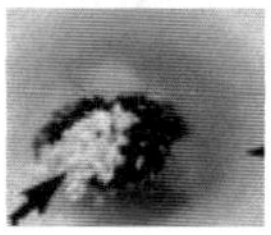

Nat. Commun. **2013**, *4*, 1777

modified commercial sponges

Energy Environ. Sci. **2012**, *5*, 7908

GO-wrapped sponge

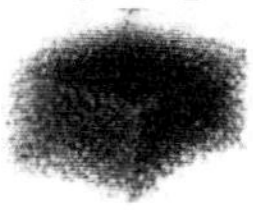

Nat. Nanotechnol. **2017**, *12*, 434

3

Catalysis Center for Energy Innovation
UNIVERSITY OF DELAWARE

Graphene Oxide Sponge (GOS)

- Explore hydrophobic and hydrophilic behaviors of GOS

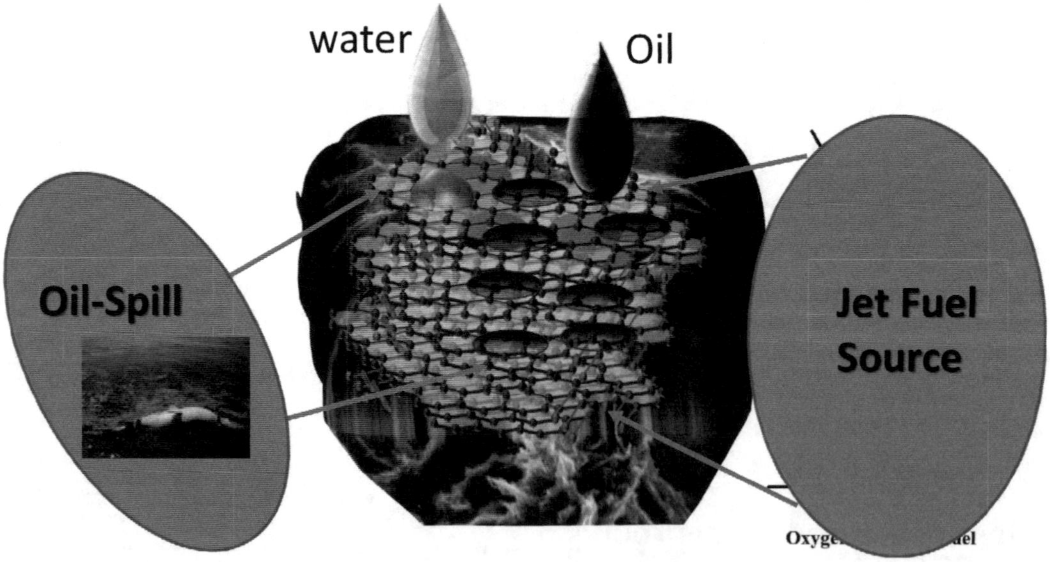

4

Catalysis Center for Energy Innovation

Hydrophilic IGO to GOS

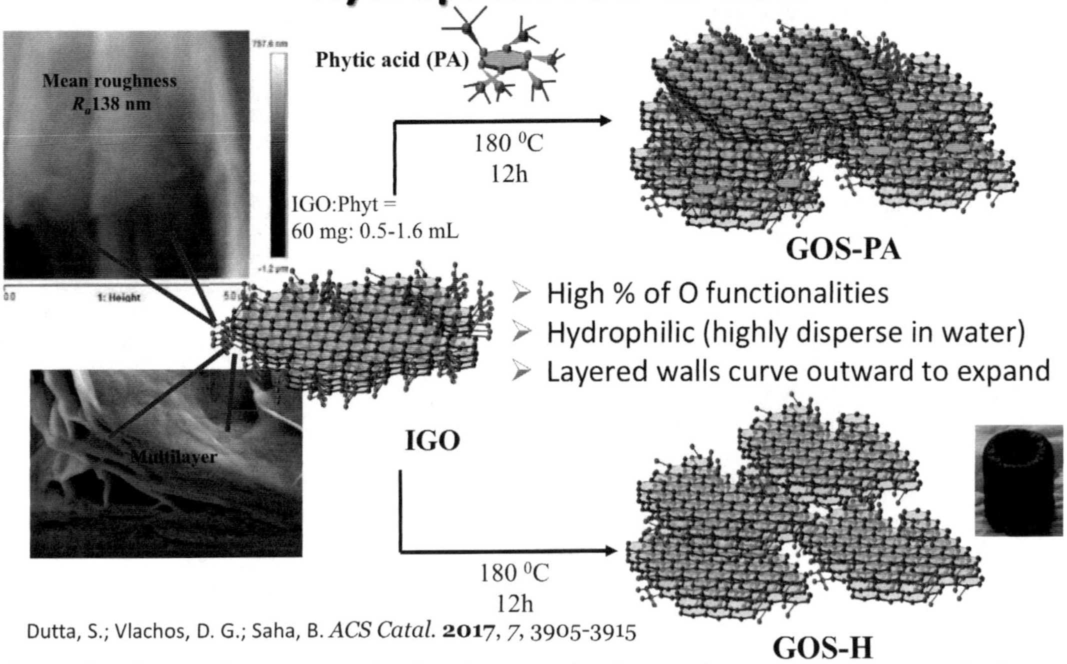

- High % of O functionalities
- Hydrophilic (highly disperse in water)
- Layered walls curve outward to expand

Dutta, S.; Vlachos, D. G.; Saha, B. *ACS Catal.* **2017**, *7*, 3905-3915

- Hydrothermal treatment is simplest method to reduce surface oxygen (gain hydrophobicity) compared to cross-linking/freeze-drying

5

Catalysis Center for Energy Innovation

Hydrophobicity of GOS from XPS C1s Peak

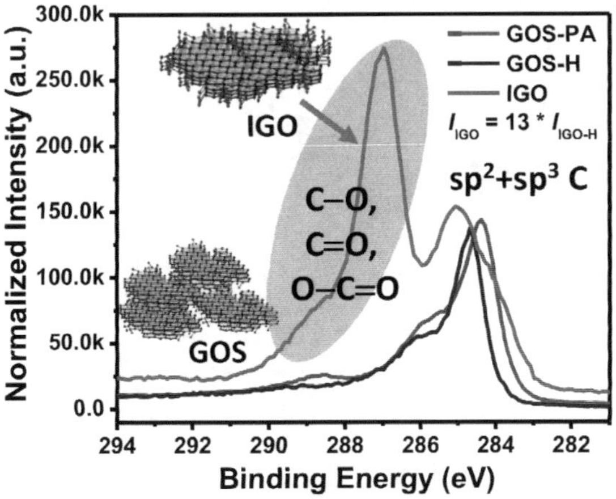

More surface oxygen (hydrophilic)

Less surface oxygen (hydrophobic)

- Increase in hydrophobicity: selective for oil/organic absorption

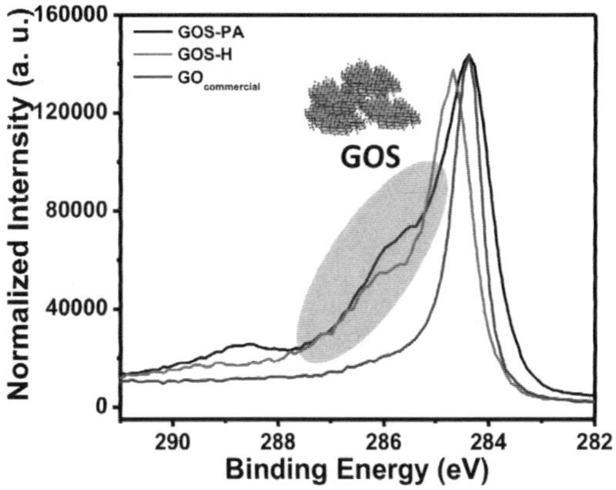

- Surface oxygen GOS > GO$_{commercial}$
- GOS more reactive

6

Catalysis Center for Energy Innovation

Defect Analysis: Raman Spectra

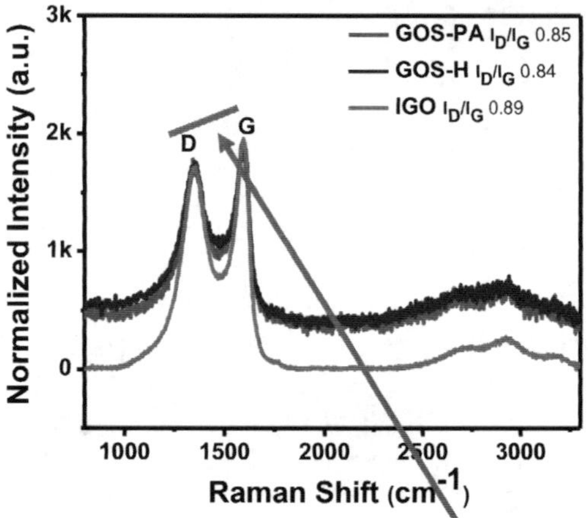

O_2 plasma (introduction of defect)

G@SiC

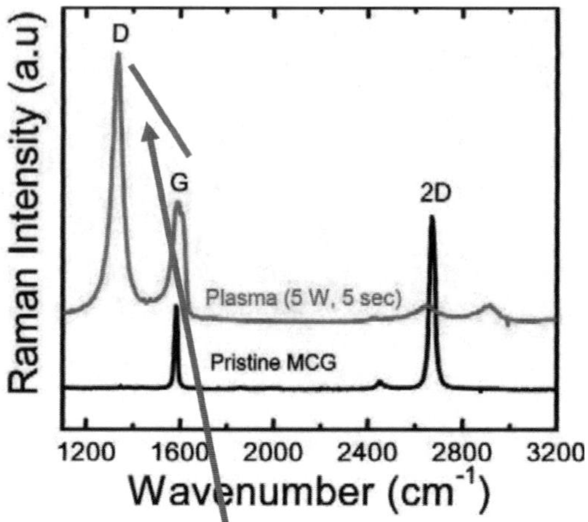

Langmuir **2010,** *26*, 3798

- GOS and IGO comparable crystalline defect
- G contains less crystalline defect before plasma treatment

7

Catalysis Center for Energy Innovation

Raman Imaging of D and G Bands

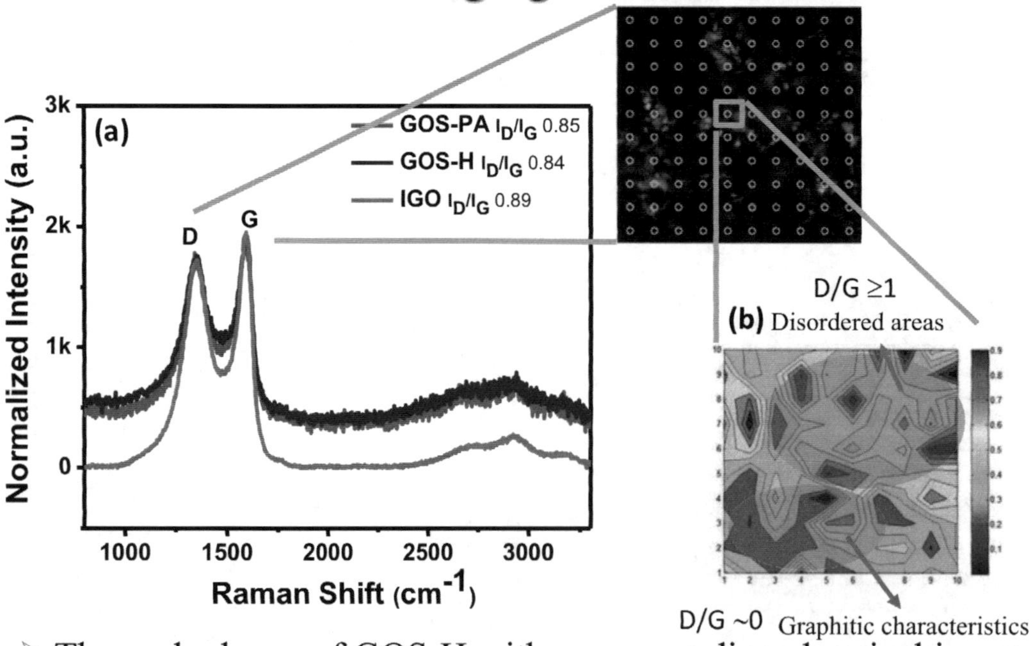

- The probed area of GOS-H with corresponding chemical image displaying the ratio of D-band and G-band intensity (b)

8

Catalysis Center for Energy Innovation

SEM Characterization of GOS-PA and GOS-H

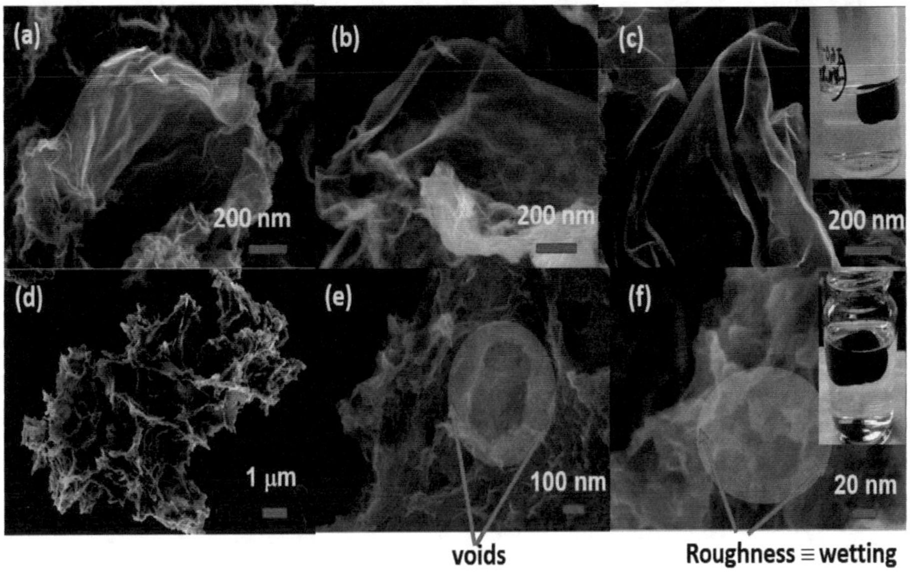

➢ GOS-H contains more rough surface as compared to GOS-PA

Catalysis Center for Energy Innovation

Water-Contact Angle of GOS

Sample	Contact Angle (Degrees)	Error (%)
HOPG (std)	67.3	3.3%
GOS-H	86.5	0.7%
GOS-PA	84.2	1.8%

➢ GOS hydrophobic with contact angle ~90°

➢ Enhancement of contact angle is countered by surface oxygen

Catalysis Center for Energy Innovation

UNIVERSITY OF DELAWARE

Oil Absorption by GOS-H

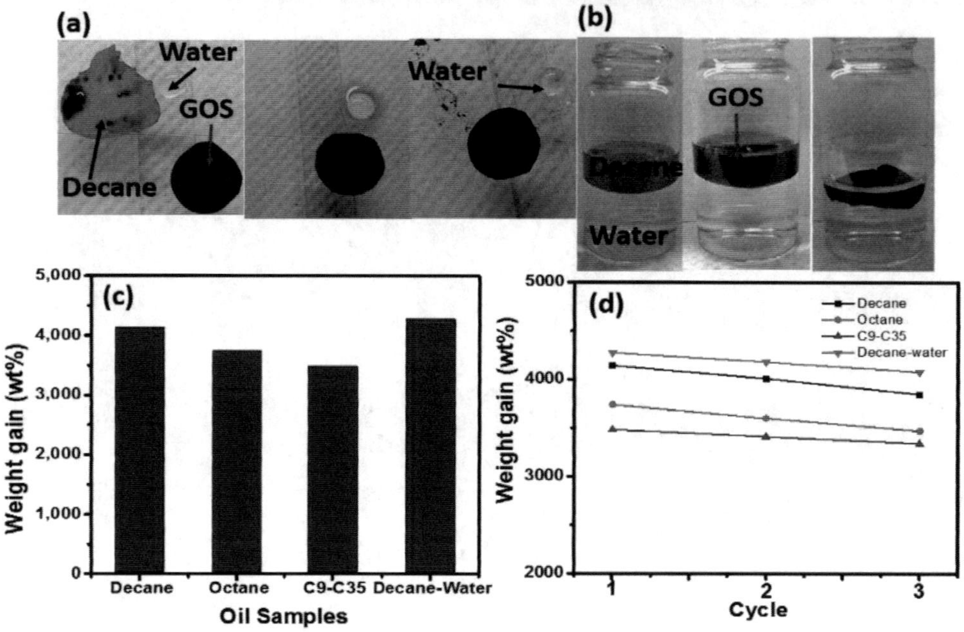

- ➢ Selective oil absorption by GOS in 15 min
- ➢ Exhibit recyclability for oil absorption

11

Catalysis Center for Energy Innovation

UNIVERSITY OF DELAWARE

Origin of Hydroxyalkylation-Alkylation (HAA) Catalysis

- ➢ Surface-oxygen active sites of IGO enables tailor the C-numbers of lignocellulosic fuel precursors

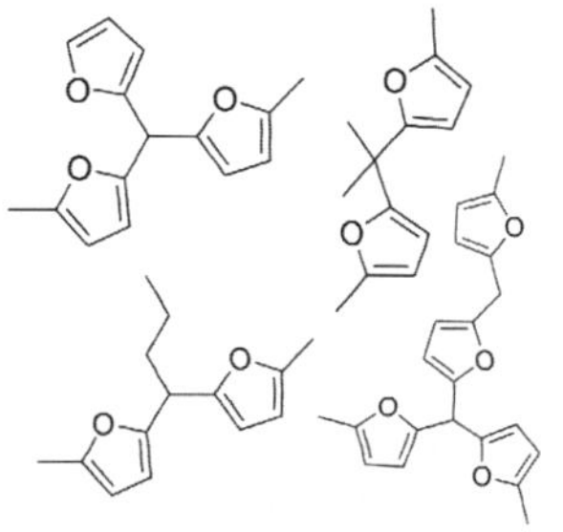

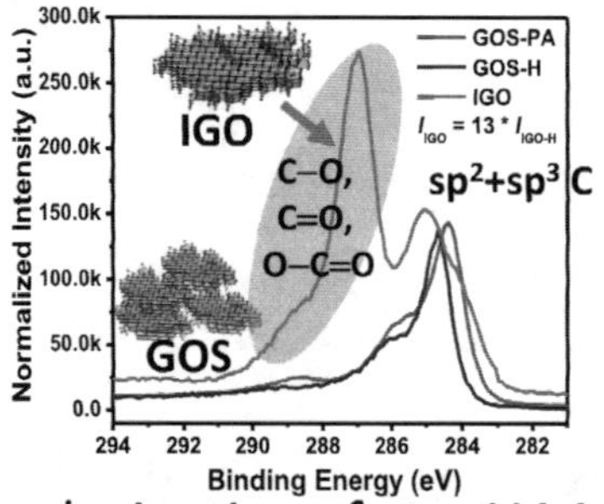

Entry	C (%)	O(%)	P(%)	S(%)
IGO-H	73.4	25.1		1.4
IGO-PA	72.5	24.2	2.1	1.2
IGO	59.2	36.6		3.1
HOPG	98.2	0.41		

- ➢ C13-C21 furan-derivatives from HAA with IGO

Dutta, S.; Saha, B. *ACS Catal.* **2017**, *7*, 5491-5499

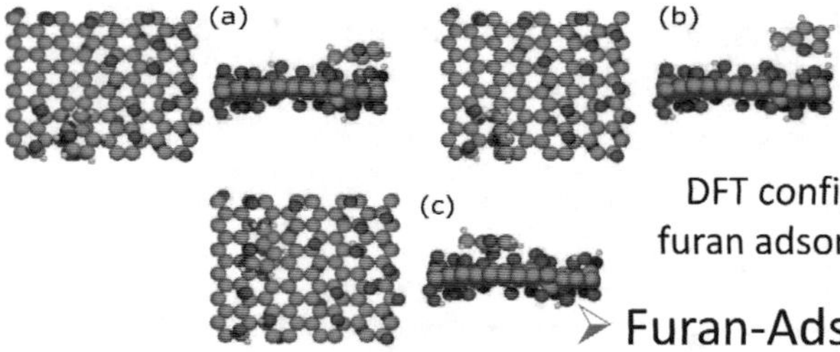

DFT configuration of furan adsorption on IGO

- ➢ Furan-Adsorption on a Oxygenated Surface

Dutta, S.; Vlachos, D. G.; Saha, B. *ACS Catal.* **2017**, *7*, 3905-3915

12

HydroxyAlkylation-Alkylation (HAA) of Furans

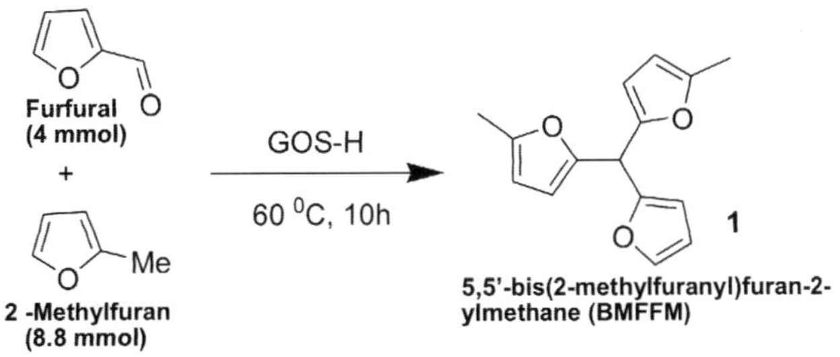

GOS-H recycle

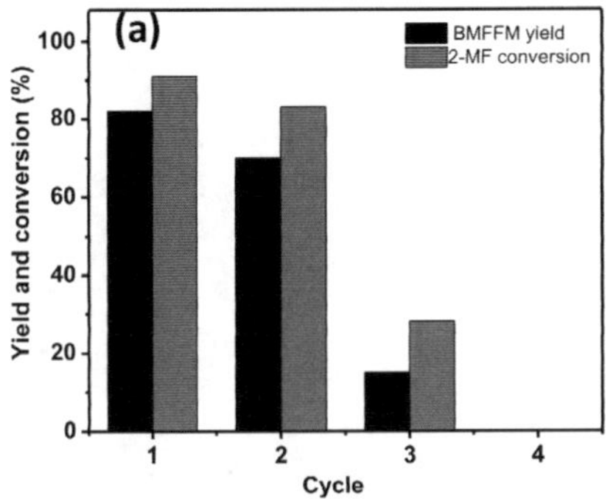

- Carbon extension of furans to desired fuel precursors
- Intermediate surface oxygen % with hydrophilic behavior GOS-H for HAA
- Loss of activity due to further reduction of surface oxygen (XPS)

Summary

- GOS derived from aqueous dispersion of IGO.
- GOS contains surface roughness responsible for wettability
- Raman imaging, XPS analysis, and SEM were used for characterization
- Percentage of surface oxygen in GOS is higher than graphene or rGO sponge
- GOS-H is effective for HAA reactions for furans
- GOS-H offers selective oil absorption from oil-water mixture

Moving Forward

- Technique for cross-linked GOS for mechanical stability

Catalysis Center for Energy Innovation

UNIVERSITY OF DELAWARE

Acknowledgements

Joseph P. Smith (Raman Spectroscopy)
Prof. Karl S. Booksh (Raman Spectroscopy)
Dr. Weiqing Zheng (SEM,TEM imaging)
Rachel Pupillo (XPS)
Dion G. Vlachos (insightful discussions)
Jeff Everhart (GC analysis)

U.S. Department of Energy, Office of Science, Office of Basic Energy Sciences, under Award No. DE-SC0001004

U.S. DEPARTMENT OF ENERGY | Office of Science

15

502391 Simultaneous Isomerization and Reactive Extraction Followed By Back Extraction of Sugars from Biomass Hydrolysate for High Purity and Yield of Ketose Sugars

Abstract:

Biomass hydrolysates are often low in sugar concentration and contain contaminants associated with the pretreatment method and sugar and lignin degradation. Cost-effective and high-yield downstream processing requires sugar streams at higher concentration, free of contaminants generated during the pretreatment process. Instead of removing contaminants from the hydrolysate, we have taken the approach of removing sugars from the hydrolysate, leaving contaminants behind. In the process of removing sugars, we enzymatically isomerize them from their aldose (glucose, xylose) to their ketose (fructose, xylulose) forms. These ketose sugars are ideally suited for production of furans â€" HMF from fructose and furfural from xylulose.

Unfortunately, the isomerization reactions do not have favorable equilibria with respect to ketose formation. To overcome this limitation, we have previously developed a method of simultaneous-isomerization-and-reactive-extraction (SIRE) to produce the ketose isomer in high yield. We have demonstrated that with a xylose/xylulose mixture, SIRE followed by 2-stage back-extraction (BE) allows high yield recovery of xylulose in nearly pure form.

For mixtures of xylose and glucose found in biomass hydrolysate, separation and purification of the ketose sugars (xylulose and fructose) from the hydrolysate will be more complex. The extraction affinities of glucose and fructose are lower than xylulose, but higher than xylose. SIRE-BE has not been previously evaluated for biomass hydrolysates with high sugar concentrations (up to 90 g/L) for recovery of a high purity, mixed C5 and C6 ketose product stream. We have developed strategies for high purity recovery of both xylulose and fructose from mixed sugar streams using SIRE-BE. We will present these strategies as well as xylulose and fructose yields and purity obtained under different experimental configurations.

494869 Development CO Sensor for Syngas Fermentation for Production of Fuels and Chemicals

Abstract:

Syngas fermentation is an emerging technology to produce fuels and chemicals from gasified biomass, coal, petcoke and solid wastes. This technology can also be used to make fuels and chemicals from gas streams that contain syngas components (CO, CO_2 and H_2). Some acetogens can convert these gases to organic acids and alcohols via the acetyl-CoA pathway. Challenges for commercialization of syngas fermentation technology include mass transfer limitations and low alcohol productivity associated with low cell mass concentration in the bioreactor and inhibition effects of high CO concentrations on key enzymes such as hydrogenase. Our recent results showed that the control of syngas feed rate to match the kinetics requirements of cells resulted in a stable operation of syngas fermentation (US Patent Application Publication, US 2016/0281114 A1 and International Patent Application No. PCT/US2015/60720). However, this required the use of gas chromatography and mathematical models to estimate the dissolved CO concentration in the medium, which was time consuming. Direct monitoring of dissolved CO and H_2 concentrations in the fermentation medium is a key to improve productivity and process stability. A real-time monitoring system for the dissolved CO and H_2 concentrations is being developed, which includes a built in-house dissolved CO sensor, a sample handling unit, a central control unit and an off-the-shelf dissolved H_2 sensor. The dissolved CO sensor was developed based on the spectral responses of CO in the mid infrared range (MIR). The CO sensor was housed in a designed cage with a careful alignment between a MIR LED as a light source and a photodetector as a signal receiver. A preamplifier and a lock-in-amplifier were used to preprocess the raw signal before being collected by the central control unit. The CO sensor was tested with gas samples containing CO concentration range generally found in bioreactors at typical operating pressures and achieved a sensitivity of over 10 mV/ppm CO. The central control unit was also developed for data acquisition and process control. The dissolved H_2 sensor was calibrated to measure 0 to 3 ppm of dissolved H_2 in the fermentation medium. Both CO and H_2 sensors were integrated in the bioreactor control platform to allow efficient delivery of these gases to the biocatalysts to enhance conversion efficiency, productivity and operation stability of syngas fermentation. Preliminary results showed the feasibility of the use of continuous mentoring of CO and H_2 concentrations in the bioreactor. Further work is focused on improving the sensorsâ€™ performance and its integration with the bioreactor control for efficient continuous syngas fermentation.

501393 Techno-Economic Analysis (TEA) of a Novel Hybrid Enzyme- and Chemo-Catalytic Process for Producing Furans from Biomass Hydrolysate

Abstract:

Lignocellulosic biomass is a sustainable feedstock for production of furans – 5-Hydroxymethy furfural (HMF) and furfural – versatile platform molecules that can be converted to drop-in fuels and precursors for synthetic materials. However, the lack of efficient pathways from to furans from lignocellulosic sugars constitutes a major barrier to their industrial production. We have recently achieved high yields of furans from biomass hydrolysates using a novel hybrid enzyme- and chemo-catalytic technology (*Green Chem.*, **2017**, 19, 1782). The present work assesses the techno-economic feasibility of this novel technology. The process was simulated for steady-state operation of an nth plant at receiving hydrolysate containing glucose and xylose in a 3:1 ratio. The feed rate was 1000 metric tons of mixed sugars (dry-basis) per day. The equipment sizing was done based on the process-unit hold-up volumes and experimentally-determined residence times to estimate total capital costs. Using discounted cash flow analysis in CAPCOST®, a Minimum Furan Selling Price (MFSP) of $2.48/kg was estimated for a "base case" hydrolysate containing a sugar concentration of 40 g/L (Case I)). The model estimates that by processing a more concentrated mixed-sugars stream (160 g/L), from upstream high-solids processing, MFSP can be lowered to $2.33/kg due to reduction in total capital costs (Case II). Further, MFSP reduction to $2.18 can be realized by solvent optimization (Case III), and to $ 1.48/kg with on-site chemical regeneration (Case IV). Sensitivity analysis revealed that the key factors affecting MFSP are overall yield of furan (from sugar) and the hydrolysate costs. TEA performed for furan production by other state-of-the-art technologies predict a furan selling price which is 75% higher than what is possible with our recently-developed method.

496508 Electrocatalytic Hydrogenation of Lignin-Derived Bio-Oil Model Compounds Using Ruthenium on Activated Carbon Cloth to Produce Liquid Fuel Intermediates and Value-Added Products

Abstract:

Biomass fast pyrolysis (BFP) uses heat in the absence of oxygen to convert biomass to bio-oil, biochar and combustible gas. The biochar co-product can be used for the removal of pollutants in the environment and has potential for use in soil amendment and carbon sequestration. The combustible gas co-product is typically burned for process heat needed by the pyrolysis system. The major product, bio-oil, has the potential to displace liquid hydrocarbon fuels; however, bio-oil is highly oxygenated, corrosive, low in energy content and unstable during storage. Electrocatalytic hydrogenation (ECH) is employed to reduce and deoxygenate reactive compounds and improve bio-oil properties. The present study particularly focuses on the electrocatalytic stabilization of compounds derived from the pyrolytic depolymerization of lignin. As lignin is a component of biomass comprising up to 30% of the mass and 40% of the energy stored in biomass, it offers great potential as feedstock for BFP. Lignin model compounds representative of bio-oil components were subjected to ECH under mild conditions using ruthenium on activated carbon. To date, several model lignin monomers and dimers have been reduced to simpler compounds such as cyclohexanol and phenol. The effect of different electrolytes, substrate concentrations, current density, and catalyst preparation methods on conversion and current efficiency was studied. The coupling of pyrolysis and electrocatalysis for depolymerization and upgrading of lignins, that are byproducts of the biomass pretreatment and extraction methods, will enable the maximization of yields from biomass conversion to fuels and value-added products.

499405 Thermal Deconstruction Opens Biomass for Acid Hydrolysis to Sugars

Abstract:

We have explored thermal deconstruction, which involves rapidly heating then cooling biomass in an inert environment before complete pyrolysis can occur. While we produce some bio-oil and light gases, solid or liquid phase reactions greatly modify our biomass. As shown through our microscopy and chemical analysis, thermal deconstruction cracks the biopolymers and loosens their lignocellulosic structure. This change allows molecules to diffuse more readily into and out of the biomass. Additionally, we have proven that as the biopolymers fragment into oligomers they become more susceptible to further depolymerization reactions. In particular, when cellulose is thermally deconstructed it rapidly forms anhydro-oligosaccharidesâ€”oligosaccharides where the reducing end has dehydrated. We can hydrolyze these oligomers to glucose under more mild conditions than required for untreated cellulose. Dilute sulfuric acid catalyzes hydrolysisâ€”primarily in cellulose and hemicellulose oligomersâ€”producing glucose and xylose, which can be fermented to ethanol or other products. Our process allows for higher monosaccharide yields than those obtainable from fast pyrolysis.

IOWA STATE UNIVERSITY
Bioeconomy Institute

Fermentable Sugar Production from Biomass Using THF/Water and Dilute Acid Catalyst

Arpa Ghosh, Wenqin Li, Mark Mba Wright, Robert C. Brown

AIChE Annual Meeting 2017, Integrated Thermochemical and Biochemical Processing for Renewable Fuels and Chemicals Session, November 2nd, 2017

Challenges in producing cellulosic sugars via conventional enzymatic pathway

- **Cost of enzyme production**[1,2]
  - Covers 20% of total production cost of fermentable carbohydrates
  - Estimated cost contribution up to $1.47 per gallon of ethanol
- **Slow rates of bioconversion**[3]
  - It takes days or even weeks to reach glucose yields >70%
- **Achieving high concentration of sugar products**[4]
  - Sugar substrate required for fermentation is high
- **Substrate specific enzyme production could be uneconomical**

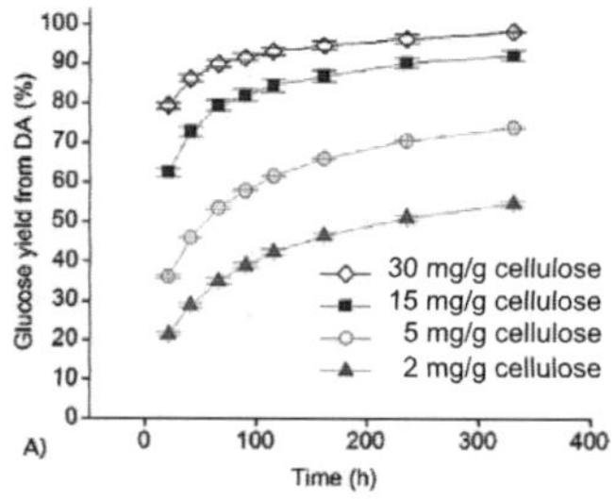

DA: Dilute acid pretreatment (common in biochemical processing of cellulosic biomass) [3]

IOWA STATE UNIVERSITY
Bioeconomy Institute

Solubilized carbohydrates production using polar aprotic solvents

- Accelerated rate of monosaccharide production from biomass in polar aprotic solvents[5,6]
- Dilution prevents secondary reactions
- Mild temperature (150-350 °C)
- Carbohydrates are solubilized
- Low consumption of acid

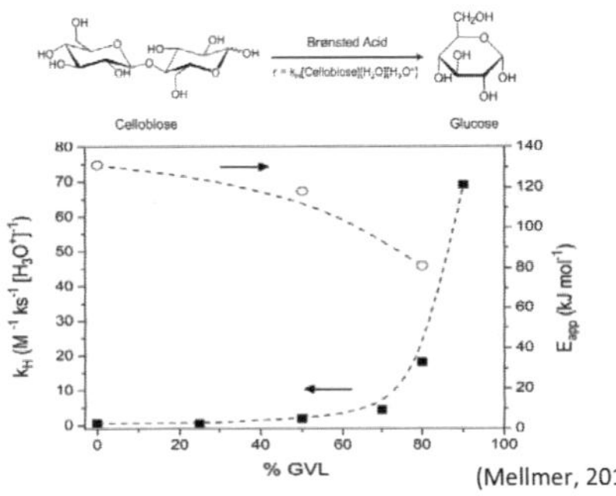

(Mellmer, 2014)

Recent progress in the field:

- Production of 68% of theoretical soluble carbohydrates from biomass at 157-217 °C using γ-valerolactone (GVL) mediated process[7]
- Other effective polar aprotic solvents: Hot, pressurized 1,4-dioxane, acetone, sulfolane, acetonitrile

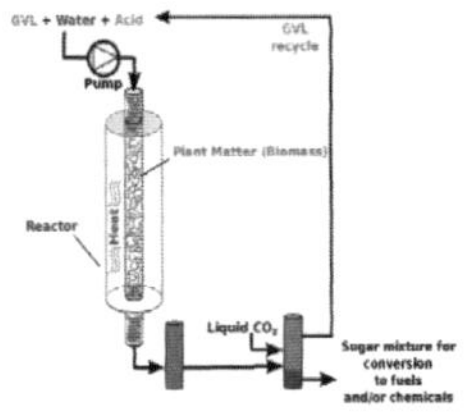

(Luterbacher, 2014)

IOWA STATE UNIVERSITY
Bioeconomy Institute

Challenges in developing cellulosic sugar production process using polar aprotic solvents

Which solvent to choose for cellulosic sugar production?

- Inadequate information on the effects of different polar aprotic solvents on solubilized carbohydrate production

Challenges with high polarity solvents

- Often expensive and thus demand quantitative recovery
- Recovery of solvent is often complicated such as the use of liquid CO_2 extraction for GVL recovery[8]

Challenges with low polarity solvents

- High temperature and high pressure processing associated with these solvents increases capital and operating costs

IOWA STATE UNIVERSITY
Bioeconomy Institute

Properties and costs of polar aprotic solvents tested

Polar aprotic solvent	Normal boiling point (°C)	Polar solubility parameter [a] ($MPa^{1/2}$)	Cost [b] ($/Kg)
1,4-Dioxane	101	2.1	105
Tetrahydrofuran (THF)	66	7.0	121
Acetone	56	13.1	80
Acetonitrile	82	18.7	153
Gamma-valerolactone (GVL)	207-208	22.1	524

[a] At reaction condition of 350 °C, 5-19 MPa; [b] From Sigma Aldrich for ACS Reagent solvents

IOWA STATE UNIVERSITY
Bioeconomy Institute

5

Use of acid catalyst removes much of the difference among solvents

No catalyst:

The higher the polar solubility parameter, the higher levoglucosan yield as a result of higher solubilization of cellulose[9]

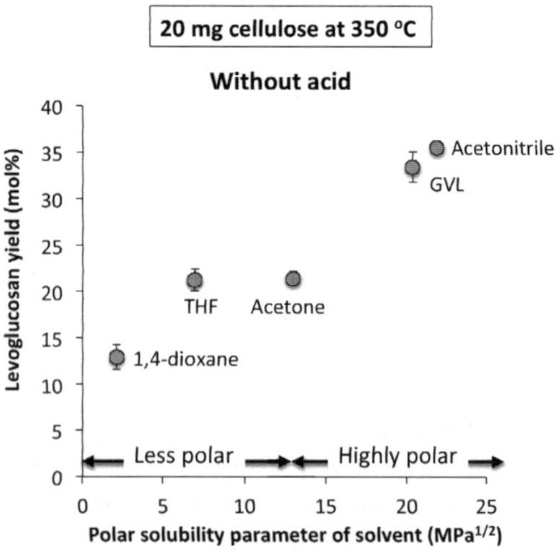

Mean yields reported based on at least duplicates.
STD error is shown in error bars.

Bioeconomy Institute

6

Use of acid catalyst removes much of the difference among solvents

No catalyst:

The higher the polar solubility parameter, the higher levoglucosan yield as a result of higher solubilization of cellulose[9]

With acid catalyst:

- Dramatically increased in low polarity solvents compared to high polarity solvents
- Levoglucosan yields equalized among solvents

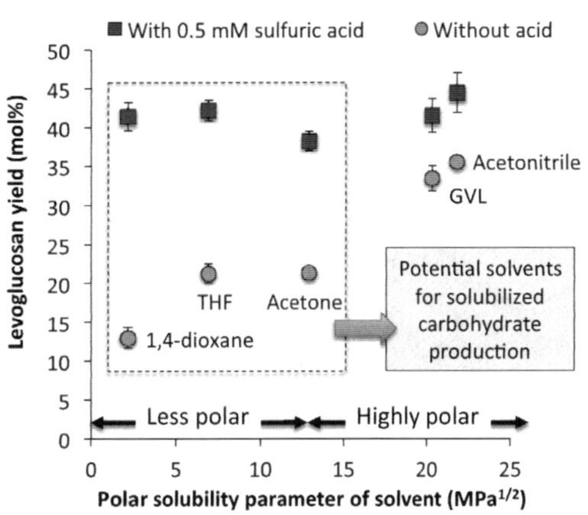

Mean yields reported based on at least duplicates.
STD error is shown in error bars.

Developing solvent liquefaction process to overcome challenges of current technologies

Goals

- Significant increase of sugar production rates with respect to enzymatic hydrolysis (>> 70% of theoretical glucose yield within 72 hours)
- Competitive solubilized carbohydrates yields with current solvent liquefaction methods (approaching 70% of theoretical yield)
- Using relatively low-cost and easily recoverable solvent

THF-Water with dilute sulfuric acid could serve as an effective solvent for sugar production

Tetrahydrofuran (THF)

Approach

Process design:

Stage-1. Solubilize lignin and hemicellulose in a mild pretreatment

- Partial removal of lignin and hemicellulose should render cellulose more susceptible to deconstruction[3]
- THF is a good solvent for dissolving lignin and decomposing hemicellulose in presence of acid catalyst[10]

Stage-2. Solubilize cellulose-rich solids in solvent liquefaction

THF is an effective solvent for acid-catalyzed cellulose depolymerization

Process economics:

THF is relatively inexpensive and has lower boiling point than water (66 °C)

9

Biomass two-step solvent process using THF/water with sulfuric acid catalyst

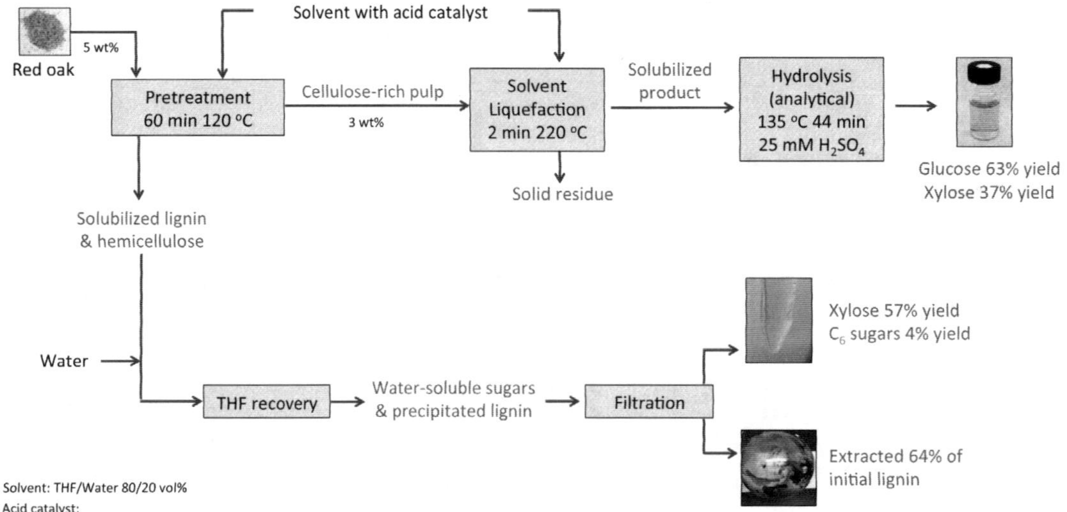

Solvent: THF/Water 80/20 vol%

Acid catalyst:

Pretreatment- 2.5-10 mM H_2SO_4 (0.5-2 wt% of red oak), Solvent liquefaction- 2.5 mM H_2SO_4

10

Prolonged pretreatment at high acid loading gives delignified biomass and solubilized hemicellulose keeping cellulose intact

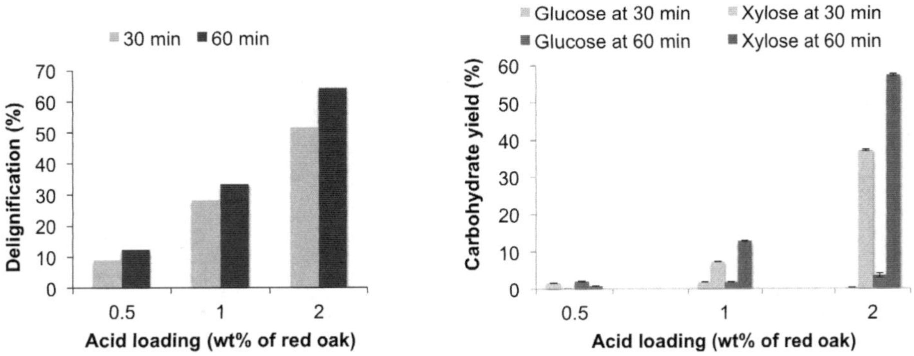

- **64% lignin extracted from water-insoluble fraction of pretreatment liquid**
- **57% xylose and 4% glucose recovered in water-soluble fraction of pretreatment liquid**
- **Resulting pulp yield at 60 min pretreatment at 2 wt% acid loading: 56 wt%**

11

Effect of pretreatment on enhancing yields of solubilized carbohydrates from red oak

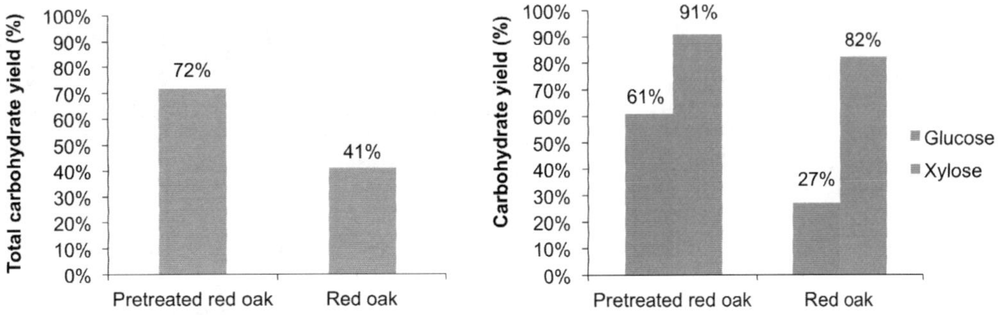

Carbohydrate yields from combined pretreatment and solvent liquefaction of red oak was significantly higher than solvent liquefaction without pretreatment

IOWA STATE UNIVERSITY
Bioeconomy Institute

Mean yields reported based on at least duplicates. STD error is within 10% of mean yield.

12

Pretreated red oak is enriched with cellulose making it easier to deconstruct by solvent liquefaction

- Lignin and xylan are reduced by 53% and 67% via pretreatment
- Glucan only reduces by 14%
- Cellulose-rich pretreated red oak could be more susceptible to deconstruction by solvent and acid

(Analysis by: Celignis Ltd.)

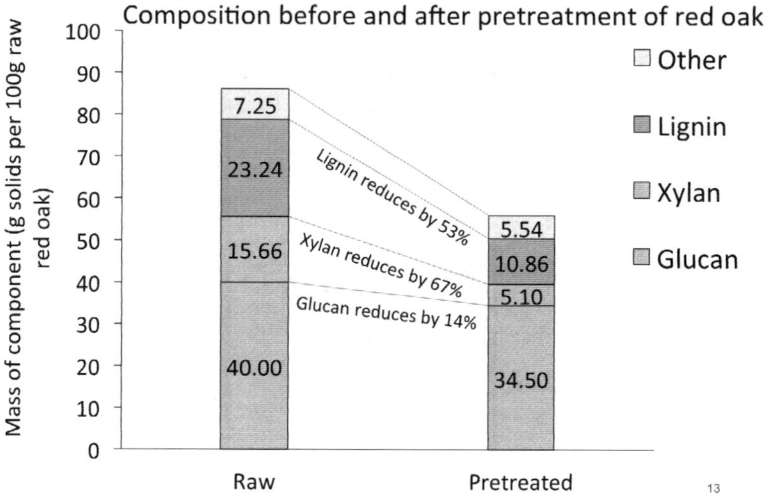

13

SEM images reveal significantly modified structure of lignocellulosics by THF/Water pretreatment

- Polymeric matrix of biomass was loosened up
- More penetrable structure with large pores and microgrooves
- Wrinkles around edges imply loss of structural strength; attributed to delignification

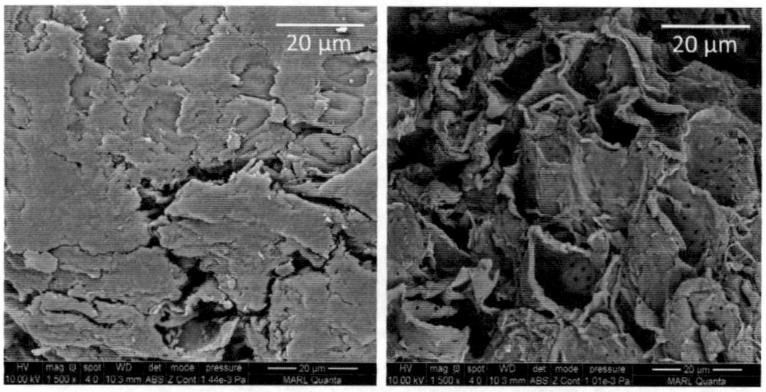

Untreated red oak **THF/Water pretreated red oak**

SEM images at 1500X magnification as taken originally

IOWA STATE UNIVERSITY
Bioeconomy Institute

14

One-step simple distillation to recover THF and sugars from solubilized products

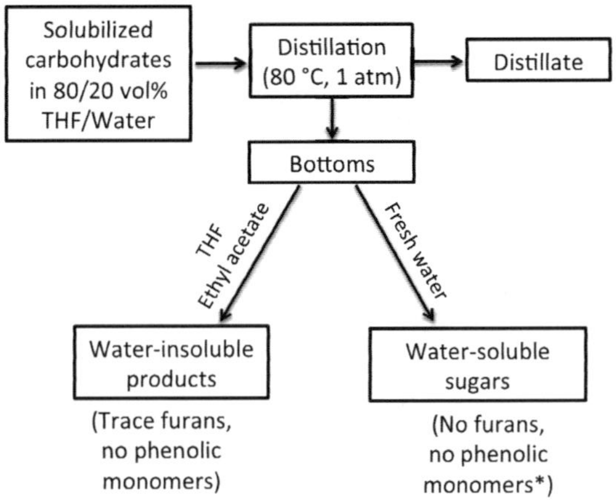

*possibly contains water-insoluble non-carbohydrate oligomers

Bioeconomy Institute

Highly concentrated sugar solution after distillation

Distillation Fraction	THF recovery (%)
Distillate (Almost pure THF, very low level of contaminants)	106.3
Bottoms (recovered products of solvent liquefaction)	0.3

Most of the sugars recovered in bottoms with no migration into distillate

Sugars in solubilized products from solvent liquefaction

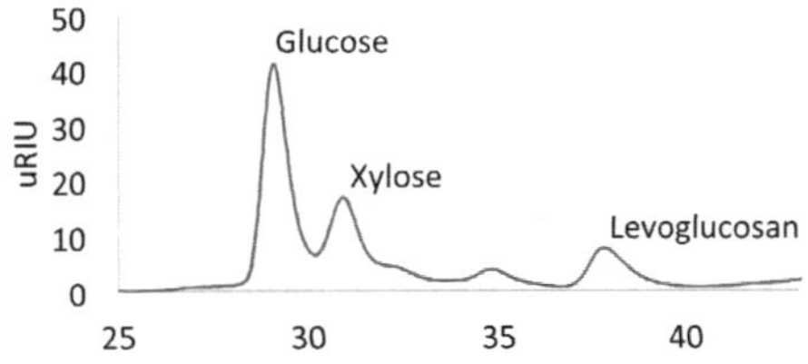

Sugars recovered by distillation of THF

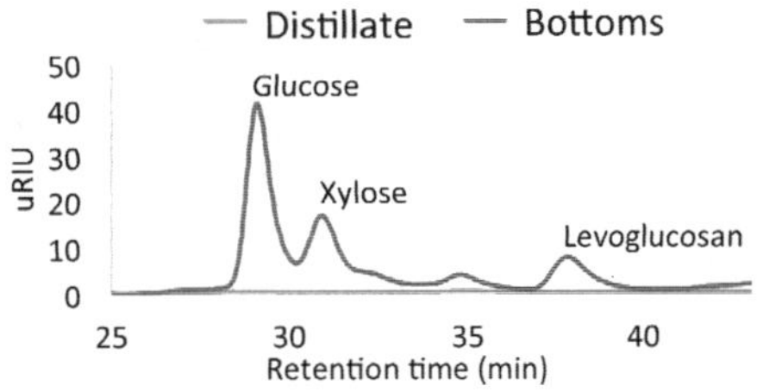

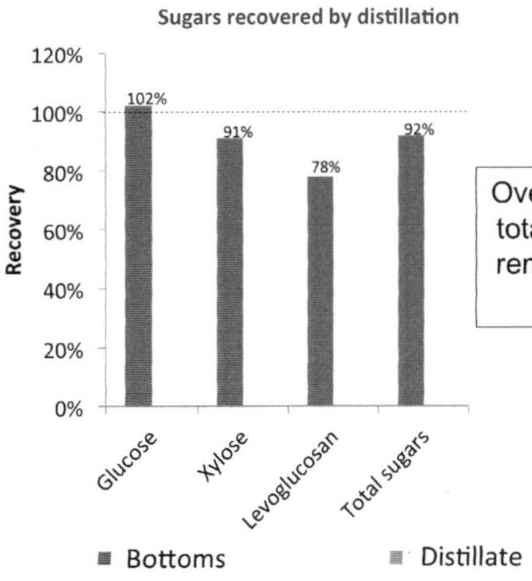

Over 90% of total sugars remained in water

16

Fermentation substrate: high concentration sugar solution as biofuel precursor at rapid rate

Process	Pretreated Biomass loading (wt%)	Glucose concentration (g/L)	Estimated glucose production (g/L/h)
Enzymatic hydrolysis[12]	**15**	**114**	0.7
GVL flow through[7]	**7**	**127**	**64**
THF batch *(this work)*	**3**	**90**	2715*
THF flow through *(predicted)*	7	215	6463

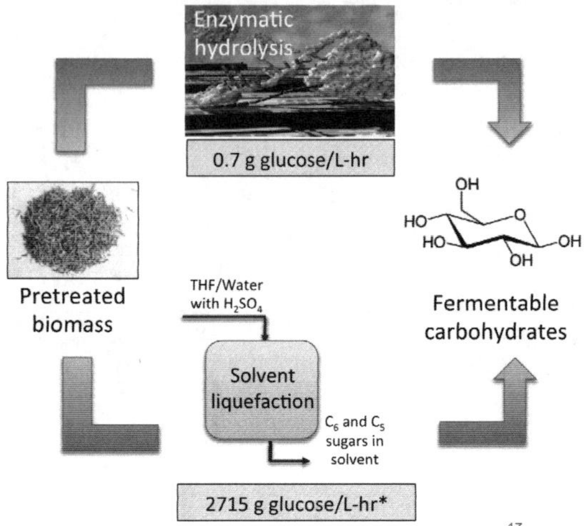

Enzymatic hydrolysis figure source: (Himmel et al., 2007)

*excluding analytical hydrolysis step of 44 min

17

Comparable or better economics of cellulosic ethanol production than current processes

- MFSP for GVL and THF processes were $3.33/gal and $2.80/gal ethanol, respectively
- Comparable/better than over both enzymatic hydrolysis and GVL processes
- Key assumptions: same feedstock cost and solvent to biomass ratio at large scale production

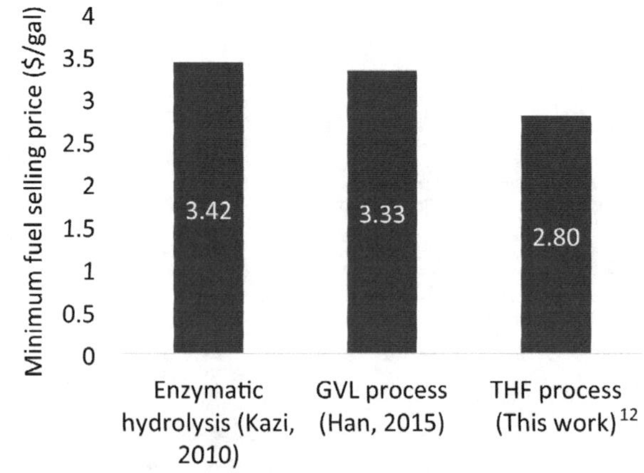

IOWA STATE UNIVERSITY
Bioeconomy Institute

18

Summary

- THF was identified as a low boiling point polar aprotic solvent that could effectively and rapidly depolymerize cellulose into solubilized carbohydrates with high yields of monomeric sugars
- A novel non-enzymatic strategy for cellulosic sugar production using THF/ water mixture shows competitive sugar yields with current solvent-mediated processes
- Preliminary techno-economic analysis indicates that THF-water process could be more economical than current processes for cellulosic ethanol production

IOWA STATE UNIVERSITY
Bioeconomy Institute

19

References

1. Lynd, L. R., Laser, M. S., Bransby, D., Dale, B. E., Davison, B., Hamilton, R., ... & Wyman, C. E. (2008). How biotech can transform biofuels. *Nature biotechnology*, *26*(2), 169.
2. Klein-Marcuschamer, D., Oleskowicz-Popiel, P., Simmons, B. A., & Blanch, H. W. (2012). The challenge of enzyme cost in the production of lignocellulosic biofuels. *Biotechnology and bioengineering*, *109*(4), 1083-1087.
3. Nguyen, T. Y., Cai, C. M., Kumar, R., & Wyman, C. E. (2015). Co-solvent Pretreatment Reduces Costly Enzyme Requirements for High Sugar and Ethanol Yields from Lignocellulosic Biomass. *ChemSusChem*, *8*(10), 1716-1725.
4. Dowe, N., & McMILLAN, J. (2001). SSF experimental protocols: lignocellulosic biomass hydrolysis and fermentation. *National Renewable Energy Laboratory (NREL) Analytical Procedures.*
5. Mellmer, M. A., Alonso, D. M., Luterbacher, J. S., Gallo, J. M. R., & Dumesic, J. A. (2014). Effects of γ-valerolactone in hydrolysis of lignocellulosic biomass to monosaccharides. *Green Chemistry*, *16*(11), 4659-4662.
6. Weingarten, R., Tompsett, G. A., Conner, W. C., & Huber, G. W. (2011). Design of solid acid catalysts for aqueous-phase dehydration of carbohydrates: The role of Lewis and Brønsted acid sites. *Journal of Catalysis*, *279*(1), 174-182.
7. Luterbacher, J. S., Rand, J. M., Alonso, D. M., Han, J., Youngquist, J. T., Maravelias, C. T., Pfleger B. F. , Dumesic, J. A. (2014). *Science, 343*, 277-280
8. Han, J., Luterbacher, J. S., Alonso, D. M., Dumesic, J. A., & Maravelias, C. T. (2015). A lignocellulosic ethanol strategy via nonenzymatic sugar production: Process synthesis and analysis. *Bioresource technology*, *182*, 258-266.
9. Ghosh, A., Brown, R. C., & Bai, X. (2016). Production of solubilized carbohydrate from cellulose using non-catalytic, supercritical depolymerization in polar aprotic solvents. *Green Chemistry*, *18*(4), 1023-1031.
10. Cai, C. M., Zhang, T., Kumar, R., & Wyman, C. E. (2013). THF co-solvent enhances hydrocarbon fuel precursor yields from lignocellulosic biomass. *Green Chemistry*, *15*(11), 3140-3145.
11. Hodge, D.B., Karim, M.N., Schell, D.J., McMillan, J.D. (2008) *Bioresour. Technol, 99*, 8940-8948
12. Li, W., Ghosh, A., Bbosa, D., Brown, R.C., Wright M.M. (2017). Techno-economic analysis of ethanol production from solvent liquefaction of lignocellulosic biomass. *ACS Sustainable. Chem. Eng.* (To be submitted)

20

Acknowledgement

- Major Professor: Prof. Robert C. Brown
- Undergraduate Research Assistants
- Bioeconomy Institute Staff, Scientists and Graduate Students
- Iowa Energy Center, Iowa EPSCoR

IOWA EPSCoR

Building Sustainable Energy Systems

IOWA STATE UNIVERSITY
Bioeconomy Institute

Thank you..

Questions?

IOWA STATE UNIVERSITY
Bioeconomy Institute

Batch mode solvent liquefaction process

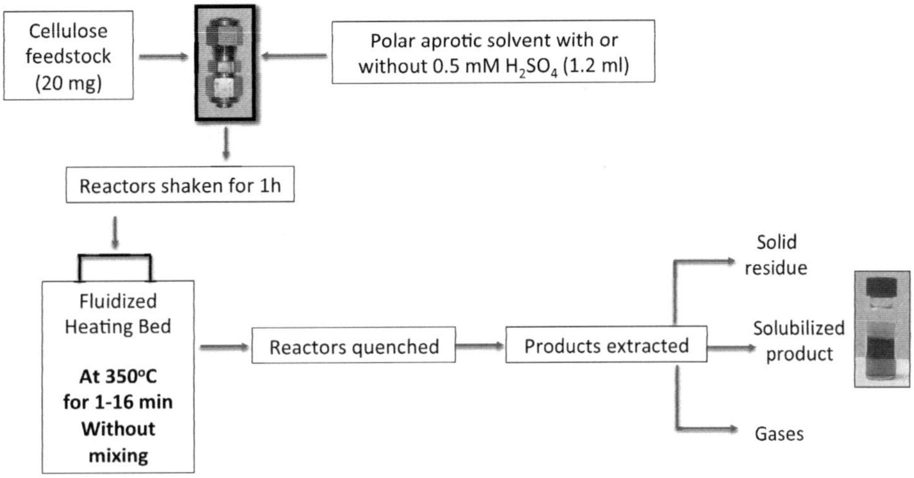

IOWA STATE UNIVERSITY
Bioeconomy Institute

Red oak composition

Composition of lignocellulose in red oak[a]

Component	Composition (wt%)	Total C_5 polysaccharides (wt%)	Total C_6 polysaccharides (wt%)	Total lignin (wt%)
Glucan	40.00	16.00	42.22	23.24
Xylan	15.66			
Mannan	1.30			
Arabinan	0.34			
Galactan	0.92			
Klason lignin	20.29			
Acid-soluble lignin	2.95			
Extractives	6.85			
Ash	0.40			

[a] Determined according to NREL procedure by Celignis, Inc.

IOWA STATE UNIVERSITY
Bioeconomy Institute

Solubilized carbohydrates quantified based on theoretical yields

$$Glucose\ yield\ (\%) = \frac{\text{concentration of glucose in solution} * \text{volume of solution}}{1.111 * \text{glucan content in red oak} * \text{initial mass of red oak}} \times 100\%$$

$$Xylose\ yield\ (\%) = \frac{\text{concentration of xylose in solution} * \text{volume of solution}}{1.136 * \text{xylan content in red oak} * \text{initial mass of red oak}} \times 100\%$$

$$Total\ carbohydrate\ yield\ (\%) = \frac{\text{glucose mass} + \text{xylose mass}}{(f_{C5} * \text{total pentosan content in red oak}) + (f_{C6} * \text{total hexosan content in red oak}) * \text{initial mass of red oak}} \times 100\%$$

Conversion factors:
f_{C5} = 1.36 for pentosan to pentose formation
f_{C6} = 1.11 for hexosan to hexose formation

IOWA STATE UNIVERSITY
Bioeconomy Institute

25

Dehydration product yield and delignification

$$\text{Dehydration product yield (wt\%)} = \frac{\text{Mass of dehydration product}}{\text{Initial mass of biomass}} \text{ X } 100\%$$

$$\text{Delignification (\%)} = \frac{\text{Mass of lignin extracted from pretreatment}}{\text{Initial lignin in biomass}} \text{ X } 100\%$$

Dehydration products: products from dehydration of carbohydrate monomers such as 5-HMF, furfural

IOWA STATE UNIVERSITY
Bioeconomy Institute

26

Mass balance and conversion of polysaccharides in two-step solvent liquefaction

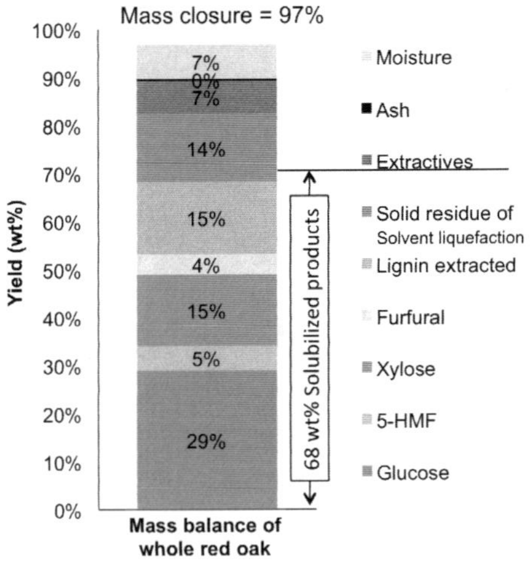

Process step	Yield (wt%)
Pretreatment	
Glucose	2
Xylose	10
Solvent liquefaction	
Glucose	28
5-HMF	5
Xylose	7
Furfural	4
Total carbohydrate derived products	56
Total	58
Initial polysaccharides	
Polysaccharide conversion*	97%

*Carbon-based molar conversion could not be done as source of furfural may both be pentosan/hexosan

27

Effect of delignification on recovery of xylose during pretreatment and glucose during subsequent solvent liquefaction

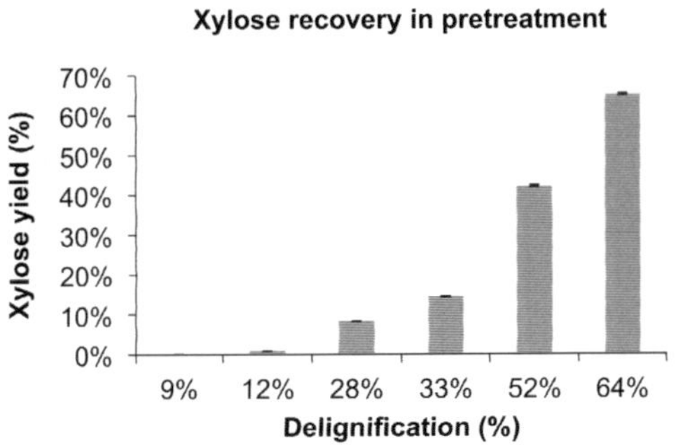

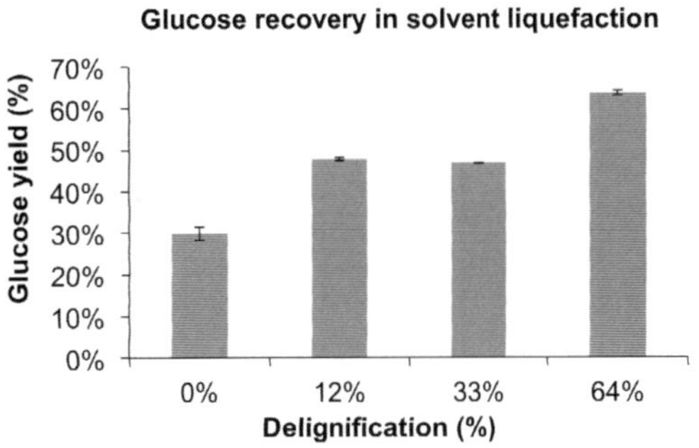

- **Delignification increases xylose recovery from hemicellulose during pretreatment**
- **Delignification increases glucose recovery from cellulose during subsequent solvent liquefaction**

IOWA STATE UNIVERSITY
Bioeconomy Institute

Mean yields reported based on at least duplicates. STD error is within 10% of mean yield.

28

Proposed block diagram for cellulosic ethanol production

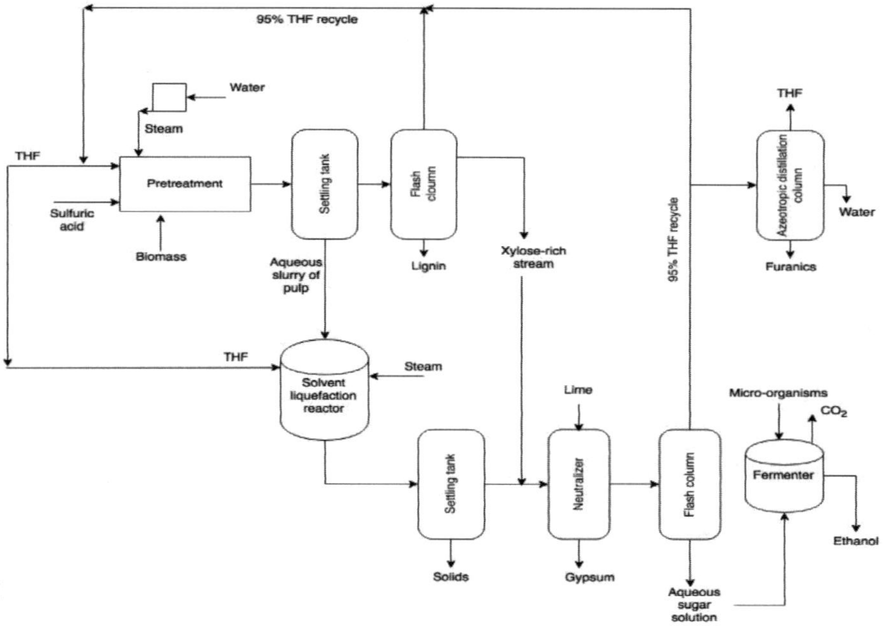

IOWA STA
Bioeconomy In

29

488549 Optimization and Performance Improvement of Bioethanol Biorefinery through the Integration of Thermochemical-Utilized Lignin Residue

Abstract:

In global economy, energy crisis has become a main issue since the global energy demand is increasing exceptionally, whilst natural reserves of crude oil are depleting as the time passes. Therefore, it is necessary to find alternative renewable energy sources. These sources include wind, solar, geothermal, hydro- and bioenergy. Bioenergy (e.g. biofuels) is the most promising renewable energy source and bioethanol is one of the main biofuels that is mostly used to be as a transportation fuel. During the production of bioethanol, a huge amount of lignin is produced as a byproduct. The global market of lignin is expected to exceed USD 6 billion by 2022, growing at 4.9% from 2015 to 2022. Due to that, strong industrial R&D has been exploring the commercialization of lignin downstream applications to produce valuable products to increase its revenue.

In this research, a biorefinery was designed to produce bioethanol by simultaneous saccharification and fermentation (SSF) of softwood (spruce) biomass after the extraction of lignin. The performance of this process is enhanced and optimized using Aspen Plus (V.2006) simulation software by investigating the potential of integrating extracted lignin back into the biorefinery. The simulation results are generally not 100% accurate but it provides a reflective prospect to process design and economic analysis of suggested processes in a limited period of time.

Firstly, the recovery of lignin from the biomass was maximised to ensure maximum production rate of bioethanol. The optimal removal rate of lignin was 20.100 tonne per hour (tph), corresponded to bioethanol production rate of 19.117 tph from a total feed charge of about 160tph.

The primary aim of this research is either to generate combined heat and power (CHP) or activated biochar. The first case was generating CHP in steam turbines that can be integrated back into the biorefinery to meet its energy requirements to minimise operating costs. The optimal turbine power of about 470 MW was produced from steam generated by the gasification of lignin. The flow rate of the steam was 2516.138 tph with turbine mechanical and isentropic efficiencies of 97% and 80% respectively. The optimal power generation was achieved by recycling steam through the gasifier to generate as much as 180 times more power than the unrecycled steam. This increase was achieved even that there was a trade-off with the gasifier power requirement, where more energy was required by the gasifier to raise the LP (low pressure) steam to superheated steam temperature. Also, the continuous recycling of steam had minimised wastewater treatment.

Heat integration was carried out to recover the waste heat from the entire system at a power duty of about 54 MW. Heat integration provided a part of the biorefinery energy requirements; reducing further its operating costs. For example, the power recovered and integrated from cooling the steam leaving the last turbine to the boiler feed water (BFW) temperature of about $105^{\circ}C$ resulted in annual cost savings of about 37 million US dollar.

The second case was producing activated biochar at a flow rate of 4.703 tph from the pyrolysis of lignin. Biochar was firstly produced at a flow rate of 7.236 tph and then thermally treated to produce activated biochar with about 35% weight loss. The activation of biochar was achieved in the presence of CO_2 at a flow rate of 1.266 tph and activation temperature of 850°C. Using steam for biochar activation showed no benefit since it reduces power generation and has more GHG emissions when compared to CO_2 activating agent. The product streams were cooled to atmospheric conditions for heat recovery. It was concluded that utilising steam generated from the pyrolysis reactor to produce activited biochar was not benefical in terms of power, cost and environment as a whole. Producing activated biochar had reduced the power generation of the whole process and minimised the potential of heat integration, which resulted in an increase in total production cost. Also the unused carbon dioxide from lignin pyrolysis was discharged to the atmostphere and more CO was formed, making the pyrolysis process harmful to the enviroment.

497847 Screening for Economically Promising Bio-Based Chemicals

Abstract:

10A. Process Design: Innovation for Sustainability

Screening for Economically Promising Bio-Based Chemicals

Wenzhao Wu, Matthew Long, Jennifer Reed, Christos T. Maravelias*

Dept. of Chemical and Biological Engineering, University of Wisconsin-Madison, Madison, WI 53706

The last decade has seen tremendous progress in metabolic engineering and synthetic biology[1,2]. These advances enable the use of engineered microorganisms such as *Escherichia coli* and yeast for the production of chemicals that are currently derived mainly from fossil fuel feedstocks[3,4]. However, it is still unclear which chemicals have the highest economic potential, which is key in developing a sustainable chemical production system. To this end, we develop a framework for the identification of economically promising chemicals that can be produced biologically using microorganisms ("bio-based chemicals").

We first examine the US high-production-volume (HPV) chemicals[5], which are manufactured in or imported into the United States in amounts equal to or greater than 454 metric tons (MT) per year. HPV chemicals include all commodity chemicals and a portion of fine chemicals. We establish an HPV chemical database (3574 chemicals) by compiling several HPV lists published by the EPA over the past two decades. Then we intersect the HPV chemical database with the KEGG database, which includes most chemicals produced by characterized reactions in biological systems, and thus 613 overlapping chemicals are found. These chemicals are then imported into a genome-scale metabolic model we develop, and 170 chemicals are identified to be producible by microorganisms, often with the addition of heterologous reactions. These 170 chemicals are the complete pool of candidate targets for bio-based production. In addition, market volume and price data is collected for them.

We first use the following two screening criteria to quantify economic potential.

Criterion 1: market volume. The market volume (MT/year) should be greater than an expected production capacity.

Criterion 2: market value. The market value ($/year) should be large enough to attract investment and recover capital cost within an expected time horizon.

The difference between the price and the production cost also plays a key role in determining the economic potential of a product. The larger the difference between a chemical's selling price and its total production cost, the larger its economic potential. However, directly implementing a third criterion based on such a difference is challenging because the downstream separation cost is highly product-dependent and difficult to estimate. To this end, we use the "separation cost margin" as Criterion 3, which is the difference between the price and the upstream cost (including raw material supply cost and bio-conversion cost). We calculate the upstream cost using base cost data from the literature as well as the theoretical yield and productivity calculated using our metabolic model. For a specific product, if the margin is negative, then the product fails Criterion 3; if the margin is positive, then we compare it with the expected separation cost. In general, the separation cost depends on (1) the product properties, e.g., separation of extracellular insoluble products tends to be easier than that of extracellular soluble ones, especially when the product concentration is low; and (2) case-specific information, e.g., whether a known solvent exists for a particular product, and special product requirement such as colorlessness and dryness.

To account for these major separation cost drivers, we introduce a general downstream bio-separation superstructure optimization framework to provide guidance in the preliminary design of separation networks[6,7]. We then apply this framework and formulate superstructure models for different classes of bio-based chemicals[8] based on their properties such as localization (intracellular/extracellular), solubility (soluble/insoluble) and density with respect to water (light/heavy). Further, we identify the optimal separation processes for base cases and the key cost drivers (such as titer and technology performances). Finally, we generate sensitivity curves (e.g. Figure 1A) and "heat maps" (e.g. Figure 1B) to show how the overall cost changes with the variation in the key cost drivers, which can help us roughly estimate the cost of separation for different classes of products.

Figure 1. Cost estimates for intracellular insoluble products under varying parameters. (A) Sensitivity curve showing cost changes with the variation in titer; (B) "Heap map" showing the change of cost and the optimal cell harvesting technology with the variation in technology performances (centrifuge efficiency and filtration retention factor). The white lines show the threshold values where the optimal technology switches. The asterisk represents the base case.

As a *preliminary* result, we identify three chemicals as the most promising ones: glutaric acid, acrylamide and propanal, by setting a uniform separation cost margin threshold for all chemicals to be greater than 1.8 $/kg (an estimated margin for a case product - polyhydroxyalkanoate), a volume greater than 31751 MT/year (an estimated production capacity), and a market size greater than 200 million $/year (to recover capital cost in ~5 years supposing a 50% market share). More accurate results can be obtained by using the aforementioned bio-separation framework with a different separation cost margin.

References

[1] Gavrilescu, M. & Chisti, Y., 2005. Biotechnology- a sustainable alternative for chemical industry. *Biotechnology advances,* Volume 23, pp. 471-499.

[2] Bornscheuer, U. T. & Nielsen, A. T., 2015. Editorial overview: chemical biotechnology: interdisciplinary concepts for modern biotechnological production of biochemicals and biofuels. *Current Opinion in Biotechnology,* Volume 35, pp. 133-134.

[3] Wilson, S. A. & Roberts, S. C., 2014. Metabolic engineering approaches for production of biochemicals in food and medicinal plants. *Current Opinion in Biotechnology,* Volume 26, pp. 174-182.

[4] Zhang, X., Tervo, C. J. & Reed, J. L., 2016. Metabolic Assessment of E. coli as a Biofactory for Commercial Products. *Metabolic Engineering,* Volume 35, pp. 64-74.

[5] US EPA, 2004. High-production-volume (HPV) chemicals status and future directions of the HPV Challenge Program. *Office of Pollution Prevention and Toxics*, Washington DC.

[6] Wu, W., Yenkie, K.M. & Maravelias, C.T., 2017. A superstructure-based framework for bio-separation network synthesis. *Computers & Chemical Engineering*, Volume 96, pp. 1-17.

[7] Yenkie, K. M., Wu, W., Clark, R. L., Pfleger, B. F., Root, T. W., & Maravelias, C. T., 2017. A roadmap for the

synthesis of separation networks for the recovery of bio-based chemicals: Matching biological and process feasibility. *Biotechnology Advances*, Volume 34(8), pp. 1362-1383.

[8] Yenkie, K. M., Wu, W., Maravelias, C.T., 2017. Synthesis and analysis of separation networks for the recovery of intracellular chemicals generated from microbial-based conversions. *Biotechnology for Biofuels*

491541 Supercritical FluidÂ Carnot Cycle for Low Temperature Waste Heat Utilization

Abstract:

Supercritical Fluid Carnot Cycle for Low Temperature Waste Heat Utilization

Many chemical processes reject low-grade heat to the environment. While it is difficult to recover low temperature waste for work production, it would be useful to design processes that, in principle, could do this. In our presentation we describe computer simulations of a process for doing this using supercritical carbon dioxide as a working fluid for the power cycle. Given the low critical temperature of carbon dioxide and the desirability of keeping the working fluid in its one phase region, CO_2 is an ideal fluid for designing such a process. Using a temperature close to its critical point for the rejection leg of the Carnot cycle, we exploit the natural flattening of the critical isotherm in this region which leads to efficient work production in Carnot engines using small temperature driving forces between the input and rejection legs of the process. The concepts appear promising as a basis for designing experimental systems for recovering low- grade waste heat for conversion into work using supercritical CO2 as the power cycleÕs working fluid.

499181 Designing Techno-Ecological Synergies While Accounting for Ecosystem Dynamics

Abstract:

Industrial processes are designed from a superstructure of multiple production pathways and the most optimal design is selected based on multiple objectives. Design objectives usually include economic goals such as maximizing net present value while simultaneously reducing impact from processes by choosing raw materials that are least toxic and identifying pathways that result in minimum emissions. However, this method towards designing industrial systems fails to address the underlying issue that the current demand for natural resources and ecosystem goods and services exceeds the rate at which these services can be supplied by ecosystems. This difference between the ecosystem service demand and supply is largely unaccounted for during design of industrial systems and ignoring the role of supporting ecosystems and the limit on ecosystem service supply has been one of the major causes for ecosystem degradation.

Originating from the concept of `unit operations', this work expands the scope of traditional process design methodologies to include ecosystem services within the design framework, by treating ecological systems as unit operations, with the goal of building process flowcharts based on the unit operation concept of ecosystems. Previous work on the use of Techno-Ecological Synergy (TES) framework for the assessment and design of systems including industrial sites [1], residential [2] and agricultural systems [3] have shown the economic and environmental benefits of using this approach. However, none of these studies consider interactions between detailed process-level decisions and the design of supporting ecosystem services. In this presentation, we show that simultaneous selection of process design variables and ecological variables while accounting for the variability and dynamics of ecosystems results in robust systems that tend towards absolute sustainability while being economically feasible. Variability and dynamics of ecological systems arises due to changes in meteorological conditions and seasonal variations. We show that, such robust systems can be designed by identifying a combination of technological and ecological alternatives.

The overall system is designed to maximize the total net present value and maximize the net ecosystem services supply with the goal of achieving absolute sustainability. A multi-period optimization problem formulated as a nonlinear program (Mo-NLP) is solved and tradeoffs are identified between maximizing the net present value of the system and the net ecosystem service supply for multiple ecosystem services for multiple time periods in a year.

Application of this framework to a biofuel production system along with a forest and wetland ecosystem indicates that systems designed by accounting for ecosystem service supply results in novel designs that have a higher economic value, and higher sustainability index. Preliminary results for the multi-period optimization problem revealed that selecting a combination of technological and ecological options during certain seasons have a higher economic and environmental performance compared to a purely technological design. Thus, coupled TES systems are expected to have economic and environmental advantages over conventional systems design, mainly due to the reduction in external inputs by moving towards a closed loop production system.

References:

[1] Gopalakrishnan, V., Bakshi, B.R. and Ziv, G., 2016. Assessing the capacity of local ecosystems to meet industrial demand for ecosystem services. AIChE Journal, 62(9), pp.3319-3333.

[2] Urban, R.A. and Bakshi, B.R., 2013. Techno-ecological synergy as a path toward sustainability of a North American residential system. Environmental science & technology, 47(4), pp.1985-1993.

[3] Hanes, R.J, Gopalakrishnan, V., Bakshi, B.R. Synergies and trade-offs in renewable energy landscapes: Balancing energy production with economics and ecosystem services. Submitted. Applied Energy.

492835 Design and Dynamic Simulation of a Solar and Natural Gas Hybrid Power Plant to Investigate the Synergies of Hybridization

Abstract:

Limited fossil energy resources and the greenhouse gas effect on the environment is creating increasing pressure over the world to look for alternative energy resources that are sustainable and ecofriendly [1]. Solar energy is currently among the popular renewable energy resources due to its recent advancements in technology and economics. However, reliability of energy supply is considered among the major drawbacks for stand-alone solar systems, as solar energy is not available during nighttime and intermittent on cloudy days. Hybridizing solar systems with fossil fuel can resolve the issue of reliability and also reduce the infrastructure cost compared to a pure solar plant [2]. Concentrated solar power (CSP) system are viable for large scale power production [3]. Parabolic trough are widely used CSP technologies as they are easy to finance and their maturity level in the market [4]. Hybridizing solar power plant with natural gas requires minimal capital cost for infrastructure [5]. In this study, we have designed and modeled a hybrid solar and natural gas plant in comparison to a stand-alone solar plant with thermal energy storage. The novel features of the hybrid design include supplemental fuel firing to ensure that high superheat temperatures are always reached in order to ensure a high cycle efficiency and the ability to operate the plant flexibly between solar and fuel modes. Another novel contribution of this work is that the design is complete with a fully dynamic model to capture the effects of rapidly changing solar conditions and an active control system to ensure that plant automation can handle these fluctuations. The hybrid system demonstrates an increase in solar-to-electric efficiency of 15 to 26% when compared to the stand-alone system. The overall efficiency also increases from 15 to 34 % in the same comparison. The hybrid system also demonstrates better environmental performance than a stand-alone system that would otherwise rely on fossil fuel backup generators at grid scale.

Reference:

[1] G. Bakos and C. Tsechelidou, "Solar aided power generation of a 300 MW lignite fired power plant combined with line-focus parabolic trough collectors field," *Renew. Energy*, 2013.

[2] Y. Zhao, H. Hong, and H. Jin, "Evaluation criteria for enhanced solar–coal hybrid power plant performance," *Appl. Therm. Eng.*, 2014.

[3] H. Price *et al.*, "Advances in Parabolic Trough Solar Power Technology," *J. Sol. Energy Eng.*, vol. 124, no. 2, p. 109, 2002.

[4] J. Peterseim, A. Tadros, U. Hellwig, and S. White, "Increasing the efficiency of parabolic trough plants using thermal oil through external superheating with biomass," *Energy Convers.*, 2014.

[5] J. H. Peterseim, S. White, A. Tadros, and U. Hellwig, "Concentrated solar power hybrid plants, which technologies are best suited for hybridisation?," *Renew. Energy*, vol. 57, pp. 520–532, 2013.

488553 The Design of Sustainable Carbon Dioxide Capture and Conversion Processes Considering Various Locations, Products and Routes

Abstract:

Due to climate change and growing concerns about global warming, methods of mitigating carbon dioxide emissions are important. The three most common approaches to mitigating emissions and reducing the concentration in the atmosphere are: improving the efficiency of processes, carbon dioxide capture and storage (CCS), and carbon dioxide capture and utilization (CCU). It is questionable, however, if they alone can satisfy the requirements; therefore, they need to be developed concurrently to provide an overall sustainable solution. Carbon dioxide utilization, which takes the captured carbon dioxide and reuses it, can provide an economic offset to the costs of capture and other new technologies in the formation of commercial products. Therefore, the sustainable design of carbon dioxide capture and utilization, especially conversion, technologies is important. To ensure this, a computer-aided, systematic framework for the synthesis-design of carbon dioxide utilization processes, inclusive of carbon dioxide capture, has been developed. This framework uses three stages, (i) process synthesis, (ii) process design and (iii) innovation, for the design of sustainable processes. In this work, the framework is implemented so that various scenarios - varying locations, prices or objective function â€“ could be investigated. The results provide insight into what processes and products should be produced via carbon dioxide conversion and how much of emissions can be mitigated in this way.

In the first stage, a superstructure of 23 steps containing over 150 intervals is optimized to find the optimal route, amongst over 30 feasible routes, to take captured carbon dioxide to a product. To aid the optimization, the developed database contains over 20 reactions, and the production of over 10 products and byproducts (methanol, dimethyl ether, ethanol, methane, dimethyl carbonate, ethylene carbonate, ethylene glycol, propylene carbonate, propylene glycol, acetic acid and succinic acid) via over 30 feasible processing routes. Optimizations for seven different scenarios are performed, in terms of prices, demands, locations and objective, to determine the impact that carbon dioxide capture and conversion can have. The seven scenarios include, (i) optimization for a fixed amount of captured CO_2 considering operating costs and capital costs (covers 2 scenarios), (ii) optimization under price uncertainty with respect to products (1 scenario), (iii) optimization with price variability for utilities, chemicals and products, to reflect sensitivity and location variability (covers 2 scenarios), (iv) optimization with variable conversion (1 scenario), and (v) optimization considering a fixed demand of the products to determine the amount of CO_2 utilized for different paths (1 scenario). The results show that if the prices of certain products changed, such as if the price of dimethyl carbonate were doubled or the price of succinic acid were decreased by half, what to produce and how to produce it is greatly influenced. Changing the parameters â€“ such as prices of utilities and materials or demands and availability of carbon dioxide sources - according to uncertainty and location, changes the optimal processing route. Considering seven scenarios, the optimal routes are found to produce succinic acid, dimethyl carbonate and methanol, amongst other value-added products.

After the optimal processing paths are found for the scenarios, they are simulated and analyzed in the second stage. A simulation library containing over 30 different capture and conversion processes to various products has been developed, including the optimal routes determined in the scenarios. In Stage 3, the production of dimethyl carbonate via different carbon dioxide utilization routes and the production of succinic acid are selected for further improvement as they are among the highest rated processes. For these cases, the targeted sustainability indicators can be improved; in the case of dimethyl carbonate production via ethylene carbonate, for example, the net carbon dioxide emissions (considering also energy production), operating costs and capital costs can be reduced by more than 10%. For already innovative processes, this is not always necessary. However, for processes which already exist or are not already carbon dioxide reducing, this can be used to find more sustainable processes.

This presentation will highlight the framework and the corresponding models, methods and tools, especially the database and simulation library. In addition, the application of this framework will be highlighted through a case study, which will encompass superstructure optimization of a large network of capture and utilization processes for all seven scenarios. The detailed design and analysis will be shown for the results of these scenarios. Finally, some conclusions about the role of capture and utilization as a bridge between current and future technology will be discussed.

491405 Optimal Integrated Plant for Waste to Biodiesel Production

Abstract:

Current developed societies generate large volumes of waste. The origin can vary from sludge from water treatment, urban waste, forest residues and manure mostly. Apart from the volume, their composition is complex and dangerous. Anaerobic digestion presents a technological option to stabilize the waste while generating added value in the form of methane and digestate that can be used as fertilizer. For decades, the aim of processing the waste has focused on power production [1]. However, biogas is an interesting carbon source, not only because of its content in methane, but also due to the amount of CO2 available. Dry reforming can be used to process this biogas into methanol [2]. A step further allows the production of biodiesel from waste. The production of biodiesel from algae requires nutrients for algae growing [3], that can be provided by the digestate. Next, the methanol produced from the biogas is used for the transesterification of the oil.

In this work we have optimized an integrated facility for the production of fully renewable biodiesel from waste via anaerobic digestion. On the one hand, biogas is produced from cattle and pig manure. After its purification, it is dry reformed to obtain syngas. Methanol is later synthesized from the syngas. On the other hand, the digestate remaining is used as nutrient to grow algae. The effect of the nutrient composition on the algae growth is considered. The algae are harvested and the oil is extracted using mechanical press and a solvent, that it is later recovered. The oil is transesterified using the methanol produced from the biogas. The problem is formulated as a large NLP for the optimal operating conditions, biogas composition and digestate nutrient content. Dry reforming, methanol and biodiesel synthesis reactors are modelled to account for the effect of the operating conditions on the yield. Distillation columns are modelled using short cut methods but including constraints on the reflux ratio [2,4], the operating pressures and temperatures from the literature and the rigorous simulation of the column in CHEMCAD [4].Optimal HEN and water networks are designed afterwards.

3110kt/y of manure are processed for the production of almost 90Mgal/y of biodiesel and 46.7 kt/y of methanol. The process uses part of the biogas to fuel the reformer but still needs 1.5 MJ/gal of FAME and 1.22 gal of water per galon of biodiesel. The production cost of biodiesel is competitive with previous work [2], 0.31€/gal, but the investment is higher due to the need to produce the biogas, 193 M€.

References

1.-León E., Martín, M. (2016) Optimal production of power in a combined cycle from manure based biogas Energ. Conv. Manag. 114 89–99

2.-Hernández, B, Martín, M. Optimal composition of the biogas for dry reforming in the Production of Methanol. IECR. 10.1021/acs.iecr.6b01044

3. -Xin, L., Hong_ying, H., Ke, G., Ying-xue, S. (2010) Effects of different nitrogen and phosphorus concentrations on the growth, nutrient uptake, and lipid accumulation of a freshwater microalga Scenedesmus sp Bioresource Technology 101 5494–5500

4.-Martín, M., Grossmann, I.E. (2012) Simultaneous optimization and heat integration for biodiesel production from cooking oil and algae. Ind. Eng. Chem Res. 51 (23), 7998–8014

498740 Overcoming the Challenges of Using Corn Stover As Feedstock in Autothermal Pyrolysis

Abstract:

Corn stover biomass is abundant across the Midwestern United States and is considered a potentially promising feedstock from which biorenewable fuels and chemicals could be produced. However, significant challenges to economically and efficiently convert the lignocellulose fractions into valuable products remain. Fast pyrolysis is an endothermic process capable of thermally deconstructing biomass into bio-oil vapors, non-condensable gases (NCG's), and solid biochar at an elevated temperature (500°C). Autothermal operation differs from traditional pyrolysis with the addition of a small amount of air introduced in the reactor to generate sufficient exothermic energy to drive pyrolytic reactions and achieve a neutral energy balance. Pilot scale testing in a bubbling fluidized bed reactor resulted in low equivalence ratios (ER) during autothermal operation that ranged from 8-11% of the theoretical air requirements for stoichiometric combustion; these ER's are almost half of those compared to autothermal gasifiers. Significant process intensification was reached when air-blown autothermal operation was achieved using 100% air as the fluidizing gas and a maximum biomass feed of 20 kg hr^{-1} was obtained. Bio-oil yields greater than 50 wt. % were achieved along with biochar and NCG's yields around 20 and 30 wt. % respectively. Six bio-oil product streams were collected using a fractional recovery system consisting of three alternating pairs of condensers and electrostatic precipitators; this allowed bio-oil compounds across a range of molecular weights to be collected based on the compound's dew point. Early technical challenges of processing corn stover involved the accumulation of biochar in the reactor and degradation of bio-oil heavy ends in the first stage fraction (SF1). These were overcome by adjusting the fluidization gas velocity to promote better char particle attrition and elutriation from the reactor as well implementing a new SF1 quench tower to facilitate the condensation and collection of viscous bio-oil products. Autothermal process intensification represents a significant advancement in pyrolysis technology and removes the heat transfer bottleneck traditionally associated with thermochemical conversion. Under these conditions, the reactor's footprint was reduced by 40% while simultaneously increasing biomass throughput by a factor of four. This advancement is expected to significantly improve the scalability and process economics of corn stover pyrolysis systems.

495587 Development of a Kinetics Model for Autothermal Pyrolysis in a Fluidized Bed Reactor

Abstract:

Autothermal pyrolysis is achieved by reacting a small amount of oxygen (equivalence ratios of 0.06-0.10) with biomass in a fluidized bed reactor, which partially combusts the products of pyrolysis to overcome the endothermic enthalpy of pyrolysis. Experiments performed at Iowa State have shown that autothermal pyrolysis produces no significant loss in bio-oil yield or quality as compared to traditional non-oxidative pyrolysis. Current experimental data shows that half the heat comes from oxidation of non-condensable gases (NCGs) such as carbon monoxide, methane, ethane, and ethylene and an additional one-quarter comes from oxidation of char. However, currently available kinetic mechanisms for gas-phase combustion of NCGs do not predict significant oxidation at pyrolysis temperatures (400 - 600ËšC). On the other hand, many of the volatile organic compounds that subsequently condense to bio-oil are predicted to readily oxidize at these temperatures, thus prompting a closer examination of the gas-phase kinetics of autothermal pyrolysis.

The role of gas-phase oxidative reactions in providing the enthalpy of pyrolysis during autothermal pyrolysis are explored in this study using both experimental and computational methods. A fluidized bed operated at temperatures between 400-600ËšC was used to estimate the oxidation rates of bio-oil model compounds at equivalence ratios of 0.06. Gas analysis allowed product distributions to be determined. These results were used for modeling oxidative reactions during autothermal pyrolysis.

492975 Coupling Effects of Mass Transfer and Chemical Kinetics during the Co-Pyrolysis of Cellulose and High Density Polyethylene

Abstract:

The increasing amount of municipal solid waste (MSW) causes growing environmental concern for the humanity. Growing global population needs to secure food, energy and water resources, and with decreasing amount of available land, landfills of the waste do not constitute a long-term solution. Biomass and plastics are two of the most dominant components of MSW, and a method that can concurrently convert both waste streams into energy or useful products will be beneficial for humanity. Pyrolysis is one of the simplest ways to convert waste into high-value products. When biomass is pyrolyzed alone, however, it forms pyrolysis oil with high oxygen content, which causes low heating value, thermal instability and corrosiveness. Plastic wastes, which are rich in hydrogen and low in oxygen, could be helpful for enhancing the quality of pyrolytic oil. Many studies have shown that co-pyrolysis of biomass and plastics provides oil with higher yields and caloric value than the one produced from biomass pyrolysis alone. Despite this promise, mixed results regarding their interactions are reported in the literature, and the fundamental interactions between biomass and plastics during co-pyrolysis are still not well understood. As a result, there is a need to unravel the interactions between biomass and plastics at the molecular level. In our work, the coupling effects of the mass transfer and chemical kinetics during cellulose/plastics co-pyrolysis is studied to explain the contradictory results in the literature. An experimental apparatus was constructed to allow fast pyrolysis of cellulose, high density polyethylene (HDPE), and their mixtures. The samples with different mixing patterns but same compositions were reacted. The differences of the product yields were observed, suggesting a competition between the intraphase diffusion and interphase diffusion of the biomass-derived molecules.

495855 Fast Pyrolysis of Four Different Microalgae: Apparent Kinetic Parameters Evaluation and Product Analysis

Abstract:

Fast Pyrolysis of Four Different Microalgae: Apparent Kinetic Parameters Evaluation and Product Analysis

Ribhu Gautam*, R. Vinu

Department of Chemical Engineering and National Center for Combustion Research and Development, Indian Institute of Technology Madras, Chennai, Tamil Nadu, India

Economic growth and development is resulting in fast depletion of fossil fuel reserves. Presently, the mitigation of environmental impacts caused due to the excessive use of fossil fuels has gained importance. Over the past few years, research for replacement of conventional fossil fuels is gaining importance with focus on biomass as it has got high potential and is an abundant energy source. Compared to terrestrial biomass, marine biomass has advantages such as high growth rate, high per hectare yield and high photosynthetic efficiency. Marine biomasses have flexibility to grow in sea water, fresh water ponds, industrial effluents and waste water. Microalgae, one of the most prominent marine biomass, are abundantly distributed water plants. They consist of lipids, proteins and carbohydrates and these lipids are converted into biodiesel. Some microalgae have lipid content as high as 80 wt.%, and are hence considered as potential feedstocks for biodiesel and biofuel production.

Fast pyrolysis is a promising thermochemical processing technique to convert microalgae into liquid, gas and solid products in the absence of oxygen at moderate temperatures. High heating rates (>1000°C s^{-1}) involved in fast pyrolysis leads to faster completion of reactions within seconds. This results in high yields of liquid bio-oil as compared to slow pyrolysis. The existing kinetic models of algae pyrolysis are all based on thermogravimetric mass loss data obtained at slow heating conditions. The use of these data limits the application of these models for fast pyrolysis. Therefore there is an immense need to obtain kinetic parameters under fast pyrolysis conditions and validate them with the existing rate parameters.

This study reports a new experimental technique to evaluate the mass loss profiles of microalgae under fast pyrolysis conditions using an analytical pyrolyzer. Fast pyrolysis of four different microalgae (*Chlorella vulgaris (C. vulgaris)*, *Nannochlorpsis oculata (N. oculata)*, *Schizochytrium limacinum (S. limacinum)* and *Arthrospira platensis (A. platensis)*) was studied in the temperature range of 400-700°C and sample residence times in the range of 2-60 s. The apparent kinetic parameters of fast pyrolysis of the algae were evaluated using first order, one dimensional (1D), 2D and 3D diffusion models. Prior to the fast pyrolysis experiments the dried microalgae samples were characterized by TGA, proximate analyser, CHNS analyser, bomb calorimeter for thermal stability, proximate analysis, ultimate analysis and heating value respectively. Fast pyrolysis experiments were carried out in an analytical pyrolyzer with 4.5±0.2 mg of sample in a quartz tube. The mass loss data of the microalgae samples was generated in isothermal conditions via

gravimetry. The most conventional rate loss equation was considered for the estimation of rate constants.

Where *f(a)* is the dependence of conversion on the reaction model, *k*(T) is dependence of rate on temperature. 'a' termed as normalized conversion can be calculated with the formula a = $(m_0-m_t)/(m_0-m_8)$, where m_0, m_t and m_8 are initial sample mass, sample mass at time *t* and final mass obtained respectively. As the reaction proceeds from initial to completion, a increases from 0 to 1. This cannot be linked to specific reactions during the process, so a is the degree of overall completion of reaction. Now,

The function of *g(a)* has been approximated as $-ln(1-a)$, a^2, $[(1-a)ln(1-a)+a]$ and $[1-(1-a)^{1/3}]^2$ for first order, 1D, 2D and 3D diffusion respectively for solid state reactions. The rate constants for each reaction temperature was obtained from the slope of the *g(a)* v/s t graph. The estimation of apparent activation energy was done by fitting temperature and rate constant data in Arrhenius Equation represented below

First order kinetic model fitted the experimental data well with very high R^2 for all the four microalgae samples. The apparent activation energies (E_a) and pre-exponential factors (A) for fast pyrolysis of the algae varied in the range of 4-24 kJmol^{-1} and 0.03–1.7 s^{-1} respectively. The obtained rate parameters were validated by constructing a kinetic compensation plot, which showed that Ln(A)=0.19E_a–0.43 (R^2=0.97) justifying the kinetic compensation effect. The pyrolysates obtained from fast pyrolysis in an analytical pyrolyzer at 500°C were characterized using gas chromatograph/mass spectrometry for all the four microalgae samples. The pyrolysates were characterized in a GC/MS coupled with pyrolyzer. Fast pyrolysis *S. limacinum* yielded highest aliphatic and aromatic hydrocarbons owing to its high lipid content. A detailed kinetic analysis comprising all the kinetic models and the presence of various compounds observed in the pyrolysates will be discussed during the presentation.

498168 Overcoming Lignin Agglomeration during Pyrolytic Sugar Production in a Fluidized Bed

Abstract:

Fast pyrolysis of acid infused biomass has enabled over 300% increase in sugars derived from the biomass carbohydrate fraction. Our research goal was to investigate the problem of lignin and carbohydrate interaction after acid infusion: promoting agglomeration and ultimately inhibiting volatilization of both the sugars and the phenolics. Increasing the production of sugars during fast pyrolysis is important because the sugars can be used for fermentation to biofuels and important chemicals currently derived from petroleum. Understanding the interactions is crucial to the advancement of thermochemical conversion technology utilizing acid pretreatment for enhanced sugar production. Alkali and alkaline earth metals (AAEM) found in biomass have negative impact on sugar yields during fast pyrolysis, catalyzing the formation of low molecular weight species. Inductively coupled plasma (ICP) determination of alkali and alkaline earth metals was performed on the biomass to determine the amount of acid infusion needed to neutralize the negative effects of the AAEM. The acid infused biomass samples were pyrolyzed using a 2 kg/h fluidized bed reactor. The biomass, the equivalency ratio which influences reactor operability, the amount of acid infusion, and neutralization steps, were varied followed by evaluation using various instrumentation. Thermogravimetric Analysis (TGA) was performed on each of the biomass samples for comparison to the starting biomass. Fourier transform infrared spectroscopy (FTIR) of the biomass was utilized for each of the pretreatment approaches. Specific wavelengths were selected that are attributed to either lignin or cellulose. Net integration of the intensities of the same absorption bands were compared between experimental parameters. Both hydrolysable and water soluble sugars quantities were established between the different pretreatments by high performance liquid chromatography (HPLC), while the relative molecular weight distribution of the phenolics was determined by gel permeation chromatography (GPC). Mass balances were obtained for each reactor run. With this approach we have accomplished uninterrupted fast pyrolysis of acid infused biomass with dramatically increased sugar production.

489438 Thermochemical Methylation of Lignin to Produce High Value Aromatic Compounds

Abstract:

Thermochemical methylation of lignin to produce high value aromatic compounds

Patrick A. Johnston[a] and Robert C. Brown[a,b]

[a]Bioeconomy Institute, Iowa State University, Ames, IA 50011

[b]Department of Mechanical Engineering, Iowa State University, Ames, IA 50011

Technical lignin, the byproduct from cellulosic ethanol production, is a potential feedstock for producing coumaric and ferulic acids. These biorenewable aromatic hydroxycinnamic acids can be upgraded to high value chemicals for the manufacturing of cosmetics, flavoring agents, and sunscreens. We hypothesize that during the fast pyrolysis of lignin aromatic acid esters can be recovered from coumaric, ferulic and other hydroxycinnamic acids using a methylation agent. In this study, technical lignin from cellulosic ethanol production was pyrolyzed in a micropyrolyzer with a methylation agent to produce gas chromatography (GC) detectable aromatic acid esters. Corn stover lignin from the enzymatic hydrolysis process and the organosolv process was combined with different methylation agents to produce, identify and quantify the hydroxycinnamic acid esters. High performance liquid chromatography (HPLC) with a diode array detector (DAD) was used to separate and quantify the extracted hydroxycinnamic acids from lignin for determination of yield recoveries.

To determine the actual total theoretical yield of coumaric and ferulic acid in the lignin an alkaline extraction method was used. The theoretical yields of coumaric and ferulic in the enzymatic hydrolysis lignin was 5.6% and 1.9%, respectively. The theoretical yields of coumaric and ferulic in the organosolv lignin was 7.7% and 2.9%, respectively.

The recoveries of the hydroxycinnamic acids from enzymatic acid hydrolysis lignin with different loading of methylation agent were compared using a micropyrolyzer at 500°C. Using tetramethylammonium hydroxide (TMAH) at 1μL produced recoveries of 40% coumaric and 37% ferulic compared to the theoretical extracted amount from the lignin. The recoveries at 5μL produced coumaric and ferulic acids concentrations of 68% and 55%, respectively. The recoveries at 10μL produced coumaric and ferulic acids concentrations of 84% and 69%, respectively.

The recoveries of the hydroxycinnamic acids from organosolv lignin with different loading of methylation agent were compared using a micropyrolyzer at 500°C. Using tetramethylammonium hydroxide (TMAH) at 1μL produced recoveries of 35% coumaric and 27% ferulic compared to the theoretical extracted amount from the lignin. The recoveries at 5μL produced coumaric and ferulic acids concentrations of 69% and 52%, respectively. The recoveries at 10μL produced coumaric and ferulic acids concentrations of 89% and 64%, respectively.

489608 Confined Ni Nanoparticles in Mesoporous Silica Via a Polyethyleneimine-Assisted Route and Their Catalytic Performance for Methane Dry Reforming

Abstract:

The major greenhouse gases of CH_4 and CO_2 were converted into CO and H_2 synthesis gas via dry reforming of methane with the minimization of the carbon coking on the nickel confined mesoporous silica catalyst. To immobilize nickel nanoparticles in the ordered channels of the mesoporous silica, polyethyleneimine (PEI) was used as a stabilizer, reducing, and capping agent of nanoparticles. The nickel-PEI chelated complexes were formed by dissolving PEI and the nickel precursor into an aqueous solution. Then, the positively charged NH_2 terminal groups of the complexes were strongly interacted with the negatively charged surface silanol groups of the mesoporous silica. From this strong interaction, the nickel-PEI chelates were anchored in the porous framework of their parent support during the impregnation steps. A long hydrocarbon chain of PEI and its steric hindrance prevented the aggregation of the nickel-PEI complexes during the drying and calcination steps. Therefore, nickel nanoparticles were highly dispersed inside the porous framework even after the reduction step, suggesting the successful synthesis of the nickel well-dispersed mesoporous silica catalyst via a PEI-assisted route. Whereas the carbon coking was severely generated on the catalyst without PEI (177 g_{carbon}/g_{cat} for 40hr), the PEI-assisted catalyst showed the superior catalytic stability for dry reforming of methane with the minimization of the carbon coking (3.63 g_{carbon}/g_{cat} for 40hr). The synthesized nickel nanoparticles without PEI were placed outside of the porous structure and then, easily aggregated more than their parent support with the severe carbon coking. On the other hands, the sintering of nickel nanoparticles was more effectively hindered and the generation of carbon coking was suppressed on the confined nickel nanoparticles, since the prepared nickel particles via a PEI-assisted route were well-confined inside the pore. Therefore, combined with the simple procedure and low cost, the PEI-assisted route is one of the most promising methods for embedding nickel nanoparticles in mesoporous framework.

495924 Effects of Morphology and Site Proximity on Sorption-Enhanced Steam Methane Reforming Using Hybrid Ni-CaO Based Nanofibers

Abstract:

Hydrogen production is an essential component of the refining and petrochemical industries. Hydrogen also represents a critical energy carrier as energy portfolios shift from carbon-based to renewable technology. Currently 95% of the hydrogen in the US is produced via steam-methane reforming (SMR), with the byproducts being CO and CO_2. Sorption enhanced steam-methane reforming (SE-SMR) is an attractive adaptation of SMR technology. SE-SMR can produce a much higher hydrogen purity stream (>85%) at lower operating temperatures (500-600°C) and pressures (1-10 atm) compared to conventional steam reforming. The primary drawbacks of SE-SMR materials are the high temperatures required for sorbent regeneration and loss of sorption capacity and catalytic activity over repeated sorption-regeneration cycles. This work focuses on developing hybrid catalyst-sorbent materials that improve reaction and sorption rates by reducing the size and increasing the dispersion of metal and metal oxide domains.

Site proximity and morphology changes were achieved with nanofibrous metal oxides produced via electrospinning. Calcium oxide, nickel oxide, and magnesium oxide nanofibers were synthesized via electrospinning using calcium acetate, nickel acetate, and magnesium acetate precursors. Electrospinning solutions consisted of the salt precursor and polyvinylpyrrolidone (PVP) dissolved in an ethanol and water mixture. Electrospinning processing was conducted at 12.5-30 kV, 0.25-2 mL/hr extrusion rate, and 3-20 inch separation between the collector plate and the tip of the extrusion needle. Salt-PVP nanofibers were finally calcined at 600°C (2°C/min ramp rate) for 2 hours to decompose PVP and produce oxide nanofibers. MgO-supported Ni catalysts were also synthesized via incipient wetness impregnation of MgO powder with nickel nitrate solutions and calcined at 650°C (5°C/min ramp rate) for 5 hours.

Materials were characterized via scanning electron microscopy (SEM), x-ray diffraction (XRD), and N_2 physisorption. XRD spectra confirmed the formation of the desired metal oxide phases. SEM images indicated nanofiber diameters of approximately 80-200 nm, 65-95 nm, and 100-300 nm for calcined calcium oxide, nickel oxide, and magnesium oxide, respectively. The nanofibrous materials exhibited higher porosity, smaller crystallites, and smaller metal oxide domain sizes than their supported particle analogs.

The reactivity and catalytic activity of the nanofibrous materials were tested using CO_2 thermogravimetry, temperature-programmed reduction, and steam-methane reforming experiments. Calcium oxide nanofibers exhibited higher rates and higher saturation CO_2 capacities (78% vs 31%) compared to bulk calcium oxide powder. The increased rate and conversion of the calcium oxide nanofibers result form their smaller more reactive domains. Temperature-programmed reduction experiments were used to characterize Ni-based materials. Commercial Ni-based reforming catalyst exhibited reduction peaks at 500°C and 720°C, however, nickel oxide nanofibers exhibited a single reduction peak centered at 350°C. The much lower reduction temperature results from smaller NiO domains that are more evenly dispersed throughout the nanofiber compared to the particle analogs. The performance of absorber materials for SE-SMR was tested in a plug flow reactor at 500-650°C, 12-80 kPa of H_2O, 12-80 kPa of CH_4, and 1-5 atm total pressure. Catalyst-sorbent beds comprised of nanofibrous materials operated for longer periods of time and could be regenerated at milder conditions compared to beds consisting of supported Ni-catalysts and bulk CaO powder.

The results of this work indicate a promising opportunity for the application of nanofibrous materials in SE-SMR. Smaller, more reactive metal oxide domains allow for these materials to achieve stoichiometric CO_2 capacities without mass transfer restricted sorption rates. These hybrid catalyst-sorbent materials also retain their maximum capacity and catalytic activity over multiple cycles because they are less susceptible to particle agglomeration and morphology changes during thermal regeneration. These improvements in SE-SMR catalyst-sorbent materials has the potential to reduce costs associated with reforming and potentially lead to economically competitive distributed H_2 generation by (i) replacing expensive alloys required for high temperatures and pressures with less expensive materials of construction, (ii) eliminating water-gas shift reactors and downstream hydrogen purification, (iii) reducing carbon deposition in the reactor, and (iv) reducing the size of heat exchange equipment.

497081 Methane Partial Oxidation and Dry Reforming to Syngas Using the $La_{0.9}Ca_{0.1}FeO_{3-Î´}$Â Mixed-Conductor

Abstract:

One of the most promising technologies for carbon dioxide (CO_2) Capture and Sequestration is oxyfuel combustion, i.e. the use of pure oxygen in the combustion of fossil fuels. The major energy penalty in oxyfuel combustion comes from oxygen production. The current technology uses cryogenic methods and this process is energy intensive; it can lead to approximately 25% reduction in the net efficiency of a power plant.

Mixed ionic-electronic conducting (MIEC) membranes can serve as a better substitute. These materials have the potential of decreasing the energy consumption for oxygen separation significantly given that they operate under an oxygen chemical potential difference between the two sides. In addition, due to both oxygen separation and reaction (process intensification), chemical processes become simpler and this leads to a lower operating cost. MIEC materials are becoming popular in the energy sector, especially for chemical conversion processes.

A stable oxygen-conducting membrane under reducing conditions is $La_{0.9}Ca_{0.1}FeO_{3-Î´}$ (LCF). Our study investigates the performance of LCF in applications combining oxyfuel combustion, CO_2 reuse and syngas production under a reactive environment using methane-carbon dioxide mixtures in the oxygen lean side. In the presence of fuel, experimental measurements show that the oxygen flux through the membrane increases by one order of magnitude compared to the non-reactive case due to heterogeneous reactions of hydrogen and carbon monoxide with lattice oxygen ions. These surface reactions produce additional steam and carbon dioxide which further increase the syngas yield by reforming methane in the gas-phase. The aforementioned performance enhancement is accompanied by significant reuse of CO_2. Our results suggest that the performance of LCF is limited by the surface reactions on the fuel side and hence, we show that addition of nickel nanoparticles on the fuel side increase the oxygen permeation and the methane performance significantly by reducing the surface polarization resistance.

494091 Probing the Catalytically Active Phase of Cobalt Fischer-Tropsch Catalysts from First Principles: Predicting High Coverage Surface Hydroxyl Conformations Under Reaction Conditions

Abstract:

The reaction mechanism of Fischer-Tropsch (FT) synthesis is a continual source of intrigue and debate. While two classes of mechanisms have been proposed (namely, carbide vs. CO-insertion), the composition and configuration of the catalytically active phase has received a comparatively small amount of attention. Chemical transient kinetic (CTK) experiments performed in our group on model Co FT catalysts have yielded evidence for a CO-insertion mechanism, but just as importantly, these experiments have also provided evidence for a "crowded" reactive surface, where oxygen, hydrogen, and carbon bind to the surface of the catalyst in quantities exceeding the monolayer limit [1]. Because of the considerable delay in water formation in CTK experiments it is considered likely that the catalytically active phase contains substantial amounts of surface hydroxyl (OH). Such OH may therefore be considered a candidate precursor for constructing the "most abundant surface intermediate" ("masi") necessary for inducing hydrocarbon chain lengthening. Irrespective of what the mechanistic pathway for constructing the "masi" looks like, we present here a DFT-based investigation of an OH-rich surface, anticipating that such information will provide guidance for more complete mechanistic studies addressing chain lengthening initiation and growth.

By utilizing the Alloy Theoretic Automated Toolkit (ATAT) [2,3] to generate a large variety of extended Co (111) surfaces with different coverages and configurations of OH, we constructed a large library of OH/Co(111) structures. The face centered cubic facet is chosen because we wish to model smaller particle size Co, which typically transitions from its native hcp phase to the fcc phase. From this library we find that proximal OH have a large tendency to hydrogen bond as seen in Figure 1. These structures are quite common in the calculations, which suggests they are likely relevant to their stable surface phases. However, to actually test this notion, a corresponding lattice gas (LG) model using these structures will be constructed and subsequent grand canonical Monte Carlo simulations will be performed. In this way, we will elucidate the nature of OH patches on Co catalysts under FT synthesis conditions and their susceptibility to an interaction with CO molecules. A prevalence of surface OH was previously considered key in the CO dissociation on Co catalysts during FT [4], but may also react with CO molecules to provide an oxygen- and hydrogen containing C1 "masi". Perhaps more importantly, however, this result, combined with our previous CTK evidence [1], suggests that modeling the FT reaction over purely metallic Co—a common practice—is ill-advised if one wishes to properly test the mechanism of the reaction

Figure 1. Example of predicted ground state structures of OH on Co(111) at 0.5 ML (left panel) and 0.75 ML (right panel). All ground state structures below 0.5 ML do not exhibit hydrogen bonding and compression like that seen above. However, all ground state structures above 0.5 ML do. Blue spheres are Co, red spheres are O, and white spheres are H.

CITATIONS

1. Schweicher, J., A. Bundhoo, and N. Kruse, Hydrocarbon Chain Lengthening in Catalytic CO Hydrogenation: Evidence for a CO-Insertion Mechanism. Journal of the American Chemical Society, 2012. 134(39): p. 16135-16138.

2. van de Walle, A., M. Asta, and G. Ceder, The Alloy Theoretic Automated Toolkit: A User Guide. Calphad - Computer Coupling of Phase Diagrams and Thermochemistry, 2002. 26: p. 539-553.

3. van de Walle, A. and G. Ceder, Automating First-Principles Phase Diagram Calculations. Journal of Phase Equilibria, 2002. 23: p. 348.

4. Gunasooriya, G.T.K.K., et al., Key Role of Surface Hydroxyl Groups in C–O Activation during Fischer–Tropsch Synthesis. ACS Catalysis, 2016. 6: p. 3660-3664.

496731 Promoted Mixed Oxides for "Low-Temperature" Methane Partial Oxidation in Absences of Gaseous Oxidants

Abstract:

Promoted Mixed Oxides for "Low-Temperature" Methane Partial Oxidation in Absences of Gaseous Oxidants

Reforming of methane to syngas is an important but energy intensive operation in the synthesis of chemical and liquid fuels from natural gas. Steam reforming does not give an ideal H_2 to CO ratio for liquid fuel or chemical synthesis and conventional methane partial oxidation (POx) approaches require costly air separation. Chemical looping reforming (CLR), which partially oxidizes methane into H_2 and CO in the absence of steam or gaseous oxygen, offers a simpler and potentially more efficient route for syngas generation. This is achieved by cyclic removal and replenishment of active lattice oxygen in oxygen carrier particles also known as redox catalysts. Selecting a redox catalyst particle with high selectivity and facile oxygen transport is very important for CLR. Several redox catalysts have been reported in literature, but their activities toward methane POx, at low temperatures (<800 °C), are often limited due to the high activation energy for the migration and removal of lattice oxygen. Moreover, syngas selectivity is often low, as metal oxide surface species that catalytically activate methane at these temperatures tend to be non-selective. To address these limitations, we investigated the effects of promoting catalytic activity of oxide surfaces for a number of mixed metal oxide based redox catalysts. Our findings indicate that surface promotion can lower the onset temperature of methane POx by as much as 300 °C while achieving >90% syngas selectivity.

While the highly dynamic nature of the system makes mechanistic studies challenging, understanding the mechanism is important to aid rational catalyst design. We report pulse reaction and isotope studies indicating that the selectivity of the redox catalysts is related to the relative rates of lattice oxygen (O^{2-}) conduction to the surface and the surface oxygen removal by the gas-solids reactions. This is particularly the case at the early stage of the reaction. Surface promotion can increase the oxygen removal rate from the surface, thereby changing the type oxygen species present on the surface. This increase in oxygen removal is caused by the enhancement of surface methane activation. It is also shown that the rate of bulk lattice oxygen diffusion to the surface can be notably affected by surface promotion of the redox catalysts. This results from the change in driving force for O^{2-} conduction.

498172 A Highly Selective Route from Syngas to Ethanol: Tandem Catalysis Unconstrained By Anderson-Schulz-Flory Distribution

Abstract:

Selective higher alcohols synthesis (HAS) from $CO/CO_2/H_2$ mixtures (syngas) has been motivated by the demand of such products in the chemical and energy sectors, as well as the versatility and economics of syngas production. Research efforts in catalyst development for HAS have spanned decades, and in many cases drew inspiration from methanol synthesis and Fischer-Tropsch technologies. However, unlike the latter processes, insufficient selectivity and activity in HAS from syngas have prevented commercialization of chemocatalytic routes. Thus, further improvement in both metrics is required.

In many cases of HAS catalyzed by single catalysts, the reaction is believed to proceed by concerted hydrogenation and chain growth (C-C coupling) steps, resulting in co-production of aldehydes, alkenes, and alkanes, with chain-length approximately following the Anderson-Schulz-Flory (ASF) distribution. Such a wide spread in products increases separation costs for targeted HAS, and results in lower-value products (such as methane) competing for feed utilization. Advances in the engineering of the catalyst surface composition and structure have led to significant improvements in selectivities, but, to date, the aforementioned limitations have not been completely eliminated. One approach to break away from the constraints of the ASF distribution in HAS is to spatially separate the hydrogenation and chain-growth steps of the process by pairing catalysts that, individually, selectively catalyze only one of the steps, but, in a multi-bed tandem catalysis scheme, enable both.

In this work, we developed a combination of precious-metal-free, inexpensive, heterogeneous catalysts that, under differential conversions of CO/H_2 feed, are able to achieve upwards of 80% selectivity towards ethanol, on a CO_2-free carbon basis, with <1% methane selectivity. We will discuss the response of this system to process variables, and its stability. Additionally, some of the thermodynamic and kinetic considerations that are important in the design of such tandem-catalytic schemes for HAS will be highlighted.

494763 Control of Metal-Support Interaction of Fe@CNTs By Surface Modification of CNT and Its Application to Direct Olefin Synthesis from Syngas

Abstract:

Rapid economic development in China arouses the intense interest to produce increasing amounts of light olefins (e.g., ethylene, propylene) in both academia and industry. Traditionally, steam cracking of naphtha with dwindling and non-renewable crude oil is the main route for olefin synthesis [1,2]. Most recently, direct synthesis of light olefins from coal-/biomass-based syngas is regarded as the most promising non-petroleum route due to low-energy consumption, low-cost, high-efficiency, etc [1,3,4]. Generally, the catalyst design strategy for direct syngas route is mainly from the aspect of support modification, promoter doping, metal precursor synthesis and construction of multifunctional active sites [1,3,4]. Regardless of multifunctional catalyst, Fe is considered as the most promising active metal for FTO with _-Fe_5C_2 commonly recognized as active phase [1,5,6]. The key to high olefin selectivity is to modulate the Fe catalyst with appropriate capability for CO dissociation and hydrogenation.

Carbon materials possess high surface area, chemical inertness, mechanical strength, fruitful surface chemistry and electronic properties via covalent modifications. The type and quantities of surface chemical groups determined the interactions between carbon support and metal species [7,8]. Previous work demonstrated that iron-carbon interactions played an important role in Fischer-Tropsch process. Comparing with other conventional supports, the iron species on carbon materials are generally more reducible and can be transformed to iron carbides more facilely in the FTS reaction. In this work, the interaction between Fe and CNT is regulated by different surface modifications on CNT. The effect of the chemical and physical properties of the different CNT supports on the FTO performance of the Fe/CNTs was discussed.

Table 1 FTO performance of various carbon materials supported Fe catalysts

The CNTs were prepared by surface modification under different conditions. Fe supported catalysts were prepared by incipient wetness impregnation of the modified CNTs after drying and calcination. Structural and electronic properties of Fe/CNTs were characterized by BET, FTIR, Raman, XPS, XRD, TEM, TPR, TPD, XAFS, etc.

Figure 1 Catalytic performance of Fe catalysts supported on CNTs with different modifications

We evaluated first the FTO performance on multiple carbon materials supported Fe catalysts and found that CNT is the most favorable carbon support for olefin synthesis (Table 1). Moreover, we found that there existed large difference in performance of Fe catalysts supported on CNT with different modification methods (Figure 1). In similar conversion level, defect-rich HTCNT supported Fe showed highest initial conversion and olefin selectivity. Besides, OCNT with large amounts of O-functional groups supported Fe catalyst exhibited the lowest olefin selectivity but most stable activity. Based on the performance results, we tried to carry on various characterizations to elucidate the intrinsic effect of CNT modification on Fe-CNT interaction. BET, zeta potential and point of zero charge suggested that defect-rich HTCNT possessed highest surface area and less negative charge, while OCNT is the opposite. Raman and FTIR spectra evidenced the large amounts of O-contained groups (-C=O, -COOH) and rich defect on OCNT and HTCNT, respectively. In addition, there existed obvious electron transfer between Fe and modified CNTs, which is also demonstrated by XPS, in situ XRD, TPR, XAFS, etc.

Figure 2 Possible structure-performance relationship for FTO over different kinds of Fe NPs immobilized on CNT

The presence/absence of surface functionalities can directly affect the catalytic behavior of the active phases. The defects on CNT benefits to the initial activity and selectivity to light olefins. High concentration of O groups enhanced the stability of Fe during FTO. Moreover, the performance strongly depends on the degree of intimacy between CNT and Fe NPs. The surface modification caused the difference in Fe-CNT interface property, which significantly affected the electronic environment of Fe, ultimately altering the FTO performance. An optimal surface chemistry also facilitates catalyst conversion into its catalytically active state.

References

[1] H.M. Torres Galvis, J.H. Bitter, C.B. Khare, M. Ruitenbeek, A. I. Dugulan, K. P. de Jong, Science, 335 (2012) 835-838.

[2] L. Zhong, F. Yu, Y. An, Y. Zhao, Y. Sun, Z. Li, T. Lin, Y. Lin, X. Qi, Y. Dai, L. Gu, J. Hu, S. Jin, Q. Shen, H. Wang, Nature, 538 (2016) 84-87

[3] F. Jiao, J. Li, X. Pan, J. Xiao, H. Li, H. Ma, M. Wei, Y. Pan, Z. Zhou, M. Li, S. Miao, J. Li, Y. Zhu, D. Xiao, T. He, J. Yang, F. Qi, Q. Fu, X. Bao, Science, 351 (2016) 1065-1068

[4] K. Cheng, B. Gu, X. Liu, J. Kang, Q. Zhang, Y. Wang, Angew. Chem. Intern. Ed., 128 (2016) 4803-4806

[5] Y. Cheng, J. Lin, K. Xu, H. Wang, X. Yao, Y. Pei, S. Yan, M. Qiao, B. Zong, ACS Catal., 6 (2016) 389-399

[6] D. Fu, W. Dai, X. Xu, W. Mao, J. Su, Z. Zhang, B. Shi, J. Smith, P. Li, J. Xu, Y.-F. Han ChemCatChem, 7 (2015) 752-756

[7] T.O. Eschemann, W.S. Lamme, R.L. Manchester, T. E. Parmentier, A. Cognigni, M. R¿nning, K. P. de Jong, J Catal., 328 (2015) 130-138

[8] M. Oschatz, J.P. Hofmann, T. W. van Deelen, W. S. Lamme, N. A. Krans, E. J. M. Hensen, K. P. de Jong, ChemCatChem, 9 (2017) 620-628

492789 Process Control for Sustainability and Life Cycle Inventory (LCÎ™) Monitoring: Application to Biomass/Coal Co-Gasification System

Abstract:

In recent years, due to the increasing global energy demand and the fast depletion of fossil fuels, biomass has received considerable attention. Biomass consists of a renewable feedstock alternative to produce energy and chemicals with potential for ameliorating the environmental effects of using fossil fuels. However, the utilization of biomass is facing some challenges, such as limited and intermittent supply, as well as low energy density. Biomass/coal co-gasification can be an economically attractive and technically viable alternative to convert biomass to energy and chemicals while improving over the sustainability performance of a coal gasification process. To investigate the optimal combination of biomass and coal in terms of economics and sustainability, process systems engineering approaches are needed.

A few process systems contributions have been made to optimize the sustainability of chemical processes by incorporating sustainability indicators into process design and optimization. However, an efficient and systematic framework for the integration of advanced process control with sustainability assessment tools is yet to be developed. To fill this gap, a process control strategy for sustainability has been initially proposed considering the evaluation of different sustainability indicators [1, 2]. In such strategy, only performance variables associated with process efficiency were considered in the controller formulation. In this presentation, the novel addition of multiple sustainability objectives to the process controller is explored, focusing on systems characterized by high-dimensional and nonlinear chemical process models. In particular, the developed framework considers the combination of multi-objective optimization and advanced process control to automatically take the chemical process to a sustainable operating point. This framework will be illustrated via a biomass/coal co-gasification process for syngas production with the end goal of acetic acid manufacturing, including the sustainability analysis for different scenarios.

The process control for sustainability framework includes four main parts: (i) steady-state and dynamic modeling; (ii) multi-objective optimization formulation; (iii) implementation of the advanced control strategy; and (iv) Life Cycle Inventory (LCI) monitoring. In this application, the whole process model is developed in Aspen Plus and Aspen Plus Dynamics based on existing literature information [3-4]. With the established models in Aspen and a link for data communication between Aspen and MATLAB, the multi-objective problem is solved to maximize profit and optimize sustainability performance, employing a genetic algorithm-based approach developed in MATLAB. To reach the optimal operating point defined by the optimization algorithm, an advanced biologically-inspired optimal control strategy (BIO-CS) [1, 5] is implemented to handle the multiple and conflicting objectives associated with economic and sustainability performance. Finally, LCI data is collected for monitoring of steady-state and transient scenarios. In this presentation, the details on the application results of this novel framework for sustainable process control are discussed, focusing on the tradeoffs between using coal and biomass for the sustainable production of chemicals.

The views expressed in this presentation are those of the authors and do not necessarily represent the views or policies of the U.S. Environmental Protection Agency.

References:

1. Li, S.; Mirlekar, G.; Ruiz-Mercado, G.J.; Lima, F.V. Development of Chemical Process Design and Control for Sustainability. *Processes* 2016.
2. Lima, F.V.; Li, S.; Mirlekar, G.V.; Sridhar, L.N.; Ruiz-Mercado, G.J. Modeling and advanced control for sustainable process systems In *The analysis, synthesis and design of chemical engineering processes*, Ruiz-Mercado, G.J.; Cabezas, H. (eds.), Eds. Elsevier, 2016.
3. Robinson, P.J.; Luyben, W.L. Simple Dynamic Gasifier Model That Runs in Aspen Dynamics. Ind. Eng. Chem. Res. 2008.
4. Smith, R.L.; Ruiz-Mercado, G.J.; Meyer, D.E. et al. Coupling Computer-Aided Process Simulation and Estimations of Emissions and Land Use for Rapid Life Cycle Inventory Modeling. *ACS Sustainable Chemistry & Engineering* 2017.
5. Mirlekar, G.V.; Li, S.; Lima, F.V. Design and implementation of a biologically-inspired optimal control strategy (BIO-CS) for chemical process control. Submitted for Publication.

501293 Changing Global Demand for Fossil Based Electricity with Adoption of Renewables at Urban Scale

Abstract:

Urbanization is happening at a rapid pace on a global level. In 2009, the percentage of the worldâ€™s population living in urban areas surpassed 50%, and it is expected to reach 70% by 2050. The changing infrastructure design, urban sprawl and variations in climate affect the energy consumption patterns in cities. We use earlier developed statistical models to estimate change in energy demand in urban areas globally in the categories of heating, cooling, electricity and transportation for 2030 [1-4]. The results show a probable large rise in demand worldwide, particularly tied to decreasing urban density and expected population growth. Particularly, In the US all the categories studied involve an increase except for heating. This growth in energy demand will further increase the need for increasing energy production capacity in and around urban areas. In this work, we study the portfolio of energy production change under different scenarios of adoption of renewable energy adoption. For adoption scenarios of various energy supply sources, we use a genetic algorithm based optimization framework for the potential inclusion of renewable and other distributed energy sources while minimizing costs and environmental impacts. We focus on the ten largest US cities for building this prototype, which reflect an array of supply options. Based on our analysis we provide a preliminary estimate for changes in the demand of conventional energy needs (such as coal and NG based power consumption) for various scenarios of renewable energy adoption in urban areas of the US.

[1] Singh, S., & Kennedy, C. (2015). Estimating future energy use and CO 2 emissions of the world's cities. Environmental Pollution, 203, 271-278.
[2] Kennedy, C.A., Stewart, I., Facchini, A., Cersosimo, I., Mele, R., Chen, B., Uda, M., Kansal, A., Chiu, A., Kim, K.G. and Dubeux, C., 2015. Energy and material flows of megacities. Proceedings of the National Academy of Sciences, 112(19), pp.5985-5990.
[3] Isaac, M., & Van Vuuren, D. P. (2009). Modeling global residential sector energy demand for heating and air conditioning in the context of climate change. Energy policy, 37(2), 507-521.
[4] McNeil, M. A., & Letschert, V. E. (2008). Future air conditioning energy consumption in developing countries and what can be done about it: the potential of efficiency in the residential sector. Lawrence Berkeley National Laboratory.

493868 Conceptual Design of a Novel Efficient Hydrogen Production Process from Natural Gas Using the Systematic "G-H" Methodology

Abstract:

Hydrogen is a clean fuel and is thus expected to play an important role as an energy carrier in a future decarbonized energy scenario that has a high share of renewable energy. However, global hydrogen production is currently dominated by fossil fuel-based processes with significant inefficiencies. Quantifying the source of these inefficiencies and designing novel efficient processes is essential to realizing a hydrogen-based energy economy. Further, these novel processes have to include carbon dioxide capture and storage (CCS).

Natural gas (containing predominantly methane) is a common feedstock for hydrogen production processes that use the steam-methane reforming (SMR) reaction followed by a water-gas shift (WGS) reaction. Methane is combusted in a furnace to provide heat for the endothermic reforming reaction. We simulate a state-of-the-art hydrogen production process in Aspen HYSYS, and then perform an exergy analysis of the model in order to pinpoint the unit operations responsible for the largest sources of inefficiency. The exergy analysis includes physical as well as mixing and chemical exergy parameters that are obtained using user-defined subroutines programmed into Aspen HYSYS. The results of the analysis show that the source of largest exergy destruction (and hence inefficiency) is the furnace (43.8%), followed by the steam generation heat exchanger (20.88%) and the reformer unit (13.73%). Using insight from the exergy analysis, we deduce that the cause of inefficiencies is incorrect operating conditions (temperature and pressure) of different unit operations such as the furnace, SMR and WGS reactors. However, it is not practical to design a more efficient process by merely changing the operating conditions of one unit operation since all interconnected unit operations would be affected by this change. Thus, it is necessary to employ a systematic methodology for conceptual design of an efficient hydrogen production process. This methodology should use a "systems-level" approach, i.e. it should concurrently consider all the unit operations in the process as well as their interconnections.

The objective of this paper is to use the systematic "G-H" methodology [1] for conceptual design of a novel hydrogen production process. The G-H methodology uses information about the change in Gibbs free energy (G) and enthalpy (H) at standard conditions for chemical reactions to derive the heat and work balances of the corresponding reactor. For instance, in our project, changes in G and H are established for SMR, WGS and methane combustion reactions at standard conditions in order to set up heat and work balances for each reactor. The heat and work balances of the overall process are obtained by summing up the contributions of each reactor unit operation weighted by the extent of the reaction taking place in each operation. A desirable operating point is chosen for the overall process and this fixes the operating conditions for the unit operations as well. The operating temperature for each unit is given by its "Carnot temperature" or "reversible temperature", which is defined as the temperature at which the reaction work requirements are provided exactly by the heat supplied such that both requirements are met concurrently. Units operating at their Carnot temperatures are reversible and hence are efficient since they do not destroy any exergy. Herein lies the value of the G-H methodology: It uses the concept of reversibility to provide a target for the operating conditions that give the highest efficiency from the enormous choice of reactor unit operating conditions. In addition, the G-H methodology is applied at the systems-level ensuring that the entire process is highly efficient rather than focusing on the efficiency of individual unit operations.

The Carnot temperatures of the SMR, WGS and methane combustion reactions in the furnace are calculated to be 960.83 K, 974.67 K and 145,866.25 K respectively, though the furnace is designed to operate at a more realistic temperature of 1000.00 K. The flow compositions of the material streams are determined from the extent of reaction of the corresponding unit operations, and a block diagram of the flowsheet is shown in Figure 1 below.

Figure 1: Block diagram of a process for hydrogen production from natural gas

While the G-H methodology ensures a high efficiency, it does not take conversion of the reactions into consideration. In this paper, we extend the G-H methodology by investigating alternative ways of achieving maximum conversion. Specifically, we study two options: separation and recycling of the unreacted feed and integrated separation and reaction in a hydrogen membrane reactor. For the first case, we model the process with separation and recycle of unreacted products for both the SMR and WGS reaction in Aspen HYSYS. We examine how the conversion of each reaction varies with the separation split ratio, and design a flowsheet that gives complete conversion.

For the second case, we formulate a novel procedure to determine the conversion of a hydrogen membrane reactor given the reactor operating conditions and membrane parameters. The problem can be stated as follows: Assume that the SMR and WGS membrane reactors run at the operating conditions obtained from the G-H methodology; what is the expected range of achievable conversion for a particular set of hydrogen membrane characteristics - permeability, thickness of the membrane layer, hydrogen partial pressure profiles at both the feed and retentate side, hydrogen flux and the hydrogen permeation rate-determining step. The procedure involves dividing the membrane reactor into a series of reaction steps followed by membrane separation steps. In each reaction step, the change in conversion is calculated analytically for the membrane reactor operating temperature, and the products of this reaction step are the feed of the next separation step. We assumed a palladium membrane for hydrogen permeation and modelled the separation step according to Sievert's law [2]. Each separation step is further divided into a number of sub-steps in which the feed partial pressure of hydrogen is updated. The final retentate concentration forms the feed of the next reaction step. The results of this procedure give an attainable region of conversion for the SMR and WGS reactions. This attainable region is used to develop a framework for estimating required membrane areas by plotting area contours for different reactor operating pressures. Finally, we show how this procedure fits into the G-H methodology.

Thus, we present two alternative flowsheets showing conceptual design of novel hydrogen production processes from natural gas. These processes are more efficient than the state-of-the-art showing the value of the systematic G-H methodology.

References:

[1] James Alistair Fox, Diane Hildebrandt, David Glasser, and Bilal Patel. "A Graphical Approach to Process Synthesis and its Application to Steam Reforming." *AIChE Journal* 59, no. 10 (2013): 3714-3729.

[2] Samhun Yun, and S. Ted Oyama. "Correlations in palladium membranes for hydrogen separation: a review." *Journal of membrane science* 375, no. 1 (2011): 28-45.

499900 CO_2 Capture and Conversion to Chemicals Via Syngas: Reactor Modeling, Process Synthesis and Optimization

Abstract:

CO_2 emissions from electricity generation increased by 50% between 2000 and 2013 [1]. CO_2 capture and storage (CCS) has drawn wide attention in the recent past but a full-scale power plant with an integrated CCS facility would cost much more than a conventional power plant [2]. CO_2 conversion for utilization can be an alternative to CCS. CO_2 is a source of both carbon and oxygen and, if combined with a hydrogen source, can produce valuable hydrocarbons. Methane, from various sources, such as natural gas and shale gas, has been considered for this purpose for the conversion of CO_2 to synthesis gas. This approach to capture and convert CO_2 has several possibilities that can be analyzed in a systematic manner by proposing a superstructure network with all the alternatives embedded in it. The challenge is to characterize each process in the superstructure with an accurate model for a holistic analysis. We have approached this challenge by trying to strike a balance between model accuracy and complexity by analyzing the trade-offs in different models and building simpler surrogate models that can retain the model predictability. Reactor models such as stoichiometric, equilibrium, pseudo-homogeneous and heterogeneous models have been considered in this analysis. Process and design variables are taken into consideration while optimizing the process synthesis model for different objective functions such as maximum possible CO_2 utilization and minimum cost of utilization. We also consider the net CO_2 utilization possible by considering the auxiliary emissions from any of the processes in the superstructure. The superstructure model is generic and robust and can accommodate additional alternatives without having to change the model. Different syngas ratios and different applications based on these ratios are discussed as well. The superstructure model is formulated as a mixed-integer nonlinear optimization (MINLP) model and is solved to optimality using the global solver ANTIGONE [3]. In this presentation, we will discuss our overall synthesis framework and the optimization results.

References

[1] I. E. A. (IEA), "CO_2 Emissions From Fuel Combustion - Highlights - 2015 Edition," 2015.

[2] Z. Yuan, M. R. Eden, and R. Gani, "Toward the Development and Deployment of Large-Scale Carbon Dioxide Capture and Conversion Processes," *Industrial & Engineering Chemistry Research,* vol. 55, pp. 3383-3419, 2016/03/30 2016.

[3] R. Misener and C. A. Floudas, "ANTIGONE: Algorithms for coNTinuous / Integer Global Optimization of Nonlinear Equations," *Journal of Global Optimization,* vol. 59, pp. 503-526, 2014.

499689 Methods for the Design of Spatially Explicit Biofuel Supply Chains

Abstract:

Methods for the Design of Spatially Explicit Biofuel Supply Chains

Rex T. L. Ng[1,2] and Christos T. Maravelias[1,2]

[1]*Department of Chemical and Biological Engineering and* [2]*DOE Great Lakes Bioenergy Research Center, University of Wisconsin-Madison, Madison, WI*

Cellulosic biomass is seen as one of the promising sources for sustainable biofuels production. In recent years, much effort has been made to design efficient biomass-to-biofuel supply chains (SCs) in terms of economic, environmental and social aspects. In many studies, biomass feedstock availability is estimated based on county-level data, assuming a harvesting site is located at the centroid of a county [1,2], or a county is divided into rectangular or square sub-areas, and a harvesting site is located at the centroid of a sub-area [3,4]. Additionally, biomass is assumed to be uniformly distributed over the area. However, unlike fossil fuels, biomass availability is highly distributed, so the aforementioned assumptions often lead to biofuel SCs that are suboptimal or even infeasible from an energy consumption point of view, that is, the primary energy required to harvest and transport biomass is higher than the energy content of the produced biofuel.

To overcome this, we develop a framework that allows us to utilize high spatial resolution biomass data in designing biomass-to-biofuel SCs. We first develop a tool based on the Agricultural Model Intercomparison and Improvement Project (AgMIP) packages to generate feedstock data using climate data from the user-specified general circulation model and representative pathway [5]. This data is then combined with a historical cropland data layer (CDL) which is provided by the U.S. Department of Agriculture [6], to generate spatial data for an area based on a user-specified resolution (down to 30 m × 30 m). Note that this generates a complex SC network with a very large number of arcs, and therefore leads to a large-scale optimization problem.

Accordingly, we then develop a preprocessing algorithm to reduce the number of arcs. We first determine the feedstock, harvesting and handling (FHH) cost up to the biorefinery gate (includes transportation and depot installation and operation costs) for different SC configurations (with or without depot and different transportation modes). If depot is selected, we determine the maximum distance for collecting distributed biomass to a potential depot given the depot capacity. Note that the maximum distance greatly depends on biomass availability at every potential depot location. Next, we compare the FHH costs for all configurations and determine the threshold distances at which one configuration becomes the most economical. For example, in a corn stover-to-bioethanol SC, depot should only be installed if the distance between a harvesting site and the biorefinery is more than 120 km. Similarly, rail transportation of densified corn stover is cheaper than truck transportation when the distance between depot and biorefinery is greater than 225 km. Based on these pre-calculated distances, we remove arcs that will never be used in an optimal solution, and thus dramatically reduce the number of defines variables.

Next, we develop a multi-period mixed-integer linear programming (MILP) model that accounts for facility location, technology selection, and capacity planning decisions in terms of depots and biorefineries, as well as auxiliary decisions such as harvesting site and biomass feedstock selection, biomass allocation, transportation mode, and inventory planning decisions.

Finally, we illustrate the applicability of the proposed model by considering a corn stover-to-bioethanol SC in Wisconsin (420 km × 500 km). We consider a one-year horizon divided into 12 periods, and two transportation modes: truck and rail. The objective is to minimize the total annual cost, which includes biomass acquisition, inventory, operating, and annualized capital costs for depots and biorefineries.

References:

[1] F. You, B. Wang, Life Cycle Optimization of Biomass-to-Liquid Supply Chains with Distributed–Centralized Processing Networks, Ind. Eng. Chem. Res. 50 (2011) 10102–10127.

[2] W. Alex Marvin, L.D. Schmidt, S. Benjaafar, D.G. Tiffany, P. Daoutidis, Economic Optimization of a Lignocellulosic Biomass-to-Ethanol Supply Chain, Chem. Eng. Sci. 67 (2012) 68–79.

[3] R.T.L. Ng, C.T. Maravelias, Design of Cellulosic Ethanol Supply Chains with Regional Depots, Ind. Eng. Chem. Res. 55 (2016) 3420–3432.

[4] R.T.L. Ng, C.T. Maravelias, Design of biofuel supply chains with variable regional depot and biorefinery locations, Renew. Energy. 100 (2017) 90–102.

[5] N.B. Villoria, J. Elliott, C. Müller, J. Shin, L. Zhao, C. Song, Rapid aggregation of global gridded crop model outputs to facilitate cross-disciplinary analysis of climate change impacts in agriculture, Environ. Model. Softw. 75 (2016) 193–201.

[6] National Agricultural Statistics Service U.S. Department of Agriculture, CropScape - Cropland Data Layer, (2016). https://nassgeodata.gmu.edu/CropScape/ (accessed April 5, 2017).

494507 Insights of Biomass Recalcitrance in *Populus* for Biomass Valorization

Abstract:

For the effective utilization of biomass, understanding of its properties, in particular, biomass recalcitrance-related properties is essential. Diverse approaches are applicable to reveal the secret of biomass recalcitrance. Among many, physicochemical characteristics of biomass feedstock do not only elucidate the biomass recalcitrance, but also provide clues to overcome its physical and chemical barriers for biomass conversion. Transcriptome analysis also gives important overview about transcriptional regulation network of biomass synthesis related gene families. *Populus* has been investigated as a promising biomass feedstock for alternative fuels and chemicals. However, biomass recalcitrance-related properties and their causable variation of gene expression are not sufficiently understood. Here, recalcitrance of *Populus* natural variants was elucidated in three aspects: (1) fermentable sugar release, (2) physicochemical properties of biomass, and (3) relative variation of gene expression. Sugar release and physicochemical characteristics of *Populus* natural variants including chemical composition, cellulose and ligninâ€™s molecular weights/degree of polymerization, cellulose crystallinity, lignin composition, and cellulose accessibility were measured by diverse analytical methods such as HPLC, GPC, 2D ^{13}C-^{1}H HSQC NMR, solid-state NMR, and others. Differential Gene expression of each variant was also investigated by RNA-seq analysis.

499309 Evaluation of Industrial Hemp Varieties As Potential Biomass Feedstock for Biofuels and Bioproducts

Abstract:

Industrial hemp is an annual herbaceous crop historically grown for fiber in the U.S. and worldwide. Despite existing applications of industrial hemp based fiber and oil products, hemp-based biofuels and bioproducts represent a new potential application area. Several key questions still remain for the technical and economic feasibility of using hemp as a bioenergy crop. Furthermore, it is necessary to screen industrial hemp species and process conditions while aiming at maximized production of biofuel and bioproducts. Hence, this study takes combined agronomy, laboratory, and economic analysis approaches to evaluate industrial hemp in comparison with other biomass feedstocks e.g. kenaf, switchgrass and biomass sorghum. Several industrial hemp varieties, currently widely cultivated in North America, Europe and Asia, were evaluated. Experimental results show an ethanol yield of 85-100 gallons /dry ton hemp stems using dilute alkali or acid pretreatment, enzymatic hydrolysis and fermentation, which is comparable to the other three tested feedstocks. Agronomy data suggest that the per hectare yield of industrial hemp stem alone was at a similar level to the other energy crops such as switchgrass and sorghum; while the hemp plants may require reduced inputs. Field trial also showed that ~1200 kg/ha hemp grain can be harvested in addition to stems. A preliminary cost analysis indicates that industrial hemp could generate higher per hectare gross profit than the other crops if both hemp grains and biofuels from hemp stem were counted. These combined evaluation results demonstrate that industrial hemp has great potential to become a promising regional commodity crop for producing both biofuels and value-added products.

503939 Thermochemical Conversion of Blended Herbaceous and Woody Biomass Feedstocks

Abstract:

The United States has a vast potential supply of renewable lignocellulosic biomass, which can be utilized as feedstocks for the production of renewable fuels and chemicals via biomass conversion technologies. In order to support a large-scale bioeconomy, it will be required that multiple types of biomass be utilized such as dedicated bioenergy crops, forestry and agricultural residues, and municipal wastes. Blending several types of biomass to produce a uniform feedstock has the potential to control feedstock quality specifications. In order to exploit the potential benefits of biomass blending, we must gain a comprehensive understanding of its effects on thermochemical conversion. In this study, we target pyrolysis as the thermochemical conversion technology of interest as it is well suited to produce liquid fuels and chemicals through thermal decomposition and condensation of pyrolysis vapors into bio-oil. The chemical composition of the feedstock used has a significant impact of the yield and quality of the bio-oil produced. In this work, we first utilized pyrolysis gas chromatography mass spectroscopy (py-GC/MS) to measure the vapor-phase pyrolysis products produced from switchgrass, two pine residue feedstocks, and binary blends of these. Next, we used an intermediate-scale pyrolysis reactor to test the effect of blending on bio-oil yield and quality. Results indicate that the composition and content of alkali and alkaline earth metals in the feedstocks have a significant effect on the pyrolysis products, and that blending is a potential strategy to produce high quality and low variability feedstocks. Further results and discussion regarding the effects of blending on the yield, quality, and chemical composition of vapor-phase and liquid-phase pyrolysis products will be presented.

491952 Feedstock Preprocessing, Fractionation, and Blending to Improve Overall Cost, Supply, and Quality Considerations for Catalyzed and Uncatalyzed Fast Pyrolysis

Abstract:

The success of lignocellulosic biofuels and biochemical industries depends on an economical and reliable supply of high-quality biomass. As such, biomass cost, supply, and quality are critical parameters to consider when choosing feedstocks and locations for biorefineries. Biomass cost is dependent upon feedstock type, location, availability, transportation costs, and the quality specifications required by the conversion technology. Biomass quality depends on inherent feedstock properties, field and storage conditions, and any preprocessing methods used to improve quality. Biomass quantity depends on location, field conditions, annual weather, and collection and storage conditions. Because of these variables, it is necessary to develop strategies that enable supply chain improvements in biomass quality and consistency to improve lower overall conversion costs and improve yields, while also securing a less risky feedstock supply.

The presented work examines the interdependencies of these variables and how they affect the biomass blends required by biomass depots and/or biorefineries to achieve the lowest cost feedstock with sufficient quality at the quantities needed for biorefinery operation. Four biomass depots were proposed in South Carolina to each produce 200,000 tons of feedstock per year. These depots supply a centrally located 800,000 ton biorefinery that converts the feedstocks to bio-oil using either catalyzed or uncatalyzed fast pyrolysis. The four depots utilize biomass based upon availability, but the feedstock or feedstock blend still met the minimum quality requirements for the biorefinery. Costs were minimized by using waste biomass resources such as construction and demolition waste, logging residues, and forest residuals. Preprocessing methods such as air classification and water washing or acid leaching were used to upgrade biomass quality. Air classification was found to be an effective method for separating low- and high-ash fractions of woody residues. For example, a high-ash fraction of logging residues was generated that contained approximately 50 wt% of the total original ash content, but less than 8 wt% of the original organic content. Dilute-acid leaching was then effectively used to remove between 85% (25°C, 0.5 wt% acid) and 98% (90°C, 1.0 wt% acid) of the alkali and alkaline earth metals from the high-ash fraction, making it suitable for recombination with the low-ash fraction. Various combinations of such preprocessing methods were tested to produce an array of blends that were available for recombination to produce least-cost blendstocks that met various conversion specifications. For both uncatalyzed and catalyzed fast pyrolysis, all four depots produced blends that met quality and quantity specifications at a cost lower than using a single feedstock.

499592 New Approach to Ammonia Pretreatment Integrates Better Feedstock Logistics with Improved Sugar Conversion

Abstract:

The implementation of a large scale bioeconomy, where lignocellulosic residues can be used for production of commodity products (fuels, chemicals, materials, food additives, animal feed, etc.) highly depends on the ability of moving large amounts of biomass across long distances. To be able to effectively transport low density materials, such as agricultural residues, it is necessary to apply densification methods (e.g., pelletization). Though densification is expensive, it opens a range of new possibilities for feedstock logistics (e.g., ability to use various transportation methods, safer feedstock storage, etc.) and the ability to implement very large scale lignocellulosic biorefineries, as these require economic transportation of lignocellulosic feedstocks within a wide feedstock production radius. By increasing the size of the biorefinery, it is possible to take advantage of economies of scale and thus, reduce the capital cost per unit of processed biomass. To extract further value from using densified biomass, it is important to explore biomass conversion methods that benefit from having high density feedstock, which can generate higher sugar yields with lower energy inputs.

In this work, we have developed a new ammonia pretreatment strategy called Compacted Biomass with Reduced Ammonia (COBRA) pretreatment. In COBRA pretreatment, liquid ammonia is added to biomass pellets in ratios lower than 1:1, allowing native cellulose I to be converted to cellulose III at relatively low temperatures (~70 °C) and residence times of 3 to 4 hours. As pellets occupy less volume per unit mass relative loose biomass, lower ammonia volumes are required to completely submerge the solid biomass. Thus, as pellets contain low moisture content, the presence of liquid ammonia around the cellulose fibers will allow the formation of a highly digestible cellulose III. As a result, COBRA pretreated biomass can be effectively digested with 60% less enzyme relative to the traditional AFEX pretreatment. Furthermore, as COBRA pretreatment uses densified biomass, it also allows the reduction of reactor size compared to AFEX, even though it uses longer residence times. By using low ammonia loadings, small sized reactors, relatively low temperatures and pressures, low enzyme loading and taking advantage of economies of scale, COBRA pretreatment has the potential to reduce overall processing costs in the biorefinery. Here, we will present a complete optimization of COBRA pretreatment for corn stover, an enzyme cocktail optimization to maximize sugar yields and determination of sugar fermentability using *Saccharomyces cerevisiae* 424A (LNH-ST). In addition, a complete mass balance and process economics will be presented.

Screening method for selection of biomass feedstock and location in biorefneries

A. Jaksland, M. Bertran, J.D. Woodley, R. Gani

2/11 AIChE Annual Meeting 2017

Outline

- Motivation and Background
- Biomass selection
- One parameter screening method
- Case study
  - Biomass selection
  - Superstructure generation and solution
- Results
- Evaluation in three parameters
- Conclusion

Danmarks Tekniske Universitet

Motivation

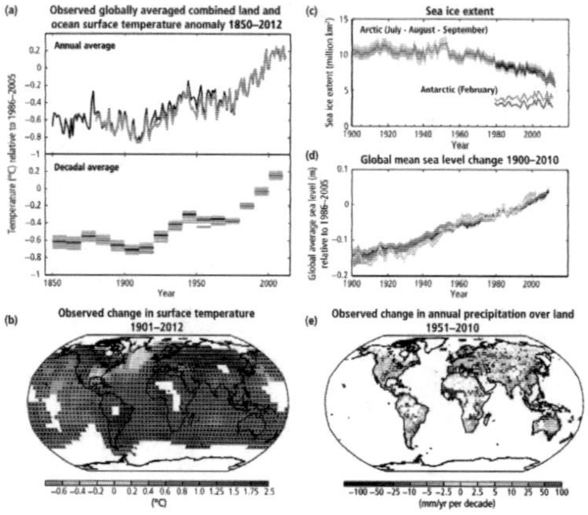

Background

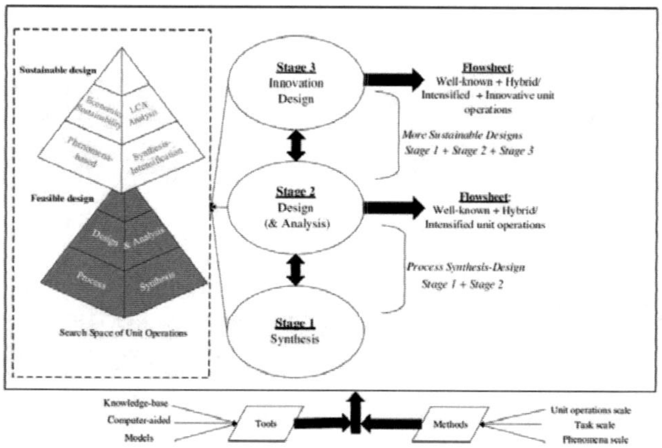

DTU Chemical Engineering
Department of Chemical and Biochemical Engineering

1) IPCC. Climate Change 2014: Synthesis Report. Contribution of Working Groups I, II and III to the Fifth Assessment Report of the intergovernmental Panel on Climate Change. 2014.
2) Deenesh K. Babi, Johannes Holtbruegge, Philip Lutze, Andrzej Gorak, John M. Woodley, and Raqul Gani. Sustainable process synthesis-intensication. Computers and Chemical Engineering, 81:218{244, 2015.

Danmarks Tekniske Universitet

Superstructure Optimisation Framework

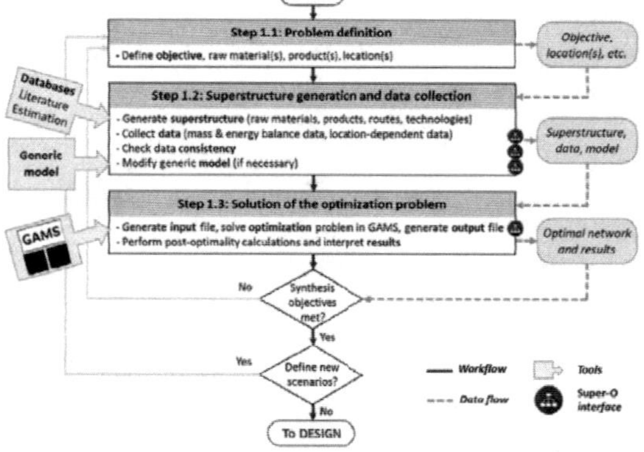

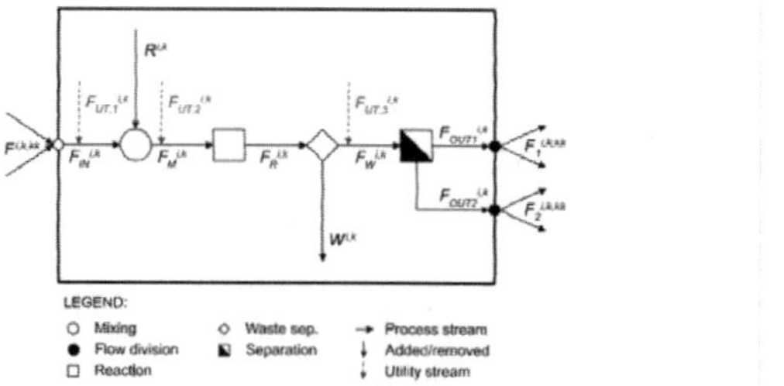

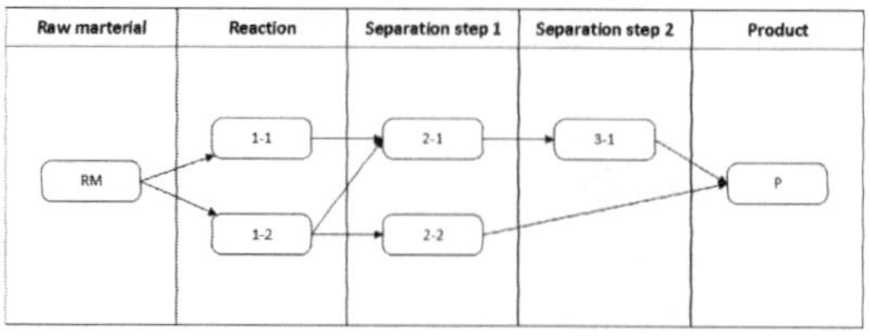

DTU Chemical Engineering
Department of Chemical and Biochemical Engineering

3) Maria Ona Bertran, Rebecca Frauzem, Lei Zhang, and Raqul Gani. A Generic Methodology for Superstructure Optimization of Different Processing Networks. Computer Aided Chemical Engineering, 38:685{690, 2016.

Biomass as Feedstock

Category	Potential biomass supply in 2050 (EJ/y)
(I) Biomass production on surplus agricultural land	0-988
(II) Biomass production on degraded lands	8-110
(III) Agricultural residues	10-32
(IV) Forest residues	10-16
(V) Animal Manure	9-25
(VI) Organic waste	1-3
(VII) Bio-materials	Minus 83-116
Total	33-1130

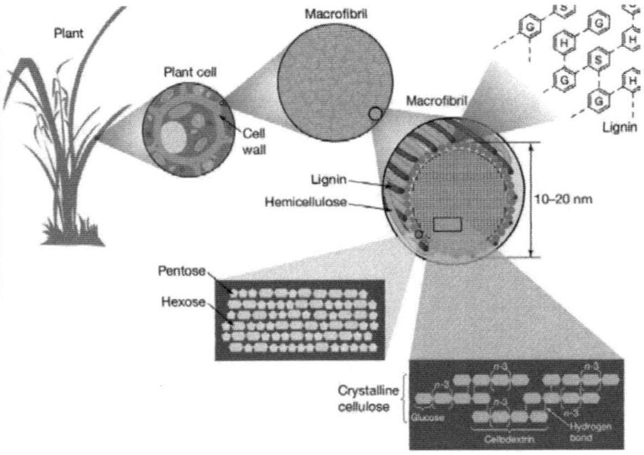

4) Moniqu Hoogwijk, Andre Faaij, Richard van den Broek, G□oran Berndes, Dolf Gielen, and Wim Turkenburg. Exploration of the ranges of the global potential of biomass for energy. Biomass and Bioenergy, 25:119(133, 2003.

Biomass Selection

Location and specie dependent variables: Composition, density, harvest field yield, transportation cost, at-field price

Three parameters: Harvest field yield, at-field price and transportation cost

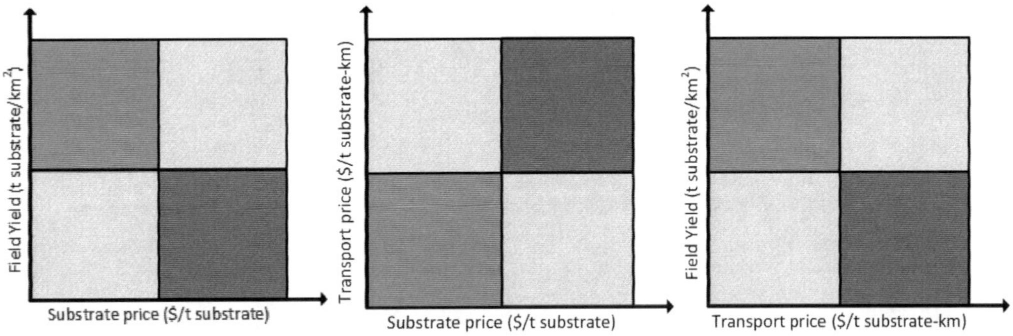

Danmarks Tekniske Universitet

Biomass Selection

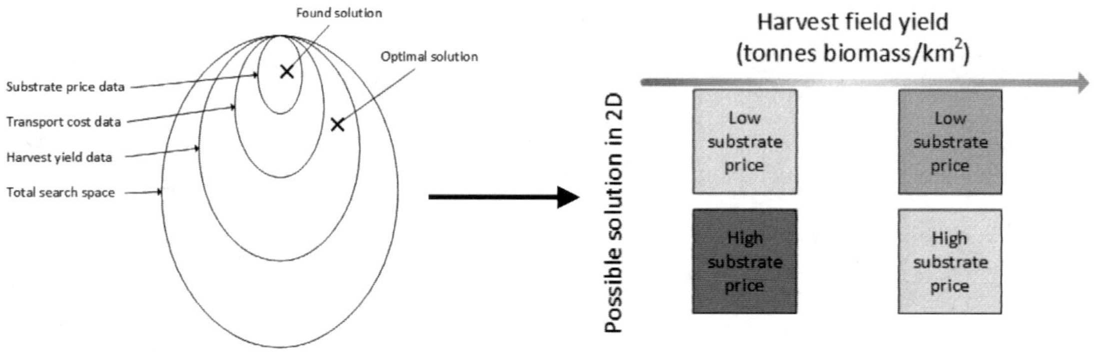

DTU Chemical Engineering
Department of Chemical and Biochemical Engineering

Danmarks Tekniske Universitet

Biomass Selection

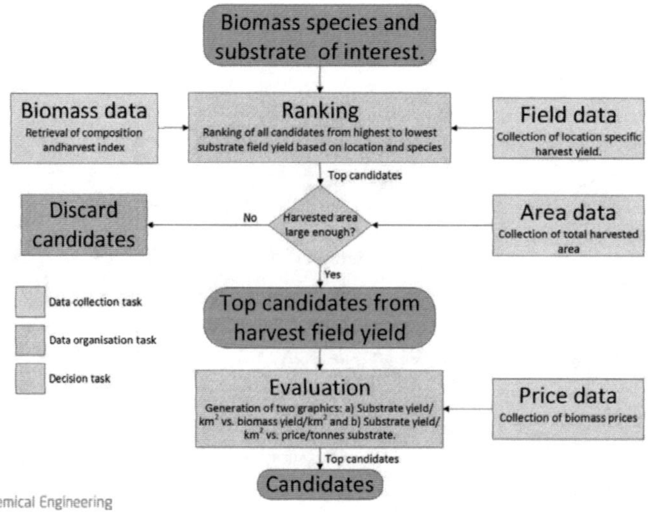

DTU Chemical Engineering
Department of Chemical and Biochemical Engineering

Case study – Step 1

- 750.000 tonnes biomass/year
- Biomass specie and location: Selected with one parameter screening
- Intermediate: glucose
- Biomass to glucose: Milling, steam explosion and enzymatic hydrolysis
- Products: 1,3-propanediol, 2,3-butanediol, succinic acid, itaconic acid
- Ideal reactions based on redox potential
- Transport and storage cost of biomass not included

Danmarks Tekniske Universitet DTU

Case study – Step 1

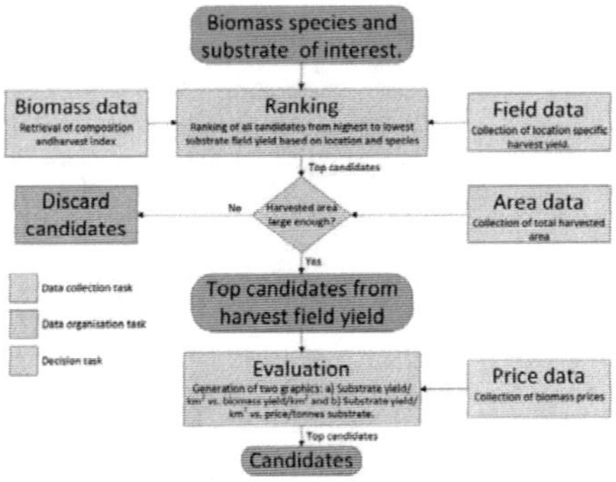

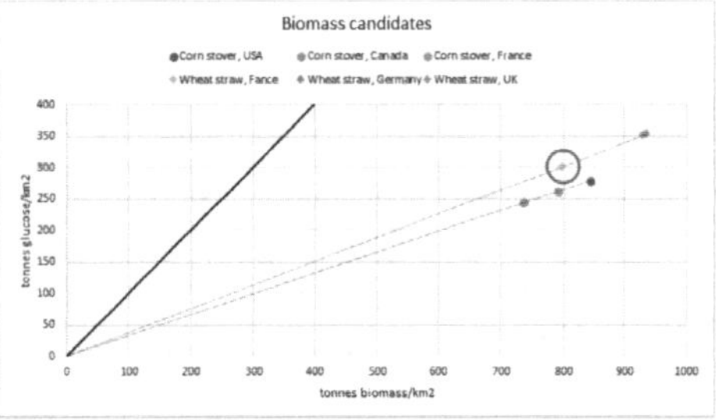

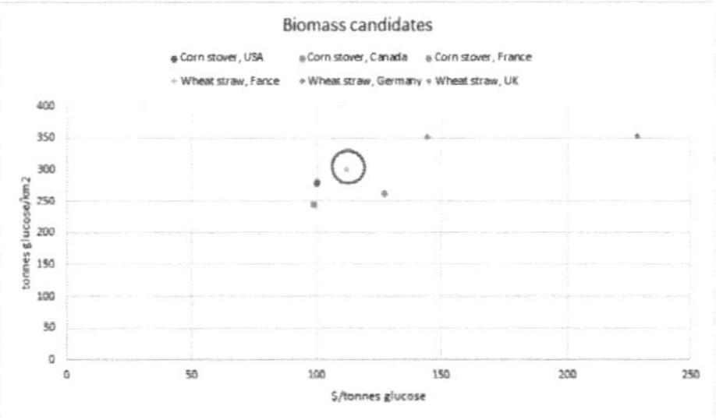

Case study Step 2 & 3

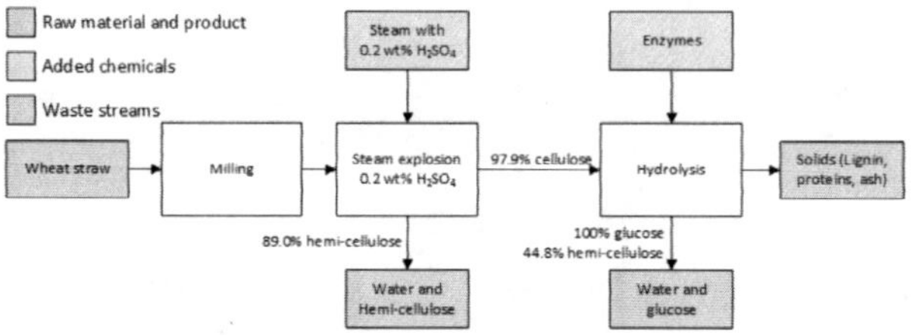

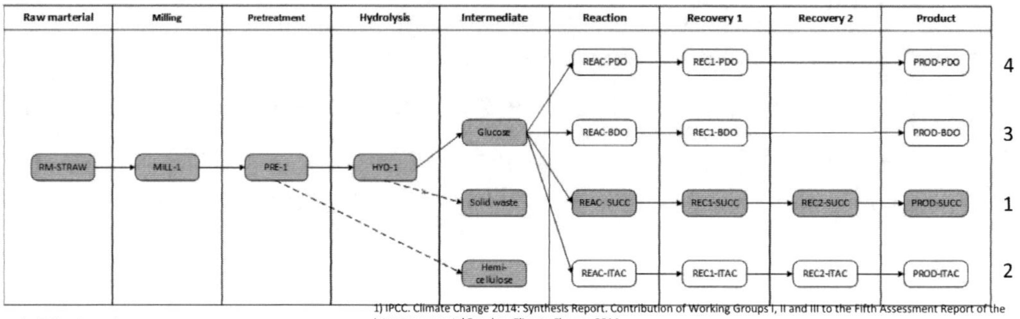

1) IPCC. Climate Change 2014: Synthesis Report. Contribution of Working Groups I, II and III to the Fifth Assessment Report of the intergovernmental Panel on Climate Change. 2014.
2) Deenesh K. Babi, Johannes Holtbruegge, Philip Lutze, Andrzej Gorak, John M. Woodley, and Raqul Gani. Sustainable process synthesis-intensication. Computers and Chemical Engineering, 81:218{244, 2015.

Results

Succinic acid		Price	Amount	Value
Raw materials	Wheat straw	0.045$/kg	85616.4	3852.7 $/hr
Added chemicals	Cellulase	15.67 $/kg	684,6	10728.1 $/hr
Products	Succinic acid	3 $/kg	29560.7	88682.1 $/hr
Utilities	Steam	0.00983 $/MJ	3680214.4 MJ/hr	36176.5 $/hr
	Cold water	0.00035 $/MJ	45830.5 MJ/hr	16.0 $/hr
	Electricity	0.0295 $/MJ	3532.6 MJ/hr	104.1 $/hr
	Total	-	3729577.5 MJ/hr	36296.7 $/hr
Operating profit				37804.6$/hr
Capital investment				59000635.0 $
Payback time				1561 hr = 65 days
Hourly profit over 20 years				37467.8 $/hr

DTU Chemical Engineering
Department of Chemical and Biochemical Engineering

Danmarks Tekniske Universitet DTU

Results – Comparison

	Product	Theo. yield	Prac. yield	Operating profit	Capital investment	Payback time
Scenario 1	1,3-propanediol	0.84 g/g	0.57 g/g	12186.8 $/hr	38.6m $	132 days
	2,3-butanediol	0.75 g/g	0.48 g/g	26856.5 $/hr	35.5m $	55 days
	Itaconic acid	0.96 g/g	0.81 g/g	29053.1 $/hr	47.0m $	67 days
	Succinic acid	1.12 g/g	0.89 g/g	37804.6 $/hr	59.0m $	65 days

Sensitivity Analysis

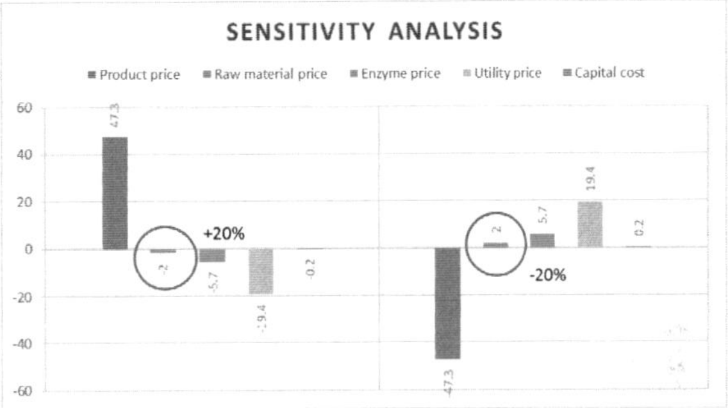

DTU Chemical Engineering
Department of Chemical and Biochemical Engineering

Danmarks Tekniske Universitet DTU

Biomass Selection in Three Parameters

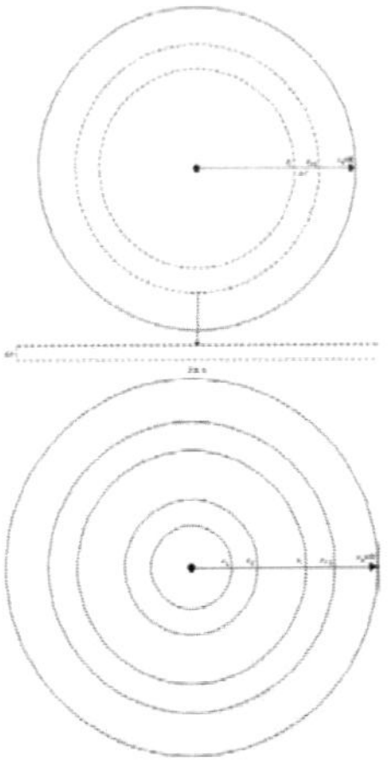

Radius as a function of capacity and yield:

$$R = \sqrt{\frac{C}{\pi Y}}$$

Transportation cost normalised by capacity as a function of radius, truck load and fuel cost:

$$\frac{T}{C} = \frac{\frac{4R^3\pi Y}{3L}F}{C} = \frac{4R^3\pi Y}{3LR^2\pi Y}F = \frac{4R}{3L}F$$

Transportation normalised by capacity as a function of capacity, yield, truck load and fuel cost

$$\frac{T}{C} = \frac{4\sqrt{\frac{C}{\pi Y}}}{3L}F = \frac{4\sqrt{C}}{3L\sqrt{Y\pi}}F$$

Symbol	Description	Unit
Y	Biomass harvest field yield	kg biomass/km^2
L	Truck load	kg biomass
R	Outer diameter	km
F	Fuel cost	$/km
C	Biomass plant capacity	kg biomass
T	Total transport cost	$
T_i	Transport cost of annulus	$
B	At-field biomass price	$/kg biomass
r	Radius (axis)	km
Δr	Differential radius piece	km

DTU Chemical Engineering
Department of Chemical and Biochemical Engineering

Biomass Selection in Three Parameters

- Transportation estimate
- France vs. Brazil
  - 796.58 t/km^2 vs. 239.27 t/km^2
  - Estimation gives: 42.3 $/t vs. 39.35 $/t
  - 941.5 km^2 vs. 3134.3 km^2

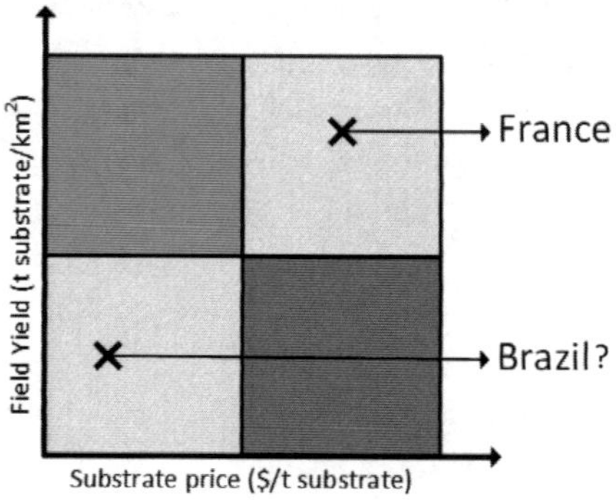

$$\frac{4\sqrt{C}}{3L\sqrt{Y_1\pi}}F + B_1 = \frac{4\sqrt{C}}{3L\sqrt{Y_2\pi}}F + B_2$$

Danmarks Tekniske Universitet DTU

Conclusion

- Location and specie dependent biomass variables can be reduced to three parameters: Harvest field yield, at-field cost and transportation cost, all based on desired substrate.
- A one-parameter based on harvest field yield has been developed.
- In a case-study a biorefinery network superstructure was developed with feedstock found from the screening.
- An operational profit of 37804.6 $/hr was found for succininc acid. The payback time was 65.
- A sensitivity analysis (+/- 20%) showed that the feedstock price only impacted the operating profit by +/- 2%.
- Expanding the potential selection of biomass to the original three parameters showed that inclusion of transport cost is small compared to at-field price.

DTU Chemical Engineering
Department of Chemical and Biochemical Engineering

Danmarks Tekniske Universitet

Thank you for listening
Questions?

DTU Chemical Engineering
Department of Chemical and Biochemical Engineering

501197 Exploring Ruthenium Metal-Support Dynamics for the Low-Temperature Partial Oxidation of Methane

Abstract:

Emmett Goodman[1], An-Chih Yang[1], Allegra Latimer[1], Liheng Wu[2], Frank Abild-Pedersen[2], and Matteo Cargnello[1], (1)Chemical Engineering, Stanford University, Stanford, CA, (2)SLAC National Accelerator Laboratory, Menlo Park, CA

Exploring Ruthenium Metal-Support Dynamics for the Low-Temperature Partial Oxidation of Methane

Department of Chemical Engineering and SUNCAT Center for Interface Science and Catalysis, Stanford University, Stanford, CA 94305, USA

Industrially, the transformation of methane to syngas stands as an invaluable reaction, the latter products being converted downstream into valuable bulk chemicals such as methanol, ammonia, and synthetic fuels. Steam or bireforming are common routes for methane conversion to syngas, but work at high temperatures because of their endothermic nature. Methane partial oxidation (CH_4 + ½ O_2 = CO + 2 H_2) is exothermic and benefits from more favorable thermodynamics. Thus, there are significant incentives to explore catalysts that can partially oxidize methane at low temperatures, with high selectivity and yield, into an industrially relevant feedstock.

Methane partial oxidation (MPO) is a complex reaction, in which cycling catalyst oxidation states dramatically influences catalyst structure, activity, and selectivity. In order to rationally design a catalyst for MPO, it is important to understand how interactions between catalyst active phases and underlying support affect catalyst structure and subsequent behavior. However, in non-uniform systems with active phases of a variety of sizes and shapes, it is often difficult to back out such structure-property relationships. By colloidally synthesizing ruthenium nanoparticles prior to support impregnation, we may make a fair comparison of activity, selectivity, and stability across supports and understand how catalyst support affects the morphological and catalytic evolution of ruthenium active phase(s). We report highly active 0.5 wt. % ruthenium catalysts prepared from monodisperse nanocrystals, which show >90% CH_4 conversion and >98% CO selectivity at ~650Â °C.

Mechanistically, at low temperatures, RuO_2 slowly catalyzes complete combustion; upon oxygen depletion in the reactor, and reduction to metallic Ru, we ignite partial oxidation, which tends toward equilibrium values. This study thus identifies two design rules for igniting low-temperature formation of synthesis gas: oxygen depletion in the reaction feed, and the effect of metal-support interaction on oxidation state. We find that selectivity and yield are strongly dependent on the support type, with a size dependence on certain â€˜activeâ€™ supports. These effects are related to how certain supports preferentially stabilize different oxidation states of ruthenium. Using this information, we can gain fundamental mechanistic insight regarding methane partial oxidation on supported ruthenium systems, and work to design more effective catalysts. Taking advantage of these design rules, we suggest the development of highly

active bifunctional catalysts, which take advantage of a palladium phase to deplete feed oxygen, and a reduced ruthenium phase to form partial oxidation products.

500783 Direct Methane Conversion to Ethylene and Ethane By Oxidative Coupling in Packed Bed and Membrane Reactors

Abstract:

The abundance of methane makes it an attractive raw material for chemical production. However, current industrial routes for converting methane to valuable chemicals typically involve a synthesis gas (CO and H_2) production step which is complex and expensive. Direct methane conversion can reduce the need for syngas production and is potentially more suitable for converting methane released from small scale sources, which is typically flared resulting in greenhouse gas emissions. Oxidative coupling of methane (OCM) is a direct route for converting methane into ethylene and ethane (C_2). Ethylene, one of the target products of OCM, is a widely-used feedstock in the petrochemical industry. However, OCM is yet to be commercialized due to the low C_2yields obtained, particularly in conventional packed bed reactors (PBRs).

It has been demonstrated that a reactor with distributed oxygen feed (i.e. an O^{2-} conducting membrane reactor) can give much higher C_2 selectivity and yield compared to a PBR. However, the relatively few membrane reactors that have been studied in OCM literature give C_2 yields that are still below the commercial viability targets[1,2]. The low performance of OCM membrane reactors is partly due to the limited oxygen flux through the membranes and low membrane surface area. Another reason for the low performance of some of these membrane reactors is the absence of a selective OCM catalyst. The integration of such catalyst in the reactor should facilitate methane activation, leading to improvements in C_2 yields. Ideally, the integrated catalyst should be ionically conductive to accept O^{2-} from the membrane, and compatible with the membrane material at the high temperatures used in fabrication and testing[1,3]

In this work, we use PBR experiments to identify O^{2-} conducting catalysts that are selective for OCM. These catalysts include lanthanum and barium containing perovskite oxides. We test these catalysts in membrane reactors that consist of a dense O^{2-} conducting membrane and a porous O^{2-}conducting layer to aid methane activation. We present our results and discuss our efforts to further improve the OCM performance of these membrane systems by increasing the membrane surface area and oxygen flux.

(1) Farrell, B. L.; Igenegbai, V. O.; Linic, S. *ACS Catal.* **2016**, *6*, 4340–4346.

(2) Wang, B.; Albarracín-Suazo, S.; Pagán-Torres, Y.; Nikolla, E. *Catal. Today* **2017**, 1–12.

(3) Farrell, B.; Linic, S. *Catal. Sci. Technol.* **2016**, *6*, 4370–4376.

493526 Sulfur As a Selective Soft Oxidant in the Catalytic Conversion of Methane to Ethylene over Metal Chalcogenides

Abstract:

Oxidative coupling of methane (OCM) in presence of an oxidant over metal oxide catalysts provides an attractive route for direct conversion of methane to ethylene.[1] The tremendous wealth of previous studies using oxygen as the oxidant have provided important insights into the mechanisms for OCM.[2] Oxygen, however, is a very strong oxidant and ultimately limits the selectivity in OCM. New coupling processes that use abundant, thermodynamically mild oxidants together with selective catalysts that reduce the thermodynamic driving force for over-oxidation and result in higher ethylene yields would offer exceptional opportunities. We thus report herein on the use of elemental sulfur as a promising soft oxidant for selective conversion of methane to ethylene. We explored the oxidative coupling of methane via sulfur (SOCM) over a range of different metal sulfide catalysts including MoS_2, RuS_2, TiS_2, PdS and Pd/ZrO_2. Experimental results and density functional theory (DFT) calculations reveal that methane conversion is directly correlated with surface metal–sulfur bond strengths. Surfaces with weakly bound sulfur are more basic and activate the C–H bonds of methane more readily. In contrast, ethylene selectivities scale inversely with surface metal–sulfur bond strengths, and surfaces with the strongest metal–sulfur bonds afford the highest ethylene selectivities. The PdS system was found to be unique in that it showed the highest rates and selectivities. More detailed characterization of the catalyst showed that the catalyst was partially reduced under reaction conditions to form $Pd_{16}S_7$. Theoretical results discussed herein show that this surface exposes both Pd as well as S sites which can more readily activate methane but also promote the C-C bond formation and the production of ethylene. We extend these results to the reactivity of oxides together with sulfur and report C_2 selectivities > 33% for Fe_3O_4. XRD and XPS analyses of Fe_3O_4 indicate that it readily converts to FeS_2. The primary products formed over these FeS_2 catalysts are ethylene, ethane and CS_2. DFT calculations show that ethylene predominantly forms in path similar to that for PdS via CH_2 coupling. The exposed Fe-S site pairs that form on the sulfided Fe_3O_4 catalyst aid in the activation of the C-H bond of methane as well as the production of ethylene. CS_2formation, on the other hand, was found likely to occur over the adsorbed sulfur sites via facile C-H bond activation steps. Ethane formation predominantly occurs via the formation of methyl fragments, their desorption from the sulfide and their recombination in the gas phase.

References:

1) Zavyalova, U., Holena, M., Schlögl, R. & Baerns, M. Statistical analysis of past catalytic data on oxidative methane coupling for new insights into the composition of high performance catalysts. ChemCatChem 3, 1935–1947 (2011).

2) Lunsford, J. H. The catalytic oxidative coupling of methane. Angew. Chem. 35, 970-980 (1995).

500461 Methane Coupling in Atmospheric Pressure Corona Discharge Microreactor

Abstract:

Methane Coupling in Atmospheric Pressure Corona Discharge Microreactor

Adam Shareghi, Ian Reddick, Justin Pommerenck, Scott Harpool, Yu Miao, Yousef Alanazi, Alexandre Yokochi, Annette von Jouanne, Nick AuYeung, Goran Jovanovic

Methane-rich landfill biogas, agricultural gas and stranded natural gas from fracking are vast energy resources that remain largely untapped. Bringing this energy to market is uneconomical with conventional technology. Converting a percentage of this unused resource into an easily transportable form represents an economic opportunity.

Methane coupling using nonthermal plasmas in microscale reactors has been demonstrated. The high temperature electrons disrupt the stable carbon-hydrogen bonds of the methane and allow the fragmented species to polymerize into higher-order hydrocarbons that can be used as liquid fuels. Nonthermal plasmas enable this chemistry to occur at ambient temperatures and pressures thus making this technology economically attractive.

Developing nonthermal plasma reactors on the microfluidic scale reduces the required materials of construction and energy input. Maximal active volume can be achieved at these smaller scales as more reactant gas is forced to move in the vicinity of the high voltage electrodes. Development and optimization of this technology to produce liquid fuels is currently the goal of the study. This presentation will summarize recent developments in this area.

497774 Non-Oxidative Methane Coupling Using Metal Hydrides As Hydrogen Storage Materials

Abstract:

Methane has experienced a meteoric rise due to its vast availability and its prominence in the constitution of natural gas, which is increasingly becoming the dominant energy source in the United States. Although highly desirable, the direct conversion of methane into useful chemicals is limited due to the molecule's high symmetry and strong C-H bonds. In this work, we seek to upgrade methane to higher hydrocarbons. Historically, the most studied method for the upgrade of methane is the oxidative coupling of methane (OCM), $2CH_4 + \frac{1}{2}O_2 \rightarrow C_2H_6 + H_2O$ [1]. However, this method has failed to yield an economically viable process due to poor carbon selectivities, which are a result of the over-oxidation of the reactive CH_3 intermediate into CO and CO_2 [2].

To avoid the over-oxidation of methane, we employ hydride-forming metals as hydrogen storage materials (HSMs) in a chemical looping approach. In the first step, hydrogen is catalytically abstracted from methane and stored within the HSM while the methyl groups are combined to form ethane. This first step follows the reaction $2CH_4 + M \rightarrow C_2H_6 + MH_2$, where M represents the hydride-forming metal. In the second step, the HSM is regenerated through thermal release, forming H_2. With the regenerated HSM, the first step can be repeated, thus creating a chemical loop.

Through both density functional theory simulations and experimental work, we have identified five Group III and IV metals (scandium, titanium, yttrium, zirconium, and hafnium) that have strong hydrogen storage capabilities and also exhibit the ability to cleave C-H bonds. In particular, titanium and yttrium have shown promise due to experimentally demonstrated rapid hydride formation kinetics and a theoretical activation energy barrier of less than 0.6 eV for the initial C-H bond scission. In addition, titanium's conversion and selectivity towards higher hydrocarbons can be tuned and improved by alloying titanium with nickel and aluminum, respectively. We have found that the use of metal hydrides and their alloys can be propitious not only for the upgrade of methane, but also for other catalytic dehydrogenation reactions.

References:

[1] Hammond, C., Conrad, S. & Hermans, *ChemSusChem* 5, (2012) 1668–86.

[2] Zaera, F, *Catal. Letters* 91, (2003) 1-10.

499404 Oxidative Coupling of Methane over Mixed Metal Oxides

Abstract:

Oxidative coupling of methane (OCM) is one of the promising routes to convert natural gas to useful chemical building blocks like ethylene. The reactions between methane and oxygen leading to C_2+ products are known to be highly exothermic and some researchers have pointed out that the hot spot temperatures within the catalyst bed in lab scale reactors could exceed the feed temperature by as much as 300°C.

We present experimental and modeling studies on the bifurcation behavior of OCM in fixed bed reactors. Feed temperature and space time are studied as bifurcation variables at various feed CH_4/O_2 ratios. Further, the bifurcation behavior (or lack of) is explained with various reactor configurations, showing the impact of heat transfer characteristics. For systems exhibiting multiple steady-states and bifurcation behavior, this work demonstrates the path dependency of the observed states and provides new insight for systems exhibiting such behavior.

497526 Experimental and Theoretical Investigation on Methane Dehydroaromatization Under Non-Oxidative Conditions

Abstract:

Advances in hydraulic fracturing technology has generated huge industrial attention towards catalytic conversion of low cost natural gas to valuable hydrocarbons. Developing molecular level understanding of catalyst active sites and its deactivation mechanism will help in the process of developing an effective commercial non-oxidative methane dehydroaromatization (MDA) catalyst. Mo_xC_y/H-ZSM-5 is one of the suitable catalyst for the MDA process. Mo/HZSM-5 contains both metal site (Mo species) and acid site (H-ZSM-5) hence act as a bi-functional catalyst to efficiently activate methane on Mo species to produce C_2H_4 which than oligomerize and aromatize to benzene and its derivatives on acidic sites of shape selective zeolite[1]-[2] . Nature of the active site for methane is suggested to be carbide or oxy-carbide but active phase structure is strongly debated. Lezcano-Gonzàlez and co-workers[3] have carried our operando X-ray method and asserted MoC_3 as possible active Mo speciation responsible for methane activation and formation of aromatics. Density functional theory calculations have been employed to study the effect of charge on catalytically active molybdenum carbide nano-cluster and its stability which will help in designing better MDA catalyst with suitable promoter species. In the study we have presented a detail investigation of effect of residual charge, Figure 1(a) showing a linear correlation between charges on MoC_3 cluster on methane activation underlying the effect of reducibility of Mo_xC_y cluster. Further, reaction mechanism insights have been studied for the methane activation and formation of ethylene, through formation of C_2H_6 and further C-H activation. Potential energy diagram for methane following scheme, Figure 1(b) shows ethylene will be formed easily on the catalyst surface, which eventually lead to formation for benzene with an intrinsic methane activation barrier of ~1.23 eV. The two resultant CH_3 groups attached to the Mo_2C_6 cluster undergo C-C coupling reaction with relatively lower barrier of +0.35eV. Following these theoretical insights, H-ZSM-5 impregnated with Mo were tested experimentally for MDA under non-oxidative conditions. In order to minimize coke deposition two different approach a) carbon dioxide (CO_2) as co-feed with methane[4] and b) in-situ topotactic transformation of MoC form hexagonally close packing to face centered cubic[5] by pre-treating catalyst with inflow of H2 at 2400ml/hr-gm GHSV for 6 hours at 350°C and then increasing the temperature to desired reaction condition with constant inflow of 30ml/min during carburization period carried out on a fixed bed reactor. Addition of CO_2 to methane feed reduces not only inert coke but also reactive coke as elucidated by TGA analysis, Figure 1(c). In the current study, the effect of variation in CO_2 concentration was studied. Results depict that addition of CO_2 enhances benzene percentage upto 35% on mole basis, Figure 1(d) in product stream as higher oxygen donor ability of CO_2 eliminate surface carbon by forming CO. Phase transformation of MoC hinders the migration of external surface Mo species to BAS present in zeolite channels resulting in retardation of inert coke deposition at pore entrance which results in enormous catalyst active with high selectivity towards aromatics, Figure 1(e) and less coke deposition resulting in high catalyst stability which is validated through coke gasification kinetics using Kissinger plot.

Figure 1. (a) Effect of residual charges on methane C-H bond activation, (b)Reaction diagram for methane activation and formation of ethylene on Mo_2C_6 nanocluster, (c) TGA analysis for Mo/HZSM-5, (d)-(e) Product selectivity during MDA process for Mo/HZSM-5.

Acknowledgement

Financial support provided by Gas Authority of India limited (GAIL) for carrying out this research work is gratefully acknowledge.

References

1. Majhi, S., Mohanty, P., Wang, H. & Pant, K. K. Direct conversion of natural gas to higher hydrocarbons: A review. *J. Energy Chem.* **22,** 543–554 (2013).

2. Gao, J. *et al.* Identification of molybdenum oxide nanostructures on zeolites for natural gas conversion. **2534,** (2015).

3. Lezcano-González, I. *et al.* Molybdenum Speciation and its Impact on Catalytic Activity during Methane Dehydroaromatization in Zeolite ZSM-5 as Revealed by Operando X-Ray Methods. *Angew. Chemie - Int. Ed.* **55,** 5215–5219 (2016).

4. Ohnishi, R. Catalytic Dehydrocondensation of Methane with CO and CO2toward Benzene and Naphthalene on Mo/HZSM-5 and Fe/Co-Modified Mo/HZSM-5. *J. Catal.* **182,** 92–103 (1999).

5. Liu, H., Bao, X. & Xu, Y. Methane dehydroaromatization under nonoxidative conditions over Mo/HZSM-5 catalysts: Identification and preparation of the Mo active species. *J. Catal.* **239,** 441–450 (2006).

499441 The Selection and Optimization of Heat Carrier Materials to Maximize Liquid Yields and Heat Recovery in Auger Pyrolyzers

Abstract:

Auger pyrolysis reactors are a promising technology in the future of commercial pyrolysis processes. Advantages of auger pyrolyzers over traditional fluidized bed pyrolyzers include the minimal requirement of an inert gas and the lack of sensitive hydrodynamic bed conditions. Additionally, auger pyrolyzers have the flexibility of different modes of heat transfer such as conventional indirect heating or directing heating via heat carrier materials to provide higher heat transfer rates. To mitigate the risk upon scale-up, a better understanding and optimization is required of auger pyrolyzer heat carrier materials. This work focuses on the selection of heat carrier materials and their ability to be recycled within the auger pyrolyzer for maximum product yields and energy recovery.

A twin-screw, lab-scale auger reactor was used to pyrolyze up to 1 kg/hr of red oak at 515°C for all trials. The effect of heat transfer properties on pyrolysis product yields was evaluated for several heat carrier materials. Stainless steel shot, silicon carbide, and sand were investigated. High liquid yields of ~65 wt. % were achieved. Sand was then chosen to further evaluate and optimize heat carrier parameters such as mass flow ratios, particle size, and long-term performance. Two sizes of sand deemed fine (250-600 μm) and coarse sand (600-1000 μm) were evaluated at heat carrier to biomass ratios of up to 20:1. The sand was then recycled at up to 5 recycles producing consistent yields of bio-oil (~60 wt. %), char (~17 wt. %) and gas (~18wt %). Regeneration of the heat carriers was found to be required trial to both remove any carbon residue and maximize heat recovery. The results from this work suggest that the selection and optimization of heat carrier for auger pyrolyzers is important in reaching high liquid yields, while minimizing processing costs attributed to attrition rates and heat recovery.

*Funding for this project was provided by Phillips 66 Company.

494870 A Kinetic Study of the Fast Micro-Pyrolysis of Hybrid Poplar

Abstract:

Hybrid poplar from the clone DN34 was studied to determine the rate of production of bio-oil vapor species as well as gas and char production with respect to time during the fast pyrolysis process. 300-660 ug samples were pyrolyzed using a micropyrolzer at 500ºC at very high heating rates and negligible vapor residence times. Individual poplar samples were run in triplicate at discrete time points ranging from 1-20 seconds in the micropyrolzer. Several bio-oil compounds from each individual sample were analyzed using GC/MS. Select bio-oil compounds, derived from the hemicellulose, cellulose, and lignin fractions of the wood, were used to determine their rates of production from thermal degradation of the biomass. These rates were quantified using standards to determine the weight percent relative to the original raw biomass with respect to time. Pyrolysis kinetic reaction models were fit to this experimental data in order to determine suitability of each model. A first order exponential decay model for degradation of the solid biomass was fit against the data along with a six-step consecutive degradation model previously developed by our group (Klinger et al. 2015). The experimental data suggests that vapor compounds derived from the hemicellulose fraction of poplar are produced at faster rates than that of the lignin or cellulose fractions, consistent with prior data from thermogravimetric analysis (TGA). When applying the first order exponential decay model, the reaction rates calculated did prove that holocellulose compounds reaction rates were approximately twice that of the lignin compounds reaction rate but did not provide the best fit. The six-step degradation model provided a better fit to all of the biooil compounds and the char data, and the stoichiometric parameters derived from the model fit showed that the first three reactions involved mostly the hemicellulose derived compounds while reaction steps 4-6 released cellulose- and lignin-derived compounds.

Reference:

Klinger, J., Bar-Ziv, E., & Shonnard, D. (2015). Unified kinetic model for torrefactionâ€"pyrolysis. Fuel Processing Technology, 138, 175-183.

497410 Experimental Investigation of Tar Recycling in Biomass Gasification

Abstract:

Tar formation is one of the major problems that hinders the development of biomass gasification process as it lowers the utilization efficiency of the syngas products and has negative effect on downstream process equipment. Extensive efforts have been made on tar control and reduction, while limited investigation have been conducted on tar recycling and reusing in biomass gasification. In this work, biomass and tar mixture at different ratios was fed to gasifiers of varied sizes to understand the impact of tar recycling on gasification. The syngas composition, tar production, ash agglomeration and reactor shut down time were studied. The experiment results show that tar recycling upgrades the syngas because of the decomposition of hydrocarbons, while lowers the syngas quality due to the extra moisture involvement in the gasification process. It is also found syngas compositions depend on the biomass/tar ratios and temperatures. The ash agglomeration problem becomes severe after the feeding of tar and makes it harder to control the reactor.

499239 A Convolutional Neural Network Model for Biomass Gasification in Fluidized Bed

Abstract:

Biomass is a renewable resource and its utilization has received great attention due to its life cycle carbon-neutrality and the potential to substitute fossil fuel to produce a variety of energy-related products. Thermochemical gasification is an important route for conversion of biomass that results in a product gas consisting H2, CO, CO2, CH4 and other light hydrocarbons that can be used as fuel gas to generate power or as well as raw material to produce variety of chemicals. Among the existing gasifiers, fluidized beds offer many advantages such as high conversion efficiency and great flexibility regarding feedstock type and size.

Modeling and simulation of FB gasifier is a problem of considerable theoretical and computational difficulty mainly due to presence of several chemical and physical processes in the reactor and the lack of proper characterization of solid fuel. The largest challenge of the existing modeling works is lack of generality and predictive capability. Here, we present a neural based (black-box) model that can predict the composition of product gas, amount of tar and solid conversion under a wide range of operating conditions and different types of feedstock.

Initially in this study, elemental composition (C, H, N, O, S), proximate analysis (Volatile matter, Fixed carbon, Ash), structural component (Cellulose, Hemicellulose, Lignin) and ash composition of 35 types of biomass were analyzed using principal component analysis (PCA), Factor analysis (FA), and a variety of machine learning approaches. A new method of biomass characterization is proposed based on composite statistical ensembles to improve understanding of nature of biomass.

Subsequently, over 200 data sets of biomass gasification in fluidized bed were collected, structured and analyzed. Multiple measures of associations between product gas characteristics, biomass constituents, and operating condition were analyzed to detect underlying features of the system that were most useful in developing the predictive model. The architecture of the neural network was founded based on the subjective knowledge of the process of biomass gasification in fluidized bed and statistical inferences of the data. The model was trained using the Bayesian regularization technique that is known for yielding more general models. It should be noted that the outputs of BBM are designed to only be used as constitutive equations and in the last step the equilibrium formulation is used as the physical framework of the model to explicitly close the mass and energy balance.

501642 Upgrading Biomass Using Low Temperature Plasma

Abstract:

Linear hydrocarbons are not easily obtained from biomass, even after thermochemical deconstruction and upgrading. The established thermochemical methods of deconstructing biomass include pyrolysis and gasification, which provide possible routes to hydrocarbons. The resulting products of biomass thermochemical deconstruction can then be upgraded by catalytic reactions such as hydrodeoxygenation of pyrolysis-oil or Fischerâ€"Tropsch reaction of biomass-derived syngas.Â A major challenge with these routes is that they produce highly heterogeneous product mixtures with low selectively. Catalyst fouling and inactivation is also problematic. A single-step process, which does not require metal catalysts, to produce oxygen-free hydrocarbons with high selectivity would be attractive. In this talk, we will report on our recent work to upgrade lignocellulosic biomass into deoxygenated hydrocarbons in a single step using a catalyst-free, low temperature plasma process. Effects of process parameters such as plasma power, pressure, gas composition (e.g. H_2 fraction), and biomass feedstock composition (e.g. lignin vs. cellulose) will be discussed. The evidence demonstrates that the plasma is capable of rapidly generating volatile deoxygenated hydrocarbons with high yield and selectivity without the need for catalysts. A preliminary technoeconomic assessment will be presented.

500694 Cellulose Pretreatment and Dissolution: Selection of Solvent and Processing Conditions

Abstract:

Efficient utilization of biomass is hindered by the recalcitrance to dissolution of semicrystalline cellulose. Pretreatment is often used to alter the structure of cellulosic biomass in order to make cellulose more accessible to solvents and enzymes. The pretreatment involves physical and/or chemical processing which affects the degree of crystallinity and size of biomass particles. We examine here the effects of (i) solvent properties, pretreatment steps and temperature, and (ii) fiber diameter and degree of crystallinity, on the kinetics of cellulose swelling and dissolution. To this end we have combined (a) experimental results on cotton fiber swelling, change in crystallinity and dissolved amount when treated under different solvent conditions, with (b) a phenomenological model that accounts for the phenomena governing the dissolution of solid cellulose, e.g., solvent penetration, transformation from crystalline to amorphous domains, specimen swelling, and polymer chain untangling. [Ghasemi, M.; Singapati, A. Y.; Tsianou, M.; Alexandridis, P., Dissolution of nanostructured polymer fibers: Numerical modeling and parametric analysis. AIChE Journal 2017, 63 (4), 1368-1383. DOI: 10.1002/aic.15615] The insights obtained from this analysis would facilitate the rational selection of solvents and the design of pretreatment processes to reduce the size and degree of crystallinity of cellulosic biomass particles, leading to enhanced biomass utilization. [Ghasemi, M.; Tsianou, M.; Alexandridis, P., Assessment of solvents for cellulose dissolution. Bioresource Technol. 2017, 228, 330-338. DOI: 10.1016/j.biortech.2016.12.049]

496185 Ionic Liquids: Solvent Design for Diverse Sustainable Biomass Valorisation

Abstract:

The processing of residual lignocellulosic biomass within the biorefinery frames impelled the use of green innovative technologies aiming at sustainable valorisation of this feedstock. The main constrain is related to the complex and recalcitrant macromolecular three-dimensional structure of biomass. The efficient and selective pre-treatment of biomass into main constituents i.e. cellulose, hemicellulose and lignin opens a possibility to produce diverse biofuels, biochemicals and biomaterials as it happens in fossil refinery. Ionic liquids (ILs) are one of the most promising solvents which offers greener solutions in biomass conversion technologies into a great variety of products.[1] The present work discloses two different scenarios of lignocellulosic biomass (wheat straw) processing with ILs, where i) alkaline or ii) acidic character of ILs remark biomass fractionation and subsequent product formation.

ILs with alkaline character demonstrated ability to dissolve the entire biomass allowing fractionation of each single fraction of biomass. Such approach produced high-purity fractions of cellulose, hemicellulose and lignin and IL was recycled and reused with success confirming the economic and environmental feasibility of the proposed method.[2] Besides, a fraction rich in phenolic compounds was also produced by employing polymeric resins and supercritical CO_2 ($scCO_2$) extraction. Silica C18 resin showed the highest extraction efficiency of phenolic compounds, while Amberlite XAD-7 was the most selective one. Nevertheless, a residual quantity of IL was found in the extracted samples and further purification was achieved by applying $scCO_2$ extraction. An extract containing vanillin, tricin, syringaldehyde, *p*-coumaric and vanillic acids, was produced with no IL presence.[3] The discussed approach is presented in Figure 1.

Figure 1. Biomass fractionation after pre-treatment with [emim][OAc] and further extraction and purification of phenolic compounds.

Another strategy of biomass valorisation was based on selective removal of the hemicellulose fraction of biomass solely using acidic ILs, i.e. 1-butyl-3-methylimidazolium hydrogen sulphate ([bmim][HSO_4]) and 1-ethyl-3-methylimidazolium hydrogen sulphate ([emim][HSO_4]). The acidity of [HSO_4]-based ILs hydrolyses hemicellulose and produce pentoses or furfural leaving a solid fraction constituted by cellulose and lignin for further valorisation in the perspective of the biorefinery concept.[4] The capacity of these ILs to hydrolyse and/or convert biomass into pentoses and furfural can be attained by controlling the reaction conditions, such as temperature, residence time and water content. A maximum pentose yield of 81.6 mol% was obtained after processing of wheat straw at 141 °C during 1.5 hours with [emim][HSO_4]/water mixture containing 58 wt. % of water. The obtained result is for now, far the best result reported in literature for biomass processing with ILs.

The present work demonstrates the great versatility of ILs to assist biomass processing for the manufacture of a wide range of products. Particularly, the ILs chemical character governing pathways in the fractionation of biomass allowing designing specific strategies towards the production of target chemicals.

Acknowledgments

IF/00471/2015 grants and by BBRI - Biomass and Bioenergy Research Infrastructure (ROTEIRO/0189/2013) and the Associated Laboratory for Sustainable Chemistry-Clean Processes and Technologies-LAQV which is financed by national funds from FCT/MEC (UID/QUI/50006/2013) and co-financed by the ERDF under the PT2020 Partnership Agreement (POCI-01-0145-FEDER-007265). This work was done in the frame of COST EUBIS TD1203 action.

References

1. R. Bogel-Lukasik, *Ionic Liquids in the Biorefinery Concept*, RSC, Cambridge, UK, 2015.

2. S. P. Magalhães da Silva, A. M. da Costa Lopes, L. B. Roseiro and R. Bogel-Lukasik, *RSC Adv.*, 2013, **3**, 16040-16050.

3. A. M. da Costa Lopes, M. Brenner, P. Fale, L. B. Roseiro and R. Bogel-Lukasik, *ACS Sustain. Chem. Eng.*, 2016, **4**, 3357-3367.

4. A. V. Carvalho, A. M. da Costa Lopes and R. Bogel-Lukasik, *RSC Adv.*, 2015, **5**, 47153-47164.

490695 Pretreatment of Biomass and Coal with Ionic Liquids for Advanced Fuel and Chemical ProductionÂ

Abstract:

Title: Pretreatment of Biomass and Coal with Ionic Liquids for Advanced Fuel and Chemical Production

Authors: C. Luke Williams[1], Chenlin Li[2], Jared C. Allen[1], and Karen Delezene-Briggs[3],
(1) Biofuels and Renewable Energy Technology, Idaho National Laboratory, Idaho Falls, ID
(2) Chemical and Radiation Measurement, Idaho National Laboratory, Idaho Falls, ID
(3) Biological and Chemical Processing, Idaho National Laboratory, Idaho Falls, ID

The United States currently faces major energy security challenges due to increasing energy demand. To reduce imports of foreign energy and fuels the development of domestic energy and materials sources, from a variety of carbon-based feedstocks, will be required. However, to efficiently transform nationally available resources, like biomass and coal, to energy and fuels will require both advanced processing techniques and catalytic upgrading. Ionic liquids are a class of solvent capable of altering the structure of many carbonaceous materials at low temperatures and atmospheric pressures. Due to the relatively mild conditions required ionic liquid pretreatment can be integrated with many advanced fuel/chemical processing techniques. In this work multiple types of biomass and several grades of coal are treated with ionic liquids. After treatment, the physical and chemical alterations in the materials are investigated using: scanning electron microscopy, X-ray diffraction, compositional analysis, enzymatic hydrolysis, and thermogravimetric analysis. It can be seen that ionic liquid pretreatment has a drastic effect on the catalytic activity of enzymatic hydrolysis for the production of biomass derived sugars, as well as altering the combustion properties of coal. Overall, ionic liquid pretreatment has the potential to improve the catalytic production of energy and materials from both biomass and fossil feedstocks by altering material structures.

499207 *In situ* Raman Microscopy to Monitor Changes in Cellulose Crystallinity During Acid Pre-Treatment

Abstract:

Dilute acid pre-treatment is a common technique for decreasing lignocellulosic biomass recalcitrance. During this process hemicellulose is removed, providing structural access for enzymes to hydrolyze cellulose. The effect of acid pre-treatment processes on the cellulose structure is well understood, in particular the change of crystallinity, partly due to uncertainty from sample preparation techniques carried out prior to *ex situ* analysis. However, cellulose crystallinity is the primary determinant of its recalcitrance to hydrolysis. Raman microscopy has been used extensively to study cellulose structural changes. Furthermore, it is a convenient and versatile tool allowing *in situ* analysis under hydrothermal environments, as water is mostly Raman transparent. Thus, understanding how cellulose crystallinity changes *in situ* would allow for improvement of pre-treatment processes by optimizing temperature and acid concentration of treatment.

In this work, we designed and built a custom view-cell reactor capable of operating at conditions of up to 220 °C and 30 bar to investigate cellulose structural changes under pre-treatment conditions. The crystallinity of ball-milled cellulose increased during the first 60 min of dilute acid pretreatment at 150 °C. After 60 min, the increasing crystallinity appeared to cease, coinciding with the appearance of a new band at 1600 cm^{-1}. To determine whether cellulose decrystallizes under pre-treatment conditions, highly crystalline cellulose was thermally treated in dilute acid. The crystallinity appeared to increase up to a temperature of 160 °C, after which it appeared to decrease. These observations could be attributed to the preferential hydrolysis of amorphous regions in the cellulose structure resulting in an increase in the relative amount of crystalline regions.

To verify the observed increase in crystallinity during dilute acid hydrolysis we carried out control tests in hydrothermal conditions in the absence of acid. At such conditions the hydrolysis rate should be rather low compared to dilute acid. Partially decrystallized cellulose started to fluoresce relatively early, within 20 minutes of reaction time at 150 °C preventing crystallinity analysis. However, similarly to reactions under acidic conditions, a new band at 1584 cm^{-1} appeared within 20 minutes of reaction time. Cool-down and washing of the cellulose with water did not completely eliminate the fluorescence and the new band at 1584 cm^{-1}, suggesting partial water solubility and surface degradation. In the absence of acid, the crystallinity of highly crystalline cellulose appeared to be relatively stable, which is consistent with low hydrolysis rate in the absence of acid.

To understand the nature of the new bands formed during acidic and hydrothermal hydrolysis, we employed IR spectroscopy. We heated cellulose in nitrogen environment and new bands appeared consistent with formation of C=O and C=C at temperatures above 230°C. Further, we compared a Raman spectrum of hydrothermal char, produced by degradation of glucose to the Raman spectra of cellulose under dilute acid and hydrothermal conditions. The G band of the hydrochar appeared at 1582 cm^{-1} similar to our observations at 1584 cm^{-1} as observed in hydrothermal treatment. These results suggest formation of degradation compounds such as HMF and humins during dilute acid and hydrothermal hydrolysis at temperatures as low as 150 °C.

In situ Ramans studies of the ball-milled and highly crystalline cellulose further understanding of the changes of cellulose structure that occur during dilute acid pre-treatment.

This presentation describes research that explores the hypothesis that simple feedstock pretreatment-digestion methods can be applied to rapidly evaluate biochemical biogas potential of anaerobic digestion in resource constrained development. Biomass feedstock, *i.e. Carica papaya* petioles were obtained from Edo State, Nigeria. Samples were analyzed with Fourier transform infrared spectroscopy after hydrochloric acid pre-treatment towards characterization of spectra with characteristic absorbance ratios indicative of physicochemical changes in biomass that include cross-linked lignin (CLL) ratio, the lateral order index (LOI), and hydrogen bond intensity (HBI) (Tables 1 & 2).

Wavenumber cm^{-1}	Raw	1%	3%	5%
897	36.1	55.4	61.5	65.5
1323	45.3	64.1	71.2	70.8
1424	43.9	60.0	66.9	67.5
1508	39.7	55.0	62.1	63.7
1599	48.0	58.8	66.3	67.6
3401	51.6	72.5	78.7	74.3

	Raw	1%	3%	5%
LOI	1.2	1.1	1.1	1.0
HBI	1.1	1.1	1.1	1.0
CLL	0.8	0.9	0.9	0.9

Figure 1 (left): Absorbance of wavenumbers associated with biomass indices indicative chemical nature of biomass for raw Carica papaya petiole, 1%, 3% and 5% hydrochloric acid pretreated samples. Wavenumber assignments: 897 cm^{-1} absorbance, i.e. glycosidic bond β-(1,4) in cellulose, 1420 -1430 cm^{-1} associated with crystalline structure of cellulose, 3400 cm^{-1}, i.e. O–H stretching H-bonds between molecules, 1320 cm^{-1}, i.e. CH rocking vibration of glucose rings, 1600cm^{-1} i.e. condensed lignin guaiacyl residues and 1508cm^{-1}, i.e. cross-linked structures of lignin guaiacyl residues.
Figure 2 (right): Summary of chemical bond ratios of Carica papaya petioles at respective pretreatment acid concentrations

Cumulative biogas yields from samples were fit to the modified Gompertz form to derive mathematical parameters related to biological reactor performance. To date, optimum sample treatment suggests that a 150% increase in biogas yield compared to non-pretreated *Carica papaya* petioles is possible. Overall, results suggest that rapid biogas potential assessment can be performed in a few days-time and provides for a visual means to characterize individual digestions in constrained research setting and collaborations (Figure 3).

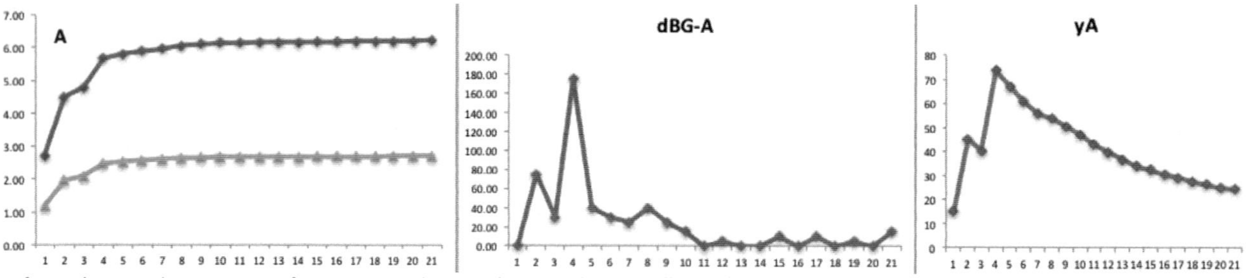

Figure 3: After selecting the Gompertz form, gas production data can be modelled either by graphing cumulative production in log or natural log forms (top and bottom lines of LEFT column), by change in the daily production (MIDDLE column), and cumulative production averaged by day of production (right column). Visual discerning maximum production (LEFT column), days of maximum production (MIDDLE column) and periods of maximum productivity (RIGHT column) are apparent.

The research reported is collaborative, in support of larger sustainability goals, and the transfer of knowledge from National lab settings such as the Department of Energy's National Renewable Energy Laboratory to biofuels research and development efforts. Overall, a goal is to assess biofuels production in more complex or integrated socio-economic models (Figure 4)

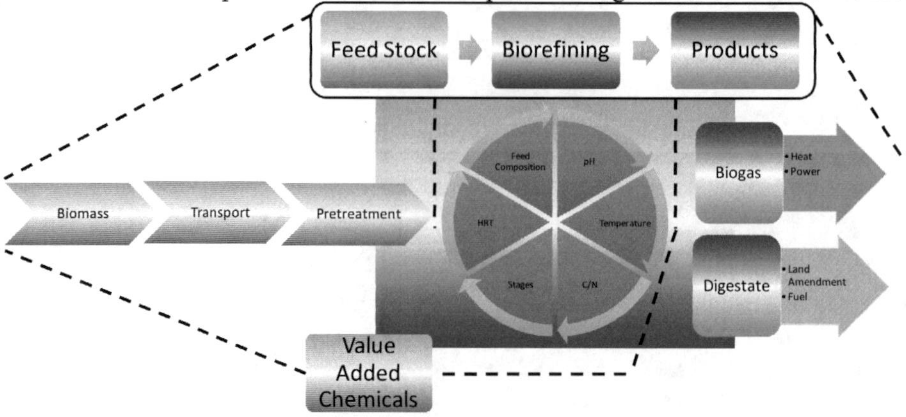

Figure 4: A framework that considers the complexity of biogas production. We propose that there are additional considerations such as: 1) natural value added chemicals, 2) socio-economic considerations, and 3) community appropriate, rural, or smaller scales to consider in addition to the traditional process specific considerations.

500004 Supercritical CO_2 Pretreatment of Wheat Straw: Hydrolysis Performance, Enzymatic Yields and Comprehensive Mass Balances

Abstract:

Wheat straw, with an estimated annual production over 170 million tons,[1, 2] has been considered the most promising agricultural residue for the production of bio-products and 2nd generation bioethanol in Europe. The processing of wheat straw into fermentable sugars represents a great challenge, mainly due to its complex and highly recalcitrant architecture. In this work, the conversion of wheat straw into pentoses using supercritical CO_2 followed by an enzymatic saccharification of cellulosic fraction is for the first time investigated. This technology benefits from the *in-situ* formation of carbonic acid. Without an additional chemical, this technology outperformed liquid hot water pretreatment allowing to produce total sugars yield as high as 84 % in comparison to 67 % with liquid hot water (and under more severe reaction conditions). Both chemical and physical effects of supercritical CO_2 resulted in a glucan to glucose conversion yield of 82 mol% representing a 26 % improvement over those achieved in liquid hot water.[3] Firstly, the effect of various process conditions, such as temperature, residence time and initial CO_2 pressure on conversion of hemicelluloses into pentoses (in both oligomeric and monomeric form) were scrutinised in an effort to define the optimal operational conditions. Then, several material characterisation techniques were used to investigate thoroughly the physical influence of supercritical CO_2on the processed solids.

Acknowledgment

IF/00471/2015 grants and by BBRI - Biomass and Bioenergy Research Infrastructure (ROTEIRO/0189/2013) and the Associated Laboratory for Sustainable Chemistry–Clean Processes and Technologies–LAQV which is financed by national funds from FCT/MEC (UID/QUI/50006/2013) and co-financed by the ERDF under the PT2020 Partnership Agreement (POCI-01-0145-FEDER-007265).

References

1. S. Kim and B. E. Dale, *Biomass Bioenerg.*, 2004, **26**, 361-375.

2. D. Montane, X. Farriol, J. Salvado, P. Jollez and E. Chornet, *J. Wood Chem. Technol.*, 1998, **18**, 171-191.

3. A. R. C. Morais, A. C. Mata and R. Bogel-Lukasik, *Green Chem.*, 2014, **16**, 4312-4322.

494899 Dilute Alkali Flowthrough Pretreatment of Softwood

Abstract:

Alkali pretreatment effectively disrupts biomass plant cell wall. In this study, dilute alkali flowthrough pretreatment of softwood was investigated at 200-270 °C for 2-10 min at a flow rate of 25 mL/min, under pH of, 8, 9, 11, and 12, respectively, comparing with water flowthrough pretreatment at pH 6.4. The composition analysis of pretreated solid residues and the recovered liquid streams showed the removal profile of biomass components and the recovery profile of sugars/chemicals as a function of pretreatment severity. 2D ^{13}C-^{1}H NMR analysis revealed that complete solubilization of biomass under high pH conditions resulted in pretreated liquid stream rich in various compounds, including monomeric and oligomeric aromatics, and carbohydrates degradation products. Results indicated that initial pH greatly affected the biomass degradation pathways and efficiency. The plausible mechanism of biomass degradation by dilute alkaline under various pretreatment severities was proposed.

498634 Production of Renewable Hydrogen from Wastewater Using Microbial Electrochemical Systems

Abstract:

Hydrogen is one of the key starting materials used in the chemical industry. Large amounts of hydrogen are also required in biorefinery schemes due to the high O/C and low H/C ratio of biomass. The H2 industry in the USA is large, but virtually all of the H2 today comes from reforming fossil fuels. Capturing renewable H2 stored in organic-rich wastewater would provide a new source of H2 that would also accomplish the wastewater treatment. Dark fermentation, a process that uses fermentative bacteria to convert organic matter to H2 without the presence of light, can achieve relatively high H2 production rates. However, the H2 yield is very low due to the formation of large quantities of side products. Microbial electrolysis cell (MEC) overcomes the fermentation barrier and provides a new approach for generating H2 from biomass. However, the key challenges for realizing the practical applications of MECs, especially for wastewater treatment include: 1) low H2 production rate; and 2) H2 uptake by H2 consumers, such as methanogens and homoacetogens. In this presentation, we will present our recent development of a hybrid microbial electrochemical system that combines the strengths of fermentation and MEC processes in one reactor for efficient H2 production. We will also present some low-cost strategies we developed against methanogens and homoacetogens for enhancing H2 recovery in the hybrid system. Future works, including scaling up and techno-economic analysis, will also be discussed.

500058 Scale-up and Process Intensification of Agricultural and Municipal Solid Waste Conversion Using Ionic Liquid Based Process

Abstract:

The sustainable production of both biofuels and chemicals depends largely on access to efficient and cost-effective biomass conversion technologies. Among the various pretreatment technologies being developed, ionic liquid (IL) pretreatment has been receiving significant recent attention as a potential "greener" process that enables fractionation of a wide range of feedstocks due to its high biomass processing efficiency. This presentation will describe recent advances in IL process development that we have demonstrated in the conversion of different types of municipal solid waste blends to fermentable sugars. We will highlight two integrated IL technologies: acidolysis with chloride based IL pretreatment and one-pot biomass conversion using a "bionic liquid" (amino acid based IL). For the acidolysis process, an acid catalyst was used to hydrolyze polysaccharides to monosaccharides following IL pretreatment, which provides an enzyme-free process with significant reductions in processing times and material costs. The one-pot biomass conversion combines the pretreatment and enzymatic saccharification into one vessel, eliminating solid-liquid separation and extensive water-washing steps with the potential to offer better economics, scalability and environmental sustainability. Process scalability and biomass conversion efficiency along with the feedstock cost and quality will be presented and discussed, as well as identifying and overcoming the remaining technical and economic challenges of these approaches to provide perspectives and opportunities for future process development and optimization.

497641 Map of Our Scrap: A GIS-Based Analysis on the Potential of Food Wastes for Bioenergy Production

Abstract:

The growth of our cities has been accompanied by the inevitable production of wastes. With the continuous increase in the amount of our wastes, their utilization has been initiated all over the world. One of the attractive utilization feedstock is food waste (FW) due to the capability of communities to segregate and direct this waste vector to established conversion technologies such as anaerobic digestion. Anaerobic digestion produces biogas consists of methane (CH_4), hydrogen (H_2), and carbon dioxide (CO_2), which are high energy-density chemicals. The digestion of food wastes typically produce 0.198 kWh/kg food waste [1]. As of now, the best sites for the installation of biogas-to-power turbine-generators are wastewater treatment plants (WWTPs) [2]. The transport of FW to WWTPs, however, consumes fuel, which may be counter-productive to a waste-to-energy initiative.

This work evaluated the energy balance associated with the digestion of food wastes at a nearest WWTP transported from the following key sources: supermarkets, hospitals, restaurants, colleges, and schools, using geographical information system (GIS) platform. The study area was Lafayette, Louisiana. The analysis was done in QGIS (Version 2.18, a.k.a. Las Palmas).

The preliminary results support the following inferences.

- This GIS-based analysis for the potential of FW as feedstock for bioenergy production showed the need for additional organic wastes to be combined with FW to possibly increase the power generation of WWTP-based biogas-to-energy turbine generators.
- Wastewater sludges produced in significant amount during water treatment in WWTPs are viable candidate organic wastes that may significantly increase biogas production. These sludges do not require transportation, hence, omitting the energy expenditure through transportation fuel.
- FWs from restaurants may significantly increase the potential bioenergy production, which may supply a utility-scale (1 MW) power plant in the city.

This city case study may be adopted to a state-wide analysis to evaluate the bioenergy potential of FW via anaerobic digestion for the state of Louisiana, USA.

Citations:

[1] Moriarty, K. (2013) Feasibility study of anaerobic digestion of food waste in St. Bernard, Louisiana. NREL, Doc. TP-7A30-57082

[2] Report: Massachusetts Department of Environmental Protection (2002) Identification, Characterization, and Mapping of Food Waste and Food Waste Generators In Massachusetts

497451 Biochar from Excelsior Residues for Plantation Production of Hybrid Poplar

Abstract:

Excelsior is a long fiber material made from wood that was used in many packing applications until the rise in the use of expanded plastics. Today, a common use of excelsior is in natural erosion control materials on construction sites. Residues from excelsior production include bark, excelsior fines, and small bits of other materials used with wood in the erosion control blankets and rolls. In this project, we investigate the potential of a three-way material pathway: making excelsior from plantation-grown hybrid poplar (instead of forest-harvested aspen), producing biochar and process heat through slow pyrolysis of excelsior residues, and using the biochar to amend soils within the hybrid poplar plantation to improve productivity. In the Southwestern U.S., many soils have elevated pH, so there is an opportunity to tailor biochar production for properties that would be advantageous for these soils, such as biochars with lower pH and lower ash contents. Results are presented from the first hybrid poplar growth study on soil amended with excelsior residue-derived biochar, as well as our initial findings related to the feasibility of the hybrid poplar-excelsior-residue biochar material pathway.

498725 Co-Pyrolysis of Plastics and Biomass Waste

Abstract:

Fast pyrolysis has been demonstrated previously for conversion of mixed waste constituents such as pine, sugarcane bagasse, corn stover, various plastics, and mixtures thereof into value-added liquid bio-oil fuel, biochar, and combustible gases. This process reduces the waste stream volume by up to 98% with low emissions and minimal impact on the environment. In an effort to understand the interactions between biomass and fossil-derived plastics, several pine/polystyrene(PS) binary mixtures were pyrolyzed in a bench-scale (1 kg/h), continuous, bubbling fluidized bed reactor at temperatures between 500 °C and 650 °C. Bio-oil properties from pyrolysis of pine/PS mixtures of 70%/30%, 80%/20%, and 90%/10% by mass were measured to determine the effect of plastic concentration on heating value, moisture content, and viscosity. Fractional condensation of heavier and lighter bio-oil fractions was demonstrated with two condensers in series, which allowed for the condensing temperature to be varied for collecting different oil fractions. The fluidizing gas temperature, residence time, reactor bed temperature, and condensing temperatures were varied to determine the effects on collection efficiency and bio-oil product characteristics such as heating value, viscosity, and water content. Samples of the bio-oil product were collected at each condenser and analyzed using gas chromatography coupled with mass spectrometry (GC/MS). Addition of plastics to the feedstock appears to have prompted a shift in bio-oil composition that indicates some interaction between the PS and pine pyrolysis intermediates. Significantly more aromatics were formed with the mixtures than when pine was pyrolyzed alone. The lower temperature fractions contained an abundance of the styrene monomer while the higher temperature fractions contained significantly heavier aromatics, including tolylacetic acid, bibenzyl, Î±-methylbibenzyl, and 6-phenyl-2-hexenal. Future work is focusing on catalytic upgrading of the heavier and lighter bio-oil fractions.

498126 Hydrothermal Liquefaction of Microalgae and Co-Product Development

Abstract:

Hydrothermal Liquefaction proves to be one of the promising process among thermochemical conversion technologies in the production of liquid transportation fuels. The red alga *Galdieria sulphuraria* was liquefied in a stainless-steel batch reactor and the influence of biomass loading, residence time and temperature was assessed. A maximum biocrude oil yield of 35.0 wt% was obtained at 300 Â°C and at a solid loading ratio of 20 % for 30 minutes residence time. Various analytical techniques were employed to analyze the valuable chemical components present in the liquefied products (bio-oil, biochar, water soluble compounds, and gaseous phase). Analysis of biochar using ICP-OES and elemental analyzer indicated the presence of macronutrients (nitrogen 6.36 mg & phosphorus 6.59 mg per gram of biochar), which are re-utilized to cultivate microalgae. Biochar was leached for macronutrients at different pH conditions (1.0 â€" 7.0) to estimate the efficiency and study the growth characteristics of microalgae. Gas Chromatography Mass spectrometry (GC-MS), ^{1}H NMR spectroscopy and calorimetric techniques were also used to analyze the products. Energy recovery was calculated through the ultimate and proximate analysis of the products.

492830 Multiphysics Simulation of Microalgae Growth in an Airlift Photobioreactor: Effect of Fluid Mixing and Shear Stress

Abstract:

Phototropic microalgae provide renewable feedstock for manufacture of biofuels, biochemicals, foods, nutrients, and pharmaceuticals. These organisms can be cultivated in photobioreactors with the aim of speeding their production. To reduce reactor design cost and improve reactor efficiency, it is necessary to simulate the key phenomena and mechanisms for biomass synthesis, as well as their mutual interactions, on reactor performance. Here, a multiphysics model has been developed with the objective to accurately simulate the effects of fluid mixing and shear stress on microalgal growth in an airlift photobioreactor. The model integrates multiphase flow dynamics, radiation transport, and algal growth kinetics using an Eulerian approach [1]. The multiphase computational fluid dynamic (CFD) model is first validated by comparing its predictions with corresponding experimental data. Subsequently, the CFD model is integrated incorporating sub-models for radiation transport and algal growth kinetics to predict reactor performance under various flow flow conditions. The simulations correctly predict the experimental trends that biomass productivity increases with increased rates of mixing, whereas it decreases with increases in shear stress. Lastly, we demonstrate the superiority of the Eulerian simulation approach by comparing its predictions not only with experimental results but by also comparing them to commonly used circulation cell and Lagrangian approaches [2].

[1] Gao, X., Kong, B. Vigil, R. D., 2017. Comprehensive computational model for combining fluid hydrodynamics, light transport and biomass growth in a Taylor vortex algal photobioreactor: Eulerian approach, *Algal Research*. 24, 1-8.

[2] Gao, X., Kong, B. Vigil, R. D., 2017. Comprehensive computational model for combining fluid hydrodynamics, light transport and biomass growth in a Taylor vortex algal photobioreactor: Lagrangian approach. *Bioresource Technology*. 224, 523-530.

496999 Techno-Economic Assessment of Microalgae Biorefineries for the Production of High Value Chemicals

Abstract:

Microalgae attract attention over the last years due to the need for alternative sources for fuels, chemicals and energy. Their major advantage over other sources of biomass is their non-competitiveness with the food sector. They require water, sunlight, nutrients and carbon dioxide to grow and their wide product portfolio renders them an exceptional source of biomass for the biorefinery concept. Furthermore, microalgae biorefineries can be integrated with current industrial applications or even other types of biorefineries.

The design of microalgae biorefineries is a challenging field and the valorization of specialty chemicals, in particular, enhances the challenge due to their unknown thermodynamic properties and the advanced technologies employed for their processing. The lack of industrial applications to draw knowledge is a significant drawback, thus extensive interaction with practitioners is of vital importance. This study capitalizes specifically on that and addresses the integrated design of a microalga biorefinery employing commercial simulation software (Aspen Plus v8.6). The design is developed under the D-Factory, an FP7 collaborative project, which aims at developing a commercial scale biorefinery process. Microalga Dunaliella is an exceptional candidate for such an incentive, since it has already been commercialized for its high value chemicals. Additional promising candidates are Nannochloropsis and Haematococcus, both offering wide range product portfolios of specialty chemicals. Nevertheless, challenges concerning the synthesis and the consequent design are evident in this study in regard to involved technologies and formulation of the product portfolio. Both advanced (e.g. high-performance countercurrent chromatography) and more conventional technologies (e.g. distillation) are employed in order to recover the desired products from the microalgal cells. The competitive chemistries present in microalgae add to the challenges related to the formulation of the product portfolio, thus the products selected should coincide with this restriction as well as with any limitations concerning the suitable selection of technologies.

For the purpose of this study, the capacity of each plant is set at 1kt per year of microalga biomass and the product portfolios consist of glycerol, carotenoids (Î²-carotene, lutein, astaxanthin etc.) and lipids (chlorophylls, EPA, DHA, PUFAs etc.). In their majority they are considered specialty chemicals and can find use in the pharmaceutical, cosmetic and nutraceutical sector. The residue after the recovery of the aforementioned compounds is rich in proteins and it is regarded as a co-product for its use as animal feed. The designed biorefineries are further studied as to their economic sustainability by calculating capital and operating costs. The cumulative cash flow diagram, starting from the construction of each plant, provides a very promising picture. Even though the evaluation is still at an early stage, as is the design of microalgae biorefineries in general, useful information is extracted which can lead to more accurate results in the future.

498705 A Process Model for Converting Low-Lipid Microalgae into Biocrude Oil Via Hydrothermal Liquefaction

Abstract:

This study aims to develop a process model for an integrated system that can simultaneously treat wild-harvested algal biomass and generate bioenergy. Previous studies focused more on high-lipid algae that normally require nitrogen depletion and contamination free culture condition, which result in low yield. On the contrary, low-lipid algae which contain large fraction of protein and carbohydrate typically have higher biomass production rate and can grow in harsher conditions. Hydrothermal liquefaction (HTL) process of biomass has shown encouraging results in converting whole algae into biocrude oil. Biocrude oil with high heating value is produced by HTL as a promising alternative for petroleum oil. After separation of biocrude oil, the aqueous phase contains large quantities of carbon, nitrogen, phosphate and other nutrients, which could be utilized in further reuse.

Investigating different configurations of the feedstock pretreatment, hydrothermal liquefaction (HTL) conditions, upgrading of biocrude oil and recycle of nutrients will help develop a sustainable and robust system for the highest net energy production and the lowest greenhouse gas emission. The present work intends to use process simulation tools to analyze the feasibility of HTL process that employs wild-harvested microalgae and cyanobacteria to produce biocrude oil and recycle the nutrients. A detailed mass balance, energy balance, techno-economic analysis will be investigated from feedstock to the final products. The simulation will be carried out by using Aspen Plus software, based on experimental results from model compounds and whole algae via HTL process.

499588 Flocculation and Vacuum Filtration of Algal-Slurry Intermediates to Enable Parallel-Algal Processing

Abstract:

Coproduction of high-value chemicals such as succinic acid is a promising route to enabling economic conversion of algal sugars and lipids to renewable fuels and chemicals. The green alga *Scenedesmus acutus* was grown at conditions specifically intended to accumulate carbohydrates, in addition to prevalent lipids. After harvesting, the algae biomass was acid pretreated in order to induce cell lysis and hydrolyze the insoluble carbohydrates to soluble sugars. The sugars were subsequently used to produce succinic acid via continuous fermentation by an immobilized culture of *Actinobacillus succinogenes*, while the lipids were extracted by hexane. The lipids are intended for upgrading to renewable-diesel blendstock, and the succinic acid is a high-value chemical intermediate that may be used to produce renewable chemicals and materials.

While immobilized-culture fermentation offers high specific productivities, it necessitates clarified sugar feedstocks. This presentation will focus on the development and use of a flocculation-enable solid-liquid separation of the pretreated algal slurries. Guided by zeta-potential measurements, trial-and-error testing of several variants of a commercial polyacrylamide flocculent (KemSep-brand, Kemira Chemicals, Helsinki, Finland) identified a suitable flocculent for pretreated algal slurries at the desired pH for fermentation. Additional dosing and mixing studies were performed, enabling vacuum filtration with acceptable permeate flux. The clarified liquor was subsequently used in a 756-hour continuous fermentation, achieving a maximum productivity of 1.2 g/L/h and a process conversion yield of 0.7 g succinic acid/g sugar.

487678 Study on Mechanism of Clinker Formation in Bamboo Combustion Process

Abstract:

In this study, we burn the bamboo powder packed layer placed on stainless steel mesh to know how the clinker is made and also we examined the possibility of clinker reduction by adding bark powder to bamboo powder. We carried out the mass change of the mesh after burning, mesh surface observation and also measured the adsorption amount and size of clinker. The following results were obtained.As a result, In the process of clinker formation, two stages were confirmed. At stage 1, about 10 Î¼m of fine white particles adsorbed onto the stainless steel mesh at the volatile component combustion of about 1000 K. In step 2, it was suggested that during combustion of char having a surface temperature of about 1200 K, these small grains melted and turned into large chunks of about 200 Î¼m.Furthermore, it was found that clinker clogging on the surface of the stainless steel mesh is reduced by mixing bamboo powder with bark and burning it. This proved that bark contains CaO with high melting point.

493264 Modeling Pyrolysis-Induced Microstructural Changes in Biomass: A Cellular Automata Approach

Abstract:

Lignocellulosic biomass is a chemically and morphologically heterogeneous material. Furthermore, this heterogeneity is in-part responsible for the vast number of thermal decomposition products seen in pyrolysis events. While modeling of biomass pyrolysis has been a subject of much research in past years at length-scales from the macro to molecular, the majority of these works have focused on a range of continuum-based approaches. Though effective at capturing global outcomes, these approaches are less tractable as frameworks for capturing microstructural effects and upscaling molecular information. This work demonstrates the use of kinetic cellular automata (kCA) as an alternative platform for the modeling and simulation of biomass pyrolysis. Asides from being effective at capturing transport and chemical processes in highly heterogeneous system, kCA is capable of modeling microstructural changes that occur as a result of chemical and physical transformations. A number of earlier benchmark trials demonstrated the convergence of the kCA to global continuum outcomes. Application of the kCA to actual biomass microstructures show promise for this platform as an intermediate length-scale tool capable of predicting char morphologies that mimic experimental outcomes at meso-scale resolution.

487672 Shrinking and Heat and Mass Transfer during Pyrolysis of Woody Biomass

Abstract:

Heat and mass transfer during pyrolysis of major components bed have been investigated numerically. For Cellulose powder, the calculation temperature and velocity vector distributions had a quasi-steady state during pyrolysis because the gas volume increased near the side wall of the packed bed and then the effective thermal conductivity decreased. The calculation temperature with the effect of volume shrinking of the packed bed agreed qualitatively with the experimental ones for all kinds of powder when the temperature plateau appeared.

493123 Plasma Gasification of Wood Pellets Using a Laboratory Scale Plasma Reactor to Produce Synthesis Gas: Effect of Increasing Temperature and Addition of Oxygen to the Process

Abstract:

A laboratory scale plasma reactor (max 15Kw power) has been used to convert wood pellets (biomass) into synthesis gas that can be used for downstream processes. A series of experiments have been conducted to see the effects of increasing temperature and adding oxygen to the process on conversion of hydrogen and carbon monoxide. The results show that increase in temperature from 400°C to 1000°C results in overall increase in biomass conversion (mass %) from 30% to 82%. In the case where no additional oxygen is supplied, the carbon and hydrogen conversion increases with increase in temperature up to 82% and 98% respectively but the molar ratio of carbon monoxide and hydrogen is unchanged at 1:1 for the temperatures studied. Addition of controlled oxygen increases the carbon monoxide yield and also results in the CO/H2 ratio of 3:2 for temperature range 800°C -1200°C. Simultaneous increase in reactor temperature and oxygen amounts to the process resulted in increased biomass conversion and a reduction of residues in the reactor.

497756 Multistage Fluidized Bed Reactor for Gasification

Abstract:

An ideal gasification process for a solid organic feedstock entails direct conversion of the feedstock into a synthesis gas, primarily CO and H_2, without any volatile or char byproducts. These byproducts coat the internal surfaces of the gasifier, rapidly degrading the performance of the gasifier and decreasing the yield of synthesis gas.

A three stage fluidized bed reactor has been developed to produce a clean synthesis gas in high yield. High quality synthesis gas can be used as a fuel for a combustion turbine or as feedstock for a chemical refinery. The first stage is the fluidized bed reactor where both the feedstock and heat transfer particles are fluidized in order to quickly heat the feedstock to a high temperature to achieve rapid pyrolysis and gasification. The second stage has a larger cross-sectional area than the first stage. This increase in area minimizes entrainment of solids in the second stage and increases the residence time to gasify char and decompose volatiles into low molecular weight gases. Typically the second stage operates at a high temperature in order to minimize condensation. The diameter of the third stage is greater than the diameter of the second stage in order to permit more char gasification and volatiles decomposition. If most of the volatiles have decomposed before entering stage three then the gases can be partially cooled in stage three. Following stage three the product stream passes through a cyclone to remove some of the entrained solids. Then the gases pass through a condenser to recover residual volatiles. Following the condenser gas samples are captured periodically for chemical analysis.

In the initial trials using this reactor, pine powder was used as the feedstock and alumina particles were used as the heat transfer medium. The temperature profile was set at 600 °C for stage 1, 800 °C for stage 2 and 300 °C in stage 3. Steady state operation was not achieved due to tar deposits on the alumina particles and the reactor walls. A close approach to steady state has been achieved by switching from alumina to zirconia particles and setting each stage at 800 °C. CO_2 was used as the fluidizing gas. Steady state was maintained for about 100 minutes and then the reactor was shut down. At this high temperature less fouling was apparent and the corresponding carbon yield in the gas fraction was about 74 %.

495901 Overall View of Lignocellulosic Biomass Decomposition in Supercritical Water

Abstract:

Supercritical water gasification is a promising technology for gasifying wet biomass. However, detail reaction kinetics of each component, cellulose, hemicellulose, and lignin have not been elucidated, and the interaction between the components has to be elucidated. In this study, mixture of cellulose, hemicellulose, and/or lignin was delivered into a continous reactor for supercritical water gasification and its reaction characteristics were determined. Reaction temperature was changed around critical point of water so that both subcritical and supercritical conditions were considered while the reaction pressure was kept at 25 MPa. Intermediate compounds were analyzed by HPLC, and comparison between avaraged single copound result and mixture was made. The interaction was clearly observed, especially in terms of hydrogen consumption. Reaction network was develped, and reaction rate constants were deteremined for each of the reaction by least square method to fit the experimental data with the calculation by the model. The Arrhenisc plot for these reactions classified ionic reaction and radical reaction. For the case of radical reacitons, Arrhenius plot showed a straight line as usual, but for the case of ionic reactions, reaction rate kinetics deviated from the straight line at high temperature, indicating ions get unstable under supercritical condition. The overall reaction kinetics was also determined, and discussed.

AUTHOR INDEX

AUTHOR INDEX

AUTHOR INDEX

AUTHOR INDEX

AUTHOR INDEX

AUTHOR INDEX

AUTHOR INDEX

AUTHOR INDEX

AUTHOR INDEX